PHYSICAL CHEMISTRY
An Advanced Treatise

Volume XIA / Mathematical Methods

PHYSICAL CHEMISTRY
An Advanced Treatise

Edited by

HENRY EYRING
*Departments of Chemistry
and Metallurgy
University of Utah
Salt Lake City, Utah*

DOUGLAS HENDERSON
*IBM Research Laboratories
San Jose, California*

WILHELM JOST
*Institut für Physikalische
Chemie der Universität
Göttingen
Göttingen, Germany*

Volume I / Thermodynamics
     II / Statistical Mechanics
     III / Electronic Structure of Atoms and Molecules
     IV / Molecular Properties
     V / Valency
     VI / Kinetics of Gas Reactions (In Two Parts)
     VII / Reactions in Condensed Phases
     VIII / Liquid State (In Two Parts)
     IX / Electrochemistry (In Two Parts)
     X / Solid State
     XI / Mathematical Methods (In Two Parts)

PHYSICAL CHEMISTRY
An Advanced Treatise

Volume XIA / Mathematical Methods

Edited by

DOUGLAS HENDERSON
IBM Research Laboratories
San Jose, California

 1975

ACADEMIC PRESS   NEW YORK / SAN FRANCISCO / LONDON
A Subsidiary of Harcourt Brace Jovanovich, Publishers

COPYRIGHT © 1975, BY ACADEMIC PRESS, INC.
ALL RIGHTS RESERVED.
NO PART OF THIS PUBLICATION MAY BE REPRODUCED OR
TRANSMITTED IN ANY FORM OR BY ANY MEANS, ELECTRONIC
OR MECHANICAL, INCLUDING PHOTOCOPY, RECORDING, OR ANY
INFORMATION STORAGE AND RETRIEVAL SYSTEM, WITHOUT
PERMISSION IN WRITING FROM THE PUBLISHER.

ACADEMIC PRESS, INC.
111 Fifth Avenue, New York, New York 10003

United Kingdom Edition published by
ACADEMIC PRESS, INC. (LONDON) LTD.
24/28 Oval Road, London NW1

Library of Congress Cataloging in Publication Data

Henderson, Douglas, Date
    Mathematical methods.

    (Physical chemistry, v. 11 A
    Includes bibliographies.
    1.   Chemistry, Physical and theoretical–Mathematics.
I. Eyring, Henry, Date   joint author.   II.   Jost,
Wilhelm, Date   joint author.   III. Title.
QD453.P55 vol. 11A, etc.   [QD455.3.M3]   541'.3'08s
ISBN 0–12–245611–4   [541'.3'0151]   74-11279

PRINTED IN THE UNITED STATES OF AMERICA

Contents

List of Contributors ix
Foreword xi
Preface xiii
Contents of Previous and Future Volumes xv

Chapter 1 / Linear Vector Spaces
R. J. Jacob

I.	Introduction: Vectors in the Physical Sciences	2
II.	Linear Vector Spaces	3
III.	Example: Three-Dimensional Euclidean Vectors—I	10
IV.	Vector Transformations	32
V.	Matrices	51
VI.	Example: Three-Dimensional Euclidean Vectors—II	64
VII.	Vector Spaces of Infinite Dimension	71
	References	78

Chapter 2 / Generalized Functions
E. W. Grundke

I.	Introduction	82
II.	Definitions	85
III.	The Algebra of Generalized Functions	98
IV.	The Calculus of Generalized Functions	108
V.	Some Singular Generalized Functions	118
VI.	Fourier Transforms	128
VII.	Laplace Transforms	143
VIII.	Conclusion	148
	References	149

Chapter 3 / Complex Variable Theory
Harris J. Silverstone

I.	Introduction	152
II.	Complex Numbers	153
III.	Analytic Functions of a Complex Variable	160
IV.	Complex Integration	167
V.	Power Series	184
VI.	Elementary Functions	193
VII.	Evaluation of Real Definite Integrals	205
VIII.	Higher Transcendental Functions	214
IX.	On Fourier Transforms	238
X.	Quantum Chemistry Integrals	241
XI.	A Formula of Lagrange and Nondegenerate Perturbation Theory	256
	References	259

Chapter 4 / Boundary-Value Problems
Douglas Henderson

I.	Introduction	262
II.	Some Typical Boundary-Value Problems	263
III.	The D'Alembert Solution of the Wave Equation	268
IV.	Separation of Variables	269
V.	Eigenvalues, Eigenfunctions, and Expansion Problems	280
VI.	Boundary-Value Problems in Cylindrical Coordinates	291
VII.	Boundary-Value Problems in Spherical Coordinates	304
VIII.	Green's Functions	316
IX.	Laplace Transform Methods	327
X.	Conformal Mapping	331
	References	335

Chapter 5 / Numerical Analysis
R. G. Stanton and W. D. Hoskins

I.	Introduction	337
II.	Approximation by Polynomial Interpolation	338
III.	Approximation by Spline Interpolation	346
IV.	Approximation by Least Squares	352
V.	Numerical Differentiation	355
VI.	Approximate Integration or Quadrature	359

VII.	Differential Equations	362
VIII.	Equations in a Single Unknown	364
IX.	Systems of Linear Equations	366
X.	Special Methods for Solving Sparse Sets of Equations.	369
	Appendix	371
	References	372

Chapter 6 / Group Theory
A. T. Amos

I.	Introduction	374
II.	Definitions	375
III.	Symmetry Operators	376
IV.	Group Representation Theory	385
V.	Some Applications in Molecular Quantum Mechanics	401
VI.	The Permutation Group and Spin	411
VII.	Continuous Groups	417
VIII.	Group Theory and the Solid State	421
	References	426

Chapter 7 / Density Matrices
F. David Peat

I.	Introduction	430
II.	The Full Density Matrix	431
III.	The Reduced Density Matrix	435
IV.	The N-Representability Problem	442
V.	The Single-Particle Reduced Density Matrix	447
VI.	The Second-Order Reduced Density Matrix	458
VII.	General Geminal Wave Functions	474
VIII.	Condensation Phenomena	480
	References	485

Chapter 8 / The Green's Function Method
C. Mavroyannis

I.	Introduction	488
II.	Double-Time Temperature-Dependent Green's Functions	490
III.	Spectral Representations	494

IV.	Properties of the Green's Functions	500
V.	The Reaction of a System to an External Perturbation	506
VI.	Calculation of the Green's Functions	517
VII.	Charge-Transfer Spectra of Molecular Crystals	527
VIII.	Perturbation Theory for the Green's Functions	539
	References	548

AUTHOR INDEX 551

SUBJECT INDEX 555

List of Contributors

Numbers in parentheses indicate the pages on which the authors' contributions begin.

A. T. Amos, Department of Mathematics, The University, Nottingham, England (373)

E. W. Grundke,* Theoretical Physics Institute, University of Alberta, Edmonton, Alberta, Canada (81)

Douglas Henderson, IBM Research Laboratories, San Jose, California (261)

W. D. Hoskins, Department of Computer Science, University of Manitoba, Winnipeg, Manitoba, Canada (337)

R. J. Jacob, Department of Physics, Arizona State University, Tempe, Arizona (1)

C. Mavroyannis, Division of Chemistry, National Research Council of Canada, Ottawa, Ontario, Canada (487)

F. David Peat, Division of Chemistry, National Research Council of Canada, Ottawa, Ontario, Canada (429)

Harris J. Silverstone, Department of Chemistry, The Johns Hopkins University, Baltimore, Maryland (151)

R. G. Stanton, Department of Computer Science, University of Manitoba, Winnipeg, Manitoba, Canada (337)

---

* Present address: Department of Physics, St. Francis Xavier University, Antigonish, Nova Scotia, Canada.

Foreword

In recent years there has been a tremendous expansion in the development of the techniques and principles of physical chemistry. As a result most physical chemists find it difficult to maintain an understanding of the entire field.

The purpose of this treatise is to present a comprehensive treatment of physical chemistry for advanced students and investigators in a reasonably small number of volumes. We have attempted to include all important topics in physical chemistry together with borderline subjects which are of particular interest and importance. The treatment is at an advanced level. However, elementary theory and facts have not been excluded but are presented in a concise form with emphasis on laws which have general importance. No attempt has been made to be encyclopedic. However, the reader should be able to find helpful references to uncommon facts or theories in the index and bibliographies.

Since no single physical chemist could write authoritatively in all the areas of physical chemistry, distinguished investigators have been invited to contribute chapters in the field of their special competence.

If these volumes are even partially successful in meeting these goals we will feel rewarded for our efforts.

We would like to thank the authors for their contributions and to thank the staff of Academic Press for their assistance.

HENRY EYRING
DOUGLAS HENDERSON
WILHELM JOST

Preface

Mathematics is at the very heart of physical chemistry. Indeed, the fact that physical phenomena are expressible in mathematical terms is the reason for the high level of development of physical science relative to other branches of science. For this reason a complete volume in this treatise has been devoted to mathematical techniques of interest to chemists. We have assumed the reader to be familiar only with those mathematical concepts, such as differentiation, integration, and ordinary differential equations, which might reasonably be encountered in an undergraduate chemistry major. The chapters here should provide the student of physical chemistry with a basic understanding of those additional mathematical techniques which are important in chemistry and should enable him to read the current literature in theoretical chemistry. The early chapters (roughly Chapters 1–6) are relatively elementary. Nonetheless, we hope that these chapters will be an addition to the literature because they are written by scientists (mainly chemists or physicists with strong chemical interests) or by mathematicians with applied interests and so should be especially useful to practicing chemists. The remaining chapters, again written by chemists and emphasizing chemical applications rather than mathematical rigor, are more advanced and specialized. Some chapters contain material never before published in a book directed particularly to chemists.

DOUGLAS HENDERSON

Contents of Previous and Future Volumes

VOLUME I

Chapter 1 / SURVEY OF FUNDAMENTAL LAWS
R. Haase

Chapter 2A / EQUILIBRIUM, STABILITY, AND DISPLACEMENTS
A. Sanfeld

Chapter 2B / IRREVERSIBLE PROCESSES
A. Sanfeld

Chapter 2C / THERMODYNAMICS OF SURFACES
A. Sanfeld

Chapter 3 / THERMODYNAMIC PROPERTIES OF GASES, LIQUIDS, AND SOLIDS
R. Haase

Chapter 4 / GAS–LIQUID AND GAS–SOLID EQUILIBRIA AT HIGH PRESSURE, CRITICAL CURVES, AND MISCIBILITY GAPS
E. U. Franck

Chapter 5 / THERMODYNAMICS OF MATTER IN GRAVITATIONAL, ELECTRIC, AND MAGNETIC FIELDS
Herbert Stenschke

Chapter 6 / THE THIRD LAW OF THERMODYNAMICS
J. Wilks

Chapter 7 / PRACTICAL TREATMENT OF COUPLE GAS EQUILIBRIUM
Max Klein

Chapter 8 / EQUILIBRIA AT VERY HIGH TEMPERATURES
H. Krempl

Chapter 9 / HIGH PRESSURE PHENOMENA
Robert H. Wentorf, Jr.

Chapter 10 / CARATHÉODORY'S FORMULATION OF THE SECOND LAW
S. M. Blinder

AUTHOR INDEX—SUBJECT INDEX

VOLUME II

Chapter 1 / CLASSICAL STATISTICAL THERMODYNAMICS
John E. Kilpatrick

Chapter 2 / QUANTUM STATISTICAL MECHANICS
D. ter Haar

Chapter 3 / CRYSTAL AND BLACKBODY RADIATION
Sheng Hsien Lin

Chapter 4 / DIELECTRIC, DIAMAGNETIC, AND PARAMAGNETIC PROPERTIES
William Fuller Brown, Jr.

Chapter 5 / ELECTRONS IN SOLIDS
Peter Gibbs

Chapter 6 / REAL GASES
C. F. Curtiss

Chapter 7 / EQUILIBRIUM THEORY OF LIQUIDS AND LIQUID MIXTURES
Douglas Henderson and Sydney G. Davison

Chapter 8 / ELECTROLYTIC SOLUTIONS
H. Ted Davis

Chapter 9 / SURFACES OF SOLIDS
L. J. Slutsky and G. D. Halsey, Jr.

AUTHOR INDEX—SUBJECT INDEX

VOLUME III

Chapter 1 / BASIC PRINCIPLES AND METHODS OF QUANTUM MECHANICS
D. ter Haar

Chapter 2 / ATOMIC STRUCTURE
Sidney G. Davidson

Chapter 3 / VALENCE BOND AND MOLECULAR ORBITAL METHODS
Ernest R. Davidson

Chapter 4 / ELECTRON CORRELATION IN ATOMS AND MOLECULES
Ruben Pauncz

Chapter 5 / ATOMIC SPECTRA
W. R. Hindmarsh

Chapter 6 / ELECTRONIC SPECTRA OF DIATOMIC MOLECULES
R. W. Nicholls

Chapter 7 / ELECTRONIC SPECTRA OF POLYATOMIC MOLECULES
Lionel Goodman and J. M. Hollas

Chapter 8 / PI ELECTRON THEORY OF THE SPECTRA OF CONJUGATED MOLECULES
G. G. Hall and A. T. Amos

Chapter 9 / IONIZATION POTENTIALS AND ELECTRON AFFINITIES
Charles A. McDowell

Chapter 10 / ELECTRON DONOR–ACCEPTOR COMPLEXES AND CHARGE TRANSFER SPECTRA
Robert S. Mulliken and Willis B. Person

AUTHOR INDEX—SUBJECT INDEX

VOLUME IV

Chapter 1 / THE VARIETY OF STRUCTURES WHICH INTEREST CHEMISTS
S. H. Bauer

Chapter 2 / ROTATION OF MOLECULES
C. C. Costain

Chapter 3 / THE VIBRATION OF MOLECULES
Gerald W. King

Chapter 4 / VIBRATIONAL SPECTRA OF MOLECULES
J. R. Hall

Chapter 5 / SPECTRA OF RADICALS
Dolphus E. Milligan and Marilyn E. Jacox

Chapter 6 / THE MOLECULAR FORCE FIELD
Takehiko Shimanouchi

Chapter 7 / INTERACTIONS AMONG ELECTRONIC, VIBRATIONAL, AND ROTATIONAL MOTIONS
Jon T. Hougen

Chapter 8 / ELECTRIC MOMENTS OF MOLECULES
A. D. Buckingham

Chapter 9 / NUCLEAR MAGNETIC RESONANCE SPECTROSCOPY
R. M. Golding

Chapter 10 / ESR SPECTRA
Harry G. Hecht

Chapter 11 / NUCLEAR QUADRUPLE RESONANCE SPECTROSCOPY
Ellory Schempp and P. J. Bray

Chapter 12 / MÖSSBAUER SPECTROSCOPY
N. N. Greenwood

Chapter 13 / MOLECULAR-BEAM SPECTROSCOPY
C. R. Mueller

Chapter 14 / DIFFRACTION OF ELECTRONS BY GASES
S. H. Bauer

AUTHOR INDEX—SUBJECT INDEX

VOLUME V

Chapter 1 / GENERAL REMARKS ON ELECTRONIC STRUCTURE
E. Teller and H. L. Sahlin

Chapter 2 / THE HYDROGEN MOLECULAR ION AND THE GENERAL THEORY OF ELECTRON STRUCTURE
E. Teller and H. L. Sahlin

Chapter 3 / THE TWO-ELECTRON CHEMICAL BOND
Harrison Shull

Chapter 5 / COORDINATION COMPOUNDS
T. M. Dunn

Chapter 6 / σ Bonds
C. A. Coulson

Chapter 7 / π Bonds
C. A. Coulson

Chapter 8 / Hydrogen Bonding
Sheng Hsien Lin

Chapter 9 / Multicentered Bonding
Kenneth S. Pitzer

Chapter 10 / Metallic Bonds
Walter A. Harrison

Chapter 11 / Rare-Gas Compounds
Herbert H. Hyman

Chapter 12 / Intermolecular Forces
Taro Kihara

Author Index—Subject Index

VOLUME VIA

Chapter 1 / Formal Kinetics
W. Jost

Chapter 2 / Survey of Kinetic Theory
C. F. Curtiss

Chapter 3 / Potential Energy Surfaces
H. Eyring and S. H. Lin

Chapter 4 / Theory of Energy Transfer in Molecular Collisions
E. E. Nikitin

Chapter 5 / Molecular Beam Scattering Experiments on Elastic, Inelastic, and Reactive Collisions
J. Peter Toennies

Chapter 6 / The Dynamics of Bimolecular Reactions
J. C. Polanyi and J. L. Schreiber

Author Index—Subject Index

VOLUME VIB

Chapter 7 / ELASTIC AND REACTIVE SCATTERING OF IONS ON MOLECULES
Arnim Henglein

Chapter 8 / COLLISION PROCESSES, THEORY OF ELASTIC SCATTERING
H. Pauly

Chapter 9 / ATOM REACTIONS
Juergen Wolfrum

Chapter 10 / RELAXATION METHODS IN GASES
A. B. Callear

Chapter 11 / UNIMOLECULAR REACTIONS, EXPERIMENTS AND THEORIES
Jürgen Troe

Chapter 12 / INTERACTIONS OF CHEMICAL REACTIONS, TRANSPORT PROCESSES AND FLOW
K. H. Hoyermann

AUTHOR INDEX—SUBJECT INDEX

VOLUME VII

Chapter 1 / THEORY OF REACTION RATES IN CONDENSED PHASES
S. H. Lin, K. P. Li, and H. Eyring

Chapter 2 / METHODS FOR THE ESTIMATION OF RATE PARAMETERS OF ELEMENTARY PROCESSES
Sidney W. Benson and David Golden

Chapter 3 / USE OF CORRELATION DIAGRAMS FOR INTERPRETATION OF ORGANIC REACTIVITY
J. Michl

Chapter 4 / PERTURBATION OF REACTIONS
Ernest Grunwald and John W. Leffler

Chapter 5 / MECHANISMS OF INORGANIC REACTIONS IN SOLUTION
R. G. Pearson and P. C. Ellgen

Chapter 6 / KINETICS OF FREE-RADICAL REACTIONS
Earl S. Huyser

Chapter 7 / HETEROGENEOUS CATALYSIS
M. Boudart

Chapter 8 / REACTIONS AT SURFACES
Milton E. Wadsworth

Chapter 9 / CHEMICAL ANNEALING REACTIONS IN SOLIDS
A. G. Maddock

Chapter 10 / REACTIONS OF SOLVATED ELECTRONS
Max S. Matheson

Chapter 11 / ISOTOPES AS PROBES IN DETERMINING REACTION MECHANISMS
Leonard D. Spicer and C. Dale Poulter

Chapter 12 / NUCLEATION IN LIQUID SOLUTIONS
M. Kahlweit

Chapter 13 / RADIATION CHEMISTRY IN CONDENSED PHASES
Asokendu Mozumder and John L. Magee

AUTHOR INDEX—SUBJECT INDEX

VOLUME VIIIA

Chapter 1 / INTRODUCTION
Robert L. Scott

Chapter 2 / STRUCTURE OF LIQUIDS
Sow-Hsin Chen

Chapter 3 / COMPUTER CALCULATION FOR MODEL SYSTEMS
F. H. Ree

Chapter 4 / DISTRIBUTION FUNCTIONS
R. J. Baxter

Chapter 5 / THE SIGNIFICANT STRUCTURE THEORY OF LIQUIDS
Mu Shik Jhon and Henry Eyring

Chapter 6 / PERTURBATION THEORIES
Douglas Henderson and J. A. Barker

AUTHOR INDEX—SUBJECT INDEX

VOLUME VIIIB

Chapter 7 / LIQUID MIXTURES
  Douglas Henderson and Peter J. Leonard

Chapter 8 / LIQUID HELIUM
  D. ter Haar

Chapter 9 / TIME-DEPENDENT PROPERTIES OF CONDENSED MEDIA
  Bruce J. Berne

Chapter 10 / CRITICAL PHENOMENA: STATIC ASPECTS
  John Stephenson

Chapter 11 / DYNAMIC CRITICAL PHENOMENA IN FLUID SYSTEMS
  H. Eugene Stanley, Gerald Paul, and Sava Milosevic

AUTHOR INDEX—SUBJECT INDEX

VOLUME IXA

Chapter 1 / SOME ASPECTS OF THE THERMODYNAMIC AND TRANSPORT BEHAVIOR OF ELECTROLYTES
  B. E. Conway

Chapter 2 / THE ELECTRICAL DOUBLE LAYER
  C. A. Barlow, Jr.

Chapter 3 / PRINCIPLES OF ELECTRODE KINETICS
  Terrell N. Andersen and Henry Eyring

Chapter 4 / TECHNIQUES FOR THE STUDY OF ELECTRODE PROCESSES
  Ernest Yeager and Jaroslav Kuta

Chapter 5 / SEMICONDUCTOR ELECTROCHEMISTRY
  Heinz Gerisher

AUTHOR INDEX—SUBJECT INDEX

VOLUME IXB

Chapter 6 / GAS EVOLUTION REACTIONS
  J. Horiuti

Chapter 7 / THE MECHANISM OF DEPOSITION AND DISSOLUTION OF METALS
John O'M. Bockris and Aleksander R. Despić

Chapter 8 / FAST IONIC REACTIONS
Edward M. Eyring

Chapter 9 / ELECTROCHEMICAL ENERGY CONVERSION
M. Eisenberg

Chapter 10 / FUSED-SALT ELECTROCHEMISTRY
G. E. Blomgren

Chapter 11 / BIOELECTROCHEMISTRY
J. Walter Woodbury, Stephen H. White, Michael C. Mackay, William L. Hardy, and David B. Chang

AUTHOR INDEX—SUBJECT INDEX

VOLUME X

Chapter 1 / DIFFRACTION OF X-RAYS, ELECTRONS, AND NEUTRONS ON THE REAL CRYSTAL
Alarich Weiss and Helmut Witte

Chapter 2 / DISLOCATIONS
P. Haasen

Chapter 3 / DEFECTS IN IONIC CRYSTALS
L. W. Barr and A. B. Lidiard

Chapter 4 / THE CHEMISTRY OF COMPOUND SEMICONDUCTORS
F. A. Kröger

Chapter 5 / CORRELATION EFFECTS IN DIFFUSION IN SOLIDS
A. D. Le Claire

Chapter 6 / SEMICONDUCTORS: FUNDAMENTAL PRINCIPLES
Otfried Madelung

Chapter 7 / SEMICONDUCTOR SURFACES
G. Ertl and H. Gerischer

Chapter 8 / ORGANIC SEMICONDUCTORS
J. H. Sharp and M. Smith

Chapter 9 / PHOTOCONDUCTIVITY OF SEMICONDUCTORS
Richard H. Bube

Chapter 10 / ORDER-DISORDER TRANSFORMATIONS
Hiroshi Sato

Chapter 11 / PRECIPITATION AND AGING
M. Kahlweit

AUTHOR INDEX—SUBJECT INDEX

VOLUME XIB

Chapter 9 / METHODS IN LATTICE STATISTICS
N. W. Dalton

Chapter 10 / PROBABILITY THEORY AND STOCHASTIC PROCESSES
Donald A. McQuarrie

Chapter 11 / NON-EQUILIBRIUM PROBLEMS—PROJECTION OPERATOR TECHNIQUES
J. T. Hynes and J. M. Deutch

Chapter 12 / SCATTERING THEORY
F. David Peat

Chapter 13 / THE SOLUTION OF INTEGRAL AND DIFFERENTIAL EQUATIONS
R. L. Somorjai

AUTHOR INDEX—SUBJECT INDEX

Chapter 1

Linear Vector Spaces

R. J. Jacob

I.	Introduction: Vectors in the Physical Sciences	2
II.	Linear Vector Spaces	3
	A. Definitions	3
	B. Schwartz's Inequality	5
	C. Linear Independence of Vectors	5
	D. Vector Components	8
III.	Example: Three-Dimensional Euclidean Vectors—I	10
	A. Vector Operations	11
	B. Orthonormal Basis and Components	15
	C. Vector Product	16
	D. Vector Fields	18
IV.	Vector Transformations	32
	A. Linear Operators	33
	B. Matrix Elements	36
	C. Eigenvectors and Eigenvalues	38
	D. Hermitian Operators	39
	E. Transformation of Basis	45
	F. Unitary Operators	48
	G. Projection Operators	49
V.	Matrices	51
	A. Basic Matrix Operations	51
	B. Special Matrices	54
	C. Determinants	55
	D. Inverse of a Matrix	56
	E. Eigenvalues of Matrices	59
	F. Basis Matrices of Small Dimension	63
VI.	Example: Three-Dimensional Euclidean Vectors—II	64
	A. Rotations of Coordinate Axes	64
	B. Tensors	69
VII.	Vector Spaces of Infinite Dimension	71
	A. Completeness in Hilbert Spaces	72
	B. Algebraic Eigenvalue Methods	75
	References	78

I. Introduction: Vectors in the Physical Sciences

The concept of a *vector* is a mathematical abstraction; it has, however, in numerous *realizations*, rendered extremely useful service to all areas of physical science. The most familiar applications of vectors are those in which various physical quantities are represented by three-dimensional *Euclidean vectors*, as for example, in Newtonian mechanics, hydrodynamics, and electrodynamics. But there are many other realizations of the notions of *vector* and *vector space* which are indispensable in applied mathematics: the generalization of the three-dimensional vectors to the non-Euclidean four-dimensional space–time vectors of the theories of special and general relativity; n-dimensional vectors used in countless situations when systems of n degrees of freedom, internal or external, are being investigated; and, indeed, even vectors of infinite dimension, as are the state functions of quantum mechanics.

A physical scientist, of course, ought to be familiar with all the vector methods at his disposal, but this information is generally obtained in a disjoint manner with the underlying unifying concepts usually lost among the details of the particular application. This is unfortunate since a unified concept of vectors can often be useful in relating apparently unrelated ideas either in fact or by analogy. An important historical example of the former was the demonstration by Schroedinger, Dirac, and von Neumann that the wave mechanics of Schroedinger and the matrix mechanics of Heisenberg are equivalent, being different realizations of quantum mechanics as expressed abstractly in an infinite-dimensional vector space called *Hilbert space*.

It is our intent here to develop the notion of *linear vector space* in such a way that the reader will become familiar with the properties used in most applications. To provide an example against which the reader can compare the abstractions, we parallel the development with a discussion of three-dimensional Euclidean vectors. Here, due to the importance of this particular type of vector, we depart long enough from the logical thread of the chapter to discuss in some depth the calculus of Euclidean vector fields.

Requirements of space force us, in keeping to our program, to avoid formality and depth in several instances, as well as limit the number of examples we can give. In such cases, we hope the references we have listed at the end of the chapter will enable the reader to pursue matters to his satisfaction.

II. Linear Vector Spaces

A. Definitions

A set of entities $\mathbf{V}_1, \mathbf{V}_2, \ldots$, constitutes a *linear vector space* $\mathscr{V}$, and the **V**'s are thus named *vectors*, if they satisfy the following requirements:

(i) There is defined a rule of combination among them called *vector addition* and written $\mathbf{V}_a \oplus \mathbf{V}_b$, such that

(a) the vector sum of any two members of $\mathscr{V}$ results in yet another member of $\mathscr{V}$ (i.e., $\mathscr{V}$ is *closed* under vector addition),

(b) vector addition is *commutative*:

$$\mathbf{V}_a \oplus \mathbf{V}_b = \mathbf{V}_b \oplus \mathbf{V}_a, \tag{2.1a}$$

(c) and *associative*:

$$(\mathbf{V}_a \oplus \mathbf{V}_b) \oplus \mathbf{V}_c = \mathbf{V}_a \oplus (\mathbf{V}_b \oplus \mathbf{V}_c), \tag{2.1b}$$

and

(d) one of the vectors in $\mathscr{V}$, **0**, behaves as a null vector, i.e.,

$$\mathbf{V}_a \oplus \mathbf{0} = \mathbf{V}_a \tag{2.1c}$$

for all $\mathbf{V}_a$.

(ii) One can combine them with complex numbers[†] by means of a *scalar multiplication*, written $\alpha \cdot \mathbf{V}$, in such a way that

(a) the product of a member of $\mathscr{V}$ with a scalar is also a member of $\mathscr{V}$ (i.e., $\mathscr{V}$ is closed under scalar multiplication),

(b) scalar multiplication is associative:

$$\alpha \cdot (\beta \cdot \mathbf{V}_a) = (\alpha\beta) \cdot \mathbf{V}_a, \tag{2.2a}$$

(c) and distributive under both scalar addition:

$$(\alpha + \beta) \cdot \mathbf{V}_a = \alpha \cdot \mathbf{V}_a \oplus \beta \cdot \mathbf{V}_a, \tag{2.2b}$$

and vector addition:

$$\alpha \cdot (\mathbf{V}_a \oplus \mathbf{V}_b) = \alpha \cdot \mathbf{V}_a \oplus \alpha \cdot \mathbf{V}_b, \tag{2.2c}$$

[†] Scalar multiplication may also be limited in its definition to the field of real numbers, as, for example, in the case of ordinary three-dimensional Euclidean vectors (cf. Section III).

and, in particular,

(d)
$$1 \cdot \mathbf{V}_a = \mathbf{V}_a \tag{2.2d}$$

for all $\mathbf{V}_a$.

It is apparent from requirements (i) and (ii) that the number of vectors in a vector space is infinite.

There must also be prescribed:

(iii) A *scalar product* between any two vectors in $\mathscr{V}$, written $(\mathbf{V}_a, \mathbf{V}_b)$, which is

(a) a complex scalar quantity,

(b) distributive in the *prefactor*:

$$(\mathbf{V}_a \oplus \mathbf{V}_b, \mathbf{V}_c) = (\mathbf{V}_a, \mathbf{V}_c) + (\mathbf{V}_b, \mathbf{V}_c), \tag{2.3a}$$

(c) and further, satisfies the relations

$$(\mathbf{V}_a, \alpha \cdot \mathbf{V}_b) = \alpha(\mathbf{V}_a, \mathbf{V}_b), \tag{2.3b}$$

$$(\mathbf{V}_a, \mathbf{V}_b) = (\mathbf{V}_b, \mathbf{V}_a)^*, \tag{2.3c}^\dagger$$

and

$$(\mathbf{V}_a, \mathbf{V}_a) \geq 0, \tag{2.3d}$$

the equality holding if and only if $\mathbf{V}_a = \mathbf{0}$.

We can see immediately from (2.3a) and (2.3b) that the scalar product is also distributive in the *postfactor*:

$$(\mathbf{V}_a, \mathbf{V}_b \oplus \mathbf{V}_c) = (\mathbf{V}_b \oplus \mathbf{V}_c, \mathbf{V}_a)^* = (\mathbf{V}_b, \mathbf{V}_a)^* + (\mathbf{V}_c, \mathbf{V}_a)^*$$
$$= (\mathbf{V}_a, \mathbf{V}_b) + (\mathbf{V}_a, \mathbf{V}_c). \tag{2.3e}$$

The analog of (2.3b) for scalars in the prefactor is also derived simply:

$$(\alpha \cdot \mathbf{V}_a, \mathbf{V}_b) = (\mathbf{V}_b, \alpha \cdot \mathbf{V}_a)^* = \alpha^*(\mathbf{V}_b, \mathbf{V}_a)^* = \alpha^*(\mathbf{V}_a, \mathbf{V}_b). \tag{2.3f}$$

We should mention that the scalar product is not always given as part of the definition of a vector space, but inasmuch as vectors are generally useless to us in applied mathematics without a scalar product, we incorporate it as part of the fundamental properties. The provision of a scalar product is said to define a linear vector space *with metric*.

---

† An asterisk (*) indicates complex conjugate.

As an exercise, the reader can show how vector subtraction, $\mathbf{V}_a \ominus \mathbf{V}_b$, may be defined in terms of vector addition and scalar multiplication. It can also be shown that the null vector is unique within a given vector space and can be obtained by scalar multiplication of any vector by zero.

The operations of vector addition and scalar multiplication should not be confused with ordinary arithmetical addition and multiplication, that is, the kind that is carried on between scalar quantities alone. Vectors are, at this point, completely abstract objects. It is to emphasize this that we introduced the symbols · and $\oplus$. The relationships between vector and arithmetical operations become apparent later when we discuss vector components. We are then able to drop the new symbols and use conventional addition and multiplication notation without fear of confusion.

B. Schwartz's Inequality

The quantity
$$\| \mathbf{V}_a \| = (\mathbf{V}_a, \mathbf{V}_a)^{1/2} \tag{2.4}$$
is called the *norm*, or sometimes loosely, the "length" of the vector $\mathbf{V}_a$. Because of (2.3d) this is always real. We will usually write $\| \mathbf{V}_a \|$ simply as V_a. A useful inequality involving the norms of vectors is *Schwartz's inequality*:
$$| (\mathbf{V}_a, \mathbf{V}_b) | \leq \| \mathbf{V}_a \| \, \| \mathbf{V}_b \| \tag{2.5}$$
where | | indicates absolute value. The equality here holds if and only if $\mathbf{V}_b$ is a scalar multiple of $\mathbf{V}_a$. We can prove (2.5) by forming the quantity
$$(\mathbf{V}_a \oplus \mu \cdot \mathbf{V}_b, \mathbf{V}_a \oplus \mu \cdot \mathbf{V}_b)$$
which is, by (2.3d), nonnegative for *any* μ. Using (2.3a), (2.3b), and (2.3f) to expand this quantity, we obtain
$$(\mathbf{V}_a, \mathbf{V}_a) + \mu^*(\mathbf{V}_b, \mathbf{V}_a) + \mu(\mathbf{V}_a, \mathbf{V}_b) + | \mu |^2 (\mathbf{V}_b, \mathbf{V}_b) \geq 0.$$
Choosing
$$\mu = -(\mathbf{V}_b, \mathbf{V}_a)/(\mathbf{V}_b, \mathbf{V}_b),$$
we arrive at the desired inequality.

C. Linear Independence of Vectors

If the only solution (set of α's) of the equation
$$\alpha_1 \cdot \mathbf{V}_1 \oplus \alpha_2 \cdot \mathbf{V}_2 \oplus \cdots \oplus \alpha_n \cdot \mathbf{V}_n = \mathbf{0} \tag{2.6}$$

is the trivial one

$$\alpha_1 = \alpha_2 = \cdots = \alpha_n = 0,$$

then the n vectors

$$\{\mathbf{V}_n\} = \mathbf{V}_1, \mathbf{V}_2, \ldots, \mathbf{V}_n$$

are said to be *linearly independent*. This means that no vector in this set can be expressed as a linear combination of some of all of the others. Now, if it is the case that there are n linearly independent vectors in $\mathscr{V}$, but not $n+1$, then $\mathscr{V}$ is said to have *dimension n*. (For the moment, let us consider n to be finite.) Consequently, *any* vector in an n-dimensional vector space $\mathscr{V}_n$ can be written as a linear combination of n or less linearly independent vectors. Another way of saying this is that a linearly independent set $\{\mathbf{V}_n\}$ *generates* the vector space $\mathscr{V}_n$, or again, the space $\mathscr{V}_n$ is the *linear manifold* of the set $\{\mathbf{V}_n\}$. Such a set is called a *basis*, and the members *basis vectors*.

It can easily be seen that a given vector space has an infinite number of basis sets. For, if $\mathbf{V}_1, \mathbf{V}_2, \ldots, \mathbf{V}_n$ form a basis of $\mathscr{V}_n$, then so do $\alpha \cdot \mathbf{V}_1 \oplus \beta \cdot \mathbf{V}_2, \alpha \cdot \mathbf{V}_1 \ominus \beta \cdot \mathbf{V}_2, \ldots, \mathbf{V}_n$, for example, because of linear independence. Thus it may be possible to select a particularly appropriate basis to use in a given situation.

1. *Orthonormal Basis*

A property that helps to simplify vector space calculations is *orthogonality*. The vectors $\mathbf{V}_a$ and $\mathbf{V}_b$ are mutually *orthogonal* if

$$(\mathbf{V}_a, \mathbf{V}_b) = 0.$$

Orthogonal vectors are linearly independent as can be seen by assuming the vectors in (2.6) to be mutually orthogonal and taking the scalar product of (2.6) with each vector in the expression in turn. Thus, for example, one obtains $\alpha_1(\mathbf{V}_1, \mathbf{V}_1) = 0$ or $\alpha_1 = 0$. The remaining α's are similarly seen to vanish. It is particularly convenient to choose as a basis of $\mathscr{V}_n$ a set of orthogonal vectors. One can add further simplification by *normalizing* the basis vectors, that is, by transforming them thus:

$$\mathbf{V}_i \to \hat{\mathbf{V}}_i \equiv (1/V_i) \cdot \mathbf{V}_i, \tag{2.7}$$

where the norm of $\hat{\mathbf{V}}_i$ is obviously unity (as is indicated from here on by a caret). If the orthogonal basis vectors are also normalized, they obey

the relationship
$$(\hat{\mathbf{V}}_i, \hat{\mathbf{V}}_j) = \delta_{ij}, \tag{2.8}$$

where δ_{ij} is the Kronecker delta:
$$\delta_{ij} = \begin{cases} 1 & \text{for } i = j \\ 0 & \text{for } i \neq j. \end{cases} \tag{2.9}$$

In the remainder of this chapter, we assume basis vectors to be *orthonormal* (orthogonal and normalized).

2. Schmidt Orthogonalization Process

The question remains as to how to form an orthonormal basis from an arbitrary basis; although orthogonal vectors are linearly independent, the converse is not generally true. The method of forming n orthonormal vectors from n linearly independent vectors is straightforward and is known as the *Schmidt orthogonalization process*. Beginning with the linearly independent set $\mathbf{V}_1, \mathbf{V}_2, \ldots, \mathbf{V}_n$, we first set

$$\mathbf{V}_1' = (1/V_1) \cdot \mathbf{V}_1. \tag{2.10}$$

We then form
$$\mathbf{V}_2' = \mathbf{V}_2 \ominus \alpha \cdot \hat{\mathbf{V}}_1'$$

where α is so chosen that $(\hat{\mathbf{V}}_1', \mathbf{V}_2') = 0$. A simple calculation gives $\alpha = (\hat{\mathbf{V}}_1', \mathbf{V}_2)$. Thus, after normalization,

$$\hat{\mathbf{V}}_2' = (V_2^2 - |(\hat{\mathbf{V}}_1', \mathbf{V}_2)|^2)^{-1/2} \cdot (\mathbf{V}_2 \ominus (\hat{\mathbf{V}}_1', \mathbf{V}_2) \cdot \hat{\mathbf{V}}_1'). \tag{2.11}$$

Proceeding in the same way, we take

$$\mathbf{V}_3' = \mathbf{V}_3 \ominus \beta \cdot \hat{\mathbf{V}}_2' \ominus \gamma \cdot \hat{\mathbf{V}}_1',$$

where $\beta = (\hat{\mathbf{V}}_2', \mathbf{V}_3)$ and $\gamma = (\hat{\mathbf{V}}_1', \mathbf{V}_3)$ assure that $(\mathbf{V}_3', \hat{\mathbf{V}}_2') = 0$ and $(\mathbf{V}_3', \hat{\mathbf{V}}_1') = 0$, respectively. One then normalizes $\mathbf{V}_3'$ and continues on through the entire set, ending up with the orthonormal set $\hat{\mathbf{V}}_1', \hat{\mathbf{V}}_2', \ldots, \hat{\mathbf{V}}_n'$. The assumed linear independence of the initial vectors in the preceding demonstration assures the existence of the vectors $\hat{\mathbf{V}}_i'$. If, for example, $\mathbf{V}_1$ and $\mathbf{V}_2$ were not linearly independent, i.e., $\mathbf{V}_2 = \mu \cdot \mathbf{V}_1$, say, then the condition $(\hat{\mathbf{V}}_1', \hat{\mathbf{V}}_2') = 0$ yields $\alpha = \mu V_1$ or $\hat{\mathbf{V}}_2' = 0$. A similar argument applies throughout the set.

D. Vector Components

Let us consider then an n-dimensional vector space generated by an orthonormal basis $\{\hat{\mathbf{V}}_i\}$. Any vector in the space can be written as a linear combination of the basis vectors:

$$\mathbf{V}_a = \sum_{i=1}^{n} a_i \cdot \hat{\mathbf{V}}_i. \tag{2.12}$$

The scalar coefficients a_i are called the *components* of the vector $\mathbf{V}_a$ in the basis $\{\hat{\mathbf{V}}_i\}$. In general, they may be complex; in fact, one can easily calculate them in terms of scalar products. Merely take the scalar product of (2.12) with the basis vector $\hat{\mathbf{V}}_k$:

$$(\hat{\mathbf{V}}_k, \mathbf{V}_a) = \left(\hat{\mathbf{V}}_k, \sum_{i=1}^{n} a_i \cdot \hat{\mathbf{V}}_i\right) = \sum_{i=1}^{n} a_i (\hat{\mathbf{V}}_k, \hat{\mathbf{V}}_i) = \sum_{i=1}^{n} a_i \, \delta_{ik} = a_k,$$

or

$$a_k = (\hat{\mathbf{V}}_k, \mathbf{V}_a) \tag{2.13}$$

for all k. Equation (2.12) can thus be rewritten as

$$\mathbf{V}_a = \sum_{i=1}^{n} (\hat{\mathbf{V}}_i, \mathbf{V}_a) \cdot \hat{\mathbf{V}}_i. \tag{2.12a}$$

We can now obtain, in terms of vector components, an understanding in an arithmetical sense for the operations of vector addition, scalar multiplication, and scalar product. Thus,

$$\mathbf{V}_a \oplus \mathbf{V}_b = \sum_{i=1}^{n} a_i \cdot \hat{\mathbf{V}}_i + \sum_{i=1}^{n} b_i \cdot \hat{\mathbf{V}}_i = \sum_{i=1}^{n} (a_i + b_i) \cdot \hat{\mathbf{V}}_i. \tag{2.14}$$

Now, if we set $\mathbf{V}_a \oplus \mathbf{V}_b = \mathbf{V}_c$ and expand $\mathbf{V}_c$:

$$\mathbf{V}_c = \sum_{i=1}^{n} c_i \cdot \hat{\mathbf{V}}_i,$$

we then have

$$\sum_{i=1}^{n} c_i \cdot \hat{\mathbf{V}}_i = \sum_{i=1}^{n} (a_i + b_i) \cdot \hat{\mathbf{V}}_i.$$

Taking the scalar product of this equation with $\hat{\mathbf{V}}_k$ gives

$$c_k = a_k + b_k \tag{2.15}$$

for all k. Thus, vector addition becomes scalar addition of components. We have a similar result for scalar multiplication:

$$\alpha \cdot \mathbf{V}_a = \mathbf{V}_b = \sum_{i=1}^{n} b_i \cdot \hat{\mathbf{V}}_i = \alpha \cdot \sum_{i=1}^{n} a_i \cdot \hat{\mathbf{V}}_i = \sum_{i=1}^{n} (\alpha a_i) \cdot \hat{\mathbf{V}}_i$$

or

$$b_i = \alpha a_i, \quad i = 1, \ldots, n. \qquad (2.16)$$

As for the scalar product,

$$(\mathbf{V}_a, \mathbf{V}_b) = \left(\sum_{i=1}^{n} a_i \cdot \hat{\mathbf{V}}_i, \sum_{j=1}^{n} b_j \cdot \hat{\mathbf{V}}_j \right) = \sum_{i=1}^{n} \sum_{j=1}^{n} a_i^* b_j (\hat{\mathbf{V}}_i, \hat{\mathbf{V}}_j)$$

$$= \sum_{i=1}^{n} \sum_{j=1}^{n} a_i^* b_j \, \delta_{ij} = \sum_{i=1}^{n} a_i^* b_i. \qquad (2.17)$$

(Note that in deriving this result we used different summation indices inasmuch as we were considering the product of two sums.) As a particular instance of (2.17), we consider the norm of $\mathbf{V}_a$:

$$V_a^2 = (\mathbf{V}_a, \mathbf{V}_a) = \sum_{i=1}^{n} |a_i|^2. \qquad (2.18)$$

Equation (2.17) is often written in the form [using (2.13)]

$$(\mathbf{V}_a, \mathbf{V}_b) = \sum_{i=1}^{n} (\mathbf{V}_a, \hat{\mathbf{V}}_i)(\hat{\mathbf{V}}_i, \mathbf{V}_b). \qquad (2.17a)$$

This is called the *closure* or *completeness* relation, and is independent of the vectors $\mathbf{V}_a$ and $\mathbf{V}_b$; it is a property solely of the orthonormal basis $\{\hat{\mathbf{V}}_i\}$. A basis is said to be *complete* if (2.17a) holds, and we are thereby guaranteed that the expansion (2.12) can be accomplished. Of course, if the basis $\{\hat{\mathbf{V}}_i\}$ in an n-dimensional space has n vectors, it is complete. We will see later that the situation is not so cut-and-dried in infinite-dimensional vector spaces.

Now that we have an understanding of vector operations in terms of the scalar components, we need no longer use the symbols $\oplus$ and $\cdot$ for vector addition and scalar multiplication, but will henceforth write $\mathbf{V}_a + \mathbf{V}_b$ and $\alpha \mathbf{V}_a$.

We conclude this section by introducing a notation that is often used in vector space calculations, especially in quantum mechanics. It was developed by P. A. M. Dirac when he used vector space concepts to unify the Schroedinger and Heisenberg forms of quantum theory. In this notation, we write $\mathbf{V}_a$ as $|a\rangle$. This is called a *ket vector*. Scalar products are written

$$(\mathbf{V}_a, \mathbf{V}_b) = \langle a | b \rangle. \tag{2.19}$$

$\langle a |$ is called a *bra vector* and is related to the ket vector $|a\rangle$ in a way that is obvious when (2.19) is compared to (2.17): the components of $\langle a |$ are the complex conjugates of those of $|a\rangle$. The *bra–ket* notation is convenient when, for example, one wishes to label a vector with several labels. Compare $\mathbf{V}_{\alpha,\mu}^{\sigma,\varrho}$ with $|\alpha\mu\sigma\varrho\rangle$. Further economy is obtained in writing such relationships as closure:

$$\langle a | b \rangle = \sum_{i=1}^{n} \langle a | i \rangle \langle i | b \rangle$$

or

$$\sum_{i=1}^{n} |i\rangle\langle i| = \mathbf{I} \tag{2.20}$$

where $\mathbf{I}$ is the *identity operator* which we introduce in Section IV.

III. Example: Three-Dimensional Euclidean Vectors—I

There are in mathematics many objects which, through a proper definition of the necessary operations, can be termed vectors. The field of complex numbers, the set of all matrices of a given dimension, and the set of all complex functions $f(x)$ defined on a real interval are realizations of vector spaces. The reader can amuse himself by inventing appropriate scalar product operations for each of these examples. Perhaps the most common application of the vector concept, however, is the *three-dimensional Euclidean vector* (TDEV). The term "vector analysis" usually means, in fact, the geometry and calculus of such vectors.

In this section, we assume that the reader has acquired some familiarity with the applications of Euclidean vectors, possibly from a study of elementary mechanics or electricity and magnetism. Thus we should be able to avoid having to introduce physical justification at each stage of our discussion. We refer to physical applications when it seems appropriate, however. Also, for the sake of brevity in this section, we usually refer to Euclidean vectors simply as *vectors*.

A TDEV is a directed straight line segment in a three-dimensional Euclidean space, i.e., in a space in which the underlying geometry is Euclidean. That the set of such directed straight line segments indeed forms a vector space is a statement we must verify by showing that the operations of vector addition, scalar multiplication, and scalar product can be defined over the set in a manner consistent with the definitions in the preceding section.

The properties of a directed straight line segment which are pertinent to the segment's being a vector are its length and its orientation with respect to a given (but arbitrary) set of Cartesian coordinate axes. (In Fig. 1 we have labeled the axes x, y, and z and have drawn the vector as an arrow in order to indicate the sense of its orientation.) Although the position of a vector is often of importance in a physical application, it is irrelevant to the specification of the vector. Any other line segment of length equal to **V**'s and parallel to **V** (with, of course, the same sense) is equivalent to **V** insofar as its being a vector is concerned. Thus, a given line segment becomes a different vector if its orientation is changed, but *not* if it is simply translated with respect to the coordinate axes.

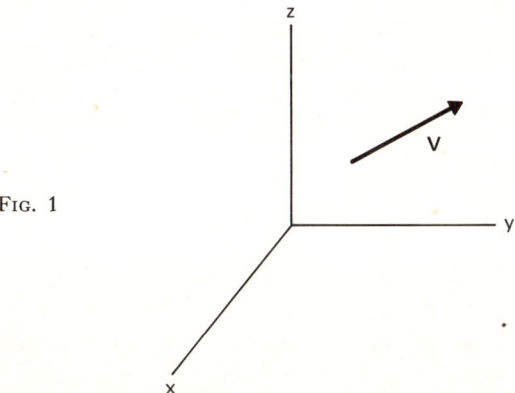

Fig. 1

A. Vector Operations

With this in mind, let us define the vector operations. (We will follow conventional use and indicate TDEVs by bold uppercase letters.)

1. *Vector Addition*

We wish to add the two vectors **A** and **B** as they appear in Fig. 2. Since we can translate a vector without altering it, we can arrange **A**

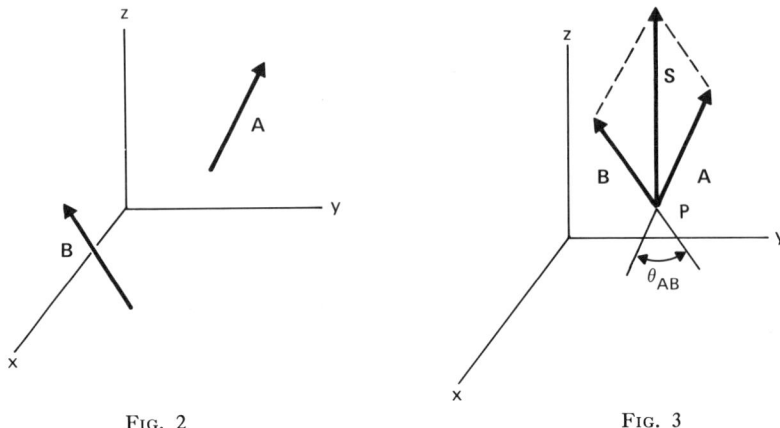

Fig. 2

Fig. 3

and **B** so that their "tails" coincide at a point P as in Fig. 3. The two vectors now define a plane in our three-dimensional space. The vector sum of **A** and **B**,

$$\mathbf{S} = \mathbf{A} + \mathbf{B},$$

is then defined by completing the parallelogram in this plane as indicated by the dashed lines, and drawing the vector **S** from P to the opposite vertex. If **A** and **B** are parallel, then **S** is simply a vector with the same direction and a length equal to the sum of their lengths, $A + B$. In general, however, the length of **S** is given by the law of cosines:

$$S^2 = A^2 + B^2 + 2AB \cos \theta_{AB} \tag{3.1}$$

where θ_{AB} is the angle of intersection of **A** and **B**.

This addition operation is clearly commutative and associative and the set of directed straight line segments is evidently closed under it. It is also easily seen that the null vector is simply a line segment of zero length, i.e. a geometrical point. If **A'** is a vector of equal length and parallel to **A** but of opposite sense, then

$$\mathbf{A'} + \mathbf{A} = \mathbf{0}.$$

Thus we may write $\mathbf{A'} = -\mathbf{A}$.

2. Scalar Multiplication

Three-dimensional Euclidean vectors are usually associated with real, rather than complex, quantities. Therefore we restrict the definition of

1. Linear Vector Spaces

B. Orthonormal Basis and Components

To construct an orthonormal basis for this vector space, we require vectors of unit length which are mutually perpendicular, i.e., $\hat{\mathbf{V}}_i, \hat{\mathbf{V}}_j, \ldots,$ where

$$\hat{\mathbf{V}}_i \cdot \hat{\mathbf{V}}_j = \cos \theta_{ij} = 0.$$

In a three-dimensional space, such a basis consists of three vectors, although there are, of course, an unlimited number of such bases. The most convenient choice is that in which the basis vectors are directed along the Cartesian axes. We call these basis vectors $\hat{\mathbf{x}}, \hat{\mathbf{y}}$, and $\hat{\mathbf{z}}$ (see Fig. 7).† The following relationships hold for them:

$$\hat{\mathbf{x}} \cdot \hat{\mathbf{x}} = \hat{\mathbf{y}} \cdot \hat{\mathbf{y}} = \hat{\mathbf{z}} \cdot \hat{\mathbf{z}} = 1, \quad \hat{\mathbf{x}} \cdot \hat{\mathbf{y}} = \hat{\mathbf{y}} \cdot \hat{\mathbf{z}} = \hat{\mathbf{x}} \cdot \hat{\mathbf{z}} = 0. \quad (3.6)$$

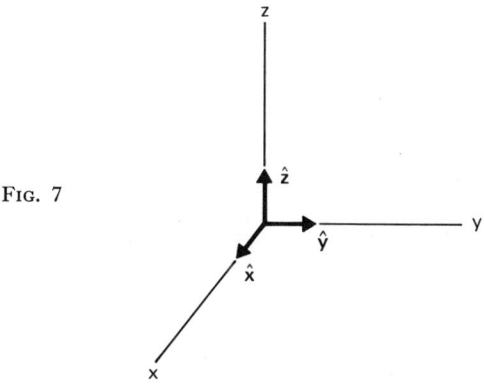

Fig. 7

As in (2.12), we can write a vector $\mathbf{A}$ in terms of the basis vectors:

$$\mathbf{A} = a_x \hat{\mathbf{x}} + a_y \hat{\mathbf{y}} + a_z \hat{\mathbf{z}}. \quad (3.7)$$

The components a_x, a_y, and a_z here are real as they represent lengths in a real space. (It is important to remember that a component of a vector is not itself a vector but a scalar.) Again, following the development of Section II.D, we can calculate a component of $\mathbf{A}$ as in (2.13), e.g.,

$$a_x = \mathbf{A} \cdot \hat{\mathbf{x}}.$$

In terms of its components, the length of a vector is [cf. Eq. (2.18)]

$$A = (\mathbf{A} \cdot \mathbf{A})^{1/2} = (a_x^2 + a_y^2 + a_z^2)^{1/2}. \quad (3.8)$$

† A more traditional notation, but one less flexible, in the opinion of the author, is $\mathbf{i}\, (= \hat{\mathbf{x}}), \mathbf{j}\, (= \hat{\mathbf{y}})$, and $\mathbf{k}\, (= \hat{\mathbf{z}})$.

The quantity
$$c_{ax} = a_x/A$$
is the direction cosine of the vector **A** with respect to the x axis, and similarly for c_{ay} and c_{az}. It is often convenient, as in Section II, to give components and basis vectors a numerical subscript notation. In doing so, we will set $a_x = a_1, a_y = a_2, a_z = a_3, \hat{\mathbf{x}} = \hat{\mathbf{x}}_1, \hat{\mathbf{y}} = \hat{\mathbf{x}}_2$, and $\hat{\mathbf{z}} = \hat{\mathbf{x}}_3$. Then,
$$\mathbf{A} = \sum_{i=1}^{3} a_i \hat{\mathbf{x}}_i.$$

The scalar product of two vectors is given in terms of their components, as in (2.17):
$$\mathbf{A} \cdot \mathbf{B} = a_1 b_1 + a_2 b_2 + a_3 b_3. \tag{3.9}$$
Thus
$$\cos \theta_{AB} = (a_1 b_1 + a_2 b_2 + a_3 b_3)/AB = c_{a1} c_{b1} + c_{a2} c_{b2} + c_{a3} c_{b3}. \tag{3.10}$$

C. Vector Product

In addition to the scalar product, there is, in TDEV spaces, a second way of combining, in a multiplicative sense, two vectors. In this case, the result is usually interpreted to be a vector (although in the strictest sense it is a tensor of rank 2; cf. Section VI.B), and the combination is called the *vector product*, or quite often, the *cross product* because of the notation **A** ✕ **B** usually given to it. Interpreted as a vector,
$$\mathbf{C} = \mathbf{A} \times \mathbf{B}$$
has a length
$$C = AB \sin \theta_{AB} \tag{3.11}$$
and a direction perpendicular to the plane of **A** and **B**. The sense of **C** is taken to be that of the advance of a right-hand threaded screw if it were turned so as to rotate the vector **A** into the vector **B** (see Fig. 8). Thus, $\mathbf{C}' = \mathbf{B} \times \mathbf{A}$ has the same length but opposite direction as **C**. The vector product is therefore not commutative:
$$\mathbf{A} \times \mathbf{B} = -\mathbf{B} \times \mathbf{A}. \tag{3.12}$$
Of course,
$$\mathbf{A} \times \mathbf{A} = \mathbf{0}. \tag{3.13}$$

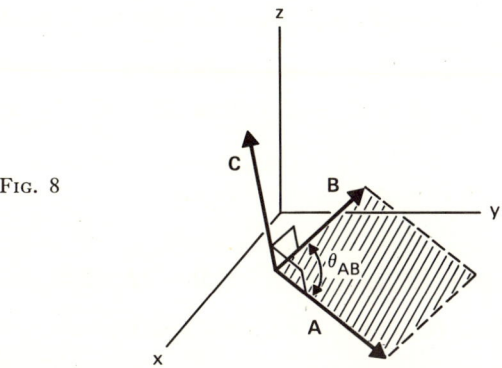

Fig. 8

The length of **C**, as given in (3.11), is equal to the area of the parallelogram formed by **A** and **B** (Fig. 8).

From the preceding definition, one can use the vector product to write the following relationships among the basis vectors:

$$\hat{x} = \hat{y} \times \hat{z} = -\hat{z} \times \hat{y}, \quad \hat{y} = \hat{z} \times \hat{x} = -\hat{x} \times \hat{z},$$
$$\hat{z} = \hat{x} \times \hat{y} = -\hat{y} \times \hat{x}. \quad (3.14)$$

The components of **C** are then easily obtained:

$$\begin{aligned}
\mathbf{C} &= c_x \hat{x} + c_y \hat{y} + c_z \hat{z} \\
&= \mathbf{A} \times \mathbf{B} \\
&= (a_x \hat{x} + a_y \hat{y} + a_z \hat{z}) \times (b_x \hat{x} + b_y \hat{y} + b_z \hat{z}) \\
&= a_x b_x \hat{x} \times \hat{x} + a_y b_y \hat{y} \times \hat{y} + a_z b_z \hat{z} \times \hat{z} \\
&\quad + (a_x b_y - a_y b_x) \hat{x} \times \hat{y} + (a_y b_z - a_z b_y) \hat{y} \times \hat{z} \\
&\quad + (a_z b_x - a_x b_z) \hat{z} \times \hat{x} \\
&= (a_x b_y - a_y b_x) \hat{z} + (a_y b_z - a_z b_y) \hat{x} + (a_z b_x - a_x b_z) \hat{y}. \quad (3.15)
\end{aligned}$$

Thus[†]

$$c_x = a_y b_z - a_z b_y, \quad c_y = a_z b_x - a_x b_z, \quad c_z = a_x b_y - a_y b_x. \quad (3.16)$$

Equation (3.15) can be written in a simple mnemonic form by means of a determinant:

$$\mathbf{C} = \begin{vmatrix} \hat{x} & \hat{y} & \hat{z} \\ a_x & a_y & a_z \\ b_x & b_y & b_z \end{vmatrix}. \quad (3.17)$$

[†] In this derivation we used the distributive property of vector products which the reader can easily obtain from the basic geometrical definition.

The following identities involving scalar and vector products are essential to vector algebra:

$$\mathbf{A} \cdot (\mathbf{B} \times \mathbf{C}) = \mathbf{B} \cdot (\mathbf{C} \times \mathbf{A}) = \mathbf{C} \cdot (\mathbf{A} \times \mathbf{B}) \quad (3.18a)$$

$$\mathbf{A} \cdot (\mathbf{B} \times \mathbf{C}) = \begin{vmatrix} a_x & a_y & a_z \\ b_x & b_y & b_z \\ c_x & c_y & c_z \end{vmatrix} \quad (3.18b)$$

$$\mathbf{A} \times (\mathbf{B} \times \mathbf{C}) = (\mathbf{A} \cdot \mathbf{C})\mathbf{B} - (\mathbf{A} \cdot \mathbf{B})\mathbf{C} \quad (3.18c)$$

$$\mathbf{A} \times [\mathbf{B} \times (\mathbf{C} \times \mathbf{A})] = (\mathbf{A} \cdot \mathbf{B})(\mathbf{A} \times \mathbf{C}) \quad (3.18d)$$

$$\mathbf{A} \times (\mathbf{B} \times \mathbf{C}) + \mathbf{B} \times (\mathbf{C} \times \mathbf{A}) + \mathbf{C} \times (\mathbf{A} \times \mathbf{B}) = 0. \quad (3.18e)$$

Equation (3.18e) is called the *Jacobi identity*.

D. VECTOR FIELDS

There exist numerous physical quantities which, by virtue of their requiring, in order to be completely determined, the specification of a magnitude *and* a direction, can be described mathematically as Euclidean vectors. The vector nature of force, velocity, momentum, and electric and magnetic field strengths, to name a few examples, is emphasized at the outset in most elementary physics courses. The equations governing these physical vectors are among the most fundamental equations of physical science. Indeed, expressing these quantities as vectors enables us to write the laws of physics more elegantly and manipulate the equations more efficiently than if we dealt with each component as a separate entity. To illustrate this, we consider the Lorentz force **F** acting on a charged particle moving with velocity **v** in an electromagnetic field of strengths **E** and **B**. In terms of components, the expression for the force has the complicated form

$$\begin{aligned} F_x &= q[E_x + (1/c)(v_y B_z - v_z B_y)] \\ F_y &= q[E_y + (1/c)(v_z B_x - v_x B_z)] \\ F_z &= q[E_z + (1/c)(v_x B_y - v_y B_x)], \end{aligned} \quad (3.19)$$

whereas with the vector form,

$$\mathbf{F} = g\mathbf{E} + (1/c)\mathbf{v} \times \mathbf{B}, \quad (3.19a)$$

and an understanding of the vector product, the nature of the Lorentz force is readily apparent. But the simplicity in the equations of physics incurred through a vector notation would be superficial [one can *learn* to recognize the meaning of the pattern in Eq. (3.19)] were it not for the algebra and calculus of vectors which we have at our disposal. With these tools, we can in most cases work directly with the vector equations to accomplish our particular task without having to consider the components separately.

We have already discussed some of the algebraic manipulations one may perform with vectors [Eqs. (3.18), for example]. In the remainder of this section, we develop the differential and integral calculus of vectors, or more precisely, of *vector fields*.

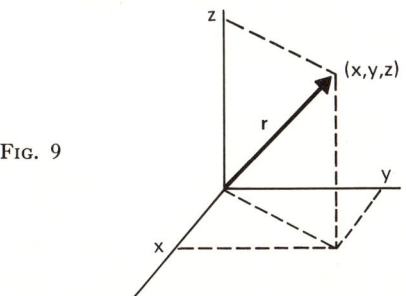

Fig. 9

The simplest example of a physical vector is the *position* or *radius vector*, which represents the position of a point in space with respect to the origin of a Cartesian coordinate system. If the coordinates of the point are x, y, and z (see Fig. 9), then the position vector can be written

$$\mathbf{r} = x\hat{\mathbf{x}} + y\hat{\mathbf{y}} + z\hat{\mathbf{z}}. \tag{3.20}^\dagger$$

The components of $\mathbf{r}$ are thus (simple, but not trivial) functions defined over the three-dimensional space. Then $\mathbf{r}$ itself can be considered a *vector function*, $\mathbf{r}(x, y, z)$, that is to say, a *mapping* of vectors in a one-to-one manner upon the points in the space. $\mathbf{r}$ is an example of a *vector field*:

A *vector field* is the ensemble of vectors prescribed by a vector function $\mathbf{A}(x, y, z)$ over a (possibly delimited) region of space.

---

† Related to the position vector is the *displacement vector*, $\Delta \mathbf{r}$, representing the separation between two points:

$$\Delta \mathbf{r} = \mathbf{r}_1 - \mathbf{r}_2 = (x_1 - x_2)\hat{\mathbf{x}} + (y_1 - y_2)\hat{\mathbf{y}} + (z_1 - z_2)\hat{\mathbf{z}}.$$

Other vector fields are the velocity vector as a function of position within a flowing substance (this gave to vector calculus much of its traditional nomenclature), the fields of gravitational or electromagnetic forces, the electromagnetic vector potential field, and the displacement field of a vibrating medium. Since the components of these vector fields are ordinary functions of the position coordinates, the mathematical concepts of existence and continuity, as found in the theory of functions, can be applied to the theory of vector fields. The equations of physics which govern these fields are, in fact, differential equations.

1. *Vector Derivatives*

The derivative of a vector field can be defined simply in terms of the derivatives of its components. Thus, if

$$\mathbf{A}(x, y, z) = A_x(x, y, z)\hat{\mathbf{x}} + A_y(x, y, z)\hat{\mathbf{y}} + A_z(x, y, z)\hat{\mathbf{z}}, \quad (3.21)$$

then we may write

$$\frac{\partial \mathbf{A}}{\partial x} = \frac{\partial A_x}{\partial x}\hat{\mathbf{x}} + \frac{\partial A_y}{\partial x}\hat{\mathbf{y}} + \frac{\partial A_z}{\partial x}\hat{\mathbf{z}}, \quad (3.22)$$

and similarly for $\partial \mathbf{A}/\partial y$ and $\partial \mathbf{A}/\partial z$. These derivatives are themselves vector fields and do or do not exist as *all three* derivatives in Eq. (3.22), for example, do or do not exist. One can easily see that (3.22) is equivalent to an alternative definition of the derivative of a vector:

$$\frac{\partial}{\partial x} \mathbf{A}(x, y, z) = \lim_{\Delta x \to 0} \frac{\mathbf{A}(x + \Delta x, y, z) - \mathbf{A}(x, y, z)}{\Delta x}. \quad (3.23)$$

The following rules of differentiation clearly hold:

$$\frac{\partial}{\partial x}(\mathbf{A} \cdot \mathbf{B}) = \mathbf{A} \cdot \frac{\partial \mathbf{B}}{\partial x} + \frac{\partial \mathbf{A}}{\partial x} \cdot \mathbf{B} \quad (3.24\text{a})$$

$$\frac{\partial}{\partial x}(\mathbf{A} \times \mathbf{B}) = \mathbf{A} \times \frac{\partial \mathbf{B}}{\partial x} + \frac{\partial \mathbf{A}}{\partial x} \times \mathbf{B} \quad (3.24\text{b})$$

$$\frac{\partial}{\partial x}(\varphi \mathbf{A}) = \varphi \frac{\partial \mathbf{A}}{\partial x} + \frac{\partial \varphi}{\partial x} \mathbf{A}, \quad (3.24\text{c})$$

where, in (3.24c), φ is a scalar field.

Physical vectors often have a dependence on time. The time derivative of a vector is defined similarly to the spatial derivative (3.22).

1. Linear Vector Spaces

An interesting and useful relationship is

$$\begin{aligned}
\mathbf{A} \cdot \frac{\partial \mathbf{A}}{\partial \xi} &= A_x \frac{\partial A_x}{\partial \xi} + A_y \frac{\partial A_y}{\partial \xi} + A_z \frac{\partial A_z}{\partial \xi} \\
&= \frac{1}{2} \frac{\partial}{\partial \xi} (A_x^2 + A_y^2 + A_z^2) = \frac{1}{2} \frac{\partial}{\partial \xi} A^2 \\
&= A \frac{\partial A}{\partial \xi},
\end{aligned} \qquad (3.25)$$

where ξ represents any of **A**'s variables, temporal or spatial. Thus, if the magnitude of **A** is to remain fixed, its differential

$$\Delta \mathbf{A} = \mathbf{A}(\xi + \Delta \xi) - \mathbf{A}(\xi)$$

must be perpendicular to **A**.

An additional complication may arise in the case of time-dependent vectors in that the time dependence may enter in two ways. First, there may be an explicit variation of the components with respect to time such as in the case of a magnetic field due to an alternating current source. But then, the source of the field, and therefore the field itself, may be in motion with respect to the coordinate system. The field at a given point then has the dependence $\mathbf{A}(t, x(t), y(t), z(t))$ and the *total* time derivative becomes

$$\frac{d\mathbf{A}}{dt} = \frac{\partial \mathbf{A}}{\partial t} + \frac{\partial \mathbf{A}}{\partial x} \frac{dx}{dt} + \frac{\partial \mathbf{A}}{\partial y} \frac{dy}{dt} + \frac{\partial \mathbf{A}}{\partial z} \frac{dz}{dt}, \qquad (3.26)$$

where the first term on the right is the explicit time derivative, i.e., the derivative with respect to the variable listed first in the argument of $\mathbf{A}(t, x, y, z)$, and dx/dt, dy/dt, and dz/dt are the components of the velocity:

$$\mathbf{v} = \frac{dx}{dt} \hat{\mathbf{x}} + \frac{dy}{dt} \hat{\mathbf{y}} + \frac{dz}{dt} \hat{\mathbf{z}}. \qquad (3.27)$$

2. The ∇ Operator

If we define the symbol ∇ to mean

$$\nabla \equiv \hat{\mathbf{x}} \frac{\partial}{\partial x} + \hat{\mathbf{y}} \frac{\partial}{\partial y} + \hat{\mathbf{z}} \frac{\partial}{\partial z}, \qquad (3.28)$$

we can write (3.26) as

$$d\mathbf{A}/dt = \partial \mathbf{A}/\partial t + (\mathbf{v} \cdot \nabla)\mathbf{A}. \qquad (3.29)$$

∇ does not represent a quantity; rather it is an example of what is called an *operator*, in this instance a *differential operator*. While we examine operators in general in the following sections, here we pursue the subject of differential operators to the extent to which they relate to Euclidean vector fields. ∇ is usually called the "del" operator, or simply "del," although its appellations often vary with the context of its use. ∇ may be applied to a scalar function to form a vector field:

$$\nabla \varphi = \frac{\partial \varphi}{\partial x} \hat{\mathbf{x}} + \frac{\partial \varphi}{\partial x} \hat{\mathbf{y}} + \frac{\partial \varphi}{\partial z} \hat{\mathbf{z}}. \tag{3.30}$$

$\nabla \varphi$ is called the *gradient* of φ and is often written grad(φ). On the other hand, when applied to a vector field in the following manner:

$$\nabla \cdot \mathbf{A} = \partial A_x/\partial x + \partial A_y/\partial y + \partial A_z/\partial z, \tag{3.31}$$

it produces a scalar field called the *divergence* of $\mathbf{A}$, or sometimes div($\mathbf{A}$). Finally, we can define the operation which we can interpret with the help of (3.17) as

$$\nabla \times \mathbf{A} = \begin{vmatrix} \hat{x} & \hat{y} & \hat{z} \\ \frac{\partial}{\partial x} & \frac{\partial}{\partial y} & \frac{\partial}{\partial z} \\ A_x & A_y & A_z \end{vmatrix}$$

$$= \left(\frac{\partial A_z}{\partial y} - \frac{\partial A_y}{\partial z}\right)\hat{\mathbf{x}} + \left(\frac{\partial A_x}{\partial z} - \frac{\partial A_z}{\partial x}\right)\hat{\mathbf{y}} + \left(\frac{\partial A_y}{\partial x} - \frac{\partial A_x}{\partial y}\right)\hat{\mathbf{z}}. \tag{3.32}$$

$\nabla \times \mathbf{A}$ is called the *curl* of $\mathbf{A}$, or just curl($\mathbf{A}$), and is a vector field. Finally, if we replace $\mathbf{A}$ in (3.31) by $\nabla \varphi$, we have

$$\nabla \cdot \nabla \varphi \equiv \nabla^2 \varphi = \frac{\partial^2 \varphi}{\partial x^2} + \frac{\partial^2 \varphi}{\partial y^2} + \frac{\partial^2 \varphi}{\partial z^2}. \tag{3.33}$$

The differential operator

$$\nabla^2 = \partial^2/\partial x^2 + \partial^2/\partial y^2 + \partial^2/\partial z^2 \tag{3.34}$$

is called the *Laplacian*.

The gradient, divergence, curl, and Laplacian have found wide application in mathematical physics and it is vital that the user of vector calculus have a grasp of their geometrical or physical meanings. To this end, we will move in a moment to the integral calculus of vectors. Before doing so, however, we give a list of identities involving these operators:

$$\mathbf{\nabla} \cdot (\varphi \mathbf{A}) = \varphi(\mathbf{\nabla} \cdot \mathbf{A}) + \mathbf{A} \cdot (\mathbf{\nabla}\varphi) \tag{3.35a}$$

$$\mathbf{\nabla} \times (\varphi \mathbf{A}) = \varphi(\mathbf{\nabla} \times \mathbf{A}) - \mathbf{A} \times (\mathbf{\nabla}\varphi) \tag{3.35b}$$

$$\mathbf{\nabla} \cdot (\mathbf{\nabla} \times \mathbf{A}) = 0 \tag{3.35c}$$

$$\mathbf{\nabla} \times (\mathbf{\nabla}\varphi) = \mathbf{0} \tag{3.35d}$$

$$\mathbf{\nabla} \cdot (\mathbf{A} \times \mathbf{B}) = \mathbf{B} \cdot (\mathbf{\nabla} \times \mathbf{A}) - \mathbf{A} \cdot (\mathbf{\nabla} \times \mathbf{B}) \tag{3.35e}$$

$$\mathbf{\nabla}(\mathbf{A} \cdot \mathbf{B}) = (\mathbf{B} \cdot \mathbf{\nabla})\mathbf{A} + (\mathbf{A} \cdot \mathbf{\nabla})\mathbf{B} + \mathbf{B} \times (\mathbf{\nabla} \times \mathbf{A}) + \mathbf{A} \times (\mathbf{\nabla} \times \mathbf{B}) \tag{3.35f}$$

$$\mathbf{\nabla} \times (\mathbf{A} \times \mathbf{B}) = (\mathbf{B} \cdot \mathbf{\nabla})\mathbf{A} - (\mathbf{A} \cdot \mathbf{\nabla})\mathbf{B} + \mathbf{A}(\mathbf{\nabla} \cdot \mathbf{B}) - \mathbf{B}(\mathbf{\nabla} \cdot \mathbf{A}) \tag{3.35g}$$

$$\mathbf{\nabla} \times (\mathbf{\nabla} \times \mathbf{A}) = \mathbf{\nabla}(\mathbf{\nabla} \cdot \mathbf{A}) - \nabla^2 \mathbf{A}. \tag{3.35h}$$

In addition, we have the following identities involving the position vector:

$$\mathbf{\nabla} \cdot \mathbf{r} = 3 \tag{3.36a}$$

$$(\mathbf{A} \cdot \mathbf{\nabla})\mathbf{r} = \mathbf{A} \tag{3.36b}$$

$$\mathbf{\nabla} \times \mathbf{r} = \mathbf{0} \tag{3.36c}$$

$$\mathbf{\nabla} \cdot (\mathbf{r}/r^3) = 0. \tag{3.36d}$$

3. *Vector Line Integration*

The ordinary Riemannian integral of a vector over a region of space is defined simply in terms of integrals of its components:

$$\int \mathbf{A} \, d\xi = \hat{\mathbf{x}} \int A_x \, d\xi + \hat{\mathbf{y}} \int A_y \, d\xi + \hat{\mathbf{z}} \int A_z \, d\xi.$$

However, much more interesting, as far as physical applications are concerned, are *line* and *surface integrals*.

Consider a curve C in our three-dimensional space, defined by the parameterization $x(\xi)$, $y(\xi)$, $z(\xi)$. The position vector corresponding to a point on this curve is thus

$$\mathbf{r} = x(\xi)\hat{\mathbf{x}} + y(\xi)\hat{\mathbf{y}} + z(\xi)\hat{\mathbf{z}}.$$

Let us divide the curve C into n segments according to the parameter

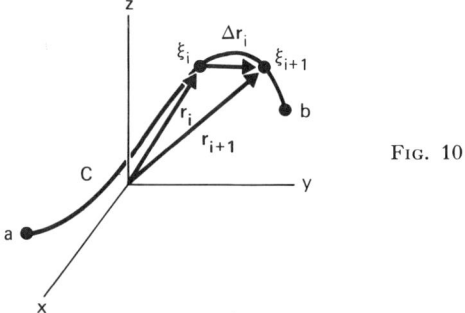

Fig. 10

intervals $\xi_0 = a < \xi_1 < \xi_2 < \cdots < \xi_i < \xi_{i+1} < \cdots < \xi_{n-1} < \xi_n = b$ (see Fig. 10). We can then define the corresponding displacement vectors:

$$\Delta \mathbf{r}_i = \mathbf{r}(\xi_{i+1}) - \mathbf{r}(\xi_i).$$

Now, consider a vector field $\mathbf{A}(x, y, z)$ defined in the space and, particularly, on C, where we may write $\mathbf{A}(x(\xi), y(\xi), z(\xi)) = \mathbf{A}(\xi)$. We let $\bar{\xi}_i$ be a value of ξ lying in the interval (ξ_i, ξ_{i+1}) and form the sum

$$I_n = \sum_{i=1}^{n} \mathbf{A}(\bar{\xi}_i) \cdot \Delta \mathbf{r}_i.$$

If we now let $n \to \infty$ in such a way that the Δr_i all approach zero, then the limit of I_n, if it exists and is unique, is called the *line integral of* $\mathbf{A}$:

$$\lim_{n \to \infty} I_n = \int_a^b (\mathbf{A} \cdot d\mathbf{r})_C \qquad (3.37)$$

where $d\mathbf{r}$ is the limiting form of the displacement vector:

$$d\mathbf{r} = \hat{\mathbf{x}} \, dx + \hat{\mathbf{y}} \, dy + \hat{\mathbf{z}} \, dz = \left(\frac{dx}{d\xi} \hat{\mathbf{x}} + \frac{dy}{d\xi} \hat{\mathbf{y}} + \frac{dz}{d\xi} \hat{\mathbf{z}} \right) d\xi. \qquad (3.38)$$

The existence and uniqueness of this integral depend on the continuity of $\mathbf{A}$ and of the slope of C. In fact, (3.37) can then be written as an ordinary Riemann integral:

$$\int_a^b (\mathbf{A} \cdot d\mathbf{r})_C = \int_a^b \left(A_x \frac{dx}{d\xi} + A_y \frac{dy}{d\xi} + A_z \frac{dz}{d\xi} \right) d\xi. \qquad (3.39)$$

If it is necessary actually to evaluate a line integral, it may be done with the help of this equation.

4. Conservative Vector Fields

Generally, the line integral of a vector from point a to point b depends on the integration path C. However, there are important instances where this is not the case, where, at least within a given region of space, the line integral is *independent* of the choice of path. This is, of course, a property of the particular vector field, which, in this case, is said to be a conservative field. Let **A** be such a field. Then,

$$\int_a^b (\mathbf{A} \cdot d\mathbf{r})_{C_1} = \int_a^b (\mathbf{A} \cdot d\mathbf{r})_{C_2} \tag{3.40}$$

where C_1 and C_2 are curves as shown in Fig. 11. If we consider the line integral around the closed curve C, formed from the segments C_1 and C_2, as in Fig. 12, we have

$$\oint_C \mathbf{A} \cdot d\mathbf{r} = 0. \tag{3.41}$$

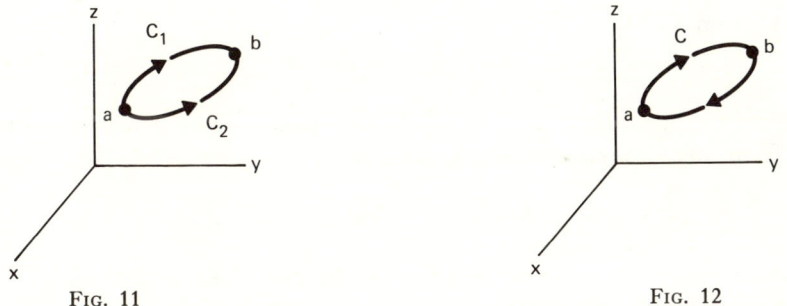

Fig. 11   Fig. 12

The terminology "conservative field" comes from mechanics. A *conservative force field* is one in which the net work done by an object while traversing a closed path is zero, i.e.,

$$\Delta W = \oint_C \mathbf{F} \cdot d\mathbf{r} = 0.$$

We will now demonstrate as a *theorem* that a necessary and sufficient condition for $\int_a^b (\mathbf{A} \cdot d\mathbf{r})_C$ to be independent of the path C is that **A** can be written as the gradient of some scalar function:

$$\mathbf{A}(x, y, z) = \boldsymbol{\nabla} \varphi(x, y, z). \tag{3.42}$$

To show sufficiency, we assume (3.42) and write

$$d\varphi = \nabla\varphi \cdot d\mathbf{r} = \mathbf{A} \cdot d\mathbf{r},$$

so that

$$\int_a^b (\mathbf{A} \cdot d\mathbf{r})_C = \int_a^b d\varphi = \varphi(x(b), y(b), z(b)) - \varphi(x(a), y(a), z(a))$$

depends only on the endpoints a and b. On the other hand, let us define the function

$$\eta(x, y, z) = \int_a^{P(x,y,z)} (\mathbf{A} \cdot d\mathbf{r})_C. \qquad (3.43)$$

If we assume $\mathbf{A}$ to be a conservative field, η depends only on the coordinates of the point P and not on the integration path. The following *directional derivative* of η is thus defined for any direction $d\hat{\mathbf{r}}$:

$$d\eta = \frac{\partial \eta}{\partial x} dx + \frac{\partial \eta}{\partial y} dy + \frac{\partial \eta}{\partial z} dz = \nabla\eta \cdot d\mathbf{r}. \qquad (3.44)$$

But, from (3.43),

$$d\eta = \mathbf{A}(x, y, z) \cdot d\mathbf{r}. \qquad (3.45)$$

Therefore, since $d\mathbf{r}$ is arbitrary (one can choose a curve which reaches P from any desired direction), we have

$$\mathbf{A}(x, y, z) = \nabla\eta(x, y, z).$$

η is the function φ we seek, and the necessity of (3.42) is proved.

From (3.35d), we see that

$$\nabla \times \mathbf{A} = \nabla \times (\nabla\varphi) = \mathbf{0}.$$

The converse is also true: If $\nabla \times \mathbf{A} = \mathbf{0}$, then (3.41) and (3.42) also hold, as the reader may prove. Hence, we have three completely equivalent conditions whereby we can determine if a vector field is conservative within a region. It is, if

1. $\oint \mathbf{A} \cdot d\mathbf{r} = 0$ for any closed path within the region, or
2. $\mathbf{A} = \nabla\varphi$, or
3. $\nabla \times \mathbf{A} = \mathbf{0}$ throughout the region.

A conservative vector field is also called *irrotational*, a term coming from fluid dynamics, for the following reason. If $\oint \mathbf{v} \cdot d\mathbf{r} \neq 0$ where $\mathbf{v}$

is the velocity field in a fluid, then a small paddle wheel placed in the fluid in the plane of **v** and **dr** at that point will be made to rotate, whereas if $\oint \mathbf{v} \cdot d\mathbf{r} = 0$, the torques delivered to the vanes of the paddle wheel will cancel and no rotation will occur.

5. Vector Surface Integrals

In Fig. 13 we have a two-dimensional surface S which is divided into n surface elements labeled ΔS_j. At the midpoint of the ith segment ΔS_i is a unit vector $\hat{\mathbf{N}}_i$, perpendicular to the surface. (We have arbitrarily decided that one side of the surface is to be the "outside" or "top" so as to fix a sense for $\hat{\mathbf{N}}_i$.) The vector field **A** is also shown at that point. We then define the *surface integral of* **A** as

$$\iint_S \mathbf{A} \cdot \hat{\mathbf{N}}\, dS = \lim_{n \to \infty} \sum_{i=1}^{n} \mathbf{A}(x_i, y_i, z_i) \cdot \hat{\mathbf{N}}_i \Delta S_i \qquad (3.46)$$

where, as with the line integral, the limit $n \to \infty$ is taken so that all $\Delta S_i \to 0$. Again, the existence of (3.46) depends on the continuity of **A** over the surface and of the tangent of the surface. The symbol $\iint$ indicates the two-dimensional nature of the integral.

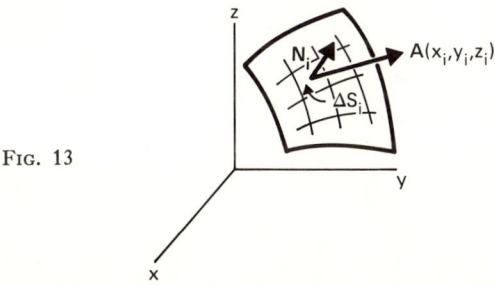

Fig. 13

6. Stokes's Theorem

The line and surface integrals we have defined above have their usefulness in applied mathematics due largely to several theorems relating vector integrals of different dimensionality. The first of these is

Stokes's Theorem Let S be a surface bounded by the curve C. Then, if the vector field $\mathbf{A}(x, y, z)$ and its derivatives are defined on the surface,

$$\oint_C \mathbf{A} \cdot d\mathbf{r} = \iint_S \mathbf{\nabla} \times \mathbf{A} \cdot \hat{\mathbf{N}}\, dS, \qquad (3.47)$$

where $\hat{\mathbf{N}}$ is the unit vector normal to the surface.

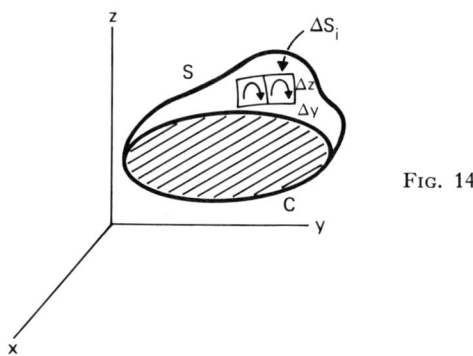

Fig. 14

To prove this, we consider the small segment of the surface, ΔS_i. We shall assume for simplicity that $\hat{\mathbf{N}}_i = \hat{\mathbf{x}}$, i.e., that ΔS_i is in the yz plane. Now, we form the line integral of $\mathbf{A}$ in the direction shown in Fig. 14:

$$\oint \mathbf{A} \cdot d\mathbf{r} = A_z\!\left(x, y, z - \frac{\Delta z}{2}\right)\Delta z + A_y\!\left(x, y, z + \frac{\Delta z}{2}\right)\Delta y$$

$$+ A_z\!\left(x, y + \frac{\Delta y}{2}, z\right)(-\Delta z) + A_y\!\left(x, y - \frac{\Delta y}{2}, z\right)(-\Delta y)$$

$$= \left[A_y\!\left(x, y, z + \frac{\Delta z}{2}\right) - A_y\!\left(x, y, z - \frac{\Delta z}{2}\right)\right]\Delta y$$

$$- \left[A_z\!\left(x, y + \frac{\Delta y}{2}, z\right) - A_z\!\left(x, y - \frac{\Delta y}{2}, z\right)\right]\Delta z$$

$$= \left[\frac{A_y(x, y, z + \Delta z/2) - A_y(x, y, z - \Delta z/2)}{\Delta z}\right.$$

$$\left. - \frac{A_z(x, y + \Delta y/2, z) - A_z(x, y - \Delta y/2, z)}{\Delta y}\right]\Delta S$$

where $\Delta S = \Delta y\, \Delta z$. In the limit $\Delta S \to 0$ this becomes

$$\lim_{\Delta S \to 0} \frac{\oint \mathbf{A} \cdot d\mathbf{r}}{\Delta S} = \frac{\partial A_y}{\partial z} - \frac{\partial A_z}{\partial y} = (\mathbf{\nabla} \times \mathbf{A})_x.$$

The generalization of this result to a surface element of arbitrary orientation $\hat{\mathbf{N}}$ is

$$\lim_{\Delta S \to 0} \frac{\oint \mathbf{A} \cdot d\mathbf{r}}{\Delta S} = (\nabla \times \mathbf{A}) \cdot \hat{\mathbf{N}}. \tag{3.48}$$

Equation (3.48) is often given as a *definition of the curl* and is entirely equivalent to our definition (3.32). If we write, then, for each of the surface elements

$$\oint_i \mathbf{A} \cdot d\mathbf{r} = (\nabla \times \mathbf{A}) \cdot \hat{\mathbf{N}}_i \Delta S_i,$$

and sum over all of them, the right-hand side becomes in the limit,

$$\iint_S (\nabla \times \mathbf{A}) \cdot \hat{\mathbf{N}} \, dS.$$

In the sum on the left-hand side, contributions from boundaries common to two surface elements will vanish, due to traversing them from opposite directions, leaving only those contributions from the surface boundary C. The sum thus becomes $\oint_C \mathbf{A} \cdot d\mathbf{r}$, and the theorem is proved.

7. *The Divergence Theorem*

Let S be a closed surface enclosing the volume V. Then, if the vector field $\mathbf{A}(x, y, z)$ and its derivatives are defined within and on the surface,

$$\iint_S \hat{\mathbf{N}} \cdot \mathbf{A} \, dS = \iiint_V \nabla \cdot \mathbf{A} \, dV. \tag{3.49}$$

To begin the proof, which proceeds similarly to the proof of Stokes's theorem, we divide the volume into several small rectangular elements. Consider the ith such element, the sides of which, for simplicity, we orient parallel to the coordinate axis (see Fig. 15). The surface integral

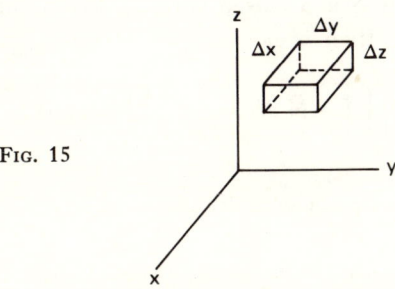

Fig. 15

over the element is

$$\iint_{S_i} \mathbf{N} \cdot \mathbf{A}\, dS = [A_x(x + \Delta x/2, y, z) - A_x(x - \Delta x/2, y, z)]\, \Delta z\, \Delta y$$
$$+ [A_y(x, y + \Delta y/2, z) - A_y(x, y - \Delta y/2, z)]\, \Delta x\, \Delta z$$
$$+ [A_z(x, y, z + \Delta z/2) - A_z(x, y, z - \Delta z/2)]\, \Delta x\, \Delta y,$$

so that

$$\lim_{\Delta V_i \to 0} \frac{\iint_{S_i} \hat{\mathbf{N}} \cdot \mathbf{A}\, dS}{\Delta V_i} = \frac{\partial A_x}{\partial x} + \frac{\partial A_y}{\partial y} + \frac{\partial A_z}{\partial z} = \mathbf{\nabla} \cdot \mathbf{A}. \quad (3.50)$$

Equation (3.50) can be considered an *integral definition* of the divergence of a vector. By writing

$$\iint_{S_i} \hat{\mathbf{N}} \cdot \mathbf{A}\, dS = \mathbf{\nabla} \cdot \mathbf{A}\, \Delta V_i,$$

summing, and taking the limit $\Delta V_i \to 0$, we obtain (3.49). Here again, we note the cancellation of contributions from adjacent volume elements in the surface sum.

A field for which $\mathbf{\nabla} \cdot \mathbf{A} = 0$ is called *solenoidal* from the geometric shape of the "flow lines" in such a field; the flow lines of solenoidal fields, such as the magnetic intensity $\mathbf{B}$, close upon themselves. If $\mathbf{\nabla} \cdot \mathbf{A} = 0$ everywhere in V, then

$$\iint_S \hat{\mathbf{N}} \cdot \mathbf{A}\, dS = 0;$$

i.e., there is no net "flow" of the vector $\mathbf{A}$ into or out of the region enclosed by S. There are no sources or sinks of the field within S. This is a special case of a physical application of the divergence theorem called:

Gauss's Law[†] If $\varrho(\mathbf{r})$ is the source distribution for an inverse square force field $\mathbf{E}$, and if S is a closed surface surrounding the field source (in part or in total), then

$$\iint_S \mathbf{E} \cdot \hat{\mathbf{N}}\, dS = 4\pi \iiint_V \varrho(\mathbf{r})\, dV, \quad (3.51)$$

where V is the volume enclosed by S.

[†] The divergence theorem is often called *Gauss's theorem*.

1. Linear Vector Spaces

Since the two known examples of inverse square force fields, the electrostatic and gravitational, obey the principle of superposition, it is sufficient to prove Gauss's law for a point source, which we will assume to be at the origin. The result (3.51) can then be obtained by adding the contributions from all point sources which make up the source density $\varrho(\mathbf{r})$. The inverse square force field due to a point source at $\mathbf{r} = \mathbf{0}$ is given by

$$\mathbf{E} = \alpha \mathbf{r}/r^3,$$

where α is the coupling strength of the particular field. Let us first consider the case where the source is *not* contained within V. Then, from the divergence theorem,

$$\iint_S \mathbf{E} \cdot \hat{\mathbf{N}} \, dS = \alpha \iiint_V \mathbf{\nabla} \cdot (\mathbf{r}/r^3) \, dV = 0$$

by (3.36d). Next, we take the case where S does surround the source at $\mathbf{r} = \mathbf{0}$. $\mathbf{E}$ is then not defined at this point within S and the divergence theorem cannot be directly applied. However, we can alter the surface by creating a spherical "bubble," S', of radius ε centered at the origin, i.e., at the source (see Fig. 16). Then, the divergence theorem can be applied to the volume between S and S' to give

$$\iint_S (\mathbf{r}/r^3) \cdot \hat{\mathbf{N}} \, dS + \iint_{S'} (\mathbf{r}/r^3) \cdot \hat{\mathbf{N}}' \, dS' = 0$$

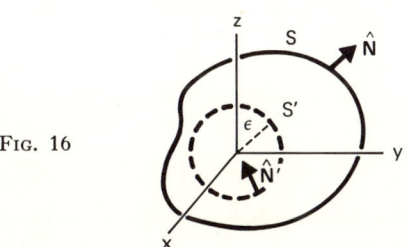

FIG. 16

as in the first case. In the second integral, however, $\hat{\mathbf{N}}'$ points *toward* the origin, that is, away from the volume. Thus,

$$\hat{\mathbf{N}}' \cdot \mathbf{r} = -\varepsilon,$$

$$\iint_{S'} (\mathbf{r}/r^3) \cdot \hat{\mathbf{N}}' \, dS' = -\frac{1}{\varepsilon^2} \iint_{S'} dS' = -4\pi$$

and
$$\iint_S \mathbf{E} \cdot \hat{\mathbf{N}}\, dS = 4\pi\alpha, \qquad (3.52)$$

which is Eq. (3.51) when $\varrho(\mathbf{r})$ is a point source.

One can use Stokes's theorem and the divergence theorem to obtain other integral identities. If φ is a scalar function and $\mathbf{C}$ a constant vector, then Stokes's theorem, applied to $\mathbf{A} = \varphi\mathbf{C}$, yields

$$\oint_C \varphi\, d\mathbf{r} = \iint_S \hat{\mathbf{N}} \times \boldsymbol{\nabla}\varphi\, dS. \qquad (3.53)$$

Substitution of $\mathbf{A} \times \mathbf{C}$ into Stokes's theorem and the divergence theorem results in

$$\oint_C d\mathbf{r} \times \mathbf{A} = \iint_S (\hat{\mathbf{N}} \times \boldsymbol{\nabla}) \times \mathbf{A}\, dS \qquad (3.54)$$

and

$$\iint_S \hat{\mathbf{N}} \times \mathbf{A}\, dS = \iiint_V \boldsymbol{\nabla} \times \mathbf{A}\, dV, \qquad (3.55)$$

respectively. The divergence theorem applied to $\mathbf{A} = \psi\boldsymbol{\nabla}\varphi$, where ψ and φ are both scalar functions, gives us *Green's first identity*:

$$\iint_S (\psi\boldsymbol{\nabla}\varphi) \cdot \hat{\mathbf{N}}\, dS = \iiint_V (\boldsymbol{\nabla}\psi \cdot \boldsymbol{\nabla}\varphi + \psi\nabla^2\varphi)\, dV. \qquad (3.56)$$

If we interchange the positions of ψ and φ in (3.56) and subtract the two equations, we obtain *Green's second identity*:

$$\iint_S (\psi\boldsymbol{\nabla}\varphi - \varphi\boldsymbol{\nabla}\psi) \cdot \hat{\mathbf{N}}\, dS = \iiint_V (\psi\nabla^2\varphi - \varphi\nabla^2\psi)\, dV. \qquad (3.57)$$

IV. Vector Transformations

We now return to the vector in its role as an abstract mathematical entity. In this section we introduce the idea of vector transformations, particularly *linear transformations*. This will bring us to the concept of a *linear operator*, the systematics of which, including the eigenvector–eigenvalue problem, we study in some depth. We also investigate the transformations of entire basis sets. The *matrix representations* of vectors and operators are introduced in preparation for our study of matrices in Section V.

Again, as in Section II, we will assume the most general situation that numbers may be complex. There are occasions, however, in specific applications, in which one may want to restrict considerations to the field of real numbers.

A. LINEAR OPERATORS

According to the defining properties of a linear vector space (as given in Section II), the processes of vector addition and scalar multiplication, or any combination of them, transform one or more vectors into other vectors. Additional prescriptions may also be given which cause one vector to be transformed into another, prescriptions which may involve a reordering or some other nonalgebraic process. Whatever the prescription, we can display such a transformation of one vector into another symbolically as

$$\mathbf{V} \to \mathbf{V}' = A\mathbf{V}, \tag{4.1}$$

where A symbolizes the particular procedure or operation and is therefore called an *operator*. The vector space throughout which the operator A is defined, that is, in which each vector has an image under the *mapping* $\mathbf{V}' = A\mathbf{V}$, is called the *domain* of A.

An operator A is said to be *linear* if

$$A(\mathbf{V}_a + \mathbf{V}_b) = A\mathbf{V}_a + A\mathbf{V}_b \tag{4.2}$$

and

$$A(\alpha \mathbf{V}) = \alpha A\mathbf{V} \tag{4.3}$$

for all vectors in its domain and for all complex scalars α. [If $B(\alpha \mathbf{V}) = \alpha^* B\mathbf{V}$ while B satisfies (4.2), then B is called *antilinear*. An example of an important antilinear operator is the time-reversal operator of quantum mechanics.] We are concerned in this chapter only with linear operators whose domains are linear vector spaces.

1. *Operator Algebra*

Operators obey the following combination rules:

$$(A + B)\mathbf{V} = A\mathbf{V} + B\mathbf{V}, \tag{4.4}$$

$$(AB)\mathbf{V} = A(B\mathbf{V}). \tag{4.5}$$

However, it is *not* generally true that

$$(AB)\mathbf{V} = (BA)\mathbf{V}. \tag{4.6}$$

If Eq. (4.6) does hold, then A and B are said to *commute*. Otherwise, for noncommuting operators, we may define the *commutator* of A and B,

$$[A, B] = AB - BA, \tag{4.7}$$

which is itself an operator. The following commutator identities are often useful in operator calculations:

$$[A, B] = -[B, A] \tag{4.8a}$$

$$[A, BC] = B[A, C] + [A, B]C \tag{4.8b}$$

$$[A, [B, C]] + [C, [A, B]] + [B, [C, A]] = \varnothing. \tag{4.8c}$$

In (4.8c), $\varnothing$ is the *null operator* and is equivalent to the scalar zero, when applied to a vector:

$$\varnothing \mathbf{V} = \mathbf{0}. \tag{4.9}$$

Equation (4.8c) is *Jacobi's identity* [cf. Eq. (3.18e)]. Equations (4.8), and all other operator equations, are only meaningful when applied to a vector; forgetting this often leads to errors in operator manipulations.

In addition to the null operator, we have another operator of trivial characteristics, namely, the *identity operator* I:

$$I\mathbf{V} = \mathbf{V} \tag{4.10}$$

for all $\mathbf{V}$ in the domain of I. We were introduced to the identity operator in Eq. (2.20). Null and identity operators are defined for every linear vector space.

If there exists an operator A_l^{-1} in the domain of A which negates the effect of A:

$$A_l^{-1} A \mathbf{V} = \mathbf{V},$$

i.e.,

$$A_l^{-1} A = I, \tag{4.11a}$$

then A_l^{-1} is called a *left inverse* of A. Similarly, a right inverse of A is such that

$$A A_r^{-1} = I. \tag{4.11b}$$

In general, $A_l^{-1} \neq A_r^{-1}$; indeed either or both may not even exist. However, it can easily be demonstrated that if *both* A_l^{-1} and A_r^{-1} exist, then they *are* equal *and unique*. Conversely, if A_l^{-1} (or A_r^{-1}) exists and is unique, then $AA_l^{-1} = I$ (or $A_r^{-1}A = I$), so that in either case, $A_l^{-1} = A_r^{-1} = A^{-1}$ exists and is unique:

$$A^{-1}A = AA^{-1} = I. \tag{4.12}$$

The operator A^{-1} is called the *inverse operator* of A, and if it exists, A is said to be *nonsingular*. An operator with no inverse (although it may have left or right inverses) is *singular*.

2. Functions of Operators

Integral powers, positive and, for nonsingular operators, negative, may be formed in the following manner:

$$A^2 = AA, \quad (AB)^2 = ABAB, \quad A^{-2} = A^{-1}A^{-1}, \tag{4.13}$$

with obvious generalizations. This permits functions of operators to be defined. Thus, if $f(x)$ is a function of the variable x which may be expanded about the point $x = 0$ in some series of positive and negative integral powers,

$$f(x) = \sum_{n=-\infty}^{\infty} a_n x^n,$$

then we can define the operator

$$f(A) = \sum_{n=-\infty}^{\infty} a_n A^n. \tag{4.14}$$

A simple but valuable example of an operator function is

$$(I - A)^{-1} = I + A + A^2 + \cdots. \tag{4.15}$$

Also of importance is the exponential operator:

$$e^A \equiv I + \tfrac{1}{1!} A + \tfrac{1}{2!} A^2 + \cdots, \tag{4.16}$$

the cosine operator:

$$\cos A \equiv I - \tfrac{1}{2!} A^2 + \tfrac{1}{4!} A^4 - \cdots + \cdots, \tag{4.17}$$

and the sine operator:

$$\sin A \equiv A - \tfrac{1}{3!} A^3 + \tfrac{1}{5!} A^5 - \cdots + \cdots. \tag{4.18}$$

Some more useful identities are as follows:

$$(AB)^{-1} = B^{-1}A^{-1} \tag{4.19a}$$

$$[A^n, B] = nA^{n-1}[A, B] \tag{4.19b}$$

$$A^{-1} - (A + B)^{-1} = (A + B)^{-1}BA^{-1} = A^{-1}B(A + B)^{-1} \tag{4.19c}$$

$$e^A B e^{-A} = B + [A, B] + \tfrac{1}{2!}[A, [A, B]] \\ + \tfrac{1}{3!}[A, [A, [A, B]]] + \cdots \tag{4.19d}$$

$$e^{A+B} = e^A e^B e^{\frac{1}{2}[B,A]} \tag{4.19e}$$

provided, in (4.19e), that $[A, [B, A]] = [B, [B, A]] = \varnothing$.

B. MATRIX ELEMENTS

Let us use the operator A to transform the vector $\mathbf{V}_a$ and take the scalar product of the result with another vector $\mathbf{V}_b$ to form the complex quantity

$$A_{ba} = (\mathbf{V}_b, A\mathbf{V}_a). \tag{4.20}$$

In the Dirac notation which we introduced at the end of Section II, this is written

$$A_{ba} = \langle b \mid A \mid a \rangle. \tag{4.20a}$$

We can, in particular, form such quantities from the members of a basis of the vector space:

$$A_{ij} = (\hat{\mathbf{V}}_i, A\hat{\mathbf{V}}_j) = \langle i \mid A \mid j \rangle. \tag{4.21}$$

In an n-dimensional vector space, there are n^2 such quantities for a given basis. They can be arranged in a square array thus:

$$\{A_{ij}\} = \begin{bmatrix} A_{11} & A_{12} & \cdots & A_{1n} \\ A_{21} & A_{22} & \cdots & A_{2n} \\ \vdots & \vdots & & \vdots \\ A_{n1} & A_{n2} & \cdots & A_{nn} \end{bmatrix}. \tag{4.22}$$

Such an array is a *matrix*; in this case a *square matrix*, for matrices may in general be nonsquare rectangular arrays, as, for example, the *column matrix* formed from the components of a vector:

$$\mathbf{V}_a = \begin{pmatrix} a_1 \\ a_2 \\ \vdots \\ a_n \end{pmatrix}. \tag{4.23}$$

The A_{ij} of Eq. (4.21) are the *matrix elements* of A in the basis $\{\hat{\mathbf{V}}_i\}$, and (4.22) and (4.23) are the *matrix representations*, in that basis, of A and $\mathbf{V}_a$, respectively. In actual practice, the quantity A_{ab} in Eq. (4.20) is also called a matrix element, although $\mathbf{V}_a$ and $\mathbf{V}_b$ are not necessarily basis vectors. We will observe stricter usage here, however, and reserve the term *matrix element* for the case where the sandwiching vectors are members of an orthonormal basis.

Having the matrix elements of an operator is much the same as having the components of a vector, in that one then has a scalar realization of an otherwise abstract object and can thus understand the basic manipulations in terms of ordinary arithmetical operations [cf. Eqs. (2.15)–(2.17)]. The "addition" of operators, Eq. (4.4), simply becomes an addition of matrix elements:

$$(A+B)_{ij} = A_{ij} + B_{ij}, \tag{4.24}$$

while the "product" of two operators, Eq. (4.5), is an operator whose matrix elements are given by[†]

$$(AB)_{ij} = \langle i | AB | j \rangle = \langle i | A \sum_k | k \rangle \langle k | B | j \rangle$$
$$= \sum_k \langle i | A | k \rangle \langle k | B | j \rangle = \sum_k A_{ik} B_{kj}, \tag{4.25}$$

where we have used the closure relationship (2.20).

The effect of an operator on any vector can be described completely in terms of a matrix representation of the operator and the vector in some basis. Expanding the new and original vectors in the basis, we obtain

$$\mathbf{V}_a{}' = \sum_i a'_i \hat{\mathbf{V}}_i = A \mathbf{V}_a = \sum_i a_i A \hat{\mathbf{V}}_i,$$

which gives, upon taking the scalar product with $\hat{\mathbf{V}}_j$,

$$a'_j = \sum_i a_i (\hat{\mathbf{V}}_j, A \hat{\mathbf{V}}_i) = \sum_i \mathbf{A}_{ji} a_i. \tag{4.26}$$

[†] We use the simplified $\sum_k$ to denote $\sum_{k=1}^n$.

C. Eigenvectors and Eigenvalues

If there are vectors $\mathbf{V}_\alpha$ in the domain of A which, when operated on by A, are altered only by a scalar multiplication:

$$A\mathbf{V}_\alpha = \alpha \mathbf{V}_\alpha, \qquad (4.27)$$

the $\mathbf{V}_\alpha$'s are called *eigenvectors* of A and the scalar α's are the associated *eigenvalues*.[†] The set of eigenvalues of an operator is called the spectrum of that operator. In terms of the Dirac ket vectors, Eq. (4.27) is written

$$A \mid \alpha \rangle = \alpha \mid \alpha \rangle. \qquad (4.27a)$$

Note that we can label an *eigenket* with its associated eigenvalue.

Solving for the eigenvalues and eigenvectors of an operator is one of the most important activities associated with physical applications of vector spaces. We now show how this may be done if we have a matrix representation of the operator. In Section VII we will encounter other methods of determining eigenvalues. Equation (4.27), the eigenvalue equation, can be written

$$(A - \alpha I)\mathbf{V}_\alpha = \mathbf{0}. \qquad (4.28)$$

If we expand $\mathbf{V}_\alpha$ in terms of basis vectors,

$$\mathbf{V}_\alpha = \sum_i a_i \hat{\mathbf{V}}_i,$$

and take the scalar product of (4.28) with $\hat{\mathbf{V}}_j$, we obtain the following set of simultaneous homogeneous equations:

$$\sum_i a_i(A_{ji} - \alpha\, \delta_{ji}) = 0, \qquad j = 1, \ldots, n, \qquad (4.29)$$

which can be solved for the a_i's *if and only if* the determinant[‡]

$$\mid A_{ji} - \alpha\, \delta_{ji} \mid = \begin{vmatrix} A_{11} - \alpha & A_{12} & \cdots & A_{1n} \\ A_{21} & A_{22} - \alpha & \cdots & A_{2n} \\ \vdots & \vdots & & \vdots \\ A_{n1} & A_{n2} & \cdots & A_{nn} - \alpha \end{vmatrix} = 0. \qquad (4.30)$$

[†] The prefix eigen is from the German and means *characteristic*.
[‡] We assume the reader to have an acquaintance with determinants. However, the rudiments of the subject are reviewed in Section V.

This equation is known as the *secular equation*. When the determinant is expanded, Eq. (4.30) becomes an nth order polynomial equation in α:

$$(-1)^n \alpha^n + C_{n-1}\alpha^{n-1} + \cdots + C_0 = 0, \qquad (4.30a)$$

where the C_i's are certain combinations of the A_{ji}. We are assured, by the fundamental theorem of algebra, that Eq. (4.30) has n solutions, or roots, $\alpha_1, \alpha_2, \ldots, \alpha_n$, which are consequently the eigenvalues of A. *In an n-dimensional space, an operator has n eigenvalues.* Some of them may, of course, be complex, although if the coefficients in (4.30a) are real, complex eigenvalues must appear in complex conjugate pairs. There may also be repetitions among the eigenvalues, with two or more being equal. If there are m equal eigenvalues, we say that there is an m-fold *degeneracy*, and the corresponding eigenvectors (if they exist) are said to be mutually degenerate.

One can, by solving Eq. (4.30a), find eigenvalues for an operator; this does not always guarantee, however, that Eq. (4.29) will yield distinct eigenvectors, particularly when there are degeneracies. For the two principal types of operators we consider in this chapter, Hermitian and unitary operators, we *are* assured of n linearly independent eigenvectors. To find their components, we solve the n simultaneous equations

$$\sum_i a_i^{(k)}(A_{ji} - \alpha_k \delta_{ji}) = 0, \qquad j = 1, 2, \ldots, n, \qquad (4.31)$$

for each α_k in turn. The $a_i^{(k)}$ thus obtained are the components of $\mathbf{V}_{\alpha_k} = |\alpha_k\rangle$.

To summarize, then, if we know the matrix elements of an operator in some basis, it is a straightforward procedure to calculate that operator's eigenvalues and, if they exist, eigenvectors. The matrix elements of the operator and the components of the eigenvectors depend, of course, on the particular basis used. As we see later, however, *the eigenvalues are independent of the choice of basis*.

D. Hermitian Operators

We have seen that any operator in an n-dimensional vector space has n eigenvalues. In general, though, the eigenvalues and eigenvectors of an operator have no systematic characteristics that lend themselves to useful application. An exception to this is the case with *Hermitian operators*.

In Eq. (4.20), the operator A transforms the postfactor (the vector to the right). We want to examine the properties of an operator, $A^\ddagger$, which, when transforming the prefactor instead, will produce the same value for the entire scalar product, i.e.,

$$(A^{\ddagger}\mathbf{V}_b, \mathbf{V}_a) = (\mathbf{V}_b, A\mathbf{V}_a) \tag{4.32}$$

for all $\mathbf{V}_a$ and $\mathbf{V}_b$. $A^{\ddagger}$ is the *Hermitian adjoint* of A, and its matrix elements in the basis $\{\hat{\mathbf{V}}_i\}$ are

$$(A^{\ddagger})_{ij} = (\hat{\mathbf{V}}_i, A^{\ddagger}\hat{\mathbf{V}}_j) = (A^{\ddagger}\hat{\mathbf{V}}_j, \hat{\mathbf{V}}_i)^* = (\hat{\mathbf{V}}_j, A\hat{\mathbf{V}}_i)^* = A_{ji}^*, \tag{4.33}$$

where we have used Eq. (2.3c) as well as the definition of $A^{\ddagger}$. Thus, the matrix representation of $A^{\ddagger}$ is formed from that of A by transposing rows and columns and taking the complex conjugate of all elements. For this reason, $A^{\ddagger}$ is also often called the *transpose conjugate* of A. Incidentally, Eq. (4.33), relating the matrix elements of $A^{\ddagger}$ to those of A, serves as proof of the existence of the Hermitian conjugate of an operator.

If $A^{\ddagger} = A$, i.e., if $A_{ba} = A_{ab}^*$ for all $\mathbf{V}_a$ and $\mathbf{V}_b$, then A is an *Hermitian operator*. (Note the distinction between *Hermitian adjoint* and *Hermitian operator*: all operators have Hermitian adjoints but most operators are not Hermitian.)

The properties of the eigenvalues and eigenvectors of Hermitian operators which we referred to above are

1. *The eigenvalues of an Hermitian operator are real.*
2. *The eigenvectors of an Hermitian operator are* (or can be adjusted to be) *orthogonal.*

These properties have their importance in that (a) the eigenvalues, being real, can represent real physical quantities, and (b) the n eigenvectors, being orthogonal, can serve as basis vectors for the domain of the operator. We can prove these characteristics of Hermitian operators by writing the eigenvalue equation of the Hermitian operator A for two eigenvectors:

$$A \mid \alpha_i \rangle = \alpha_i \mid \alpha_i \rangle \tag{4.34a}$$

$$A \mid \alpha_j \rangle = \alpha_j \mid \alpha_j \rangle. \tag{4.34b}$$

Forming the scalar product of (4.34a) and (4.34b) with $\mid \alpha_j \rangle$ and $\mid \alpha_i \rangle$, respectively,

$$\langle \alpha_j \mid A \mid \alpha_i \rangle = \alpha_i \langle \alpha_j \mid \alpha_i \rangle \tag{4.35a}$$

$$\langle \alpha_i \mid A \mid \alpha_j \rangle = \alpha_j \langle \alpha_i \mid \alpha_j \rangle, \tag{4.35b}$$

1. Linear Vector Spaces

and taking the complex conjugate of (4.35b):

$$\langle \alpha_i | A | \alpha_j \rangle^* = \langle \alpha_j | A^\dagger | \alpha_i \rangle = \alpha_j^* \langle \alpha_j | \alpha_i \rangle$$
$$= \langle \alpha_j | A | \alpha_i \rangle, \tag{4.36}$$

we have, after substracting (4.36) from (4.35a),

$$0 = (\alpha_i - \alpha_j^*)\langle \alpha_j | \alpha_i \rangle. \tag{4.37}$$

Now, if $| \alpha_i \rangle$ and $| \alpha_j \rangle$ are taken to be the same eigenvector, then, since $\langle \alpha_i | \alpha_i \rangle > 0$, $\alpha_i = \alpha_i^*$, i.e., α_i is real. On the other hand, if $| \alpha_i \rangle \neq | \alpha_j \rangle$, then, unless we have here a degeneracy, $\alpha_i \neq \alpha_j$. Consequently, from (4.37), $\langle \alpha_j | \alpha_i \rangle = 0$, which is the orthogonality condition.

In the event there is a degeneracy, it turns out that $| \alpha_i \rangle$ and $| \alpha_j \rangle$ are still linearly independent[†] and can be combined by the Schmidt process to form new eigenvectors which, as the reader may easily verify, are still degenerate with the same eigenvalue as before, but which are also orthogonal. It should be mentioned here that, in order to use the eigenvectors of an Hermitian operator as a basis, they must be normalized.

Example We will find the eigenvalues and eigenvectors of the Hermitian operator that has the matrix representation

$$A = \begin{pmatrix} 0 & 0 & -i \\ 0 & 1 & 0 \\ i & 0 & 0 \end{pmatrix}.$$

The secular equation becomes

$$\begin{vmatrix} -\alpha & 0 & -i \\ 0 & 1-\alpha & 0 \\ i & 0 & -\alpha \end{vmatrix} = -(\alpha - 1)^2(\alpha + 1) = 0.$$

The eigenvalues are thus $\alpha_1 = -1$, $\alpha_2 = \alpha_3 = 1$, and we see that we have a two-fold degeneracy. The eigenvector $\mathbf{V}_1$ for the nondegenerate eigenvalue α_1 is found by solving the simultaneous equations (4.31), which in this instance become

$$-ia_3^{(1)} + a_1^{(1)} = 0, \qquad a_2^{(1)} + a_2^{(1)} = 0, \qquad ia_1^{(1)} + a_3^{(1)} = 0.$$

[†] The proof of this statement is straightforward but uninstructive, and is omitted here.

The solution is $a_2^{(1)} = 0$ and $a_3^{(1)} = -ia_1^{(1)}$. Normalization fixes $a_1^{(1)}$ at $1/\sqrt{2}$, and

$$\hat{\mathbf{V}}_1 = \frac{1}{\sqrt{2}} \begin{pmatrix} 1 \\ 0 \\ -i \end{pmatrix}.$$

We consider next the degenerate eigenvalues α_2 and α_3. Writing (4.31) for $\alpha = 1$, we have

$$-ia_3^{(k)} - a_1^{(k)} = 0, \quad a_2^{(k)} - a_2^{(k)} = 0, \quad ia_1^{(k)} - a_3^{(k)} = 0, \quad k = 2, 3.$$

These equations are satisfied by *any* $a_2^{(k)}$ and $a_1^{(k)}$ with $a_3^{(k)} = ia_1^{(k)}$:

$$\mathbf{V}_k = \begin{pmatrix} a_1^{(k)} \\ a_2^{(k)} \\ ia_1^{(k)} \end{pmatrix}.$$

We note that whatever our choices for $a_1^{(k)}$ and $a_2^{(k)}$, $\hat{\mathbf{V}}_1$ and $\mathbf{V}_k$ are orthogonal:

$$(\hat{\mathbf{V}}_1, \mathbf{V}_k) = (1/\sqrt{2})(a_1^{(k)} + 0 + i(ia_1^{(k)})) = 0.$$

Suppose we make the arbitrary assignments $a_1^{(2)} = a_2^{(2)} = a$ and $a_1^{(3)} = -a_2^{(3)} = a$. Then,

$$\mathbf{V}_2 = a \begin{pmatrix} 1 \\ 1 \\ i \end{pmatrix} \quad \text{and} \quad \mathbf{V}_3 = a \begin{pmatrix} 1 \\ -1 \\ i \end{pmatrix}$$

are linearly independent. They are not, however, orthogonal:

$$(\mathbf{V}_2, \mathbf{V}_3) = |a|^2(1 - 1 + 1) = |a|^2 \neq 0.$$

Therefore, we apply the Schmidt process to obtain a new $\mathbf{V}_3'$ which will be orthogonal to $\mathbf{V}_2$:

$$\mathbf{V}_3' = \mathbf{V}_3 - (\mathbf{V}_2, \mathbf{V}_3)/(\mathbf{V}_2, \mathbf{V}_2) \cdot \mathbf{V}_2 = \mathbf{V}_3 - \tfrac{1}{3}\mathbf{V}_2 = \tfrac{2}{3}a \begin{pmatrix} 1 \\ -2 \\ i \end{pmatrix}.$$

($\mathbf{V}_3'$ is of course still orthogonal to $\mathbf{V}_1$ and still an eigenvector of A with eigenvalue $+1$.) Finally, normalization gives

$$\hat{\mathbf{V}}_2 = \frac{1}{\sqrt{3}} \begin{pmatrix} 1 \\ 1 \\ i \end{pmatrix} \quad \text{and} \quad \hat{\mathbf{V}}_3' = \frac{1}{\sqrt{6}} \begin{pmatrix} 1 \\ -2 \\ i \end{pmatrix}.$$

Other initial choices of $a_1^{(2)}$ and $a_2^{(2)}$ would have led to different mutually orthogonal pairs, $\hat{V}_2$ and $\hat{V}_3$.

1. *Commuting Hermitian Operators*

The problem still remains, however, of the distinguishability of degenerate eigenvectors of Hermitian operators. We have been labeling eigenvectors, literally and figuratively, by their associated eigenvectors. Degenerate eigenvectors, although orthogonal, are thus given the same label. Is there a nonarbitrary way of adding a second, distinguishing label? The answer is yes, *if* there is a second Hermitian operator whose domain coincides with that of the first and which commutes with the first.

Consider the commuting Hermitian operators A and B:

$$[A, B] = \varnothing. \tag{4.38}$$

We will take as a basis the eigenvectors of A, $\{|\alpha_i\rangle\}$. In this representation, the matrix elements of A vanish unless $i = j$:

$$A_{ij} = \langle \alpha_i | A | \alpha_j \rangle = \alpha_j \langle \alpha_i | \alpha_j \rangle = \alpha_j \delta_{ij}. \tag{4.39}$$

The matrix formed by the A_{ij} has zeros everywhere except along the diagonal going from upper left to lower right. A is consequently said to be *diagonalized* by the representation $\{|\alpha_j\rangle\}$.

Now, it is easy to show that if A has no degeneracies, then B is also diagonal in the representation $\{|\alpha_j\rangle\}$, for

$$\begin{aligned}\langle \alpha_i | (AB - BA) | \alpha_j \rangle &= 0 \\ &= (\alpha_i - \alpha_j)\langle \alpha_i | B | \alpha_j \rangle,\end{aligned} \tag{4.40}$$

so that if $i \neq j$, $B_{ij} = 0$. But this implies that the $|\alpha_j\rangle$ are also eigenvectors of B with eigenvalues $\beta_j = B_{jj}$ [we cannot tell from (4.40), of course, just what the values of B_{jj} are] and we can label the eigenvectors $|\alpha_i, \beta_i\rangle$:

$$A | \alpha_i, \beta_i \rangle = \alpha_i | \alpha_i, \beta_i \rangle, \qquad B | \alpha_i, \beta_i \rangle = \beta_i | \alpha_i, \beta_i \rangle. \tag{4.41}$$

Suppose, though, that $|\alpha_{i,1}\rangle, |\alpha_{i,2}\rangle, \ldots, |\alpha_{i,m}\rangle$ are degenerate orthonormal eigenvectors of A, with common eigenvalue α_i. Then there is an $m \times m$ submatrix of the B matrix (in the A representation) which is not necessarily diagonal, since from (4.40), $\langle \alpha_{i,k} | B | \alpha_{i,l} \rangle$ does not have to vanish if $k \neq l$. It is therefore *not* the case that the $|\alpha_{i,k}\rangle$ must also be eigenvectors of B.

Let us reverse the situation, however, and consider the eigenvectors of B. Again, where B is not degenerate, its eigenvectors coincide with A's. In fact, if B has no degenerate eigenvectors in the above degenerate subspace of A, A and B will both be diagonal in B's representation. We may therefore formulate the prescription for finding a unique, nonarbitrary labeling of the representation in which A is diagonal, as follows:

1. Find another Hermitian operator B, which commutes with A.
2. In that part of the vector space where neither A nor B is degenerate, take their mutual eigenvectors.
3. If A (B) is degenerate in a certain subspace, choose the nondegenerate eigenvectors of B (A) in that subspace.
4. If *both* A and B are degenerate in a certain subspace, any arbitrary orthonormal set of vectors in that subspace will serve as common eigenvectors. But to remove the remaining degeneracy, one must locate yet a *third* Hermitian operator which commutes with *both* A and B and repeat the process, and so forth.

When a set of mutually commuting operators is found which is large enough to have eliminated any arbitrariness in selecting and labeling a representation in which *all* are diagonal, it is called a *complete set of commuting operators*. There then exists no subspace in which all members of a complete set of commuting operators are degenerate.

The idea of complete sets of commuting operators has its greatest value in quantum mechanics where physical observables are represented by Hermitian operators, their possible (quantized) values by the eigenvalues of the operators, and the wave functions which describe specific physical states by the corresponding eigenvectors. A familiar example is the hydrogen-like atom. Here, an important observable, whose values are usually used to specify the state of the atom, is the energy E. However, it is clear from experimental evidence that specifying the energy is not adequate for a complete description of the state of the atom; one or more observables are required, in addition to the energy, for its complete specification. The question of determining which observables correspond to operators which commute with the energy operator is one of quantum physics and we do not pursue it here. We merely point out that there are several sets of three such observables (in various levels of approximation), an example of which is the following: the norm of the orbital angular momentum, L; one (and only one) vector component of the orbital angular momentum, L_i; and one (and only one) vector component of the intrinsic angular momentum (or spin) of the valence electron, S_i. The *state vector* of the atom, in this representation, is then unambiguously given by $|E, L, L_i, S_i\rangle$, where the symbols in the ket vector represent certain eigenvalues of the respective operators.

E. TRANSFORMATION OF BASIS

As we pointed out in Section II, there are, for a given n-dimensional vector space, an infinite number of different basis sets from which to choose, including the sets of eigenvectors of all noncommuting Hermitian operators. It is, for any number of reasons, often convenient to be able to transform from one basis to another. The pertinent questions here are (1) which quantities change under such a transformation, and which do not, and (2) by what prescriptions can we calculate the new values of the quantities which do change? We need of course be concerned here only with quantities such as scalar products, components, matrix elements, and so on, which are relevant to vector spaces. In answering these questions, we first determine how to effect a basis transformation and then apply it to the particular quantities.

The transformation from one basis to another is obviously a linear operation since all new basis vectors can be expressed as linear combinations of the old. Therefore, we write the transformation of the basis set $\{\hat{\mathbf{V}}_i\}$ to the basis set $\{\hat{\mathbf{V}}_i'\}$ in terms of an operator U:

$$\hat{\mathbf{V}}_i' = U\hat{\mathbf{V}}_i. \tag{4.42}$$

The orthonormality of both new and original basis sets gives

$$(\hat{\mathbf{V}}_i', \hat{\mathbf{V}}_j') = \delta_{ij} = (U\hat{\mathbf{V}}_i, U\hat{\mathbf{V}}_j) = (\hat{\mathbf{V}}_i, U^{\ddagger}U\hat{\mathbf{V}}_j),$$

or

$$U^{\ddagger}U = I. \tag{4.43}$$

If we apply the closure relationship (2.20) to (4.42), we get the expansion of $\hat{\mathbf{V}}_i'$ in the original basis:

$$\hat{\mathbf{V}}_i' = \sum_j \hat{\mathbf{V}}_j(\hat{\mathbf{V}}_j, U\hat{\mathbf{V}}_i) = \sum_j U_{ji}\hat{\mathbf{V}}_j \tag{4.44}$$

from which we obtain the values for the matrix elements of U:

$$U_{ji} = (\hat{\mathbf{V}}_j, \hat{\mathbf{V}}_i'). \tag{4.45}$$

Equation (4.43) then becomes

$$\sum_k U^*_{kj} U_{ki} = \delta_{ji}. \qquad (4.46)$$

We also have

$$\sum_k U_{jk} U^*_{ik} = \sum_k (\hat{\mathbf{V}}_j, \hat{\mathbf{V}}_k')(\hat{\mathbf{V}}_k', \hat{\mathbf{V}}_i) = (\hat{\mathbf{V}}_j, \hat{\mathbf{V}}_i) = \delta_{ij} \qquad (4.46a)$$

or

$$UU^\ddagger = I, \qquad (4.47)$$

where we have used closure for the new basis. Operators which obey *both* (4.43) and (4.47) are called *unitary*. Thus, for *unitary operators*,

$$U^{-1} = U^\ddagger. \qquad (4.48)$$

We give a more detailed account of unitary operators *per se* below, but now let us approach the questions posed earlier.

To see how a change of basis effects the components of an arbitrary vector, we expand $\mathbf{V}_a$ in both bases:

$$\mathbf{V}_a = \sum_i a_i \hat{\mathbf{V}}_j = \sum_i a_i' \hat{\mathbf{V}}_j'$$
$$= \sum_i a_i' \sum_j U_{ji} \hat{\mathbf{V}}_j = \sum_j (\sum_i U_{ji} a_i') \hat{\mathbf{V}}_j$$

or

$$a_j = \sum_i U_{ji} a_i'. \qquad (4.49)$$

After multiplying by U^*_{jk}, summing over j, and using (4.45), we obtain

$$a_k' = \sum_j a_j U^*_{jk} = \sum_j U^\ddagger_{kj} a_j. \qquad (4.50)$$

The matrix elements in different representations of an operator are also simply related:

$$A_{ij}' = (\hat{\mathbf{V}}_i', A\hat{\mathbf{V}}_j')$$
$$= \sum_k \sum_l (\hat{\mathbf{V}}_i', \hat{\mathbf{V}}_k)(\hat{\mathbf{V}}_k, A\hat{\mathbf{V}}_l)(\hat{\mathbf{V}}_l, \hat{\mathbf{V}}_j')$$
$$= \sum_k \sum_l U^*_{ki} A_{kl} U_{lj}. \qquad (4.51)$$

1. Invariants

Quantities which do not change in value under a transformation of basis are called *invariants*. They are also called, in many contexts, *scalars*. Thus, to be precise, we should use the term *scalar* in a more restrictive sense than we have so far. If we apply the name to any quantity which is invariant under a basis transformation only, then, as we have seen, matrix elements and vector components, although numbers, are not scalars. This distinction is of particular importance when dealing with three-dimensional Euclidean vector spaces and the four-dimensional Lorentzian vector space of relativistic physics. In order to avoid confusion at this point, however, we use the word *invariant*.

Although matrix elements of an operator are not invariants when the basis in which they are calculated is transformed, quantities of the general type $(\mathbf{V}_a, A\mathbf{V}_b)$ are.

$$\begin{aligned}
(\mathbf{V}_a, A\mathbf{V}_b)' &= \sum_i \sum_j a_i'^* b_j' (\hat{\mathbf{V}}_i', A\hat{\mathbf{V}}_j') \\
&= \sum_i \sum_j \sum_k \sum_l a_i'^* b_j' U_{ki}^* U_{lj} (\hat{\mathbf{V}}_k, A\hat{\mathbf{V}}_l) \\
&= \sum_k \sum_l a_k^* b_l (\hat{\mathbf{V}}_k, A\hat{\mathbf{V}}_l) \\
&= (\mathbf{V}_a, A\mathbf{V}_b).
\end{aligned} \qquad (4.52)$$

Here we have used Eqs. (4.44) and (4.50). Thus, the eigenvalues of an operator are invariants, as can be seen by letting $\mathbf{V}_a$ and $\mathbf{V}_b$ be eigenvectors of A. As another special case of the above, we have the invariance of scalar products, for upon setting $A = I$:

$$(\mathbf{V}_a, \mathbf{V}_b)' = (\mathbf{V}_a, \mathbf{V}_b). \qquad (4.53)$$

In particular, $\|\mathbf{V}_a\|$, the norm of $\mathbf{V}_a$, is invariant.

Another invariant is the trace of the matrix representation of an operator, i.e., the sum of all elements on the diagonal:

$$\begin{aligned}
\operatorname{tr}(A)' &= \sum_i A_{ii} = \sum_i \sum_k \sum_l U_{ki}^* A_{kl} U_{li} \\
&= \sum_k \sum_l \left(\sum_i U_{ki}^* U_{li} \right) A_{kl} = \sum_k A_{kk} = \operatorname{tr}(A),
\end{aligned} \qquad (4.54)$$

where we have used (4.46a). Finally, as we shall see in Section V, the invariance of A's eigenvalues implies for finite vector spaces the invariance of the determinant of the matrix $\{A_{ij}\}$.

F. Unitary Operators

Suppose we apply a unitary operator S to an arbitrary vector $\mathbf{V}_a$:

$$\mathbf{V}_a' = S\mathbf{V}_a \tag{4.55}$$

$$SS^\ddagger = S^\ddagger S = I. \tag{4.56}$$

Here, now, we are transforming a particular vector rather than an entire basis. The transformation (4.55) leaves scalar products invariant:

$$(\mathbf{V}_a', \mathbf{V}_b') = (S\mathbf{V}_a, S\mathbf{V}_b) = (\mathbf{V}_a, S^\ddagger S\mathbf{V}_b) = (\mathbf{V}_a, \mathbf{V}_b). \tag{4.57}$$

A special case of this is the invariance of $\|\mathbf{V}_a\|$.

The components of the transformed vector are

$$a'_i = (\hat{\mathbf{V}}_i, \mathbf{V}_a') = (\hat{\mathbf{V}}_i, S\mathbf{V}_a) = (\hat{\mathbf{V}}_i, S\sum_j a_j\hat{\mathbf{V}}_j) = \sum_j S_{ij}a_j. \tag{4.58}$$

Now, consider the quantity $(\mathbf{V}_a, A\mathbf{V}_b)$. If we apply the unitary transformation (4.55) to both $\mathbf{V}_a$ and $\mathbf{V}_b$, the question arises: What operator A', when sandwiched between the transformed vectors, gives the same value? We have

$$(\mathbf{V}_a', A'\mathbf{V}_b') = (\mathbf{V}_a, A\mathbf{V}_b)$$
$$= (S^{-1}\mathbf{V}_a', AS^{-1}\mathbf{V}_b') = (\mathbf{V}_a', SAS^{-1}\mathbf{V}_b').$$

If this equality is to hold for any $\mathbf{V}_a$ and $\mathbf{V}_b$, then

$$A' = SAS^{-1} = SAS^\ddagger. \tag{4.59}$$

It is simple to show that $\text{tr}(A') = \text{tr}(A)$ and (for finite-dimensional spaces) $\det(A') = \det(A)$.

We see that if one wishes a particular representation of a vector or operator, one has the choice either of changing the basis:

$$\hat{\mathbf{V}}_i' = U\hat{\mathbf{V}}_i,$$

or of retaining the same basis and transforming the vector or operator:

$$\mathbf{V}_a' = S\mathbf{V}_a, \qquad A' = SAS^\ddagger.$$

If these choices are to be equivalent, we expect that U and S should be somehow related. Indeed, if

$$a_i = \sum_j U^*_{ji} a_j = \sum_j U^\dagger_{ij} a_j = a'_i = \sum_j S_{ij} a_j,$$

1. Linear Vector Spaces

then
$$S = U^{\ddagger}. \tag{4.60}$$

The eigenvalues of a unitary operator are restricted in their magnitude to unity:

$$S|\sigma\rangle = \sigma|\sigma\rangle, \qquad \langle\sigma|S^{\ddagger}S|\sigma\rangle = |\sigma|^2\langle\sigma|\sigma\rangle = \langle\sigma|\sigma\rangle$$

or
$$|\sigma|^2 = 1. \tag{4.61}$$

Thus we can write
$$\sigma = \exp(i\alpha_\sigma), \tag{4.62}$$

where α_σ is a real number. (If S is also Hermitian, σ must be real and has only two possible values, ± 1.) Let us write (4.62) as

$$\sigma = \cos\alpha_\sigma + i\sin\alpha_\sigma.$$

If we define the operator A such that its eigenvalues are the α_σ, the operator

$$e^{iA} \equiv \cos A + i\sin A \tag{4.63}$$

has eigenvalues $\exp(i\alpha_\sigma)$. We can therefore write

$$S = e^{iA}. \tag{4.64}$$

Since A can always be found (it is that operator which is diagonal with eigenvalues α_σ), we see that any unitary operator can be written in terms of an Hermitian operator as in (4.64).

G. Projection Operators

As a useful example of an operator, we define the *projection operator*:

$$P_a \equiv \frac{1}{\langle a|a\rangle}|a\rangle\langle a| = |\hat{a}\rangle\langle\hat{a}|, \tag{4.65}$$

where $|a\rangle$ is an arbitrary vector and $|\hat{a}\rangle$ is $|a\rangle$ normalized. When applied to another vector, $|b\rangle$, P_a gives

$$P_a|b\rangle = \langle\hat{a}|b\rangle|\hat{a}\rangle$$

which is just the unit vector $|\hat{a}\rangle$ multiplied by the component of $|b\rangle$ in the "direction" of $|a\rangle$.

P_a has the peculiar property that $P_a^2 = P_a$:

$$P_a^2 | b\rangle = P_a P_a | b\rangle = P_a | \hat{a}\rangle\langle \hat{a} | b\rangle = | \hat{a}\rangle\langle \hat{a} | \hat{a}\rangle\langle \hat{a} | b\rangle$$
$$= | \hat{a}\rangle\langle \hat{a} | b\rangle = P_a | b\rangle, \qquad (4.66)$$

which is reasonable, since having once projected $| b\rangle$ along $| a\rangle$, further projections in the same direction have no further effect. Because of this, P_a is said to be *idempotent*. Projection operators can also be shown to be Hermitian.

Consider now the set of basis vectors $| i\rangle$, and the associated projection operators

$$P_i = | i\rangle\langle i |. \qquad (4.67)$$

From closure,

$$\sum_i P_i = I. \qquad (4.68)$$

When operating with P_i upon an arbitrary vector $| a\rangle$, we obtain

$$P_i | a\rangle = | i\rangle\langle i | a\rangle = a_i | i\rangle. \qquad (4.69)$$

The operator

$$Q_i \equiv I - P_i \qquad (4.70)$$

is called the *complement* of P_i and removes the ith component from $| a\rangle$.

Let us take as our basis the eigenvectors $| \alpha_i\rangle$ of the Hermitian operator A. The projection operator P_a, when expressed as a matrix in this representation, is called the *density matrix*, ϱ_a:

$$(\varrho_a)_{ij} = \langle \alpha_i | P_a | \alpha_j\rangle = \langle \alpha_i | \hat{a}\rangle\langle \hat{a} | \alpha_j\rangle. \qquad (4.71)$$

A result which is useful in many applications, particularly in quantum mechanics, is

$$\langle \hat{a} | A | \hat{a}\rangle = \sum_i \sum_j \langle \hat{a} | \alpha_i\rangle\langle \alpha_i | A | \alpha_j\rangle\langle \alpha_j | \hat{a}\rangle$$
$$= \sum_i \alpha_i \langle \hat{a} | \alpha_i\rangle\langle \alpha_i | \hat{a}\rangle$$
$$= \sum_i (\varrho_a A)_{ii} = \text{tr}(\varrho_a A) \qquad (4.72)$$

[recall (4.39)].

V. Matrices

Matrices were introduced in Section IV.B as square arrays of the quantities $\langle i | A | j \rangle$, which we then called matrix elements of the operator A. We also pointed out that the components of a vector form a columnar array called a column matrix. From these two examples we see the natural generalization of matrices to include all *rectangular arrays* of real or complex numbers leading to countless applications in all areas of mathematics.

Let A be such a rectangular array, or matrix, with m rows and n columns; the dimension of A is thus $m \times n$. In the conventional notation for the elements of A, a_{ij} or sometimes $(\mathsf{A})_{ij}$, the first subscript indices the row, and the second, the column in which the element is found:

$$\mathsf{A} = \begin{pmatrix} a_{11} & a_{12} & \cdots & a_{1j} & \cdots & a_{1n} \\ a_{21} & a_{22} & \cdots & a_{2j} & \cdots & a_{2n} \\ \vdots & \vdots & & \vdots & & \vdots \\ a_{i1} & a_{i2} & \cdots & a_{ij} & \cdots & a_{in} \\ \vdots & \vdots & & \vdots & & \vdots \\ a_{m1} & a_{m2} & \cdots & a_{mj} & \cdots & a_{mn} \end{pmatrix}. \tag{5.1}$$

We will often also indicate a matrix simply as $\mathsf{A} = \{a_{ij}\}$.

A. Basic Matrix Operations

Let us begin the discussion of the basic matrix operations by introducing the *null* and *unit* (or identity) matrices. The *null matrix* is just a matrix all of whose elements are zero: $\varnothing = \{0\}$. It may be of any rectangular dimension. The *unit matrix* is a square matrix of any dimension whose (upper left to lower right) diagonal elements are unity, the rest equaling zero, i.e.,

$$\mathsf{I} = \{\delta_{ij}\}. \tag{5.2}$$

1. *Equality of Matrices*

Two matrices are said to be equal if they are of the same dimension and are equal element by element, i.e., $\mathsf{A} = \mathsf{B}$ if $a_{ij} = b_{ij}$ for all i and j. Matrix equality is obviously transitive: If $\mathsf{A} = \mathsf{B}$ and $\mathsf{B} = \mathsf{C}$, then $\mathsf{A} = \mathsf{C}$.

2. Matrix Addition

Addition and multiplication are defined for matrices in general in a way that is consistent with operations we have given for square matrices in Eqs. (4.24) and (4.25). Thus, two matrices $\{a_{ij}\}$ and $\{b_{ij}\}$ of the same dimension can be added to yield a third matrix, $\{c_{ij}\}$, also of the same dimension, where

$$c_{ij} = a_{ij} + b_{ij}. \tag{5.3}$$

Subtraction is similarly defined. In particular, if $\mathbf{A} = \mathbf{B}$, then

$$\mathbf{A} - \mathbf{B} = \varnothing$$

where $\varnothing$ is the null matrix of appropriate dimension.

Matrix addition is commutative:

$$\mathbf{A} + \mathbf{B} = \mathbf{B} + \mathbf{A}, \tag{5.4}$$

and associative:

$$\mathbf{A} + (\mathbf{B} + \mathbf{C}) = (\mathbf{A} + \mathbf{B}) + \mathbf{C}. \tag{5.5}$$

3. Matrix Multiplication

As with addition, matrix multiplication is restricted by dimensionality requirements. Aside from one trivial exception, matrices may not be multiplied unless the number of columns of one equals the number of rows of the other. Thus, if $\mathbf{A}$ and $\mathbf{B}$ are of dimensions $l \times m$ and $m \times n$, respectively, we can form the product $\mathbf{C} = \mathbf{AB}$ to obtain a matrix of dimension $l \times n$ which has elements

$$c_{ij} = \sum_{k=1}^{n} a_{ik} b_{kj}. \tag{5.6}$$

However, if $l \neq n$, the product $\mathbf{BA}$ is not defined. Clearly, nonsquare matrices never commute, for even if $l = n$, so that both products $\mathbf{AB}$ and $\mathbf{BA}$ would be defined, their dimensions would be $n \times n$ and $m \times m$, respectively. As we saw in Section IV, square matrices also do not commute in general.

On the other hand, assuming that the respective products can be taken as indicated, matrix multiplication can be seen to be associative:

$$\mathbf{A}(\mathbf{BC}) = (\mathbf{AB})\mathbf{C}, \tag{5.7}$$

and distributive through matrix addition:

$$A(B + C) = AB + AC. \tag{5.8}$$

4. *Scalar Multiplication of Matrices*

The exception to dimensionality restrictions mentioned above is that of the 1×1 matrix: the scalar. A scalar may multiply a matrix of arbitrary dimension, yielding a matrix of the same dimension with matrix elements

$$(\alpha A)_{ij} = \alpha a_{ij}. \tag{5.9}$$

5. *Matrix Transmutations*

From a given $m \times n$ matrix A, several other matrices can be formed through simple transmutations of A (more complicated matrix transformations are discussed later in this section). We list here a few of the more useful transmutations.

A^*: the *complex conjugate* of A, dimension $m \times n$.

$$(A^*)_{ij} = a_{ij}^*. \tag{5.10}$$

$\tilde{A}$: the *transpose* of A, dimension $n \times m$.

$$(\tilde{A})_{ij} = a_{ji}. \tag{5.11}$$

$\tilde{A}$ is obtained from A by interchanging rows and columns.

$A^\ddagger$: the *Hermitian conjugate* (or *Hermitian transpose* or *transpose conjugate*) of A, dimension $n \times m$.

$$(A^\ddagger)_{ij} = a_{ji}^*, \tag{5.12}$$

i.e.,

$$A^* = (\tilde{A})^* = \tilde{A}^*. \tag{5.13}$$

The following relationships are simple to prove:

$$\widetilde{AB} = \tilde{B}\tilde{A} \tag{5.14}$$

$$(AB)^\ddagger = B^\ddagger A^\ddagger. \tag{5.15}$$

For example,

$$(\widetilde{AB})_{ij} = (AB)_{ji} = \sum_k a_{jk} b_{ki} = \sum_k (\tilde{B})_{ik} (\tilde{A})_{kj} = (\tilde{B}\tilde{A})_{ij}.$$

B. Special Matrices

Associated with the preceding matrix transmutations there is a variety of special types of matrices characterized by their properties under these and other simple operations. We list several of them in Table I. Except for the first two, all are square matrices. We have introduced here the *inverse* of $\mathbf{A}$, namely $\mathbf{A}^{-1}$, whose existence and properties are examined below. Suffice it here to say that if $\mathbf{A}^{-1}$ exists for the square matrix $\mathbf{A}$, then of course

$$\mathbf{A}\mathbf{A}^{-1} = \mathbf{A}^{-1}\mathbf{A} = \mathbf{I}. \quad (5.16)$$

We should include at this point the following definitions.

$\mathbf{A}$ is called *diagonal* if $a_{ij} = \alpha_i \delta_{ij}$ and *scalar* if $a_{ij} = \alpha\, \delta_{ij}$, i.e., $\mathbf{A} = \alpha \mathbf{I}$.

It is an often useful fact that *any square matrix can be written as the sum of a symmetric and an antisymmetric matrix or as the sum of an Hermitian and an anti-Hermitian matrix*. (For real matrices, these statements are, of course, equivalent.) This is shown by construction:

$$\mathbf{A} = \tfrac{1}{2}(\mathbf{A} + \tilde{\mathbf{A}}) + \tfrac{1}{2}(\mathbf{A} - \tilde{\mathbf{A}}) \quad (5.17a)$$

and

$$\mathbf{A} = \tfrac{1}{2}(\mathbf{A} + \mathbf{A}^\dagger) + \tfrac{1}{2}(\mathbf{A} - \mathbf{A}^\dagger). \quad (5.17b)$$

The first and second terms on the right of these equations are manifestly symmetric (Hermitian) and antisymmetric (anti-Hermitian), respectively.

TABLE I

$\mathbf{A}$ is			
Real	if $\mathbf{A}^* = \mathbf{A}$,	i.e.,	$a_{ij}^* = a_{ij}$
Imaginary	$\mathbf{A}^* = -\mathbf{A}$		$a_{ij}^* = -a_{ij}$
Symmetric	$\tilde{\mathbf{A}} = \mathbf{A}$		$a_{ji} = a_{ij}$
Antisymmetric	$\tilde{\mathbf{A}} = -\mathbf{A}$		$a_{ji} = -a_{ij}$
Hermitian	$\mathbf{A}^\dagger = \mathbf{A}$		$a_{ji}^* = a_{ij}$
Anti-Hermitian	$\mathbf{A}^\dagger = -\mathbf{A}$		$a_{ji}^* = -a_{ij}$
Idempotent	$\mathbf{A}^2 = \mathbf{A}$		$\sum_k a_{ik} a_{kj} = a_{ij}$
Orthogonal	$\tilde{\mathbf{A}} = \mathbf{A}^{-1}$		$\sum_k a_{kj} a_{ki} = \delta_{ji}$
Unitary	$\mathbf{A}^\dagger = \mathbf{A}^{-1}$		$\sum_k a_{kj}^* a_{ki} = \delta_{ji}$

C. Determinants

Before discussing the inverse and eigenvalue problems for square matrices, it is necessary for us to review the properties of determinants. We state most of the results of this section without proof inasmuch as the reader is likely to have encountered a detailed elucidation of determinant theory in a course in elementary algebra.

The determinant $|\mathbf{A}|$ of the matrix $\mathbf{A}$ is defined most succinctly by

$$|\tilde{\mathbf{A}}| = \sum_{ij\cdots v=1}^{n} \varepsilon_{ij\cdots v} a_{1i} a_{2j} \cdots a_{nv} \tag{5.18}$$

where the symbol $\varepsilon_{ij\cdots v}$ has n independent subscripts running from 1 to n and is given the values

$$\varepsilon_{ij\cdots v} = \begin{cases} +1 & \text{if } (ij \ldots v) \text{ is an even permutation of } (1\,2 \ldots n) \\ -1 & \text{if } (ij \ldots v) \text{ is an odd permutation of } (1\,2 \ldots n) \\ 0 & \text{otherwise, i.e., if any two or more subscripts are equal.} \end{cases}$$

The more common way of calculating determinants, however, is to *expand by minors*. We write

$$|\mathbf{A}| = \sum_{i=1}^{n} a_{ji} |\mathbf{A}_{(ji)}| \tag{5.19}$$

for *any* j, where $\mathbf{A}_{(ji)}$, called the *minor* of a_{ji}, is the $(n-1) \times (n-1)$ matrix which is formed by removing the jth row and the ith column from $\mathbf{A}$. Each determinant $|\mathbf{A}_{(ji)}|$ is expressed in a similar fashion, and so forth, until we have reduced the problem to finding the determinants of several 2×2 matrices. These are given by $\left|\begin{smallmatrix} a & b \\ c & d \end{smallmatrix}\right| = ad - bc$.

For larger n, though, this process becomes tedious and susceptible to errors of sign. The calculation of a determinant in a computer is usually done by a process called the *Gauss–Jordan reduction method*, which we describe below. It depends on some of the following basic properties of determinants, all of which follow more or less easily from (5.18).

(i) Interchanging two neighboring rows (or columns) of a matrix changes only the sign of the determinant.

(ii) If two rows (or columns) of $\mathbf{A}$ are equal, then $|\mathbf{A}| = 0$.

(iii) $\quad\quad\quad\quad |\tilde{\mathbf{A}}| = |\mathbf{A}|.$ (5.20)

(iv) $\quad\quad\quad\quad |\alpha \mathbf{A}| = \alpha^n |\mathbf{A}|.$ (5.21)

(v) $\quad\quad\quad\quad |\mathbf{AB}| = |\mathbf{A}||\mathbf{B}|.$ (5.22)

(vi) Multiplying one row (or column) by the constant α results in multiplying the determinant by α.

(vii) The addition of a constant multiple of a row (or column) to the corresponding elements of any other row (or column) leaves the value of the determinant unchanged.

1. Gauss–Jordan Reduction

This method of calculating the determinant of a matrix consists of reducing the matrix to triangular form, i.e., all zeros below (or above) the diagonal, by means of operations of the type described in (vii) above, under all of which the determinant is invariant. From (5.18) or (5.19) it is easy to see that the determinant of a triangularized matrix is just the product of the diagonal elements. This process is most easily appreciated through an example. Let us calculate the determinant of

$$\begin{pmatrix} 4 & 2 & -1 \\ 1 & 0 & 5 \\ 2 & 3 & -2 \end{pmatrix}.$$

We leave the first row as it is. To reduce the second row, we substract from it $\frac{1}{4} \times$ the first row. Similarly, the third row undergoes partial reduction by subtracting $\frac{2}{4} \times$ the first row from it. At this point the matrix has become

$$\begin{pmatrix} 4 & 2 & -1 \\ 0 & -\frac{1}{2} & \frac{21}{4} \\ 0 & 2 & -\frac{3}{2} \end{pmatrix}.$$

The third row is now completely reduced by adding to it $4 \times$ the new second row. We are left with the triangularized matrix

$$\begin{pmatrix} 4 & 2 & -1 \\ 0 & -\frac{1}{2} & \frac{21}{4} \\ 0 & 0 & \frac{39}{2} \end{pmatrix},$$

the determinant of which is $4 \times (-\frac{1}{2}) \times \frac{39}{2} = -39$.

D. Inverse of a Matrix

We can now approach the problem of how to calculate the inverse of a square matrix. This is done formally, and for small matrices, by techniques differing from those used in computers for large matrices. We begin the discussion of the former by proving the following:

Theorem If $\mathbf{A}^{-1}$ exists, its matrix elements are given by

$$(\mathbf{A}^{-1})_{ij} = |\mathbf{A}_{(ji)}| / |\mathbf{A}| \qquad (5.23)$$

where $\mathbf{A}_{(ji)}$ is the minor of a_{ji}.

Before giving the proof, we point out that $\mathbf{A}^{-1}$ will exist if and *only if* the determinant of $\mathbf{A}$ does not vanish! A matrix whose determinant vanishes is called *singular*; thus the adjective *singular* is used for matrices without inverses and *nonsingular* otherwise.

We will now prove (5.17) for a right inverse; that $\mathbf{A}^{-1}$ is then also a left inverse is shown in an exactly similar manner. From (5.23) we have

$$(\mathbf{A}\mathbf{A}^{-1})_{ij} = \sum_k a_{ik} \frac{|\mathbf{A}_{(jk)}|}{|\mathbf{A}|} = \frac{1}{|\mathbf{A}|} \sum_k a_{ik} |\mathbf{A}_{(jk)}|. \qquad (5.24)$$

When $i = j$, this becomes unity by virtue of (5.19). On the other hand, when $i \neq j$, each term in $\sum_k a_{ik}|\mathbf{A}_{(jk)}|$ will have a factor $a_{ik}a_{il}$, $k \neq l$, for all k and l, since $\mathbf{A}_{(jk)}$ is the minor of an element in the kth column other than a_{ik}. Therefore, this sum is actually the determinant of a matrix that has two rows, namely the ith and jth, equal. It consequently vanishes according to property (ii) of determinants given earlier. Hence,

$$(\mathbf{A}\mathbf{A}^{-1})_{ij} = \delta_{ij}.$$

When we combine this with the same result for $(\mathbf{A}^{-1}\mathbf{A})_{ij}$, the theorem is proved. The reader may show that $\mathbf{A}^{-1}$ so defined is unique.

As a corollary to (5.24), we have

$$\sum_k a_{ik}|\mathbf{A}_{(jk)}| = |\mathbf{A}|\delta_{ij}, \qquad \sum_k a_{ik}|\mathbf{A}_{(il)}| = |\mathbf{A}|\delta_{kl}. \qquad (5.25)$$

1. Calculating the Inverse

To use the formula (5.24) to calculate the inverse of a matrix requires the calculation of n^2 $(n-1) \times (n-1)$ determinants $|\mathbf{A}_{(ij)}|$ as well as the combination of n of them to form $|\mathbf{A}|$. For large values of n, this can become a very inefficient use of time. Among many alternatives are the Cayley–Hamilton method (to be described later) and an extension of the Gauss–Jordan elimination method which serves as the basis for most computer matrix inversion techniques. The algorithm for the latter is demonstrated as follows for the matrix whose determinant we calculated earlier.

We begin by writing the matrix and the 3×3 unit matrix side by side for convenience:

$$A = \begin{pmatrix} 4 & 2 & -1 \\ 1 & 0 & 5 \\ 2 & 3 & -2 \end{pmatrix}, \quad I = \begin{pmatrix} 1 & 0 & 0 \\ 0 & 1 & 0 \\ 0 & 0 & 1 \end{pmatrix}.$$

Any subsequent operation performed on A is to be performed in exactly the same way on I. The object of the game is to reduce A, by means of multiplying and adding rows, to the unit matrix.

(a) Divide the first row by 4. (This step is called normalization and leaves the diagonal element equal to 1. If $a_{11} = 0$, we can start with another row.) Under this operation

$$A \to \begin{pmatrix} 1 & \tfrac{1}{2} & -\tfrac{1}{4} \\ 1 & 0 & 5 \\ 2 & 3 & -2 \end{pmatrix}, \quad I \to \begin{pmatrix} \tfrac{1}{4} & 0 & 0 \\ 0 & 1 & 0 \\ 0 & 0 & 1 \end{pmatrix}.$$

(b) Substract the first row from the second and twice the first row from the third to obtain

$$A \to \begin{pmatrix} 1 & \tfrac{1}{2} & -\tfrac{1}{4} \\ 0 & -\tfrac{1}{2} & \tfrac{21}{4} \\ 0 & 2 & -\tfrac{3}{2} \end{pmatrix}, \quad I \to \begin{pmatrix} \tfrac{1}{4} & 0 & 0 \\ -\tfrac{1}{4} & 1 & 0 \\ -\tfrac{1}{2} & 0 & 1 \end{pmatrix}.$$

(c) Normalize the second row by multiplying by -2.

$$A \to \begin{pmatrix} 1 & \tfrac{1}{2} & -\tfrac{1}{4} \\ 0 & 1 & -\tfrac{21}{2} \\ 0 & 2 & -\tfrac{3}{2} \end{pmatrix}, \quad I \to \begin{pmatrix} \tfrac{1}{4} & 0 & 0 \\ \tfrac{1}{2} & -2 & 0 \\ -\tfrac{1}{2} & 0 & 1 \end{pmatrix}.$$

(d) Subtract $\tfrac{1}{2} \times$ the second row from the first and twice the second row from the third.

$$A \to \begin{pmatrix} 1 & 0 & 5 \\ 0 & 1 & -\tfrac{21}{2} \\ 0 & 0 & \tfrac{39}{2} \end{pmatrix}, \quad I \to \begin{pmatrix} 0 & 1 & 0 \\ \tfrac{1}{2} & -2 & 0 \\ -\tfrac{3}{2} & 4 & 1 \end{pmatrix}.$$

(e) Normalize the third row by multiplying by $\tfrac{2}{39}$.

$$A \to \begin{pmatrix} 1 & 0 & 5 \\ 0 & 1 & -\tfrac{21}{2} \\ 0 & 0 & 1 \end{pmatrix}, \quad I \to \begin{pmatrix} 0 & 1 & 0 \\ \tfrac{1}{2} & -2 & 0 \\ -\tfrac{3}{39} & \tfrac{8}{39} & \tfrac{2}{39} \end{pmatrix}.$$

(f) Subtract 5 × the third row from the first and $-\frac{21}{2}$ × the third row from the second. This gives finally

$$\mathbf{A} \to \begin{pmatrix} 1 & 0 & 0 \\ 0 & 1 & 0 \\ 0 & 0 & 1 \end{pmatrix}, \quad \mathbf{I} \to \tfrac{1}{39} \begin{pmatrix} 15 & -1 & -10 \\ -12 & 6 & 21 \\ -3 & 8 & 2 \end{pmatrix}.$$

The assertion is that the matrix on the right is $\mathbf{A}^{-1}$, as the reader can check.

That this procedure works in general is seen by realizing that each of the operations performed on $\mathbf{A}$ can be expressed as an $n \times n$ matrix multiplying $\mathbf{A}$. Thus, step (a) is carried out by the matrix

$$\mathbf{S}_a = \begin{pmatrix} \tfrac{1}{4} & 0 & 0 \\ 0 & 1 & 0 \\ 0 & 0 & 1 \end{pmatrix}$$

and step (b) by the matrix

$$\mathbf{S}_b = \begin{pmatrix} 1 & 0 & 0 \\ -1 & 1 & 0 \\ -2 & 0 & 1 \end{pmatrix},$$

and so forth. The complete transformation is given by the product of these matrices which is itself a matrix. This matrix is that which, when multiplying $\mathbf{A}$, gives $\mathbf{I}$; to wit, $\mathbf{A}^{-1}$. But we have multiplied $\mathbf{I}$ by the same matrix, obtaining thereby $\mathbf{A}^{-1}$.

If $\mathbf{A}^{-1}$ does not exist, the procedure will ultimately lead to a point where there are no nonzero diagonal elements about which reductions can still be made.

E. EIGENVALUES OF MATRICES

The eigenvalue problem for matrices is exactly as given in the discussion of the eigenvalue problem for linear operators in Section IV. Indeed, there, as an example, we solved for the eigenvalues of a 3×3 matrix. For completeness here, however, we review the essentials.

The matrix $\mathbf{A}$ will have the quantity α as an eigenvalue if the determinant of the matrix $\mathbf{A} - \alpha \mathbf{I}$ vanishes:

$$|\mathbf{A} - \alpha \mathbf{I}| = 0 \tag{5.26}$$

[cf. Eq. (4.30)]. The column matrices which in the aforementioned example were called eigenvectors may also be called *eigencolumns* by some authors. They are obtained by solving the set of n homogeneous equations represented in matrix form by

$$\mathbf{A}\mathbf{V}_\alpha = \alpha \mathbf{V}_\alpha \tag{5.27}$$

[cf. Eq. (4.31)] for each α.

We expand the secular equation (5.26) to obtain the polynomial equation

$$(-1)^n |\mathbf{A} - \alpha \mathbf{I}| = \alpha^n + a_1 \alpha^{n-1} + \cdots + a_n = 0. \tag{5.28}$$

If $\alpha_1, \ldots, \alpha_n$ are the roots (i.e., the eigenvalues of $\mathbf{A}$), we can also write

$$(-1)^n |\mathbf{A} - \alpha \mathbf{I}| = (\alpha - \alpha_1)(\alpha - \alpha_2) \cdots (\alpha - \alpha_n). \tag{5.29}$$

Comparison of the coefficients of the individual powers of α in these equations gives

$$\begin{aligned} a_1 &= -(\alpha_1 + \alpha_2 + \cdots + \alpha_n) \\ a_2 &= \alpha_1 \alpha_2 + \alpha_1 \alpha_3 + \cdots + \alpha_{n-1}\alpha_n \\ &\vdots \\ a_n &= (-1)^n \alpha_1 \alpha_2 \cdots \alpha_n. \end{aligned} \tag{5.30}$$

Of particular interest are a_1 and a_n. Comparing (5.28) with (4.30), we see that

$$a_1 = -\sum_i a_{ii}$$

or

$$\sum_i \alpha_i = \text{tr}(\mathbf{A}). \tag{5.31}$$

Setting $\alpha = 0$ in (5.28) gives $a_n = (-1)^n |\mathbf{A}|$. Consequently,

$$|\mathbf{A}| = \alpha_1 \alpha_2 \cdots \alpha_n. \tag{5.32}$$

1. *Diagonalization of Hermitian Matrices*

In our discussion in Section IV of transformations of basis vectors, we saw that if such a transformation were effected by the unitary operator U,

$$\hat{\mathbf{V}}_i' = U\hat{\mathbf{V}}_i, \tag{4.42}$$

then the matrix elements of the operator A would be transformed according to

$$A'_{ij} = \sum_k \sum_l U^*_{ki} A_{kl} U_{lj}. \tag{4.51}$$

In matrix form, Eq. (4.51) is written

$$\mathsf{A}' = \mathsf{U}^\ddagger \mathsf{A} \mathsf{U} = \mathsf{U}^{-1} \mathsf{A} \mathsf{U} \tag{5.33}$$

[not to be confused with Eq. (4.59) in which an operator rather than a basis is transformed]. The particular question we have in mind here is What unitary operator U will perform the transformation from the original basis to that provided by the eigenvectors of an Hermitian operator? The answer is given by Eq. (4.45) as that operator whose matrix elements *in the original basis* are

$$U_{ij} = (\hat{\mathbf{V}}_i, \hat{\mathbf{V}}_{\alpha_j}). \tag{5.34}$$

Now, in terms of matrices, the question above becomes, What unitary matrix will diagonalize the Hermitian matrix A? And the answer is, as seen from (5.34), that matrix whose columns are the normalized eigencolumns of A.

2. *The Cayley–Hamilton Theorem*[†]

If A is Hermitian it obeys its own secular equation.

In order to prove this, we require first:

Lemma If A and B are related by a unitary transformation

$$\mathsf{B} = \mathsf{U}^\ddagger \mathsf{A} \mathsf{U}, \qquad \mathsf{U}^\ddagger = \mathsf{U}^{-1}, \tag{5.35}$$

and if $f(x)$ is a function which can be written as a series expansion in nonnegative powers of x (negative powers are allowed if the inverses of A and B exist), then

$$f(\mathsf{B}) = \mathsf{U}^\ddagger f(\mathsf{A}) \mathsf{U}. \tag{5.36}$$

For,

$$\mathsf{B}^2 = (\mathsf{U}^\ddagger \mathsf{A} \mathsf{U})(\mathsf{U}^\ddagger \mathsf{A} \mathsf{U}) = \mathsf{U}^\ddagger \mathsf{A}^2 \mathsf{U}$$

$$\mathsf{B}^n = (\mathsf{U}^\ddagger \mathsf{A} \mathsf{U})(\mathsf{U}^\ddagger \mathsf{A} \mathsf{U}) \cdots (\mathsf{U}^\ddagger \mathsf{A} \mathsf{U}) = \mathsf{U}^\ddagger \mathsf{A}^n \mathsf{U} \qquad (n \text{ factors})$$

[†] This theorem actually holds for *any* $n \times n$ matrix with n linearly independent eigenvectors.

so that
$$f(\mathbf{B}) = \sum_n a_n(\mathbf{U}^\ddagger \mathbf{A}\mathbf{U})^n = \mathbf{U}^\ddagger(\sum_n a_n \mathbf{A}^n)\mathbf{U} = \mathbf{U}^\ddagger f(\mathbf{A})\mathbf{U}.$$

Now, the Cayley–Hamilton theorem is proved by showing that $\mathbf{A}'$, the diagonal matrix whose diagonal elements are the eigenvalues of $\mathbf{A}$, obeys the secular equation for $\mathbf{A}$. Let the secular equation (5.28) be written as

$$\varphi(\alpha) = 0, \tag{5.37}$$

where $\varphi(\alpha) = \alpha^n + a_1\alpha^{n-1} + \cdots + a_n$. Then consider

$$\varphi(\mathbf{A}') = \mathbf{A}'^n + a_1\mathbf{A}'^{n-1} + \cdots + a_n\mathbf{I}.$$

Since $(\mathbf{A}')_{ij} = \alpha_i \delta_{ij}$, we have

$$(\mathbf{A}'^n)_{ij} = \alpha_i^n \delta_{ij},$$

giving
$$(\varphi(\mathbf{A}'))_{ij} = \varphi(\alpha_i) \delta_{ij} = 0.$$

Thus,
$$\varphi(\mathbf{A}') = \varnothing$$

and, by the lemma,

$$\varphi(\mathbf{A}) = \mathbf{U}\varphi(\mathbf{A}')\mathbf{U}^\ddagger = \varnothing. \tag{5.38}$$

The Cayley–Hamilton theorem leads to algorithms which simplify, in some instances, calculations of the eigenvalues and inverse of a matrix. For example, one can calculate the secular equation without expanding the determinant. Let $\mathbf{V}_0$ be any arbitrary column matrix and form $\mathbf{V}_1 = \mathbf{A}\mathbf{V}_0$, $\mathbf{V}_2 = \mathbf{A}^2\mathbf{V}_0 = \mathbf{A}\mathbf{V}_1$, ... We write the secular equation for $\mathbf{A}$:

$$\mathbf{A}^n + a_1\mathbf{A}^{n-1} + a_2\mathbf{A}^{n-2} + \cdots + a_n\mathbf{I} = \varnothing, \tag{5.38a}$$

and multiply on the right by $\mathbf{V}_0$, obtaining

$$\mathbf{V}_n + a_1\mathbf{V}_{n-1} + \cdots + a_n\mathbf{V}_0 = \varnothing. \tag{5.39}$$

Equation (5.39) is simply a set of n simultaneous homogeneous equations which may be solved (by Gaussian reduction, for example) for the a_1's.

We may also use (5.39), once we have the a_i's, to calculate $\mathbf{A}^{-1}$. A simple manipulation of (5.38a) gives

$$\mathbf{I} = -(1/a_n)\mathbf{A}^n - (a_1/a_n)\mathbf{A}^{n-1} - \cdots - (a_{n-1}/a_n)\mathbf{A},$$

where we assume $a_n \neq 0$. (If $a_n = 0$, $|\mathbf{A}| = 0$ and $\mathbf{A}^{-1}$ does not exist.) Upon multiplication by $\mathbf{A}^{-1}$, this equation becomes

$$\mathbf{A}^{-1} = -(1/a_n)\mathbf{A}^{n-1} - (a_1/a_n)\mathbf{A}^{n-2} - \cdots - (a_{n-1}/a_n)\mathbf{I}. \quad (5.40)$$

F. Basis Matrices of Small Dimension

The idea of linear independence as applied to vectors can be extended to matrices. (In fact, the set of $m \times n$ complex matrices constitutes a linear vector space. The reader is invited to work out the details.) We therefore have the possibility of establishing basis sets of matrices. The basis sets for 2×2 and 3×3 matrices are of particular interest in applied matrix theory and we give here a special example of each.

It is easy to see that a complete basis for the set of $n \times n$ complex matrices consists of n^2 such matrices. They are usually taken to be the unit matrix and $n^2 - 1$ traceless matrices.

1. 2×2

$$\mathbf{I} = \begin{pmatrix} 1 & 0 \\ 0 & 1 \end{pmatrix}, \quad \sigma_1 = \begin{pmatrix} 0 & 1 \\ 1 & 0 \end{pmatrix}, \quad \sigma_2 = \begin{pmatrix} 0 & -i \\ i & 0 \end{pmatrix}, \quad \sigma_3 = \begin{pmatrix} 1 & 0 \\ 0 & -1 \end{pmatrix}. \quad (5.41)$$

The σ_i are called the *Pauli matrices* and obey the following relationships:

$$\sigma_i^2 = \mathbf{I}, \quad i = 1, 2, 3 \quad (5.42)$$

$$[\sigma_i, \sigma_j] = 2i\varepsilon_{ijk}\sigma_k \quad (5.43)$$

$$\sigma_i \sigma_j = i\sigma_k, \quad (5.44)$$

where, in (5.44), the subscripts i, j, and k are even permutations of 1, 2, and 3; when not a subscript, $i = \sqrt{-1}$. In addition, the σ_i are Hermitian and have determinant $+1$.

2. 3×3

Here we have nine members of the basis: the 3×3 unit matrix and

$$\mathbf{F}_1 = \begin{pmatrix} 0 & 1 & 0 \\ 1 & 0 & 0 \\ 0 & 0 & 0 \end{pmatrix}, \quad \mathbf{F}_2 = \begin{pmatrix} 0 & -i & 0 \\ i & 0 & 0 \\ 0 & 0 & 0 \end{pmatrix}, \quad \mathbf{F}_3 = \begin{pmatrix} 1 & 0 & 0 \\ 0 & -1 & 0 \\ 0 & 0 & 0 \end{pmatrix}$$

$$\mathbf{F}_4 = \begin{pmatrix} 0 & 0 & 0 \\ 0 & 0 & 1 \\ 0 & 1 & 0 \end{pmatrix}, \quad \mathbf{F}_5 = \begin{pmatrix} 0 & 0 & 0 \\ 0 & 0 & -i \\ 0 & i & 0 \end{pmatrix}, \quad \mathbf{F}_6 = \begin{pmatrix} 0 & 0 & 1 \\ 0 & 0 & 0 \\ 1 & 0 & 0 \end{pmatrix} \quad (5.45)$$

$$\mathbf{F}_7 = \begin{pmatrix} 0 & 0 & -i \\ 0 & 0 & 0 \\ i & 0 & 0 \end{pmatrix}, \quad \mathbf{F}_8 = \frac{1}{\sqrt{3}} \begin{pmatrix} 1 & 0 & 0 \\ 0 & 1 & 0 \\ 0 & 0 & -2 \end{pmatrix}.$$

The commutation relations for these matrices are

$$[\mathbf{F}_i, \mathbf{F}_j] = 2if_{ijk}\mathbf{F}_k, \quad (5.46)$$

where the f_{ijk} are antisymmetric with respect to their indices and

$$f_{123} = +1$$
$$f_{147} = f_{246} = f_{257} = f_{345} = f_{571} = f_{376} = f_{165} = \tfrac{1}{2} \quad (5.47)$$
$$f_{458} = f_{678} = \sqrt{3}/2,$$

all others, except those whose indices are merely permutations of the above, being zero.

The generation of matrix bases of higher dimension is properly a function of group representation theory and the interested reader is referred to Chapter 6.

VI. Example: Three-Dimensional Euclidean Vectors—II

A. Rotations of Coordinate Axes

We have seen in Section V how the effect of a transformation of basis upon the components of a vector can be represented by a unitary matrix. As a simple but important example of this we examine here the behavior of three-dimensional Euclidean vectors under rotations of the coordinate axes. Since we are dealing with real vectors, the transformation matrices are *orthogonal*.

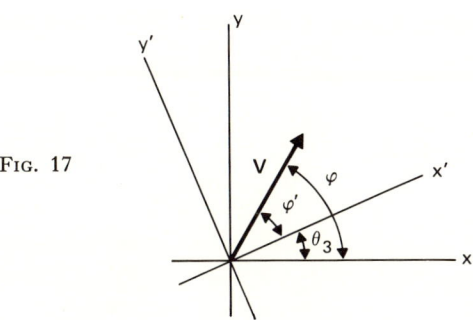

Fig. 17

In general, the specification of the orientation of a set of coordinate axes in three dimensions requires three independent parameters. To begin with, however, we will simplify matters by fixing the z axis and consider only rotations in the xy plane. The situation is as illustrated in Fig. 17 where the z axis is normal to the plane of the page and directed outward. In the original (unprimed) basis, the components of **V** are

$$V_1 = V \cos \varphi \quad \text{and} \quad V_2 = V \sin \varphi,$$

whereas in the new (primed) basis,

$$V_1' = V \cos \varphi' \quad \text{and} \quad V_2' = V \sin \varphi'.$$

Using the fact that $\varphi = \theta_3 + \varphi'$ along with various trigonometric identities to eliminate V, we obtain

$$V_1' = V_1 \cos \theta_3 + V_2 \sin \theta_3, \quad V_2' = -V_1 \sin \theta_3 + V_2 \cos \theta_3. \quad (6.1)$$

In matrix form this becomes

$$\begin{pmatrix} V_1' \\ V_2' \end{pmatrix} = \begin{pmatrix} \cos \theta_3 & \sin \theta_3 \\ -\sin \theta_3 & \cos \theta_3 \end{pmatrix} \begin{pmatrix} V_1 \\ V_2 \end{pmatrix}. \quad (6.2)$$

If our vector also had a z component, it would not have been affected by the rotation $V_3' = V_3$. The entire situation is then given by the *rotation matrix*

$$\mathsf{R}_3(\theta_3) = \begin{pmatrix} \cos \theta_3 & \sin \theta_3 & 0 \\ -\sin \theta_3 & \cos \theta_3 & 0 \\ 0 & 0 & 1 \end{pmatrix}. \quad (6.3a)$$

The orthogonality of R_3, $\sum_{k=1}^{3} r_{ki} r_{kj} = \delta_{ij}$, obtains from the identity $\cos^2 \theta_3 + \sin^2 \theta_3 = 1$.

One can accomplish similar rotations of the basis about the x and y axes by the respective matrices R_1 and R_2:

$$R_1(\theta_1) = \begin{pmatrix} 1 & 0 & 0 \\ 0 & \cos\theta_1 & \sin\theta_1 \\ 0 & -\sin\theta_1 & \cos\theta_1 \end{pmatrix} \quad (6.3b)$$

$$R_2(\theta_2) = \begin{pmatrix} \cos\theta_2 & 0 & -\sin\theta_2 \\ 0 & 1 & 0 \\ \sin\theta_2 & 0 & \cos\theta_2 \end{pmatrix}. \quad (6.3c)$$

In the above, we have defined angles to be positive if they represent right-handed rotations, i.e., x into y, y into z, and z into x.

1. Euler Angles

An arbitrary rotation of coordinate axes can be built up from individual rotations about the three original axes in turn. The entire rotation is then represented by the product $R(\theta_1, \theta_2, \theta_3) = R_3 R_2 R_1$. A more useful and conventional description of rotations is that in terms of the *Euler angles*. In this scheme (see Fig. 18) one rotates first through an angle θ_3 about

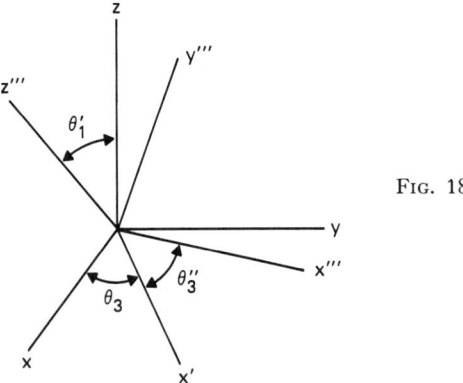

Fig. 18

the original z axis to give the intermediate x and y axes, x' and y', whereas $z' = z$. The second rotation is through an angle θ_1' about the x' axis, giving a new z axis, z'', and yet another y axis, y''; $x'' = x'$. Finally, a rotation about the z'' axis through the angle θ_3'' finishes the process, giving the final set of coordinate axes, x''', y''', and z'''. The more conventional notation for the rotation angles is $\theta_3 = \varphi$, $\theta_1' = \theta$, $\theta_3'' = \psi$. In terms of these, the rotation matrices take the form

$$R(\varphi, \theta, \psi) = R_3(\psi)R_1(\theta)R_3(\varphi)$$

$$= \begin{pmatrix} \cos\psi & \sin\psi & 0 \\ -\sin\psi & \cos\psi & 0 \\ 0 & 0 & 1 \end{pmatrix} \begin{pmatrix} 1 & 0 & 0 \\ 0 & \cos\theta & \sin\theta \\ 0 & -\sin\theta & \cos\theta \end{pmatrix}$$

$$\times \begin{pmatrix} \cos\varphi & \sin\varphi & 0 \\ -\sin\varphi & \cos\varphi & 0 \\ 0 & 0 & 1 \end{pmatrix}. \tag{6.4}$$

2. Cayley–Klein Parameters

If we introduce the imaginary unit $i = \sqrt{-1}$, we may write a 2×2 rather than a column matrix representation for a Euclidean vector:

$$\mathbf{V} = \begin{pmatrix} V_3 & V_1 - iV_2 \\ V_1 + iV_2 & -V_3 \end{pmatrix}. \tag{6.5}$$

The modulus of **V** is then the negative of the determinant of this matrix. We can expand **V** in terms of the Pauli matrices [Eq. (5.41)]:

$$\mathbf{V} = V_1\sigma_1 + V_2\sigma_2 + V_3\sigma_3. \tag{6.6}$$

Now, a rotation of the coordinate axes will be represented as a *unitary* transformation of the vector matrix **V**:

$$\mathbf{V}' = \mathbf{U}\mathbf{V}\mathbf{U}^\ddagger. \tag{6.7}$$

In terms of the Euler angles,

$$\mathbf{U} = \begin{pmatrix} \cos\frac{\theta}{2}\exp(i(\varphi+\psi)/2) & i\sin\frac{\theta}{2}\exp(-i(\varphi-\psi)/2) \\ i\sin\frac{\theta}{2}\exp(i(\varphi-\psi)/2) & \cos\frac{\theta}{2}\exp(-i(\varphi+\psi)/2) \end{pmatrix}. \tag{6.8}$$

The matrix elements of **U** are called the *Cayley–Klein parameters* and have their greatest application in the quantum mechanical description of spin $\tfrac{1}{2}$ particles. For the individual rotations making up the Euler sequence,

$$\mathbf{U}_3(\varphi) = \begin{pmatrix} e^{i\varphi/2} & 0 \\ 0 & e^{-i\varphi/2} \end{pmatrix} = \mathbf{I}\cos(\varphi/2) + i\sigma_3\sin(\varphi/2) \tag{6.9a}$$

$$\mathbf{U}_1(\theta) = \begin{pmatrix} \cos\theta/2 & i\sin\theta/2 \\ i\sin\theta/2 & \cos\theta/2 \end{pmatrix} = \mathbf{I}\cos(\theta/2) + i\sigma_1\sin(\theta/2). \tag{6.9b}$$

In general, a rotation through the angle α about an arbitrary axis defined by the unit vector $\hat{\mathbf{a}}$ is represented by

$$U_{\hat{a}}(\alpha) = I\cos(\alpha/2) + i\boldsymbol{\sigma} \cdot \hat{\mathbf{a}}\sin(\alpha/2), \qquad (6.9c)$$

where the meaning of $\boldsymbol{\sigma} \cdot \hat{\mathbf{a}}$ should be clear to the reader.

3. Vector Rotations

We recall the fact that one may accomplish the same end by fixing the basis and rotating the vector rather than holding the vector fixed in space and transforming the coordinate axes. Thus, in Fig. 19 we obtain a vector with the components V_1' and V_2' of Eq. (6.1) simply by rotating the vector about the z axis through the angle $\theta = -\theta_3$. If we write

$$\mathbf{V}' = \mathbf{Q}_3 \mathbf{V}, \qquad (6.10)$$

then

$$\mathbf{Q}_3(\theta) = \mathbf{R}_3(-\theta) = \tilde{\mathbf{R}}_3(\theta). \qquad (6.11)$$

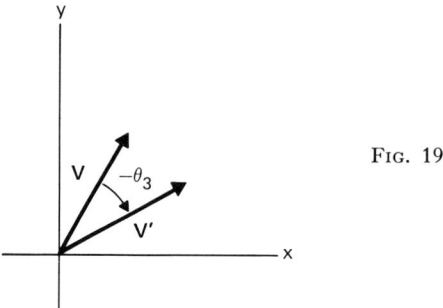

Fig. 19

For a general rotation,

$$\mathbf{V}' = \mathbf{Q}(\varphi, \theta, \psi)\mathbf{V}, \quad \mathbf{Q}(\varphi, \theta, \psi) = \mathbf{R}(-\varphi, -\theta, -\psi) = \tilde{\mathbf{R}}(\varphi, \theta, \psi). \quad (6.12)$$

In terms of the Cayley–Klein matrices,

$$\mathbf{V}' = \mathbf{S}\mathbf{V}\mathbf{S}^{\ddagger} \qquad (6.13)$$

$$\mathbf{S}(\varphi, \theta, \psi) = \mathbf{U}^{\ddagger}(\varphi, \theta, \psi) \qquad (6.14)$$

[cf. Eq. (4.60)].

B. Tensors

A quantity, such as $\mathbf{A} \cdot \mathbf{B}$, whose value does not change under a rotation of the coordinate frame is called an invariant, or more conventionally in Euclidean vector spaces, a *scalar* (cf. Section IV). A vector's coordinates are not invariants, however, and transform according to

$$a_j' = \sum_l r_{il} a_l, \qquad (6.15)$$

where $\{r_{il}\} = \mathsf{R}$, the rotation matrix. Furthermore, the elements of the matrix representation of an operator in the vector space transform according to

$$a_{ij}' = \sum_{lm} r_{il} r_{jm} a_{lm} \qquad (6.16)$$

[cf. Eqs. (4.49) and (4.51)].

We can generalize this to sets of quantities having an arbitrary number of subscripts and define a *tensor of rank n* as being such a set whose elements $a_{ij\cdots k}^{(n)}$ have n independent indices (ranging in value, in the case of three-dimensional Euclidean spaces, from 1 to 3) and which transform under rotations of the coordinate system according to the rule

$$a_{ij\cdots k}^{(n)'} = \sum_{lm \cdots q} r_{il} r_{jm} \cdots r_{kq} a_{lm \cdots q}^{(n)}. \qquad (6.17)$$

From this definition we see that scalars, vectors, and operators, or more precisely, their matrix representations, are tensors of rank 0, 1, and 2, respectively.

The reader should note that an arbitrary 3×3 matrix is not necessarily a tensor. To be a tensor, the matrix elements must have a defined relationship with the underlying geometry of the space, such as is the case with the matrix elements of an operator, and this relationship must admit to the transformation rule given in Eq. (6.16). Tensors are thus recognized fundamentally by their transformation properties.

Tensors of various ranks may be combined to form tensors of yet other ranks. A simple example of this is the process of *contraction*. By setting two indices equal and summing over them, we produce a tensor of rank $n - 2$ from a tensor of rank n:

$$a_{ij\cdots l}^{(n-2)} = \sum_k a_{ij\cdots kk\cdots l}^{(n)}.$$

That the object on the left is a tensor can be proved from Eq. (6.17) and the orthogonality of the transformation matrix **R**. One may also combine the tensors $a^{(m)}_{ij\cdots l}$ and $b^{(n)}_{pq\cdots s}$ to create a tensor of rank $m + n - 2$:

$$c^{(m+n-2)}_{ij\cdots lpq\cdots s} = \sum_k a^{(m)}_{ij\cdots k\cdots l} b^{(n)}_{pq\cdots k\cdots s}.$$

Here, we have equated an index of $a^{(m)}$ to one of $b^{(n)}$ and summed over it.

Tensors of rank higher than 2 are of primary interest to the theory of groups and group representations. We do not pursue them further in this chapter.

1. *The Inertia Tensor*

We complete this section with an example from the dynamics of rotating bodies which illustrates the appearance of tensors as well as the matrix eigenvalue problem in classical mechanics.

Consider a rigid body which is situated and oriented in some arbitrary fashion with respect to a bodyfixed system of coordinate axes. The *moments of inertia* I_{11}, I_{22}, and I_{33} and the *products of inertia* I_{ij} $i, j = 1, 2, 3, i \neq j$, are calculated from the formulas

$$I_{ii} = \int \varrho(\mathbf{r})(r^2 - x_i^2)\, dv, \qquad I_{ij} = -\int \varrho(\mathbf{r}) x_i x_j\, dv, \qquad i \neq j, \qquad (6.18)$$

where r is the length and x_i is the ith component of the vector **r** as shown in Fig. 20, $\varrho(\mathbf{r})$ is the mass density at the position **r**, and the integral is over the volume of the body. Note that $I_{ij} = I_{ji}$. Now suppose that the body revolves with angular velocity **ω** about an axis directed through the origin of the body-fixed coordinate system. The angular momentum

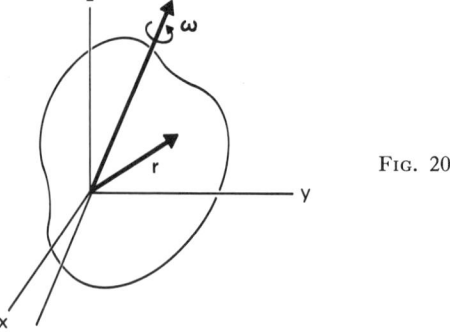

FIG. 20

of the object is then given by

$$L_1 = I_{11}\omega_1 + I_{12}\omega_2 + I_{13}\omega_3$$
$$L_2 = I_{21}\omega_1 + I_{22}\omega_2 + I_{23}\omega_3 \qquad (6.19)$$
$$L_3 = I_{31}\omega_1 + I_{32}\omega_2 + I_{33}\omega_3,$$

or, in matrix form,

$$\mathbf{L} = \mathscr{I}\boldsymbol{\omega}. \qquad (6.19a)$$

$\mathscr{I}$ is called the *inertia tensor*, and correctly so inasmuch as its elements transform according to Eq. (6.16).

The description of the rotational motion of an object is quite cumbersome when given as in Eq. (6.19). Matters would be much simpler if the products of inertia were all zero, i.e., if $\mathscr{I}$ were diagonal. We know by now, of course, that such a state of affairs is easy to arrive at; we merely have to diagonalize $\mathscr{I}$ and transform the body-fixed coordinate system accordingly. Indeed, the new axes will be oriented in the directions of the eigenvectors of $\mathscr{I}$ (recall that $\mathscr{I}$ is real and symmetric, and that its eigenvectors are therefore orthogonal). These are called the *principal axes* of the object with respect to the (unchanged) origin of the coordinate system. The (real) eigenvalues of $\mathscr{I}$, i.e., the diagonal elements in the *principal axes* coordinate frame, are called the *principal moments of inertia*. Finding the principal axes and moments of inertia, I_i, reduces Eqs. (6.19) to

$$L_i' = I_i \omega_i', \qquad i = 1, 2, 3, \qquad (6.20)$$

where the primes indicate, as before, the transformed quantities. If in fact the rotation is about one of the principal axes, the angular momentum in then in the direction of the axis of rotation, the situation most often encountered in elementary physics textbooks.

VII. Vector Spaces of Infinite Dimension

The extension of the ideas in Sections II and IV to linear vector spaces of infinite dimension is, in most respects, straightforward, and the greater part of the results of these sections can be applied directly without amendment. One must, however, stop to reconsider the notion of *completeness* as a matter of principle and the determination of eigenvalues and eigenvectors as a matter of calculational expediency, matrices of infinite dimension being cumbersome, if not impossible, to manipulate. Since both of these considerations are of importance in quantum mechanics, which finds its formal underpinnings in the theory of infinite-dimensional linear vector spaces called *Hilbert spaces*, we devote the final section of this chapter to them.

A. COMPLETENESS IN HILBERT SPACES

The idea of completeness in a vector space of infinite dimension is based on the convergence of sequences of vectors. By the convergence of a vector sequence we mean, rather loosely, that if we have an infinite set of vectors placed by some scheme in a one-to-one correspondence with the positive integers,

$$\mathbf{V}_1, \ \mathbf{V}_2, \ldots, \mathbf{V}_n, \ldots, \tag{7.1}$$

then, by choosing n large enough, we can find a vector $\mathbf{V}_n$ which is arbitrarily "close" to some limit vector $\mathbf{V}$. Exactly what we mean by "close" to the limit vector is specified in the definitions of two types of convergence: *strong* and *weak convergence*.

(a) *Strong convergence.* The vector sequence (7.1) converges strongly to the limit vector $\mathbf{V}$ if for any $\varepsilon > 0$, there exists an $N(\varepsilon)$ such that

$$\| \mathbf{V}_n - \mathbf{V} \| < \varepsilon \tag{7.2}$$

for all $n > N(\varepsilon)$.

Strong convergence measures closeness by the length of the difference vector. The second method of defining convergence utilizes the limits of scalar products.

(b) *Weak convergence.* The vector sequence (7.1) converges weakly to the limit vector $\mathbf{V}$ if

$$\lim_{n \to \infty} (\mathbf{V}_n, \mathbf{U}) = (\mathbf{V}, \mathbf{U}) \tag{7.3}$$

for every vector $\mathbf{U}$ in the vector space.

It follows immediately from the Schwartz inequality (2.5) that strong convergence implies weak convergence. The implication is not reciprocated, however.

1. Linear Vector Spaces

Now consider the countably infinite orthonormal set of vectors $\{\hat{\mathbf{V}}_i\}$. We can use any other vector $\mathbf{U}$ to form a vector sequence:

$$\mathbf{U}_1 = a_1\hat{\mathbf{V}}_1$$
$$\mathbf{U}_2 = a_1\hat{\mathbf{V}}_1 + a_2\hat{\mathbf{V}}_2$$
$$\vdots \quad\quad (7.4)$$
$$\mathbf{U}_n = \sum_{i=1}^{n} a_i\hat{\mathbf{V}}_i$$

where

$$a_i \equiv (\hat{\mathbf{V}}_i, \mathbf{U}). \quad (7.5)$$

We are now in a position to define the *completeness* property precisely:

The set $\{\hat{\mathbf{V}}_i\}$ is *complete* if the vector sequence $\mathbf{U}_1, \mathbf{U}_2, \ldots, \mathbf{U}_n, \ldots$ converges strongly to the limit $\mathbf{U}$ for every $\mathbf{U}$ in the vector space.

If such a complete set exists, the vector space $\mathscr{V}$ then qualifies as a *Hilbert space* and $\{\hat{\mathbf{V}}_i\}$ is a basis for the space.

Alternative conditions for completeness as derived from the definition stated above are given in a fundamental theorem for Hilbert spaces. Before stating and proving it though, we must consider the following lemma:

Lemma If the vector $\mathbf{V}$ in $\mathscr{V}$ is orthogonal to each of the members of the complete basis $\{\hat{\mathbf{V}}_i\}$, it is the null vector.

To prove this, we take the scalar product

$$\left(\mathbf{V}, \mathbf{V} - \sum_{i=1}^{N(\varepsilon)} b_i\hat{\mathbf{V}}_i\right),$$

where

$$b_i \equiv (\hat{\mathbf{V}}_i, \mathbf{V}).$$

Applying the Schwartz inequality and strong convergence, we have

$$\left|\left(\mathbf{V}, \mathbf{V} - \sum_{i=1}^{N(\varepsilon)} b_i\hat{\mathbf{V}}_i\right)\right| \leq \|\mathbf{V}\| \left\|\mathbf{V} - \sum_{i=1}^{N(\varepsilon)} b_i\hat{\mathbf{V}}_i\right\| \leq \|\mathbf{V}\|\varepsilon.$$

But, since $\mathbf{V}$ is orthogonal to each $\hat{\mathbf{V}}_i$, $b_i = 0$ for all i, and $\|\mathbf{V}\|^2 \leq \|\mathbf{V}\|\varepsilon$ or $\|\mathbf{V}\| \leq \varepsilon$ for any $\varepsilon > 0$. Thus $\|\mathbf{V}\| = 0$, which can be the case only if $\mathbf{V} = \mathbf{0}$.

Theorem In order for $\{\hat{\mathbf{V}}_i\}$ to be complete, each of the following is a necessary and sufficient condition:

(a) It is always true that for any arbitrary vector $\mathbf{V}_a$ in $\mathscr{V}$, the equation

$$\mathbf{V}_a = \sum_{i=1}^{\infty} a_i \hat{\mathbf{V}}_i \tag{7.6}$$

holds, where

$$a_i = (\hat{\mathbf{V}}_i, \mathbf{V}_a) \tag{7.7}$$

[cf. Eqs. (2.12) and (2.13)].

(b) The space generated by $\{\hat{\mathbf{V}}_i\}$, i.e., the linear manifold of $\{\hat{\mathbf{V}}_i\}$, equals $\mathscr{V}$.

(c) One can write the equation

$$(\mathbf{V}_a, \mathbf{V}_b) = \sum_{i=1}^{\infty} (\mathbf{V}_a, \hat{\mathbf{V}}_i)(\hat{\mathbf{V}}_i, \mathbf{V}_b) \tag{7.8}$$

for any $\mathbf{V}_a$ and $\mathbf{V}_b$ in $\mathscr{V}$ [cf. Eq. (2.17a)].

We prove this theorem by showing first that condition (a) follows from the definition of completeness. We form the vector difference

$$\mathbf{W} = \mathbf{V}_a - \sum_{i=1}^{\infty} a_i \hat{\mathbf{V}}_i, \qquad a_i = (\hat{\mathbf{V}}_i, \mathbf{V}_a)$$

and take the scalar product of $\mathbf{W}$ with $\hat{\mathbf{V}}_j$ to obtain

$$(\hat{\mathbf{V}}_j, \mathbf{W}) = (\hat{\mathbf{V}}_j, \mathbf{V}_a) - \sum_{i=1}^{\infty} a_i (\hat{\mathbf{V}}_j, \hat{\mathbf{V}}_i) = a_j - a_j = 0$$

for each j. By the previous lemma then, $\mathbf{W} = \mathbf{0}$, and no part of $\mathbf{V}_a$ is left out of the sum. We show next that condition (b) follows from condition (a). Since we can write

$$\mathbf{V}_a = \sum_{i=1}^{\infty} (\hat{\mathbf{V}}_i, \mathbf{V}_a) \hat{\mathbf{V}}_i \tag{7.9}$$

for arbitrary $\mathbf{V}_a$ in $\mathscr{V}$, $\mathscr{V}$ is contained in the space generated by $\{\hat{\mathbf{V}}_i\}$. On the other hand, the basis vectors $\hat{\mathbf{V}}_i$ are all contained in $\mathscr{V}$, and consequently so are all linear combinations of them, including, in particular, the limits of sequences. $\mathscr{V}$ is therefore identical to the manifold of $\{\hat{\mathbf{V}}_i\}$.

If we can now demonstrate that condition (b) implies completeness, we will have come full circle and shown that conditions (a) and (b) are each necessary and sufficient. Suppose, then, that $\mathbf{V}_a$ is orthogonal to all the $\mathbf{\hat{V}}_i$. It is then orthogonal to all linear combinations of the $\mathbf{\hat{V}}_i$ and hence, by (b), to all of $\mathscr{V}$, including $\mathbf{V}_a$ itself. Thus, $(\mathbf{V}_a, \mathbf{V}_a) = 0$ and $\mathbf{V}_a = \mathbf{0}$. But, in order for the orthogonality of $\mathbf{V}_a$ with the $\mathbf{\hat{V}}_i$ to imply that $\mathbf{V}_a$ is the null vector, $\{\mathbf{\hat{V}}_i\}$ must, as seen in the proof of the lemma, be a complete set.

Finally, we prove the necessity and sufficiency of condition (c) for the completeness of $\{\mathbf{\hat{V}}_i\}$. For $\mathbf{V}_a = \mathbf{V}_b = \mathbf{V}$, Eq. (7.8) becomes

$$(\mathbf{V}, \mathbf{V}) = \sum_{i=1}^{\infty} (\mathbf{V}, \mathbf{\hat{V}}_i)(\mathbf{\hat{V}}_i, \mathbf{V}).$$

If $\mathbf{V}$ is orthogonal to all the $\mathbf{\hat{V}}_i$, then $(\mathbf{V}, \mathbf{V}) = 0$ and $\mathbf{V} = \mathbf{0}$. The completeness of $\{\mathbf{\hat{V}}_i\}$ then follows just as above. Conversely, if we assume completeness, we have, from strong and weak convergence,

$$(\mathbf{V}_a, \mathbf{V}_b) = \lim_{n \to \infty} \left(\sum_{i=1}^{n} (\mathbf{\hat{V}}_i, \mathbf{V}_a)\mathbf{\hat{V}}_i, \sum_{j=1}^{n} (\mathbf{\hat{V}}_j, \mathbf{V}_b)\mathbf{\hat{V}}_j \right)$$

$$= \lim_{n \to \infty} \sum_{i,j=1}^{n} (\mathbf{\hat{V}}_i, \mathbf{V}_a)^*(\mathbf{\hat{V}}_j, \mathbf{V}_b)(\mathbf{\hat{V}}_i, \mathbf{\hat{V}}_j)$$

$$= \lim_{n \to \infty} \sum_{i=1}^{n} (\mathbf{V}_a, \mathbf{\hat{V}}_i)(\mathbf{\hat{V}}_i, \mathbf{V}_b) = \sum_{i=1}^{\infty} (\mathbf{V}_a, \mathbf{V}_i)(\mathbf{V}_i, \mathbf{V}_b).$$

This completes the proof of the theorem.

In applications, the completeness of a particular vector space can be determined by appealing to one of the conditions in the theorem. On the other hand, completeness may be imposed upon a vector space as an additional assumption of a theory, as is done in quantum mechanics. It should then be a demonstrated property of any realization of the vector space which arises from the theory. To summarize, then, completeness for an infinite-dimensional linear vector space is an additional property which, if possessed by the vector space, enables one to write the expansions (7.6) and (7.8). A complete infinite-dimensional linear vector space is called a *Hilbert space*.

B. ALGEBRAIC EIGENVALUE METHODS

The calculation of eigenvalues in the case of infinite-dimensional vector spaces cannot, in general, be done by diagonalization of matrix representations. Matrix techniques are useful here only when there are certain simple repetitive properties of the infinite-dimensional matrices which allow us to treat finite submatrices independently. Usually one is left with two possibilities: find a realization, other than matrix, for the operator and its domain so that the eigenvalue equation obtains a form tractable by other techniques, such as differential or integral equations; or, utilize the basic properties (Hermiticity, commutation relations, etc.) of the operators themselves. There are numerous instances of the use of this latter *algebraic* technique, again mostly from quantum mechanics, and we give here only one simple, but ubiquitous, illustration. Examples of the integral–differential methods are copious throughout applied mathematics and are dealt with in other chapters of this book. We leave the discussion of *function space* problems to them.

Consider then the operator a and its Hermitian conjugate $a^\dagger$. We will assume in this example that they obey the commutation relationship

$$[a, a^\dagger] = I. \tag{7.10}$$

The operator

$$N \equiv aa^\dagger \tag{7.11}$$

is Hermitian and must, therefore, have real eigenvalues and orthogonal eigenvectors. We label the eigenvalues n_i and write the eigenvalue equation as

$$N \mid n_i \rangle = n_i \mid n_i \rangle, \tag{7.12}$$

where

$$\langle n_i \mid n_j \rangle = \delta_{ij}. \tag{7.13}$$

The problem is to determine the spectrum of N and the corresponding eigenvectors; we will do this using the commutation relationship (7.10).

If we substitute (7.11) into (7.12),

$$aa^\dagger \mid n_i \rangle = n_i \mid n_j \rangle,$$

operate on the left with $a^\dagger$,

$$(a^\dagger a) a^\dagger \mid n_i \rangle = n_i a^\dagger \mid n_i \rangle,$$

and use (7.10), we obtain

$$(aa^\dagger - I) a^\dagger \mid n_i \rangle = n_i a^\dagger \mid n_i \rangle$$

or

$$N(a^\dagger \mid n_i \rangle) = (n_i + 1)(a^\dagger \mid n_i \rangle). \tag{7.14}$$

The vector $a^\dagger |n_i\rangle$ is thus an eigenvector of N with eigenvalue $n_i + 1$, i.e.,

$$a^\dagger |n_i\rangle = c_i |n_i + 1\rangle \tag{7.15}$$

where c_i is a normalization factor included in order to retain the normalization (7.13). We determine its value later. Because of its ability to "boost" an eigenvector to one of a higher eigenvalue, $a^\dagger$ is called a *raising*, *boosting*, or *creation* operator. Similarly, a is a *lowering* or *destruction* operator, for

$$a^\dagger a |n_i\rangle = (N - I)|n_i\rangle = (n_i - 1)|n_i\rangle$$
$$(aa^\dagger)a|n_i\rangle = (n_i - 1)a|n_i\rangle$$

or

$$N(a|n_i\rangle) = (n_i - 1)(a|n_i\rangle)$$

so that

$$a|n_i\rangle = d_i |n_i - 1\rangle \tag{7.16}$$

where, again, d_i is a normalization factor. Now, from property (2.3d), we have

$$n_i = \langle n_i | N | n_i \rangle = \langle n_i | aa^\dagger | n_i \rangle > 0. \tag{7.17}$$

This condition on n_i is sufficient to establish the spectrum of N as the positive integers, for, if n_k were a positive number lying between the integers $m - 1$ and m, then m repeated applications of a to $|n_k\rangle$ would yield an eigenvector with eigenvalue less than 0. We thus have $n_i = i$, $i = 1, 2, 3, \ldots$.

To construct the eigenvectors, we first consider $|1\rangle$. Obviously, in order to avoid negative eigenvalues,

$$a|1\rangle = 0. \tag{7.18}$$

A representation for $|1\rangle$ can be determined from (7.18) if one has a realization for the operator a. Here we will construct the remainder of the $|n_i\rangle$ in terms of $|1\rangle$. Our problem is only that of determining the c_i in (7.15), and this is done by taking the scalar product of each side of Eq. (7.15) with itself:

$$\langle n_i | aa^\dagger | n_i \rangle = |c_i|^2 \langle n_i + 1 | n_i + 1 \rangle = |c_i|^2 = n_i \langle n_i | n_i \rangle = n_i.$$

Thus, $c_i = (n_i)^{1/2}$ where we have arbitrarily taken the argument of c_i to vanish. One can likewise show that $d_i = (n_i - 1)^{1/2}$, which is consistent with Eq. (7.18) for n_1.

The situation is summarized then as follows:

$$\mathsf{N} \mid n_i \rangle = n_i \mid n_i \rangle, \tag{7.12}$$

where $\mathsf{N} = \mathsf{a}\mathsf{a}^\ddagger$,

$$n_i = i \quad \text{for} \quad i = 1, 2, 3, \ldots, \tag{7.19}$$

and $\langle n_i \mid n_j \rangle = \delta_{ij}$. Furthermore,

$$\mathsf{a}^\ddagger \mid n_i \rangle = (n_i)^{1/2} \mid n_i + 1 \rangle, \tag{7.15a}$$

$$\mathsf{a} \mid n_i \rangle = (n_i - 1)^{1/2} \mid n_i - 1 \rangle, \tag{7.15b}$$

and

$$\mid n_i \rangle = [(n_i - 1)!]^{-1/2} (\mathsf{a}^\ddagger)^{i-1} \mid 1 \rangle. \tag{7.20}$$

General References

Linear Vector Spaces (General)

Dennery, P., and Krzywicki, A. (1967). "Mathematics for Physicists," Chapter II. Harper, New York.

Halmos, P. R. (1958). "Finite-Dimensional Vector Spaces." Van Nostrand Reinhold, Princeton, New Jersey.

Matson, F. A. (1970). "Vector Spaces and Algebras for Chemistry and Physics." Holt, New York.

Three-Dimensional Euclidean Vector Spaces, Vector Fields, and Applications from Mechanics and Electrodynamics

Goldstein, H. (1950). "Classical Mechanics." Addison-Wesley, Cambridge, Massachusetts.

Lass, H. (1957). "Elements of Pure and Applied Mathematics," Chapter 2. McGraw-Hill, New York.

Morse, P. M., and Feshbach, H. (1957). "Methods of Theoretical Physics," Chapter 2. McGraw-Hill, New York.

Panofsky, W. K. H., and Phillips, M. (1962). "Classical Electricity and Magnetism." Addison-Wesley, Reading, Massachusetts.

Wylie, C. R. (1960). "Advanced Engineering Mathematics," Chapter 11. McGraw-Hill, New York.

Matrices and Determinants, Numerical Methods

Irving, J., and Mullineux, N. (1959). "Mathematics in Physics and Engineering," Chapter 5. Academic Press, New York.

Kuo, S. S. (1965). "Numerical Methods and Computers," Chapters 8 and 9. Addison-Wesley, Reading, Massachusetts.

Wylie, C. R. (1960). "Advanced Engineering Mathematics," Chapter 1. McGraw-Hill, New York.

Hilbert Spaces and Applications to Quantum Mechanics

Dirac, P. A. M. (1958). "The Principles of Quantum Mechanics." Oxford Univ. Press, London and New York.

Merzbacher, E. (1970). "Quantum Mechanics," Chapter 14. Wiley, New York.

Schmiedler, W. (1965). "Linear Operators in Hilbert Spaces." Academic Press, New York.

Chapter 2

Generalized Functions

E. W. Grundke

    I. Introduction 82
   II. Definitions 85
       A. Test Functions 85
       B. Generalized Functions 88
       C. Sequences of Generalized Functions 93
       D. An Alternative Approach 96
  III. The Algebra of Generalized Functions 98
       A. Introduction 98
       B. Spatial Transformations 99
       C. Products 101
       D. Division 104
  IV. The Calculus of Generalized Functions 108
       A. Differentiation 108
       B. Integration 114
   V. Some Singular Generalized Functions 118
       A. Definitions of Inverse Powers and Related Functions 118
       B. Applications 123
  VI. Fourier Transforms 128
       A. Fourier Transforms of Ordinary Functions 128
       B. The Spaces S and S' 130
       C. Fourier Transforms of Generalized Functions in S' 134
       D. Particular Results 136
       E. Transforms of Periodic Generalized Functions 139
       F. Several Dimensions 142
 VII. Laplace Transforms 143
       A. Laplace Transforms of Ordinary Functions 143
       B. Laplace Transforms of Generalized Functions 145
VIII. Conclusion 148

       References 149

I. Introduction

When Dirac (1926, 1930) introduced the δ function to expedite the formalism of quantum mechanics, he defined it to be a highly singular function having the properties

$$\delta(x) = \begin{cases} \infty, & x = 0, \\ 0, & x \neq 0, \end{cases} \tag{1.1}$$

$$\int_{-\infty}^{\infty} \delta(x)\phi(x)\,dx = \phi(0), \tag{1.2}$$

where ϕ is any function continuous at $x = 0$. Unfortunately, it is easily shown that no function, in the usual sense of the word, can have the properties of δ; yet, like Heaviside's operational calculus (1893), $\delta(x)$ proved to be an extremely useful tool in the hands of physical scientists although lacking rigorous mathematical justification.

Such difficulties are best regarded as *shortcomings* of classical functional analysis. (In a similar vein, a great deal of progress is made once one regards the lack of a real solution of $x^2 + 1 = 0$ as a shortcoming of the real number system.) Indeed, other difficulties abound in ordinary analysis: functions fail to have derivatives, functions have divergent definite integrals, many theorems have very tedious necessary conditions, and so forth.

The *generalized function* (GF), or *distribution*, is an extension of the usual notion of a function that provides solutions, in many instances, for the difficulties mentioned above. Although this concept began to emerge in the mathematical literature before the work of Laurent Schwartz, it is he who laid the foundations for GF theory in his papers of 1944–1948.[†] The theory has been accepted rapidly, stimulating research in other areas of mathematics (such as differential equations, operational calculus, and harmonic analysis) as well as the theoretical physical sciences.

This chapter aims to provide an elementary introduction to GF theory for the physical scientist. Our approach will emphasize the parallelism between GFs and ordinary functions. The reader is assumed to have mastered undergraduate "advanced calculus" and to have some familiarity with the theories of complex variables and linear spaces. Strict mathematical rigor is not our aim; however, in Section II we occasionally deal with more mathematical details than our goals would justify so that the

---

[†] See Schwartz (1966) and references quoted therein.

interested reader can make an easier transition to the more specialized literature if he wishes.

An extensive literature on GFs exists even at the textbook level. Although many works[†] are written primarily for an audience of mathematicians, there is an abundance of rigorous yet very readable accounts, including Bremermann (1965), Marchand (1962), Liverman (1964), and the fine treatise of five volumes by Gel'fand *et al.* (1964). Temple (1955), using a method due to Mikusinski, introduces the subject in a way that is much less abstract than the original approach of Schwartz, and Lighthill (1958) follows this route in his book, which is probably better known among scientific workers than the other references mentioned. The work of de Jager (1964, 1970) is written specifically for physicists. The accounts by Erdélyi (1961) and Saltzer (1958) appear in engineering-oriented publications, and some engineering texts (e.g., Papoulis, 1962) are including appendices on GFs. Zemanian (1965) has written a good text for engineering and mathematics undergraduates. Finally, to complete the circle begun by Dirac, we find that quantum mechanics texts, such as those by Messiah (1966) and Capri (1975), are including a discussion of GFs to justify their use of $\delta(x)$. In this connection, we might also refer to the Wightman axiomatic formulation of relativistic quantum field theory (Schweber 1961; Streater and Wightman 1964), an approach relying heavily on GFs. Having quoted these general references, we shall dispense, for the most part, with detailed references in the remainder of the chapter.

Before defining GFs rigorously in Section II, it may be helpful to introduce the subject heuristically by discussing some properties and uses of $\delta(x)$. Of course $\delta(x)$ is typically used for describing physical phenomena like impulsive forces, point sources, and the like. "Derivatives" of δ are used to represent point dipole sources and higher order multipoles. One even sees $\delta(x)$ used as an approximation to other functions (even continuous functions) in order to make the mathematics of the problem more manageable![‡]

[†] Here we may quote Schwartz (1966), Halperin (1952), Treves (1967), Donoghue (1969), Gårding and Lions (1959), Jantscher (1971), Beltrami and Wohlers (1966), Friedman (1963), Mikusinski (1959), Erdélyi (1962), Korevaar (1955), Arsac (1966), and Bouix (1964). The subject is also treated in the more recent books on functional analysis, such as that by Edwards (1965).

[‡] For example, Baxter (1968) replaces the attractive portion of the molecular pair interaction by a negative δ function. As a result he is able to obtain an analytic solution for the Percus–Yevick equation, and hence an analytic equation of state, for a fluid of these "sticky spheres."

Notice that in these examples $\delta(x)$ invariably appears under an integral sign; it is an "input" to the problem and not a final answer. Its very definition, (1.1) and (1.2), really gives it meaning only when integrated with another function ϕ. We shall refer to such a function ϕ as a *test function*, giving the connotation that δ is being tested or evaluated with ϕ. As another example, consider the theory of Green's functions associated with partial differential equations which is discussed by Henderson in Chapter 4. There again, the δ function always occurs as a source term and is effectively under an integral sign as far as the final solution is concerned.

To summarize: $\delta(x)$ *is defined in terms of its evaluation using some test function* $\phi(x)$. The same will be true of all GFs. In fact, our first task in Section II is to establish suitable sets of test functions, and then we proceed to define GFs in terms of these test functions. This is the approach of Schwartz. The reader has undoubtedly also seen $\delta(x)$ defined by a limiting procedure such as

$$\delta(x) = \lim_{m \to \infty} \frac{m}{\pi(1 + x^2 m^2)}. \tag{1.3}$$

This type of definition is related to the approach of Temple, Lighthill, and others mentioned above, and is discussed briefly in Section II.D. In Sections III and IV we develop the algebra and calculus of GFs; we shall find that most operations performed on ordinary functions can be extended to GFs. Section V defines some specific GFs related to inverse powers and demonstrates some simple applications. Section VI discusses Fourier transforms of GFs, treating Fourier series as a special case. In Section VII we briefly define the Laplace transform for GFs, and end with some closing remarks in Section VIII.

Although we give the fundamental definitions in n dimensions, many subsections deal with just the one-dimensional case without a serious loss of generality. To denote position in the ordinary Euclidean n-dimensional space we shall use the vector $\mathbf{r}$, its components being x_1, $x_2, \ldots, x_n$. For $n = 1$ we shall dispense with the vector notation and simply use the symbol x, as above.

It should be mentioned that there is another important extension of the usual concept of function, namely the theory of *convolution quotients*. This is not a functional approach but rather an algebraic one, being based on the properties of rings of continuous functions. We touch on the subject very briefly in Section III.C.2.

II. Definitions

A. Test Functions

1. *Preliminaries*

In Section I it was pointed out that generalized functions are characterized by their "evaluation using some test function." Let us make this more precise. We shall shortly define a generalized function as a *mapping* from some appropriate set of *functions* (the *domain* of the mapping) to the set of real numbers (the *range* of the mapping). Such mappings are called *functionals*; they are to be distinguished from ordinary functions, which have as their domains some set of *numbers* such as the real line, the points in a plane, the integers, and the like.

For example, the energy E in many physical fields $\phi(r)$ (e.g., elastic or electromagnetic fields) is given by the volume integral

$$E\langle\phi\rangle = \int |\nabla\phi(\mathbf{r})|^2 \, dV. \tag{2.1}$$

Here $\phi(\mathbf{r})$ is a function of the position $\mathbf{r}$, while $E\langle\phi\rangle$ is a functional of ϕ. We use angular brackets to enclose the argument of a functional. Notice that $\mathbf{r}$ is really a dummy variable in (2.1).

2. *The Set D*

Many different sets of functions can be used as *test functions*, or domains on which to define generalized functions. Although we introduce other test functions in Sections VI and VII, we begin with a set of test functions, denoted by D_n, whose members have the following properties:

1. They are real-valued functions of n real variables. They are defined for all real values of their arguments, and are infinitely differentiable, i.e., they have partial derivatives of all orders (including all mixed partial derivatives).

2. They are identically zero outside some bounded region (or some finite interval, in the case of D_1).

The closure of the set on which a member of D_n is not zero is called its *support*. Different functions in D_n may have different supports. We shall use the symbol D to discuss test functions in one of the D_n, $n = 1, 2, \ldots$, when the number of variables is not important.

Clearly the functions in D are extremely smooth, and because of their bounded support they can give no divergence problems when integrated over all space. In fact, many of the pleasing properties of generalized functions follow directly from the strict requirements placed upon the test functions. One is entitled to wonder whether such incredibly smooth functions even exist at all. The answer is yes; in one dimension the classic example is a function of the form

$$\varrho_a(x) \equiv \begin{cases} C_a^{-1} \exp(x^2/a^2 - 1)^{-1}, & 0 < |x| < a, \\ 0, & |x| \geq a, \end{cases} \quad (2.2)$$

where C is a constant that normalizes the area under ϱ_a to unity. See Fig. 1. The support of ϱ_a is the finite interval $-a \leq x \leq a$, and it is easily shown that ϱ_a and all its derivatives $\varrho_a^{(k)}$, $k = 1, 2, 3, \ldots$, approach zero as $|x| \to a_-$; clearly ϱ_a belongs to D_1.

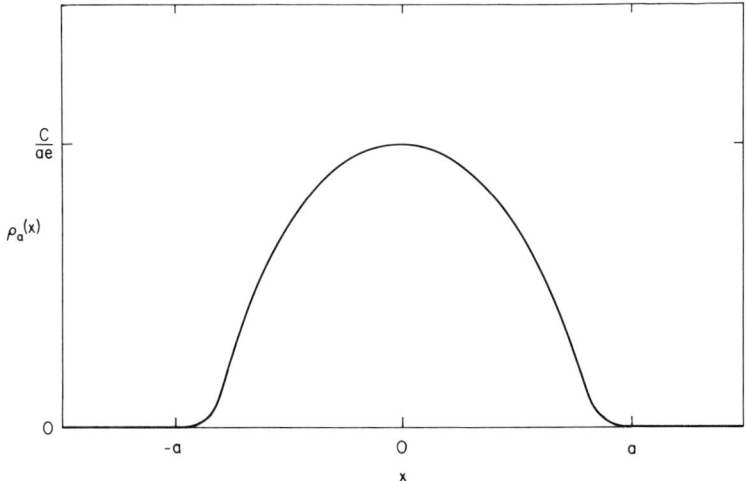

FIG. 1. The function $\varrho_a(x)$ defined by (2.2).

There is in fact an abundance of functions in D. Any continuous function may be approximated as closely as one wishes by functions in D. (See Section II.D). An important special case in D_1 is the function

$$\sigma_a(x; b, c) \equiv \int_{-\infty}^{x} [\varrho_a(t - b) - \varrho_a(t - c)] \, dt, \quad (2.3)$$

which is identically equal to 1 when $b + a \leq x \leq c - a$, that is, when x lies in the interval between the supports of the two ϱ_a's.

2. Generalized Functions

Several other properties of the members of D are worth noting. If ϕ belongs to D_n, then all (partial) derivatives of ϕ also belong to D_n and have their support contained in that of ϕ. Further, D is closed under multiplication and addition: If ϕ_1 and ϕ_2 are in D_n, then $\phi_1\phi_2$ and $\phi_1 + \phi_2$ are in D_n also. In fact, multiplication by an infinitely differentiable function α that may not have bounded support is also permitted: $\alpha\phi$ is in D_n if ϕ is in D_n. If ϕ is chosen to be σ_a, then $\alpha\sigma_a$ is a test function equaling a prescribed function α in the interval where $\sigma_a = 1$.

3. D as a Linear Space

The functions in D_n form a linear space. This important statement means simply that the members of D_n satisfy the postulates listed under (i) and (ii) in Section II.A of Chapter 1 (by Jacob) of this volume (omitting the word *vector* wherever it occurs). The linear combination $a\phi_1 + b\phi_2$ of two members, ϕ_1 and ϕ_2, of D_n is defined in the ordinary way, and the postulates need no elaboration here.

The main reason for mentioning the notion of a linear space at this point is to introduce the definition of *convergence* of a sequence of functions in D. To give any space such as D a mathematical structure, one must define a *topology*, giving rise to a notion of "distance" or "closeness" in the space. On the real line, the "distance" between two points is ordinarily defined to be the absolute value of their difference. That is why the usual "ε and δ" definitions of limits involve absolute values of differences.

How best to define the "distance" between ϕ_1 and ϕ_2, two members of D_n, is not immediately clear. One could use the quantity

$$\int |\phi_1(\mathbf{r}) - \phi_2(\mathbf{r})| \, dV, \tag{2.4}$$

or perhaps the maximum value of this integrand. However, it turns out to be far better to involve the derivatives of the ϕ's as well as their values. Specifically, we shall say that a sequence $\{\phi_m\}$, $m = 1, 2, 3, \ldots$, of functions in D_n converges to the test function that is identically zero if

1. the ϕ_m *and all their derivatives* converge *uniformly* to zero as $m \to \infty$, and
2. the supports of the ϕ_m are contained in some fixed bounded region.

As in any linear space, the distance between two elements ϕ_1 and ϕ_2 is defined as the distance of $\phi_1 - \phi_2$ from zero. Thus $\{\phi_m\}$ converges to ϕ (written $\phi_m \to \phi$ as $m \to \infty$) if $\{\phi_m - \phi\}$ converges to zero. Written algebraically, condition 1 says that corresponding to any $k(= 0, 1, 2, \ldots)$ and any $\varepsilon > 0$ there exists an integer M_k so that $m > M_k$ guarantees that $|\phi_m^{(k)}(\mathbf{r})| < \varepsilon$ for all $\mathbf{r}$. We shall state without proof that if $\phi_m \to \phi$, then the function ϕ is also in D.

A few examples will help to visualize this process of convergence in D. Referring to (2.2), the sequence

$$\{m^{-1}\varrho_a(x)\}, \qquad m = 1, 2, 3, \ldots \tag{2.5}$$

will converge to the zero function. However,

$$\{m^{-1}\varrho_a(x - m)\}, \qquad m = 1, 2, 3, \ldots \tag{2.6}$$

fails to converge because condition 2 above is not met. Neither will the sequence

$$\{m^{-1}\exp(-m^2x^2)\varrho_a(x)\}, \qquad m = 1, 2, 3, \ldots \tag{2.7}$$

converge to zero in spite of the fact that $\phi_m \to 0$ uniformly. The reason is that

$$\phi_m''(0) = -2m\varrho_a(0) + m^{-1}\varrho_a''(0), \tag{2.8}$$

which actually diverges as $m \to \infty$.

B. Generalized Functions

1. Regular Generalized Functions

Let us consider the set of real-valued *locally integrable* functions of n real variables (i.e., their definite integral over every bounded region exists). Corresponding to any such function $f(\mathbf{r})$ we form the functional

$$f\langle\phi\rangle \equiv \int f(\mathbf{r})\phi(\mathbf{r})\,dV, \tag{2.9}$$

where ϕ is in D_n and the integration extends over all space (or equivalently, over the support of ϕ). Under our assumptions, (2.9) will always exist, and we call $f\langle\phi\rangle$ the *regular generalized function* (or *regular GF*) defined on D_n by $f(\mathbf{r})$.

Consider some examples:

1. If $f(\mathbf{r})$ is a constant, say c, then

$$f\langle\phi\rangle = c\langle\phi\rangle = c\int \phi(\mathbf{r})\,dV. \tag{2.10}$$

In particular, if $c = 0$ we have the zero GF, whose value at any ϕ is zero.

2. If $f(x) = |x|$, then

$$f\langle\phi\rangle = |x| \langle\phi(x)\rangle = -\int_{-\infty}^{0} x\phi(x)\,dx + \int_{0}^{\infty} x\phi(x)\,dx. \quad (2.11)$$

Note that we allow ourselves the convenience of using the symbol $|x|$ for the GF as well as the ordinary function, even though x is strictly a dummy variable in the GF case. The analog of (2.11) for n dimensions is

$$|\mathbf{r}|\langle\phi\rangle = \int \left(\sum_{k=1}^{n} x_k^2\right)^{1/2} \phi(\mathbf{r})\,dV. \quad (2.12)$$

3. The sign function

$$\operatorname{sgn} x \equiv \begin{cases} -1, & x < 0 \\ 0, & x = 0 \\ 1, & x > 0 \end{cases} \quad (2.13)$$

gives rise to the GF

$$(\operatorname{sgn} x)\langle\phi\rangle = -\int_{-\infty}^{0} \phi(x)\,dx + \int_{0}^{\infty} \phi(x)\,dx. \quad (2.14)$$

For future reference let us define the Heaviside step function $H(x)$ in terms of $\operatorname{sgn} x$ as

$$H(x) \equiv \tfrac{1}{2}(1 + \operatorname{sgn} x). \quad (2.15)$$

4. Although $\ln |x|$ has a singularity at $x = 0$, it is integrable there, and we have the regular GF

$$(\ln |x|)\langle\phi\rangle = \int_{-\infty}^{\infty} \ln |x|\,\phi(x)\,dx. \quad (2.16)$$

5. The function x^{-1} is not integrable over any interval including the origin and hence does not define a regular GF.

The usefulness of definition (2.9) lies in the following result, which we state without proof and which holds with one minor qualification (see below): The regular GFs $f_1\langle\phi\rangle$ and $f_2\langle\phi\rangle$ that correspond to two *different* functions $f_1(\mathbf{r})$ and $f_2(\mathbf{r})$ cannot be the same, i.e., there will be some ϕ in D for which $f_1\langle\phi\rangle \neq f_2\langle\phi\rangle$. Thus the functional (2.9) is sufficiently sensitive to the behavior of f, and there is a sufficient variety of ϕ's in D to enable (2.9) to "tell the difference" between different f's. In short, regular GFs on D are faithful representations of locally integrable functions. (This result justifies the casual use of the same symbol above for a function and its regular GF; the context will make it clear which is meant.)

The qualification mentioned in the last paragraph is that $f_1(\mathbf{r})$ and $f_2(\mathbf{r})$ must not differ merely on a set of measure zero (such as a set of isolated points). For example, if we had assigned any finite number other than zero to sgn(0) in (2.13), the GF (2.14) would have remained unaltered. This qualification will cause us no trouble at all—it is actually an advantage to have the theory intrinsically suppress such physically unimportant features in a function, and makes the enunciation of many theorems more straightforward.

It is important to emphasize that in making the transition from a function to its GF, we have lost the pointwise description of the function. The function is specified by its value $f(\mathbf{r})$ at every point $\mathbf{r}$ in Euclidean space; the GF is specified by its value $f\langle\phi\rangle$ at every function ϕ in D—or, if you wish, by some sort of "average value" on the region of Euclidean space that is the support of ϕ. Thus one can sketch a graph of an ordinary function of one variable on a piece of paper, but for GFs this is not possible (except on an open interval where the GF "equals" an ordinary function; see Section II.B.3).

The loss of the pointwise description is hardly shocking to the modern scientist, who has grown accustomed to living with the uncertainty principles of quantum mechanics. Furthermore, on a more macroscopic scale, if the functions involved are related to the outcome of some experimental procedure, one may regard the evaluation of a GF at some ϕ as the result of a measurement with an apparatus (ϕ) whose resolution is finite and large enough to "smear" the "ideal" output. We return to some of these points in Section II.B.3 after first widening our concept of GFs.

2. Singular Generalized Functions

We have not yet succeeded in defining the Dirac δ function or, for that matter, a GF to correspond to x^{-1} and other nonintegrable functions. We must enlarge the definition of a GF to include functionals of the ϕ's which are not of the form (2.9) (i.e., which are *non*regular or *singular* GFs). To that end, let us examine the essential mathematical properties of (2.9).

First, regular GFs are *linear* functionals, since, for any real constants a and b, and for any ϕ_1 and ϕ_2 in D, (2.9) shows that

$$f\langle a\phi_1 + b\phi_2\rangle = af\langle\phi_1\rangle + bf\langle\phi_2\rangle. \tag{2.17}$$

This may be compared with Section IV.A of Chapter 1 (by Jacob); also see (2.1) above for an example of a *non*linear functional.

Second, regular GFs are *continuous* functionals on D with respect to our earlier definition of convergence in D. In other words, it follows from (2.9) that if $\phi_m \to \phi$ (in the sense of convergence in D), then the sequence of real numbers $\{f\langle\phi_m\rangle\}$ will converge to the value $f\langle\phi\rangle$. For example, the functional $f\langle\phi\rangle = \text{sgn}(\phi(\mathbf{0}))$ is neither linear nor continuous.

Our generalization of (2.9) will consist of admitting any functional obeying the two conditions above. Thus we define a generalized function on D to be any real-valued *continuous linear functional* defined for all ϕ in D. The set of all GFs on D is denoted by D' and includes regular GFs as a proper subset. In the same spirit as before, we use D_n' to denote the set of GFs defined on D_n.

We can now quote the δ function (or more properly, the δ *functional*) as an example of a singular GF:

$$\delta\langle\phi\rangle \equiv \phi(\mathbf{0}). \tag{2.18}$$

This functional is clearly both continuous and linear, and hence qualifies as a GF. Another example is the functional $\partial\phi(\mathbf{0})/\partial x_i$, which is also a GF; we shall encounter this one again when discussing the "derivatives" of δ. (GFs like this last one motivate the inclusion of the derivative in our definition of convergence in D.) A third example of a singular GF is the line integral

$$\int_C \phi(\mathbf{r})\,d\mathbf{r} \tag{2.19}$$

taken along a fixed path C in n dimensions, $n \geq 2$. This functional is a GF, and is singular since it cannot be written as a volume integral in n dimensions. Later we shall find that (2.19) is closely related to $\delta(\phi)$. A discussion of x^{-1}, the other example mentioned earlier, is postponed to Section III.D.

An important consequence of linearity is that *any* GF evaluated at the zero test function gives zero:

$$f\langle 0\rangle = f\langle 0 \cdot \phi\rangle = 0 \cdot f\langle\phi\rangle = 0, \tag{2.20}$$

where f and ϕ are both arbitrary.

The reader should be warned that our nomenclature is by no means universal. The terms *generalized function*, *regular generalized function*, *distribution*, and others are used by various authors with different meanings, making it imperative to consult the definition in each case.

Similarly, many different GF notations are in use. One finds $f\langle\phi\rangle$ written variously as $f(\phi)$, (f, ϕ), $(f\phi)$, or $\langle f, \phi\rangle$, to name just a few. We shall also use the mathematical "slang" $\int f(\mathbf{r})\phi(\mathbf{r})\, dV$ and also $f(\mathbf{r})\langle\phi(\mathbf{r})\rangle$ even when f is singular and the symbol $f(\mathbf{r})$ *per se* has no meaning at all. These notations are useful because they suggest the linearity and continuity conditions that f does in fact satisfy.

3. Equality of GFs. Support of a GF

Suppose we have two n-dimensional GFs, f_1 and f_2, defined in some way, and we can show that

$$f_1\langle\phi\rangle = f_2\langle\phi\rangle \tag{2.21}$$

for *all* ϕ in D_n. Then we say that f_1 and f_2 are *equal*, writing $f_1 = f_2$. This remark would be trivial, were it not for a useful refinement of the notion of equality.

Suppose (2.21) is satisfied not by all ϕ in D_n, but only those ϕ whose support lies inside some open region R of the n-dimensional Euclidean space. Then we say f_1 equals f_2 *on the open set* R. If, furthermore, f_1 happens to be a regular GF, we may say that the (singular) GF f_2 equals the *function* $f_1(\mathbf{r})$ on R. In a sense, this concept gives us a partial recovery of the pointwise description as discussed earlier.

Compare, for example, $|x|$ in (2.11) and

$$x\langle\phi\rangle = \int_{-\infty}^{\infty} x\phi(x)\, dx \tag{2.22}$$

and notice that the GFs x and $|x|$ are equal in the open region $x > 0$. This is true because if the support of ϕ lies in $x > 0$, the integrals over negative x in (2.22) and (2.11) vanish and the remaining integrals are equal. A more interesting example is the singular GF $\delta(\mathbf{r})$ (2.18) which equals the ordinary function zero in the open region consisting of all $\mathbf{r} \neq \mathbf{0}$. Notice that we can talk of equality in *open* regions only; to include closed regions would admit single points, and the nature of our test functions prevents us from inquiring about the value of a GF at single points, as mentioned earlier.

As in the case of $\delta(\mathbf{r})$, it may happen that a GF equals zero on some open region R. We define the *support* of a GF to be the complement of R. The support of $\delta(\mathbf{r})$ is the single point $\mathbf{r} = 0$. (In fact, no GF except δ and its derivatives, which we define later, has a single point as its support.) For x and $|x|$, on the other hand, R is empty and their support is the whole real axis.

Notice that the support of a GF, like the support of a test function, is always a closed set. If the support of a GF, f, has no points in common with the support of some test function ϕ, then $f\langle\phi\rangle$ is zero. More generally, for any f and ϕ, one can modify ϕ outside the support of f without changing the value of $f\langle\phi\rangle$.

C. Sequences of Generalized Functions

1. *Convergence of Sequences*

The two simplest operations on the set D', the set of all GFs on D, are the addition of two GFs, f_1 and f_2, and the multiplication of f_1 by a real constant, c. These are defined in the obvious way by the corresponding operations on the real numbers $f_1\langle\phi\rangle$ and $f_2\langle\phi\rangle$:

$$(f_1 + f_2)\langle\phi\rangle \equiv f_1\langle\phi\rangle + f_2\langle\phi\rangle \tag{2.23}$$

and

$$(cf_1)\langle\phi\rangle \equiv c \cdot f_1\langle\phi\rangle. \tag{2.24}$$

Clearly $f_1 + f_2$ and cf_1 are also GFs. With these definitions D' itself becomes a linear space.

It is useful to give a definition of convergence in the space D', just as we did in D. Therefore let us construct a *sequence* of GFs in D_n', $\{f_m\}$, $m = 1, 2, 3, \ldots$; it is just a set of (regular or singular) GFs labeled by an integer index m.

The definition of convergence in D_n' is simple: $\{f_m\}$ converges to some limit f as $m \to \infty$ if, for all ϕ in D_n, the sequence of real numbers $\{f_m\langle\phi\rangle\}$ converges to the value $f\langle\phi\rangle$ in the ordinary sense of convergence on the real axis. Clearly $f\langle\phi\rangle$ is some kind of functional on D_n, but on the face of it the definition does not guarantee that f is also a GF (i.e., that it is also continuous and linear). It is true nevertheless: We state without proof that if $\{f_m\}$ converges in D_n', the limit f is also in D_n'.

A number of results, which are analogs of theorems on sequences of ordinary functions, follow immediately from the definition of convergence in D'. For example, if $f_m \to f$ and $g_m \to g$, then $(f_m + g_m) \to (f + g)$; if $f_n \to f$ and c is a real constant, then $(cf_m) \to cf$; and so on.

Some examples are in order.

1. The sequence of regular GFs formed from the ordinary functions

$$f_m(x) = \begin{cases} 0, & |x| > m^{-1} \\ m/2, & |x| < m^{-1} \end{cases} \tag{2.25}$$

converges to $\delta(x)$ as $m \to \infty$, since

$$\tfrac{1}{2} \int_{-m^{-1}}^{m^{-1}} m\phi(x)\, dx \to \phi(0) \qquad \text{as} \quad m \to \infty. \tag{2.26}$$

However, the sequence $\{f_m^2\}$ fails to converge.

2. The sequence of singular GFs defined by

$$g_m \langle \phi \rangle = \phi(m^{-1}) \tag{2.27}$$

also converges to δ as $m \to \infty$. The g_m are in fact translated δ's, as we shall see in Section III.B.1.

3. If a sequence of ordinary locally integrable functions converges uniformly to some limit, the corresponding sequence of regular GFs will converge in D', but this condition is by no means a necessary one, as example 1 shows. A more striking example is $\{\sin mx\}$, which fails to converge in the ordinary sense as $m \to \infty$. However, the corresponding GF converges to zero, since

$$\int_{-\infty}^{\infty} \sin mx \phi(x)\, dx = m^{-1} \int_{-\infty}^{\infty} \cos mx \phi'(x)\, dx \leq m^{-1} \int_{-\infty}^{\infty} |\phi'(x)|\, dx, \tag{2.28}$$

using integration by parts.

4. Notice that in (2.28) we could have integrated by parts as often as desired, because of the differentiability of the test functions. Therefore, for any positive number p, the sequence $\{m^p \sin mx\}$ also converges to zero in D_1'! Clearly the requirements for convergence in D' are much less stringent than ordinary uniform convergence—a very useful feature of GF theory.

2. Series of Generalized Functions

It is a simple extension of the ideas discussed above to turn to infinite series of GFs. A series is formed by adding the members of a sequence:

$$f_1 + f_2 + f_3 + \cdots. \tag{2.29}$$

2. Generalized Functions

We say that the series (2.29) converges in D' to a limit f if the sequence g_m of partial sums

$$g_m = \sum_{k=1}^{m} f_k \tag{2.30}$$

converges to f (in the sense of convergence defined above) as $m \to \infty$. We then write

$$= \sum_{k=1}^{\infty} f_k. \tag{2.31}$$

From the preceding subsection it follows that f is also in D'. Also the usual results on series hold; e.g., if $\sum_m f_m = f$ and $\sum_m g_m = g$, then $\sum_m (f_m + g_m) = f + g$, and so forth.

As with sequences, we can find series that converge in D' but diverge badly as ordinary pointwise functions. For example, all the terms in a Fourier series are locally integrable and define a series of regular GFs:

$$\sum_{m=0}^{\infty} \int_{-\infty}^{\infty} (a_m \cos mx + b_m \sin mx)\phi(x)\, dx. \tag{2.32}$$

This series converges in D' provided a_m and b_m are dominated by some power of m, i.e., $(|a_m| + |b_m|)/m^p$ is bounded for some p, which may be positive! To prove this we simply integrate (2.32) by parts 4ν times, picking ν so that $4\nu > p + 2$. This gives

$$\sum_{m=1}^{\infty} m^{-4\nu} \int_{-\infty}^{\infty} (a_m \cos mx + b_m \sin mx)\phi^{(4\nu)}(x)\, dx, \tag{2.33}$$

which cannot exceed a constant times $\sum m^{-2}$ and hence (2.32) converges. Not only that, but differentiating the Fourier series term by term any finite number of times still leaves the GF (2.32) convergent, since each differentiation merely increases the value of p by one! [This result is not restricted to Fourier series; after defining the derivative of a GF, we shall see that the sequence of first derivatives of any convergent sequence (or series) is still convergent.]

3. Parametric Generalized Functions

There is no need for the index m of a sequence $\{f_m\}$ to be restricted to integer values. One may equally well consider a *parametric GF*, which is a set of GFs $\{f_\gamma\}$ labeled by a parameter γ that may take on a continuous set of real values (or even complex values). As before, f_γ is said to converge to some limit f as $\gamma \to \gamma_0$ if the quantity $f_\gamma \langle \phi \rangle$ converges to $f \langle \phi \rangle$ for all ϕ. Here γ_0 may be finite or infinite. As before, f is also in D'.

Consider the following example. Suppose that a locally integrable function $f(x)$ is represented numerically by its values $f(x_m)$ on some grid of points x_m. To be specific, let us take $x_m = m\Delta$, $m = 0, \pm 1, \pm 2, \ldots$, where Δ is a constant. The following reasoning shows that we are justified in regarding the numerical representation as a row of Dirac δ's

$$g_\Delta(x) = \sum_{m=-\infty}^{\infty} f(x_m)\, \delta(x - x_m) \qquad (2.34)$$

where the translated δ is defined by

$$\delta(x - x_m)\langle \phi(x) \rangle \equiv \phi(x_m). \qquad (2.35)$$

Really g_Δ is a parametric GF with parameter Δ, and $\Delta g_\Delta \to f$ as $\Delta \to 0$:

$$\lim_{\Delta \to 0} \Delta g_\Delta \langle \phi \rangle = \lim_{\Delta \to 0} \sum_m f(x_m)\phi(x_m)\Delta = \int_{-\infty}^{\infty} f(x)\phi(x)\, dx = f\langle \phi \rangle \qquad (2.36)$$

by the definition of a Riemann integral.

For a fixed ϕ, $f_\gamma \langle \phi \rangle$ is simply an ordinary function of the real variable γ, and we may speak of the continuity, differentiability, and so on, of f with respect to γ. In fact, we can define a new parametric GF, the derivative of f with respect to the parameter γ, according to

$$\left(\frac{\partial f}{\partial \gamma}\right)_{\gamma_0} \langle \phi \rangle \equiv \frac{\partial}{\partial \gamma}(f_\gamma \langle \phi \rangle)_{\gamma=\gamma_0}, \qquad (2.37)$$

provided the derivative on the right exists. Likewise the definite integral of f_γ with respect to γ is the GF

$$\left(\int_{\gamma_1}^{\gamma_2} f_\gamma\, d\gamma\right)\langle \phi \rangle \equiv \int_{\gamma_1}^{\gamma_2} f_\gamma \langle \phi \rangle\, d\gamma, \qquad (2.38)$$

which is not parametric, assuming γ_1 and γ_2 to be fixed. The integral (2.38) is an analog of the summation in (2.31). We shall find a number of occasions to use parametric GFs as well as sequences in following sections of this chapter.

D. AN ALTERNATIVE APPROACH

Although our definition of GFs as continuous linear functionals is probably the most common one, other equivalent definitions are possible. It is useful to indicate one alternate definition using *fundamental sequences*,

2. Generalized Functions

which is the approach taken by a number of authors (Temple, 1955; Lighthill, 1958; Liverman, 1964). For the sake of simplicity we confine the discussion to the case of one dimension.

Suppose we have a sequence of piecewise continuous[†] functions $\{f_m(x)\}$, $m = 1, 2, 3, \ldots$, with the property that

$$\lim_{m \to \infty} \int_{-\infty}^{\infty} f_m(x)\phi(x)\,dx \qquad (2.39)$$

exists for all ϕ in D_1, although the limit of the f_m themselves need not exist. If, in addition, the limit (2.39) is a continuous functional of ϕ, we say that $\{f_m\}$ is a *fundamental sequence*.

Clearly the functional (2.39) defines a GF, since it is linear in ϕ and continuity has been assumed. We shall refer to the limit (2.39) as the *value* of the fundamental sequence at ϕ. Of course more than one fundamental sequence may have the same value for all ϕ, and in this case they are said to be *equivalent*. The crux of the "alternative approach" is to define a GF *to be* a class of equivalent fundamental sequences. (However, we reserve the term GF for functionals defined as in Section II.B.) For example, both (1.3) and (2.25) belong to the class of fundamental sequences for $\delta(x)$.

We should now demonstrate that the set of all classes of equivalent fundamental sequences is actually identical to D'. We saw above that each class of equivalent fundamental sequence defines a unique GF, so we need only prove that each GF in D' can be represented by at least one fundamental sequence. This can be done by simply showing how such a fundamental sequence may be constructed, as follows.

Given f, an arbitrary GF in D_1', we form a parametric family of GFs according to the rule

$$f_\gamma(x)\langle\phi(x)\rangle = f(x)\langle\phi(x-\gamma)\rangle. \qquad (2.40)$$

As discussed in Section II.C.3, such an expression is an ordinary function of γ, say $g(\gamma)$, once ϕ is fixed. (In this case g is even infinitely differentiable.) For ϕ let us choose one of the functions $\varrho_{1/m}(x)$, $m = 1, 2, 3, \ldots$, defined in (2.2), giving a sequence of ordinary functions

$$g_m(\gamma) = f(x)\langle\varrho_{1/m}(x-y)\rangle. \qquad (2.41)$$

[†] The properties required here vary from author to author, ranging from piecewise continuity to infinite differentiability. In a given situation one may select the functions with a view to the operations that are to be performed on them.

Now $\{g_m\}$ is a fundamental sequence for f. This follows from the linearity and continuity of f: Given any $\psi(\gamma)$ in D_1,

$$\int_{-\infty}^{\infty} g_m(\gamma)\psi(\gamma)\, d\gamma = \int f(x)\langle \varrho_{1/m}(x-\gamma)\rangle \psi(\gamma)\, d\gamma$$

$$= \int f(x)\langle \psi(\gamma)\varrho_{1/m}(x-\gamma)\rangle\, d\gamma$$

$$= f(x)\left\langle \int \psi(\gamma)\varrho_{1/m}(x-\gamma)\, d\gamma \right\rangle$$

$$\to f(x)\langle \psi(x)\rangle \quad \text{as} \quad m \to \infty. \tag{2.42}$$

where the last step follows from a form of the mean value theorem for integrals. The trivial extension of this proof to n dimensions establishes the equivalence of D' with the classes of fundamental sequences.

Incidentally, there is an interesting special case in the proof just completed. If f is actually a regular GF, $g_m(x)$ is a sequence of infinitely differentiable functions converging pointwise to $f(x)$, provided f is continuous at x. This supports a remark made in Section II.A.2.

Having indicated this alternative approach, we now resume our original development. We return to a discussion of fundamental sequence in Section V.B.3 to illuminate the character of some singular GFs. The reader would also find it rewarding to examine some of the intervening material from the viewpoint of fundamental sequences. In many instances this viewpoint may enhance one's intuitive understanding of the subject significantly.

III. The Algebra of Generalized Functions

A. INTRODUCTION

We have now established a new mathematical creature, the generalized function (GF) on D, but our manipulations with it are still quite rudimentary. Sections III and IV will widen our repertoire of operations with GFs.

In defining new operations on GFs we shall often be guided by the regular GF

$$\int f(\mathbf{r})\phi(\mathbf{r})\, dV, \tag{3.1}$$

corresponding to the locally integrable function $f(\mathbf{r})$, since we want our

B. Spatial Transformations

1. Translation

For ordinary functions, *translation* through a vector **a** is an operation that gives the "shape" of the function a displacement **a**. This is achieved by writing $f(\mathbf{r} - \mathbf{a})$ instead of $f(\mathbf{r})$. For regular GFs, (3.1) shows it is immaterial whether f is displaced by **a** or ϕ by $-\mathbf{a}$. This leads to the following definition of $f(\mathbf{r} - \mathbf{a})$ for GFs in general:

$$f(\mathbf{r} - \mathbf{a})\langle\phi(\mathbf{r})\rangle \equiv f(\mathbf{r})\langle\phi(\mathbf{r} + \mathbf{a})\rangle. \tag{3.2}$$

[Notice that $\phi(\mathbf{r} + \mathbf{a})$ is in D if $\phi(\mathbf{r})$ belongs to D, so that the right side is meaningful.] For example, the Dirac δ shifted by **a** is

$$\delta(\mathbf{r} - \mathbf{a})\langle\phi(\mathbf{r})\rangle = \delta(\mathbf{r})\langle\phi(\mathbf{r} + \mathbf{a})\rangle = \phi(\mathbf{a}) \tag{3.3}$$

as anticipated in (2.35). Thus nuclear positions in a molecular configuration are often written as a "number density" using a GF of the form

$$\sum_k \delta(\mathbf{r} - \mathbf{a}_k). \tag{3.4}$$

It may happen that a GF is invariant under some translation, that is, $f(\mathbf{r} - \mathbf{a}) = f(\mathbf{r})$ for some **a**. In this case f is called *periodic*, with a period **a**.

Actually, $f(\mathbf{r} - \mathbf{a})$ is an example of a parametric GF with the (vector-valued) parameter **a**. In Section III.C.2 we shall want to regard GFs in one dimension in this light. In fact, $f(x - a)$ is continuous in a, and all its derivatives with respect to a exist:

$$\left(\frac{d^k}{da^k} f(x - a)\right)\langle\phi(x)\rangle \equiv \frac{d^k}{da^k} \left(f(x - a)\langle\phi(x)\rangle\right)$$

$$= \frac{d^k}{da^k} \left(f\langle\phi(x + a)\rangle\right) = f\langle\phi^{(k)}(x + a)\rangle. \tag{3.5}$$

Finally, if f has a bounded support, $f(x - a)$ has another useful property. For a fixed ϕ, $f(x - a)\langle\phi\rangle$ is zero when a lies outside some finite range of values and hence is itself a test function in D. Some of these ideas were already implicit in the discussion concerning (2.40)–(2.42).

Further, by treating the displacement **a** as a parameter, we can create a "fence" of δ's in two or more dimensions by performing a line integral that lets **a** assume values along some specified contour C. Evaluating this GF we get

$$\left(\int_C \delta(\mathbf{r} - \mathbf{a})\,d\mathbf{a}\right)\langle\phi(\mathbf{r})\rangle = \int_C \delta(\mathbf{r} - \mathbf{a})\langle\phi(\mathbf{r})\rangle\,d\mathbf{a} = \int_C \phi(\mathbf{a})\,d\mathbf{a}, \quad (3.6)$$

which reproduces the example in (2.19). The fence can be given a non-uniform "height" by including an **a**-dependent factor under the integral sign.

2. Other Changes of Variables

Translation is only one of many possible "spatial transformations" or changes of the independent variable of the test functions. In each case we examine the behavior of (3.1) under the transformation in question, and then define the transformation for singular GFs accordingly. For a simple "change of scale," $\mathbf{r} \to a\mathbf{r}$, regular GFs behave according to

$$\int f(a\mathbf{r})\phi(\mathbf{r})\,dV = |a|^{-n}\int f(\mathbf{r})\phi(a^{-1}\mathbf{r})\,dV \quad (3.7)$$

in n dimensions, so we define

$$f(a\mathbf{r})\langle\phi(\mathbf{r})\rangle \equiv |a|^{-n}f(\mathbf{r})\langle\phi(a^{-1}\mathbf{r})\rangle \quad (3.8)$$

for all GFs.

An interesting special case of (3.8) occurs when $a = -1$, for then

$$f(-\mathbf{r})\langle\phi(\mathbf{r})\rangle = f(\mathbf{r})\langle\phi(-\mathbf{r})\rangle. \quad (3.9)$$

In one dimension, it may happen that $f(ax) = a^\lambda f(x)$ for $a > 0$ and all ϕ; then f is called *homogeneous* of degree λ. For example, x is homogeneous of degree 1 and $\delta(x)$ of degree -1. If $f(-x) = f(x)$ [or $-f(x)$], f is called *even* [or *odd*]. All of these terms are obvious extensions of the terminology for ordinary functions. Incidentally, an even (odd) GF gives zero when evaluated on an odd (even) test function.

In one dimension it is easy to generalize (3.8) to the change of variable $x = x(u)$, where x is a strictly monotonic and infinitely differentiable function of a new real variable, u. (At least these conditions must hold throughout the support of f.) Given $f(x)\langle\phi(x)\rangle$, we seek to define the new GF $h(u) = f(x(u))$. Performing this transformation in (3.1) suggests the definition

$$h(u)\langle\phi(u)\rangle = f(x(u))\langle\phi(u)\rangle \equiv f(x)\left\langle \left|\frac{dx}{du}\right|^{-1}\phi(u(x))\right\rangle. \quad (3.10)$$

(The derivative is the Jacobian of this transformation.) In one dimension (3.8) is now seen to be a special case of this definition, with $x(u) = au$. If f is the Dirac δ,

$$\delta(x(u))\langle\phi(u)\rangle = \delta(x)\langle \mid x' \mid^{-1}\phi(u(x))\rangle$$
$$= \mid x'(u_0) \mid^{-1}\phi(u_0) = \mid x'(u_0) \mid^{-1}\delta(u - u_0)\langle\phi(u)\rangle, \quad (3.11)$$

justifying a result used frequently in the literature. Here u_0 is the solution of $x(u) = 0$.

C. Products

1. *Multiplication by an Ordinary Function*

Considerable difficulty has been encountered in attempts to define the product of two GFs analogous to the product $f_1(\mathbf{r})f_2(\mathbf{r})$ or two ordinary functions f_1 and f_2. Even (3.1) is of no help, since the product of two locally integrable functions need not be locally integrable. From the viewpoint of fundamental sequences, for example, (2.25) is a fundamental sequence for δ, but its square diverges, indicating that δ^2 cannot be defined this way.

However, in one special case a product can always be defined. If f is in D' and α is an infinitely differentiable function, the product αf is the GF defined by

$$(\alpha f)\langle\phi\rangle \equiv f\langle\alpha\phi\rangle. \quad (3.12)$$

This is consistent with (3.1) and is always meaningful because $\alpha\phi$ belongs to D for any ϕ in D. The multiplication defined by (3.12) has many properties of the multiplication of ordinary functions: It is associative and distributive [i.e., $\alpha_1(\alpha_2 f) = (\alpha_1\alpha_2)f$, $\alpha(f_1 + f_2) = \alpha f_1 + \alpha f_2$, and $(\alpha_1 + \alpha_2)f = \alpha_1 f + \alpha_2 f$], even and odd factors combine to produce even or odd products as usual, and so forth. Multiplication of a GF by a constant is clearly a special case of (3.12).

As an example, consider

$$(\alpha(\mathbf{r})\,\delta(\mathbf{r}))\langle\phi(\mathbf{r})\rangle = \delta(\mathbf{r})\langle\alpha(\mathbf{r})\phi(\mathbf{r})\rangle = \alpha(\mathbf{0})\phi(\mathbf{0}), \qquad (3.13)$$

which shows that actually

$$\alpha(\mathbf{r})\,\delta(\mathbf{r}) = \alpha(\mathbf{0})\,\delta(\mathbf{r}). \qquad (3.14)$$

In particular, $x\,\delta(x) = 0$. Notice that the numerical representation of a function in (2.34) could have been written

$$\sum_m f(x_m)\,\delta(x - x_m) = \sum_m f(x)\,\delta(x - x_m) = f(x)\sum_m \delta(x - x_m), \qquad (3.15)$$

provided f is infinitely differentiable at the x_m.

2. Convolution

Let us consider the one-dimensional *convolution* product

$$(f * g)(x) \equiv \int_{-\infty}^{\infty} f(x')g(x - x')\,dx' \qquad (3.16)$$

of two ordinary functions, f and g. For GFs (3.1) suggests the definition

$$\begin{aligned}
(f * g)\langle\phi\rangle &= \int_{-\infty}^{\infty} \phi(x) \int_{-\infty}^{\infty} f(x')g(x - x')\,dx'\,dx \\
&= \int_{-\infty}^{\infty}\int_{-\infty}^{\infty} f(x')g(x'')\phi(x' + x'')\,dx'\,dx'' \\
&= f(x')\langle g(x'')\langle\phi(x' + x'')\rangle\rangle.
\end{aligned} \qquad (3.17)$$

However, some caution is required. We must be sure that the quantity in the outer angular brackets is actually a test function before we adopt (3.17) as a definition for convolutions of GFs. According to the discussion of Section III.B.1, this requirement will be met if the support of g is bounded. With this qualification, let us take $f(x)\langle g(x')\langle\phi(x + x')\rangle\rangle$ as our definition of $(f * g)\langle\phi\rangle$ for GFs. If f also has bounded support, $g * f$ is also defined and can be shown to equal $f * g$. The extension of this definition to more than one dimension presents no problems.

The Dirac δ has bounded support, so that for any ϕ in D_1' we may evaluate

$$(f * \delta)\langle\phi\rangle = f(x)\langle\delta(x')\langle\phi(x + x')\rangle\rangle = f(x)\langle\phi(x + 0)\rangle = f\langle\phi\rangle. \qquad (3.18)$$

Therefore δ displays the role of an identity element for the operation of convolution. (This result is used in constructing the solution of a partial differential equation from its Green's function.) Let us convolute f and $\delta(x - x_0)$, obtaining

$$(f * \delta(x - x_0))\langle\phi\rangle = f(x)\langle\delta(x' - x_0)\langle\phi(x + x')\rangle\rangle = f(x)\langle\phi(x + x_0)\rangle, \tag{3.19}$$

which is equivalent to a translation:

$$f * \delta(x - x_0) = f(x - x_0). \tag{3.20}$$

Consider, as another example, $f * \alpha$ where α is the regular GF corresponding to a bounded function with bounded support. In that case our definition is equivalent to

$$(f * \alpha)\langle\phi\rangle = \left[\int \alpha(x')f(x - x')\,dx'\right]\langle\phi(x)\rangle, \tag{3.21}$$

where the right side is an integral with respect to the parameter x' (Section II.C.3).

The question of an inverse operation for convolution has been a fruitful one. Mikusinski (1959) has developed a theory of *convolution quotients* which achieves many of the same generalizations over ordinary functions as does the theory outlined in this chapter. In a loose sense, convolution quotients provide another route to generalized functions, but the correspondence with the present theory is not one to one. For instance, Mikusinski's theory does not deal with more than one dimension nor with "generalized functions" of unbounded support. The interested reader is referred to Mikusinski (1959), Erdélyi (1961, 1962), and Marchand (1962) for further details.

3. *Other Products*

Other products of ordinary functions are commonly employed. One of the most important is the *direct product*, in which two functions of n_1 and n_2 variables, respectively, are multiplied to give a function of $n_1 + n_2$ variables; for example,

$$h(x, y, z) = f(x)g(y, z). \tag{3.22}$$

Such an operation can also be defined for GFs. In fact, in a more formal development one usually define convolutions in terms of direct products. However, the direct product is not needed for our purposes in this chapter.

D. Division

1. *Division by x*

Let us consider the inverse of the problem of multiplying by an infinitely differentiable function $\alpha(\mathbf{r})$. Given f in D', find a GF g satisfying

$$\alpha g \langle \phi \rangle = f \langle \phi \rangle. \tag{3.23}$$

In the simplest case, α has no zeros so that α^{-1} is again infinitely differentiable. The solution is just $g = \alpha^{-1} f$.

If α does have one or more zeros, the problem is not trivial. Let us specialize to one dimension and take $\alpha(x) = x$, and return to the more general problem after solving

$$xg(x)\langle \phi(x) \rangle = f(x)\langle \phi(x) \rangle. \tag{3.24}$$

We know from the start that g cannot be unique, since $g + c\delta$ must satisfy (3.24) if g does, in view of (3.14); c is an arbitrary constant.

The whole problem is that $x^{-1}\phi(x)$ is not in D_1 [unless $\phi(0) = 0$]. Corresponding to any ϕ, therefore, let us construct the "related" test function ψ:

$$\psi(x) = \phi(x) - \phi(0)\phi_0(x). \tag{3.25}$$

Here $\phi_0(x)$ is any fixed member of D_1 satisfying $\phi_0(0) = 1$. Now $\psi(0) = 0$, and a Taylor expansion of ψ to one term in x shows immediately that $x^{-1}\psi(x)$ is also in D_1. In fact, $g\langle \phi \rangle = f\langle x^{-1}\psi \rangle$ is actually a solution of (3.24). This can be verified by substitution:

$$xg\langle \phi \rangle = g\langle x\phi \rangle = f\langle x^{-1}[x\phi(x) - [x\phi(x)]_{x=0}\phi_0(x)]\rangle = f\langle \phi \rangle. \tag{3.26}$$

[The quantity in the outer square brackets is the test function related to $x\phi(x)$ by (3.25).] It is easy to show that $c\delta$ is the *only* solution of $xg\langle\phi\rangle = 0$, and therefore the *general* solution of (3.24) is

$$g\langle \phi \rangle = f\left\langle \frac{\phi(x) - \phi(0)\phi_0(x)}{x} \right\rangle + c\,\delta\langle\phi\rangle, \tag{3.27}$$

where c is an arbitrary constant and ϕ_0 is any function in D_1 such that $\phi_0(0) = 1$.

Notice that there has been no restriction whatever on f. *Any* GF in D_1' can be divided by x, yielding another GF as the solution. As a practical matter, notice that $xg_1 = xg_2$ does not imply $g_1 = g_2$, but rather $g_1 = g_2 + c\,\delta$. This is a more satisfactory alternative to the usual restriction "whenever $x \neq 0$" in ordinary analysis.

As an example of division by x, let us find $x^{-1}\,\delta$:

$$x^{-1}\,\delta\langle\phi\rangle = \delta\left\langle\frac{\phi(x) - \phi(0)\phi_0(x)}{x}\right\rangle + c_1\,\delta\langle\phi\rangle. \qquad (3.28)$$

The first δ requires the use of l'Hospital's rule to obtain the value of the argument at $x = 0$. The result is

$$x^{-1}\,\delta\langle\phi\rangle = \phi'(0) - \phi(0)\phi_0'(0) + c_1\phi(0) = \phi'(0) + c_2\,\delta\langle\phi\rangle. \qquad (3.29)$$

This example illustrates a more general point. If one exercises the freedom in the choice of ϕ_0, the effect is simply to shift c, which is arbitrary anyway. Thus we may fix ϕ_0 once and for all to remove the redundant arbitrariness in (3.27).

2. The Generalized Function x^{-1}

Having solved the division problem, we are now in a position to investigate the singular GF that corresponds to the ordinary function x^{-1}. We shall want to select x^{-1} to be one of the solutions of

$$xg\langle\phi\rangle = 1\langle\phi\rangle = \int_{-\infty}^{\infty} \phi(x)\,dx. \qquad (3.30)$$

These are given by

$$g\langle\phi\rangle = \int_{-\infty}^{\infty} x^{-1}[\phi(x) - \phi(0)\phi_0(x)]\,dx + c\,\delta\langle\phi\rangle, \qquad (3.31)$$

using the symbols defined in the preceding subsection.

Now we must select a unique solution from the set (3.31) and call it x^{-1}. To make a unique choice let us demand that x^{-1} be odd; this will be true if $c = 0$ and ϕ_0 is even. Then x^{-1} is uniquely specified in spite of the remaining arbitrariness in ϕ_0, since any two choices of ϕ_0 differ by an even function that vanishes at the origin, and such a difference contributes nothing to the integral in (3.31). A very convenient choice for ϕ_0 results from replacing $\phi(0)\phi_0(x)$ by $\frac{1}{2}(\phi(x) + \phi(-x))$, giving

$$x^{-1}\langle\phi\rangle = \tfrac{1}{2}\int_{-\infty}^{\infty} x^{-1}(\phi(x) - \phi(-x))\,dx$$
$$= \int_{0}^{\infty} x^{-1}(\phi(x) - \phi(-x))\,dx. \qquad (3.32)$$

This definition gives x^{-1} many of the properties of the ordinary function x^{-1}. On any open interval excluding the origin, $x^{-1}\langle\phi\rangle$ equals the ordinary function x^{-1} in the sense of Section II.B.3. We have already ensured that $xx^{-1} = 1$ and that x^{-1} is odd; in fact, x^{-1} is homogeneous of degree -1. Finally, after discussing the differentiation of GFs in Section IV.A, we find that (3.32) is actually the derivative of the regular GF $\ln|x|$, and later we define other inverse powers of x by further differentiation.

In effect, (3.32) has "doctored up" the function $\phi(x)/x$ so that it can be "integrated." The reader may have begun to suspect a connection with the Cauchy principal value (denoted by P) of a divergent integral. If the integrand $f(x)$ has a singularity at x_0, where $a < x_0 < b$, one defines

$$P\int_a^b f(x)\,dx = \lim_{\varepsilon \to 0_+} \left(\int_a^{x_0-\varepsilon} + \int_{x_0+\varepsilon}^b \right) f(x)\,dx, \qquad (3.33)$$

if the limit exists. The point of (3.33) is that x_0 is approached *at the same rate* from both sides. For $f(x) = \phi(x)/x$, we can rearrange the integral as follows:

$$P\int_{-\infty}^{\infty} x^{-1}\phi(x)\,dx = \tfrac{1}{2} P \int_{-\infty}^{\infty} x^{-1}(\phi(x) - \phi(-x))\,dx. \qquad (3.34)$$

But if ϕ is in D, this integrand is not singular, so the P symbol may be dropped. Therefore

$$x^{-1}\langle\phi\rangle = P\int_{-\infty}^{\infty} x^{-1}\phi(x)\,dx. \qquad (3.35)$$

The Cauchy principal value arises in the calculus of residues (see Chapter 3 by Silverstone) when a singularity lies on the contour of integration; it is pleasing that this important concept emerges from GF theory in a perfectly natural manner. Along with the definition of a regular GF it encourages us to think of $f\langle\phi\rangle$ as the "value" of the integral

$$\int_{-\infty}^{\infty} f(x)\phi(x)\,dx \qquad (3.36)$$

even when this diverges as an ordinary integral. Note, incidentally, that the GF division by x can always be performed, even if the corresponding principal value fails to exist, as for $x^{-1}\operatorname{sgn} x$, for example.

3. Division by Other Functions

By now the reader may have scoffed that division by x is trivial compared with the problem of division by an arbitrary infinitely differentiable function. This is not the case. Consider the necessary properties of the divisor α in the one-dimensional case in order for (3.23) to have a solution. Clearly it may not be identically zero over a finite interval since that would amount to division by zero. In fact, there must be only a finite number of zeros in any finite interval, and these must be of finite order. Suppose these zeros occur at $x_1, x_2, \ldots, x_k$ (for a given test function we need to consider only a finite number of them), and they are of order $m_1, m_2, \ldots, m_k$. That is, near x_j,

$$\alpha(x) \sim c(x - x_j)^{m_j}. \tag{3.37}$$

Then the function

$$\beta(x) = \alpha(x) \prod_{j=1}^{k} (x - x_j)^{-m_j} \tag{3.38}$$

has no zeros, and (3.23) can be replaced by

$$\prod_{j=1}^{k} (x - x_j)^{m_j} g\langle\phi\rangle = \beta^{-1} f\langle\phi\rangle. \tag{3.39}$$

Clearly the division by factors linear in x is of fundamental importance.

Division by $(x - x_1)$ is a simple extension of division by x. The solution of

$$(x - x_1) g\langle\phi\rangle = f\langle\phi\rangle \tag{3.40}$$

is

$$g\langle\phi\rangle = f\left\langle \frac{\phi(x) - \phi(x_1)\phi_1(x)}{(x - x_1)} \right\rangle + c\,\delta(x - x_1)\langle\phi\rangle \tag{3.41}$$

where c is arbitrary and ϕ_1 is an arbitrary test function except that $\phi_1(x_1) = 1$. Similarly, the solution of

$$(x - x_1)(x - x_2) g\langle\phi\rangle = f\langle\phi\rangle \tag{3.42}$$

turns out to be

$$g\langle\phi\rangle = f\left\langle \frac{\phi(x) - \phi_{12}(x)[(x - x_1)\phi(x_2) - (x - x_2)\phi(x_1)]/(x_2 - x_1)}{(x - x_1)(x - x_2)} \right\rangle$$
$$+ c_1\,\delta(x - x_1) + c_2\,\delta(x - x_2), \qquad x_1 \neq x_2, \tag{3.43}$$

where c_1 and c_2 are arbitrary constants and ϕ_{12} is any test function with $\phi_{12}(x_1) = \phi_{12}(x_2) = 1$. Division by factors like $(x - x_1)^m$ can be accomplished by applying (3.41) m times. The result contains $\delta(x - x_1)$ through $\delta^{(m-1)}(x - x_1)$ multiplied by arbitrary coefficients; see (4.13) and (4.14).

In several dimensions a division may have not only isolated zeros but also lines, surfaces, or hypersurfaces of zeros. The solution follows the spirit of the one-dimensional case. A test function equal to 1 at the zeros is used to avoid singular results, and an arbitrary fence or sheet of δ's [as in (3.6)] becomes a part of the solution.

An example of division by a sinusoidal function can be found in Section VI.E.1.

IV. The Calculus of Generalized Functions

A. Differentiation

1. A Single Variable

In this section we define the GF analogs of differentiation and integration, again using the regular GF to lead us to useful definitions. We begin with the derivative of a GF in D_1'.

Suppose that $f(x)$ is not only a locally integrable function but also has a locally integrable derivative f'. Then the corresponding regular GF

$$f'\langle\phi\rangle = \int_{-\infty}^{\infty} f'(x)\phi(x)\,dx, \tag{4.1}$$

can be integrated by parts to give

$$f'\langle\phi\rangle = f(x)\phi(x)\Big|_{-\infty}^{\infty} - \int_{-\infty}^{\infty} f(x)\phi'(x)\,dx = -\int_{-\infty}^{\infty} f(x)\phi'(x)\,dx. \tag{4.2}$$

The integrated part equals zero since ϕ is assumed to be in D and hence has bounded support. Notice that f' is required to be locally integrable only in the first step of deriving (4.2). Let us generalize this further, and *define* f', the derivative of an arbitrary member f of D_1', to be

$$f'\langle\phi\rangle \equiv -f\langle\phi'\rangle. \tag{4.3}$$

Equation (4.3) is always meaningful because ϕ' is in D if ϕ is in D. Furthermore, f' is easily shown to be a linear and continuous functional of ϕ, and hence it is a GF itself. (Continuity follows from the inclusion of derivatives in the definition of convergence in D.) Differentiation may be applied repeatedly, giving the mth-order derivative

$$f^{(m)}\langle\phi\rangle = (-1)^m f\langle\phi^{(m)}\rangle. \tag{4.4}$$

Each of the $f^{(m)}$ is also a GF; in other words, *every GF in D_1' has GF derivatives of all orders*! The stringent demands made of the test functions in Section II.A.2 are paying handsome dividends.

Before turning to some examples, let us obtain (4.3) by an interesting alternative route. By analogy with the definition of ordinary derivatives, consider

$$\lim_{\varepsilon\to 0} \varepsilon^{-1}[f(x+\varepsilon) - f(x)]\langle\phi(x)\rangle. \tag{4.5}$$

Using the definition of translation and the linearity and continuity of f, this limit can be written as

$$\lim_{\varepsilon\to 0} \varepsilon^{-1} f\langle\phi(x-\varepsilon) - \phi(x)\rangle = f\langle\lim_{\varepsilon\to 0} \varepsilon^{-1}[\phi(x-\varepsilon) - \phi(x)]\rangle$$
$$= -f\langle\phi'\rangle, \tag{4.6}$$

so that (4.3) is equivalent to

$$f' = \lim_{\varepsilon\to 0} \varepsilon^{-1}[f(x+\varepsilon) - f(x)], \tag{4.7}$$

just as if f were a function instead of a GF.

2. Examples. Rules of Differentiation

Let us work out the derivatives of some specific GFs. The derivative of $|x|$ [see (2.11)] is

$$|x|'\langle\phi\rangle = \int_{-\infty}^{0} x\phi'(x)\,dx - \int_{0}^{\infty} x\phi'(x)\,dx \tag{4.8}$$

which simplifies, through integration by parts, to become

$$-\int_{-\infty}^{0} \phi(x)\,dx + \int_{0}^{\infty} \phi(x)\,dx = \text{sgn}(x)\langle\phi(x)\rangle, \tag{4.9}$$

as expected. Now let us differentiate again:

$$(\operatorname{sgn} x)'\langle\phi\rangle = \int_{-\infty}^{0} \phi'(x)\,dx - \int_{0}^{\infty} \phi'(x)\,dx$$

$$= \phi(x)\Big|_{-\infty}^{0} - \phi(x)\Big|_{0}^{\infty} = 2\phi(0) = 2\delta\langle\phi\rangle. \qquad (4.10)$$

Similarly, $H' = \delta$, where H is defined in (2.15). Thus we see that δ enters the theory in a perfectly way: The regular GF corresponding to a function with a jump discontinuity of height C at $x = x_0$ will produce, on differentiation, a term $C\,\delta(x - x_0)$.

Like all members of D_1', δ itself must have derivatives of all orders. They are

$$\delta^{(m)}\langle\phi\rangle = (-1)^m\,\delta\langle\phi^{(m)}\rangle = (-1)^m\phi^{(m)}(0). \qquad (4.11)$$

Multiplying these by an infinitely differentiable function $\alpha(x)$ and using the Leibnitz formula for $(\alpha\phi)^{(m)}$ gives the identity

$$\alpha\,\delta^{(m)} = \sum_{k=0}^{m} (-1)^k C_k^m \alpha^{(m)}(0)\,\delta^{(m-k)}, \qquad (4.12)$$

where $C_k^m \equiv m![k!(m-k)!]^{-1}$. This generalizes (3.13). In particular

$$x^k\,\delta^{(m)} = \begin{cases} 0, & k > m \\ (-1)^k C_k^m\,\delta^{(m-k)} & k \leq m, \end{cases} \qquad (4.13)$$

which shows why the arbitrary term in the result of a division by x^k is

$$\sum_{m=0}^{k-1} c_m\,\delta^{(m)}. \qquad (4.14)$$

From these examples it is clear that many rules of ordinary differentiation are expected to be valid for GF differentiation. Specifically, we have rules for the derivatives of:

1. The sum of two GFs:

$$(f_1 + f_2)'\langle\phi\rangle = -(f_1 + f_2)\langle\phi'\rangle = -f_1\langle\phi'\rangle - f_2\langle\phi'\rangle$$
$$= f_1'\langle\phi\rangle + f_2'\langle\phi\rangle = (f_1' + f_2')\langle\phi\rangle. \qquad (4.15)$$

2. A GF multiplied by a constant:

$$(Cf)'\langle\phi\rangle = -(Cf)\langle\phi'\rangle = -f\langle(C\phi)'\rangle = Cf'\langle\phi\rangle. \qquad (4.16)$$

2. Generalized Functions

Rules 1 and 2 establish GF differentiation as a linear operation; it is also easily verified to be continuous.

3. A constant GF:

$$C'\langle\phi\rangle = -C\langle\phi'\rangle = -C\int_{-\infty}^{\infty}\phi'(x)\,dx = -C\phi\Big|_{-\infty}^{\infty} = 0. \quad (4.17)$$

Conversely, it will follow from the discussion of Section IV.B below that any GF whose derivative is zero must be a constant GF.

4. A GF multiplied by an infinitely differentiable function α:

$$(\alpha f)'\langle\phi\rangle = -\alpha f\langle\phi'\rangle = -f\langle\alpha\phi'\rangle = -f\langle(\alpha\phi)' - \alpha'\phi\rangle$$
$$= f'\langle\alpha\phi\rangle + \alpha'f\langle\phi\rangle = (\alpha f' + \alpha'f)\langle\phi\rangle. \quad (4.18)$$

Rule 2 is really a special case of this.

5. A convolution of two GFs:

$$(f*g)'\langle\phi\rangle = -(f*g)\langle\phi'\rangle = -f(x_1)\langle g(x_2)\langle\phi'(x_1+x_2)\rangle\rangle$$
$$= f(x_1)\langle g'(x_2)\langle\phi(x_1+x_2)\rangle\rangle = (f*g')\langle\phi\rangle. \quad (4.19)$$

In particular, $f*\delta^{(m)} = (f*\delta)^{(m)} = f^{(m)}$, so that convolution with $\delta^{(m)}$ is identical to differentiating m times.

6. A composite function: Using the notation of (3.10) and writing d/dx or d/du instead of primes, we have

$$\frac{d}{du}f(x(u))\langle\phi(u)\rangle = -f(x(u))\left\langle\frac{d\phi(u)}{du}\right\rangle$$
$$= -f(x)\left\langle\left|\frac{dx}{du}\right|^{-1}\frac{d\phi(u(x))}{du}\right\rangle$$
$$= -f(x)\left\langle\left|\frac{dx}{du}\right|^{-1}\frac{dx}{du}\frac{d\phi(u(x))}{dx}\right\rangle$$
$$= \frac{df(x)}{dx}\left\langle\left|\frac{dx}{du}\right|^{-1}\frac{dx}{du}\phi(u(x))\right\rangle$$
$$= \frac{dx}{du}\frac{df(x(u))}{dx}\langle\phi(u)\rangle. \quad (4.20)$$

This is the analog of the chain rule for ordinary differentiation. Inserting $x(u) = \pm u$ shows that the derivative of an even GF is odd, and vice versa.

7. A homogeneous function: If f is homogeneous of degree λ, then f' is homogeneous of degree $\lambda - 1$, but the converse is false.

In addition to these general rules, various formulas for derivatives of particular functions hold for the analogous GFs. Thus we have

$$(x^n)' \langle \phi \rangle = n x^{n-1} \langle \phi \rangle, \tag{4.21}$$

$$(\sin x)' \langle \phi \rangle = \cos x \langle \phi \rangle, \tag{4.22}$$

and so forth. In particular, it is interesting that $(\ln |x|)'$ equals the GF x^{-1} as defined by (3.32):

$$\begin{aligned}
(\ln |x|)' \langle \phi \rangle &= -\ln |x| \langle \phi' \rangle = -\int_{-\infty}^{\infty} \ln |x| \phi'(x) \, dx \\
&= -\int_{0}^{\infty} \ln x [\phi(x) - \phi(-x)]' \, dx \\
&= -\ln x [\phi(x) - \phi(-x)] \Big|_{0}^{\infty} + \int_{0}^{\infty} x^{-1} [\phi(x) - \phi(-x)] \, dx \\
&= x^{-1} \langle \phi \rangle.
\end{aligned} \tag{4.23}$$

At this stage in an analogous development of ordinary differential calculus one would probably introduce the study of maxima and minima of a function. This cannot in general be done for GFs since the underlying notion of "value at a point" has been lost. Notice also that the differentiation of GFs is not related to the functional differentiation encountered in variational calculus. GF derivatives are taken with respect to the independent variables of the ϕ's. In a functional derivative, on the other hand, the ϕ's themselves are being varied, and the derivative is being taken in the space D. Since GFs are linear functionals, such a procedure would be trivial and uninteresting in the present case.

3. *Several Variables*

It is a simple extension of Section IV.A.1 to introduce partial derivatives of GFs in D_n', $n > 1$. One may proceed via either (4.2) or (4.7), using the divergence theorem (Chapter 1, Section III, by Jacob) as a generalization of integration by parts in the former case. Either way one is led to define $\partial f / \partial x_k$ as

$$(\partial f / \partial x_k) \langle \phi \rangle \equiv -f \langle \partial \phi / \partial x_k \rangle \tag{4.24}$$

where f is in D_n' and $k = 1, 2, \ldots, n$ identifies one of the Cartesian coordinates. Again the resulting functional is always a GF.

2. Generalized Functions

Higher order and mixed partial derivatives are defined by repeatedly applying (4.24). Thus the second-order derivatives of f are

$$\frac{\partial^2 f}{\partial x_k \, \partial x_j} \langle \phi \rangle = -\frac{\partial f}{\partial x_j} \left\langle \frac{\partial \phi}{\partial x_k} \right\rangle = +f \left\langle \frac{\partial^2 \phi}{\partial x_j \, \partial x_k} \right\rangle. \tag{4.25}$$

We conclude that all GFs have partial derivatives of all orders. Furthermore, the mixed derivatives [$k \neq j$ in (4.25), for example] are always independent of the order in which the various differentiations are performed, thanks to the properties of the ϕ's. The rules mentioned in the preceding subsection generalize to several variables as expected.

As an example, consider the Laplacian operator ∇^2:

$$\nabla^2 f \equiv \sum_{k=1}^{n} (\partial^2 f / \partial x^2_k). \tag{4.26}$$

Let us apply ∇^2 to the regular GF defined in $D_2{}'$ by the function

$$U(x, y) = -2\lambda \ln(x^2 + y^2)^{1/2}. \tag{4.27}$$

The reader familiar with electrostatics, for instance, will immediately recognize U as the solution of Laplace's equation for a line charge of constant linear charge density λ lying along the z axis. Therefore, since the problem has been posed in two dimensions, we expect to find $\nabla^2 U = 4\pi\lambda \, \delta(\mathbf{r})$. [A *derivation* of (4.27) using GFs is given in Section V.B.2.]

Applying (4.26) to U gives

$$(\nabla^2 U) \langle \phi(x, y) \rangle = U \langle \nabla^2 \phi \rangle = -2\lambda \iint \ln(x^2 + y^2)^{1/2} \nabla^2 \phi \, dx \, dy. \tag{4.28}$$

The integration extends over the support of ϕ. A simplification results by applying Green's theorem in the plane (an analog of integration by parts) to obtain

$$(\nabla^2 U) \langle \phi \rangle = -2\lambda \iint (\nabla \ln(x^2 + y^2)^{1/2} \cdot \nabla \phi \, dx \, dy \tag{4.29}$$

where ∇ is the gradient operator (Chapter 1, Section III.D.2.) This integral is readily evaluated by converting it to plane polar coordinates:

$$(\nabla^2 U) \langle \phi \rangle = -2\lambda \int_0^{2\pi} \int_0^\infty (\nabla \ln r) \cdot (\nabla \phi) r \, dr \, d\theta$$

$$= -2\lambda \int_0^{2\pi} \int_0^\infty \frac{\partial \phi}{\partial r} dr \, d\theta = -2\lambda \int_0^{2\pi} \phi \bigg|_{r=0}^{r=\infty} d\theta = 4\pi\lambda\phi(0)$$

$$= 4\pi\lambda \, \delta \langle \phi \rangle, \tag{4.30}$$

which is the anticipated result. As an ordinary function, U is not differentiable at the origin, of course, and the conventional proofs that $\nabla^2 U = 4\pi\lambda\,\delta$ require special care at this point. However, in the GF formalism the proof is quite straightforward.

4. Differentiation of Sequences

Before directing our attention to integration, let us examine the differentiation of sequences of GFs. Suppose $f_m \to f$ as $m \to \infty$, as defined in Section II.C.1, where the f's are in D_1' for simplicity. Since $f_m^{(k)}\langle\phi\rangle$ is simply $(-1)^{(k)} f_m\langle\phi^{(k)}\rangle$, it is trivial that $f_m^{(k)}$ also converges, since $f_m \to f$ for *any* test function, including $\phi^{(k)}$. Thus $f_m^{(k)} \to f^{(k)}$. The same is clearly true of series: A convergent series differentiated term by term k times converges to the kth derivative of the original limit. Needless to say, this is an enormous improvement over the state of affairs for ordinary functions.

As an extension of these ideas, consider a family f_γ of parametric GFs. If $f_\gamma \to f$ as $\gamma \to \gamma_0$, it follows by the same reasoning as above that $f_\gamma^{(k)} \to f^{(k)}$. Furthermore, differentiation with respect to x commutes with differentiation with respect to γ:

$$\left(\frac{\partial f_\gamma}{\partial \gamma}\right)'\langle\phi\rangle = -\frac{\partial f_\gamma}{\partial \gamma}\langle\phi'\rangle = -\frac{\partial}{\partial \gamma}(f_\gamma\langle\phi'\rangle)$$
$$= \frac{\partial}{\partial \gamma}(f_\gamma'\langle\phi\rangle) = \frac{\partial}{\partial \gamma}(f_\gamma')\langle\phi\rangle. \qquad (4.31)$$

B. Integration

1. The Indefinite Integral

It is natural at this point to seek a definition of the indefinite integral, or primitive, of a GF of one variable. In other words, corresponding to any f in D_1', we want to find another GF, $f^{(-1)}$, whose derivative is f:

$$(f^{(-1)})'\langle\phi\rangle = f\langle\phi\rangle. \qquad (4.32)$$

From another viewpoint one might say we are solving a simple first-order differential GF equation.

Comparing (4.32) and (4.3) suggests that the solution might be

$$f^{(-1)}\langle\phi\rangle = -f\!\left\langle\int_{-\infty}^{x}\phi(t)\,dt\right\rangle, \qquad (4.33)$$

but here is a trap for the unwary: The argument of f on the right side of (4.33) need not be in D_1 since the integral may not approach zero for large x. A definition like (4.33) would only be valid in a subspace of D_1. The similarity of this difficulty with the problem of division (Section III.D.1) is striking.

As in the case of division, we can decompose ϕ in a suitable manner and obtain a solution. Let us construct

$$\psi(x) = \phi(x) - 1\langle\phi\rangle\phi_0(x) \tag{4.34}$$

where $1\langle\phi\rangle$ is the constant that results from evaluating the constant unit GF [see (3.30) or (2.10)] at ϕ, and ϕ_0 is any arbitrary member of D_1 that has $1\langle\phi_0\rangle = 1$. Once ϕ_0 is fixed, this decomposition is unique; what it accomplishes is to extract from ϕ a test function, namely ψ, whose primitive

$$\psi^{(-1)}(x) = \int_{-\infty}^{x} \psi(t)\, dt \tag{4.35}$$

is also in D_1. If $\phi^{(-1)}$ is already in D, then $\psi = \phi$ since $1\langle\phi\rangle = 0$ in this case.

Now we can state that the solution of (4.32) is

$$f^{(-1)}\langle\phi\rangle \equiv -f(x)\langle\psi^{(-1)}(x)\rangle = -f(x)\left\langle \int_{-\infty}^{x}[\phi(t) - 1\langle\phi\rangle\phi_0(t)]\, dt\right\rangle. \tag{4.36}$$

That this is a solution may be verified by differentiation:

$$(f^{(-1)})'\langle\phi\rangle = -f^{(-1)}\langle\phi'\rangle = f(x)\left\langle\int_{-\infty}^{x}[\phi'(t) - 1\langle\phi'\rangle\phi_0(t)]\, dt\right\rangle$$
$$= f(x)\langle\phi(x) - 0\rangle = f\langle\phi\rangle, \tag{4.37}$$

using the fact that $1\langle\phi'\rangle = 0$. Notice that the solution (4.36) actually reduces to (4.33) if $\phi^{(-1)}$ is in D. Furthermore, it is easily verified that (4.36) gives the correct answer for the special case of a regular GF.

The freedom in the selection of ϕ_0 is precisely that of an arbitrary additive constant, as we might expect. This can be shown as follows. Let ϕ_0 be altered by an amount $\Delta\phi_0$. Clearly $1\langle\Delta\phi_0\rangle$ must be zero so that the constraint $1\langle\phi_0 + \Delta\phi_0\rangle = 1$ can be met, and hence $\Delta\phi_0^{(-1)}$ is in D_1. The change in $f^{(-1)}\langle\phi\rangle$ brought about by $\Delta\phi_0$ is $f\langle\Delta\phi_0^{(-1)}\rangle 1\langle\phi\rangle$, according to (4.36); but this is just $c\langle\phi\rangle$, where c is an arbitrary constant. Therefore we may fix ϕ_0 once and for all, and write the primitive as the family of GFs

$$f^{(-1)}\langle\phi\rangle = -f\langle\psi^{(-1)}\rangle + c\langle\phi\rangle. \tag{4.38}$$

Actually, it is easy to show that the *only* solution of $f' = 0$ is a constant, and therefore (4.38) and (4.36) are general solutions of (4.32).

It is not a trivial point that $f^{(-1)}$ exists for any f and is itself a GF, implying that any GF may be integrated any number of times. In fact, we state without proof a much stronger result: Every GF is locally the finite-order derivative of some regular GF. In other words, on a given bounded interval, no GF is so singular that a finite number of integrations will not yield a regular GF (and then of course one further integration gives the regular GF corresponding to a continuous function). This important result emphasizes that the theory of GFs is in effect a theory of differentiation of ordinary functions.

By way of an example, let us find the primitive of $\delta(x)$. It is

$$\begin{aligned}\delta^{(-1)}\langle\phi\rangle &= -\delta\left\langle\int_{-\infty}^{x}[\phi(t) - 1\langle\phi\rangle\phi_0(t)]\,dt\right\rangle \\ &= -\int_{-\infty}^{0}\phi(t)\,dt + 1\langle\phi\rangle\int_{-\infty}^{0}\phi_0(t)\,dt \\ &= H\langle\phi\rangle + c\langle\phi\rangle. \end{aligned} \tag{4.39}$$

Like differentiation, integration is a linear operation: If c_1 and c_2 are real constants and f_1 and f_2 are in D_1', then

$$(c_1 f_1 + c_2 f_2)^{(-1)} = c_1 f_1^{(-1)} + c_2 f_2^{(-1)}, \tag{4.40}$$

bearing in mind that there are arbitrary additive constant GFs associated with these integrals. As in the case of differentiation, a convergent sequence integrated term by term converges to a primitive of the original limit, and the other remarks of Section IV.A.4 apply to integration as well.

2. The Definite Integral

It would be useful to define a GF operation corresponding to the definite integral

$$\int_a^b f(x)\,dx \tag{4.41}$$

of an ordinary function $f(x)$. We may expect some difficulty in doing this, since the fundamental theorem of integral calculus states that (4.41) equals $f^{(-1)}(b) - f^{(-1)}(a)$, whereas for GFs we cannot in general speak

2. Generalized Functions

of the value at a point. However, when a GF equals an integrable ordinary function in some open interval containing a and another containing b, progress can be made easily.

Comparing (4.41) with the definition of a regular GF, we see the need for a testing function equaling unity between a and b and zero elsewhere. Such a function is not in D, of course, but may be approximated in the following sense:

$$\lim_{\varepsilon \to 0_+} \sigma_\varepsilon(x; a, b) = \begin{cases} 1, & a < x < b \\ \tfrac{1}{2}, & x = a, \quad x = b \\ 0, & \text{otherwise,} \end{cases} \quad (4.42)$$

where σ is defined in (2.3). Here "lim" stands for a limit in the pointwise sense; in D the limit does not exist, but this need not prevent $f\langle \phi_\varepsilon \rangle$ from having a well-defined limit as $\varepsilon \to 0$. (By analogy the sequence $\{n\}$, $n = 1, 2, 3, \ldots$, diverges on the real axis, but $\{f(n) = 1/n\}$ converges to zero.) In fact, let us define the real quantity

$$\int_a^b f(x) \, dx \equiv \lim_{\varepsilon \to 0_+} f(x)\langle \sigma_\varepsilon(x; a, b) \rangle, \quad (4.43)$$

if it exists, to be the definite integral of the GF $f\langle \phi \rangle$ from a to b. We can expect this limit to exist whenever f equals a regular GF in neighborhoods of the endpoints a and b. This represents an improvement over ordinary definite integrals, whose existence depends on the behavior of the integrand throughout the interior of interval (a, b) as well.

When (4.43) exists, the fundamental theorem of integral calculus applies to GFs, since

$$f\langle \sigma_\varepsilon \rangle = (f^{(-1)})'\langle \sigma_\varepsilon \rangle = -f^{(-1)}\langle \sigma_\varepsilon' \rangle$$
$$= f^{(-1)}(x)\langle \varrho_\varepsilon(x - b) - \varrho_\varepsilon(x - a) \rangle, \quad (4.44)$$

so that

$$\int_a^b f(x) \, dx = \lim_{\varepsilon \to 0_+} f^{(-1)}(x)\langle \varrho_\varepsilon(x - b) - \varrho_\varepsilon(x - a) \rangle$$
$$= f^{(-1)}(b) - f^{(-1)}(a). \quad (4.45)$$

The last step assumes $f^{(-1)}$ to be a continuous function at a and b, which is true if the restriction on f mentioned above holds.

As in ordinary calculus, (4.45) is useful in evaluating integrals. For example, using (4.39) we have

$$\int_a^b \delta(x) \, dx = H(b) - H(a) = 1, \quad (4.46)$$

provided $a < 0 < b$. If $[a, b]$ does not include the origin, (4.46) equals zero, and if $a = 0$ or $b = 0$, then the integral is not defined in this approach. We can use (4.23) to integrate the GF x^{-1} across the origin. For example,

$$\int_{-1}^{+1} x^{-1}\, dx = \ln |x| \Big|_{-1}^{+1} = 0, \tag{4.47}$$

as expected from (3.35).

Definite integrals of GFs enjoy many of the elementary properties of ordinary definite integrals, although it will not be necessary to elaborate them in detail. In particular, the chain rule can be combined with (4.45) to show that the usual change-of-variable procedure under the integral sign holds for GFs. In many cases this permits (4.43) to be extended to the case of the improper integral, where a or b or both are not finite. [See Section V.B.2 for an example and Gel'fand *et al.* (1964) for further details.]

As with differentiation, convergent sequences and series may be integrated term by term. For parametric GFs, integration commutes with both integration and differentiation with respect to the parameter. Thus

$$\frac{\partial}{\partial \gamma} \int_a^b f_\gamma(x)\, dx = \int_a^b \frac{\partial f_\gamma}{\partial \gamma}\, dx \tag{4.48}$$

and

$$\int_{\gamma_1}^{\gamma_2} \left\{ \int_a^b f_\gamma(x)\, dx \right\} d\gamma = \int_a^b \left\{ \int_{\gamma_1}^{\gamma_2} f_\gamma(x)\, d\gamma \right\} dx. \tag{4.49}$$

We shall have occasion to use (4.48) in Section V.B.2.

V. Some Singular Generalized Functions

A. Definitions of Inverse Powers and Related Functions

1. *Integral Powers*

We have already defined x^{-1} in Section III.D.2, and the calculus developed in Section IV now permits us to discuss other inverse power GFs. The GFs defined in this subsection correspond to the ordinary functions x^{-n}, where n is a positive integer.

2. Generalized Functions

Several subsections hence we discuss certain GFs whose definitions are not consistent from author to author. To avoid ambiguities it is important at the outset to state explicitly what properties we shall expect a GF named x^{-n} to possess:

1. Equality, in the GF sense (Section II.B.3), with the corresponding ordinary function when $x \neq 0$.
2. Uniqueness, i.e., the definition should contain no arbitrary parameters.
3. $(x^{-n})' = -nx^{-n-1}$ [and $(\ln |x|)' = x^{-1}$, which is already satisfied].
4. Homogeneity of degree $-n$ (Section III.B.2).
5. $xx^{-n} = x^{-n+1}$, with $xx^{-1} = 1$ for $n = 1$ as we have already seen.

Let us try the following definition for x^{-n} based directly on property 3:

$$x^{-(n+1)}\langle\phi\rangle \equiv \frac{(-1)^n}{n!}(x^{-1})^{(n)}\langle\phi\rangle = \frac{1}{n!}\int_0^\infty \frac{\phi^{(n)}(x) - \phi^{(n)}(-x)}{x}\,dx. \tag{5.1}$$

This automatically satisfies requirement 2 since derivatives are uniquely defined; it also satisfies 4 because of the behavior of homogeneity under differentiation (Section IV.A.2) and because x^{-1} is homogeneous of degree -1. It is easy to prove that condition 5 is met:

$$xx^{-(n+1)}\langle\phi\rangle = x^{-(n+1)}\langle x\phi\rangle = \frac{1}{n!}\int_0^\infty x^{-1}\left[\frac{d^n}{d\xi^n}\xi\phi(\xi)\right]\Big|_{\xi=-x}^{\xi=x}\,dx$$

$$= \frac{1}{n}\int_0^\infty \{\phi^{(n)}(x) - \phi^{(n)}(-x)$$

$$+ nx^{-1}[\phi^{(n-1)}(x) - \phi^{(n-1)}(-x)]\}\,dx = x^{-n}\langle\phi\rangle. \tag{5.2}$$

Condition 1 is also met. This can be demonstrated explicitly, but follows more easily from 3 and the fact that x^{-1} satisfies 1.

Clearly definition (5.1) gives us the family of GFs, x^{-n}, that we have sought. Let us defer applications of these to Section V.B, and continue by defining other closely related GFs. Incidentally, it is trivial that the five conditions are also met by the positive powers of x, which are all regular GFs.

2. Nonintegral Powers. Logarithms

The extension from integral powers to real powers is not so simple as it may appear. The difficulty is that the ordinary function x^γ has complex values for $x < 0$ and γ nonintegral. An apparent remedy is to work instead with $|x|^\gamma$ or $|x|^\gamma \, \text{sgn}\, x$. Both of these functions can be written in terms of $H(x)|x|^\gamma$, upon which we shall therefore focus our attention. Let us adopt the simpler notation

$$x_+^\gamma = \begin{cases} x^\gamma, & x > 0, \\ 0, & x < 0, \end{cases} \qquad (5.3)$$

and seek a GF to correspond to x_+^γ.

Our criteria for a GF x_+^γ will be the five requirements listed in the preceding subsection, with x^{-n} replaced by x_+^γ and $\ln|x|$ by $\ln x_+$, which is defined by

$$\ln x_+ = \begin{cases} \ln x, & x > 0, \\ 0, & x < 0. \end{cases} \qquad (5.4)$$

In addition, let us try to impose a sixth requirement:

6. x_+^γ should be continuous in the parameter γ.

First of all, notice that x_+^γ defines a regular GF for $\gamma > -1$:

$$x_+^\gamma \langle \phi \rangle \equiv \int_0^\infty x^\gamma \phi(x)\, dx, \qquad \gamma > -1. \qquad (5.5)$$

However, this integral diverges as $\gamma \to -1$ [unless we happen to have $\phi(0) = 0$] and it is already clear that requirement 6 cannot be satisfied at $\gamma = -1$. Nevertheless, for $-2 < \gamma < -1$ we can follow requirement 3 and define

$$x_+^\gamma \langle \phi \rangle = (\gamma + 1)^{-1} (x_+^{\gamma+1})' \langle \phi \rangle, \qquad -2 < \gamma < -1. \qquad (5.6)$$

Again this result diverges as $\gamma \to -2$. We shall have to impose the restriction $\gamma \ne -1, -2, -3, \ldots$ in defining x_+^γ. (This is not inconsistent with the preceding subsection. There we were not dealing with these "half-range" functions.) The procedure of differentiation can be continued, leading to the general definition

$$x_+^\gamma \equiv \prod_{j=1}^m (\gamma + j)^{-1} (x_+^{\gamma+m})^{(m)}, \qquad -m-1 < \gamma < -m, \qquad (5.7)$$

where m is a positive integer. In this definition, we are again relying ultimately on (5.5). The reader may easily verify that (5.7) does indeed satisfy requirements 1–5 (but of course not 6).[†]

[†] By analogy with the development in this paragraph, one can define r^γ in n dimensions, where r is defined in (2.12). For $\gamma > -n$ the function r^γ defines a regular GF, and for $\gamma < -n$ one lets the relation $\nabla^2 r^{\gamma+2} = (\gamma + 2)(\gamma + n) r^\gamma$ play the role of (5.6).

2. Generalized Functions

Defining
$$x_-^\gamma \langle \phi(x) \rangle \equiv x_+^\gamma \langle \phi(-x) \rangle, \tag{5.8}$$

we can now form the GFs
$$|x|^\gamma = x_+^\gamma + x_-^\gamma \tag{5.9}$$

and
$$|x|^\gamma \operatorname{sgn} x = x_+^\gamma - x_-^\gamma. \tag{5.10}$$

These GFs have the properties one would expect: (5.9) and (5.10) are even and odd, respectively, and

$$(|x|^\gamma)' = \gamma |x|^{\gamma-1} \operatorname{sgn} x, \tag{5.11}$$

$$(|x|^\gamma \operatorname{sgn} x)' = \gamma |x|^{\gamma-1}, \tag{5.12}$$

$$x|x|^\gamma = |x|^{\gamma+1} \operatorname{sgn} x, \tag{5.13}$$

$$x|x|^\gamma \operatorname{sgn} x = |x|^{\gamma+1}, \tag{5.14}$$

and so on. In spite of the earlier restrictions on γ, there is a cancellation of singularities so that $|x|^\gamma$ does *not* diverge as γ approaches a negative *even* integer, and likewise $|x|^\gamma \operatorname{sgn} x$ exists for negative *odd* integers. In these cases one simply obtains the GFs x^{-n} of the preceding subsection; for example, as $\gamma \to -1$, (5.10) clearly approaches x^{-1} as defined in (3.32).

For the ordinary functions one has the identity $\partial x_+^\gamma / \partial \gamma = x_+^\gamma \ln x$, and this can be exploited to define several new GFs involving powers of logarithms:

$$x_+^\gamma (\ln x)^m \equiv \frac{\partial^m}{\partial \gamma^m} x_+^\gamma, \qquad \gamma \neq -1, -2, -3, \ldots. \tag{5.15}$$

Likewise,

$$x_-^\gamma (\ln |x|)^m \equiv \frac{\partial^m}{\partial \gamma^m} x_-^\gamma, \qquad \gamma \neq -1, -2, -3, \ldots, \tag{5.16}$$

$$|x|^\gamma (\ln |x|)^m \equiv \frac{\partial^m}{\partial \gamma^m} |x|^\gamma, \qquad \gamma \neq -1, -3, -5, \ldots, \tag{5.17}$$

and

$$|x|^\gamma \operatorname{sgn} x (\ln |x|)^m \equiv \frac{\partial^m}{\partial \gamma^m} (|x|^\gamma \operatorname{sgn} x), \qquad \gamma \neq -2, -4, -6, \ldots. \tag{5.18}$$

For $x \neq 0$ these GFs equal the ordinary function which the notation suggests. In the cases where γ approaches an integer, $x^{-n}(\ln|x|)^m$ can be defined using (5.17) or (5.18) for even or odd n, respectively.

3. x_+^{-n}

The restriction $\gamma \neq -1, -2, -3, \ldots$ in the definition of x_+^γ is awkward, since x_+^{-1}, $x_+^{-1} \ln x$, and the like would be desirable GFs to define. We can make progress only by relaxing some of our six criteria—or rather five, since 6 has already gone by the board.

Consider the divergence of $x_+^\gamma \langle \phi \rangle$ at $\gamma = -n$, where n is a positive integer, as a function of γ for some fixed arbitrary ϕ. The singularities are all simple poles ("$1/x$-type" singularities) since $(\gamma + n) x_+^\gamma$ is well behaved at $\gamma = -n$:

$$\lim_{\gamma \to -n_-} (\gamma + n) x_+^\gamma \langle \phi \rangle = \lim_{\gamma \to -n_-} \left\{ \prod_{j=1}^{n-1} (\gamma + j) \right\}^{-1} (x_+^{\gamma+n})^{(n)} \langle \phi \rangle$$

$$= \left\{ \prod_{j=1}^{n-1} (-n + j) \right\}^{-1} H^{(n)} \langle \phi \rangle$$

$$= \frac{(-1)^{n-1}}{(n-1)!} \delta^{(n-1)} \langle \phi \rangle = \frac{\phi^{(n-1)}(0)}{(n-1)!}. \quad (5.19)$$

If we think of $x_+^\gamma \langle \phi \rangle$ in terms of a Laurent expansion (Chapter 3, Section V.E, by Silverstone) about $\gamma = -n$, clearly the leading term is of order $(\gamma + n)^{-1}$. Let us then define x_+^{-n} as the next term of the expansion (i.e., the "constant" term):

$$x_+^{-n} \langle \phi \rangle \equiv \lim_{\gamma \to -n_-} \frac{\partial}{\partial \gamma} (\gamma + n) x_+^\gamma \langle \phi \rangle. \quad (5.20)$$

Defined in this way, x_+^{-n} has several of the properties we seek. For $n = 1$,

$$x_+^{-1} \langle \phi \rangle = \lim_{\gamma \to -1_-} \frac{\partial}{\partial \gamma} (\gamma + 1) x_+^\gamma \langle \phi \rangle = \lim_{\gamma \to -1_-} \frac{\partial}{\partial \gamma} (x_+^{\gamma+1})' \langle \phi \rangle$$

$$= -\frac{\partial}{\partial \gamma} \int_0^\infty x^{\gamma+1} \phi'(x) \, dx \bigg|_{\gamma = -1}$$

$$= -\int_0^\infty \ln x \, \phi'(x) \, dx = (\ln x_+)' \langle \phi \rangle, \quad (5.21)$$

just as demanded by requirement 3. However, for $n \neq 1$, condition 3 is not met. Instead, (5.20) leads to

$$(x_+^{-n})' = -nx_+^{-(n+1)} + \frac{(-1)^n}{n!} \delta^{(n)}. \tag{5.22}$$

This can be written as

$$x_+^{-(n+1)} = \frac{(-1)^n}{n!} \left\{ (x_+^{-1})^{(n)} + \delta^{(n)} \sum_{k=1}^{n} \frac{1}{k} \right\} \tag{5.23}$$

which, together with (5.21), is often a more convenient expression than the definition (5.20). Conditions 1 and 2 hold for x_+^{-n} but 4 is easily shown to fail. Finally, 5 holds, with

$$xx_+^{-1}\langle\phi\rangle = H\langle\phi\rangle \tag{5.24}$$

for $n = 1$.

In summary, definition (5.20) satisfies criteria 1, 2, and 5. A different subset of the six conditions could have been satisfied by using a definition slightly different from (5.20), but in any case some of the conditions will always be violated. In particular, x_+^{-n} cannot be a special case of the GF x_+^ν discussed earlier.

By analogy with the treatment in the preceding subsection, one can define x_-^{-n} and then proceed to $|x|^{-n}$ and $|x|^{-n} \operatorname{sgn} x$. If n is even (or odd), the cancellation of the singularities again takes place and $|x|^{-n}$ (or $|x|^{-n} \operatorname{sgn} x$, respectively) becomes identical with x^{-n} as defined in Section V.A.1 and satisfies requirements 1–5.

Finally, we can also pick out the constant term in the Laurent expansion of (5.15) and define

$$x_+^{-n}(\ln x)^m = \lim_{\gamma \to -n_-} \frac{\partial}{\partial \gamma}(\gamma + n)\frac{\partial^m}{\partial \gamma^m} x_+^\nu \tag{5.25}$$

and other related GFs.

B. Applications

1. *Finite Part of Divergent Integrals*

Direct applications of the GFs of Section V.A are not hard to find. Let us begin with some interesting and useful results obtained by evaluating definite integrals (Section IV.B.2) of the above GFs.

Consider, for example, the classically divergent integral

$$I_1 = \int_{-1}^{1} x^{-3} \, dx. \tag{5.26}$$

We may interpret the integrand as the GF x^{-3}, and the integration can then be performed since the latter equals the ordinary function x^{-3} in neighborhoods of $+1$ and -1. By (5.1) and (4.45) we have

$$I_1 = -\tfrac{1}{2}x^{-2}\Big|_{-1}^{1} = 0. \tag{5.27}$$

Clearly this is just the Cauchy principal value of (5.26).

A divergent integral need not have a Cauchy principal value in order for GF theory to assign it a value. For example, consider

$$I_2 = \int_0^a x^{-3/2}\,dx, \qquad a > 0, \tag{5.28}$$

which we may interpret and integrate as follows:

$$I_2 = \int_{-\infty}^a \cdot x_+^{-3/2}\,dx = -2x_+^{-1/2}\Big|_{-\infty}^{a} = -2/\sqrt{a}. \tag{5.29}$$

Actually, this result is just the Hadamard "finite part" (Hadamard, 1923) of the integral (5.28). Apparently the Hadamard finite part of an integral is a special case of the GF definite integral.

As a third example, consider the following generalization of (5.26). If $n > 1$ is an integer,

$$I_3 = \int_{-1}^1 x^{-n}\,dx = -(n-1)^{-1}x^{-n+1}\Big|_{-1}^{1} = \begin{cases} -2/(n-1), & n \text{ even,} \\ 0, & n \text{ odd.} \end{cases} \tag{5.30}$$

For the sake of comparison, one may integrate the function z^{-n} of the complex variable z from $z = -1$ to $z = 1$ along any contour that avoids the origin. For $n > 1$ the residue (see Chapter 3, Section VII.A, by Silverstone) of z^{-n} at the origin is zero. Thus the result is independent of the path and in fact is identical to (5.30). In other words, GF theory evaluates integrals like (5.30) consistently with complex variable theory but without the necessity of bypassing the singularity.

This last example is indicative of the close connection between GFs and complex variables; see, for example, Bremermann (1965). In general it is pleasing that GF theory fits into the structures of classical mathematics so well.

2. Physical Examples

The integrals mentioned in Section V.B.1 are not artificially devised examples. We can turn to electrostatics for perhaps the simplest physical examples.

Consider the problem of finding the electric field due to a (mathematical idealized) uniform line source of charge extending along the whole x axis. The potential in esu at $(0, y, z)$ due to an element of charge dq located at $(x, 0, 0)$ is

$$dU = dq/(r^2 + x^2)^{1/2} \tag{5.31}$$

where $r^2 = y^2 + z^2$. Taking the negative gradient of dU gives the electric field element of magnitude

$$dE = dq/(r^2 + x^2). \tag{5.32}$$

Now we can perform a vector integration on dE over all x, giving an electric field of magnitude

$$E = 2\lambda/r \tag{5.33}$$

where λ is the (constant) linear charge density. (Of course E points outward from the x axis). From (5.33) it is obvious that a potential

$$U = -2\lambda \ln r + \text{constant} \tag{5.34}$$

would give a correct description of the field, yet U cannot be evaluated from (5.31) directly since the integral

$$U = \int_{-\infty}^{\infty} \frac{\lambda \, dx}{(r^2 + x^2)^{1/2}} = 2\lambda \int_0^{\infty} \frac{dx}{(r^2 + x^2)^{1/2}} \tag{5.35}$$

diverges. The reader is probably aware that such a divergence arises in all electrostatics problems involving sources of infinite extent. The difficulty can be avoided, of course, by proceeding to the limit of infinite extent after, rather than before, performing the integration.

The theory of GFs provides a more elegant way of evaluating U directly. We could consider $(r^2 + x^2)^{-1/2}$ to be a GF of x with a parameter r. Since integration with respect to x commutes with differentiation with respect to a parameter [see (4.48)], we should be able to integrate (5.35) as a GF to obtain (5.34) and thus (5.33). First, however, let us change variables to $\alpha = x^{-1}$, giving

$$U = 2\lambda \int_0^{\infty} \alpha^{-1}(r^2\alpha^2 + 1)^{-1/2} \, d\alpha \tag{5.36}$$

which we interpret as the GF definite integral

$$U = 2\lambda \int_a^\infty \alpha_+^{-1}(r^2\alpha^2 + 1)^{-1/2}\,d\alpha, \qquad a < 0. \tag{5.37}$$

The integrand is simply a smooth multiplying function times the GF α_+^{-1} of Section V.A.3. Its primitive is the regular GF

$$\ln \alpha_+ - \ln(1 + (r^2\alpha^2 + 1)^{1/2})_+ + \text{constant}, \tag{5.38}$$

where, as in (5.3), the subscript "+" denotes that the function is evaluated as shown for positive α but equals zero for negative α. Finally, using (4.45), (5.37) becomes

$$U = 2\lambda \lim_{\alpha \to \infty} \ln \frac{\alpha}{1 + (r^2\alpha^2 + 1)^{1/2}} = 2\lambda \ln |r|^{-1} = -2\lambda \ln r, \tag{5.39}$$

which agrees with (5.34), as expected. We have already seen in Section IV.A.3 that (5.39) satisfies the appropriate Laplace's equation.

The case of the uniform infinite sheet of charge at $z = 0$ is similar to the example above. In that case the integral analogous to (5.35)–(5.37) is

$$U = 2\pi\lambda \int_0^\infty (x\,dx/(r^2 + x^2)^{1/2}) = 2\pi\lambda \int_0^\infty \alpha^{-2}(r^2\alpha^2 + 1)^{-1/2}\,d\alpha$$

$$= 2\pi\lambda \int_a^\infty \alpha^{-2}(r^2\alpha^2 + 1)^{-1/2}\,d\alpha, \qquad a < 0. \tag{5.40}$$

The primitive in this case is $-\alpha_+^{-1}(r^2\alpha^2 + 1)^{1/2} - \delta$, so that

$$U = -2\pi\lambda |r|, \tag{5.41}$$

which gives the field expected from elementary considerations. Incidentally, setting $r = 0$ in the first form of (5.40) gives the GF integral

$$\int_0^\infty dx = 0.$$

3. Illustration by Fundamental Sequences

We can use the fundamental sequences introduced in Section II.D to examine in detail how the classically divergent integrals we have been evaluating are handled in the GF theory. We use the fact that if a fundamental sequence $\{f_m\}$ represents a GF f, then $\{f_m'\}$, provided it exists and is also fundamental, represents f'. Furthermore, if $\int_a^b f\,dx$ can be defined, then it equals the limit of $\int_b^a f_m\,dx$ as $m \to \infty$.

Consider the definite integral of x^{-2} from -1 to $+1$, which equals -2, according to (5.30). Now x^{-2} is the second derivative of $-\ln x$, which is a regular GF. We can therefore choose a suitable fundamental sequence for $-\ln x$ and examine its second derivative in order to get a feeling for the behavior of x^{-2}. A convenient choice is

$$f_m = \begin{cases} -\ln x, & |x| \geq m^{-1}, \\ \ln m + \tfrac{1}{2}(1 - m^2 x^2), & |x| < m^{-1}. \end{cases} \quad (5.42)$$

Several members of $\{f_m\}$ are shown in Fig. 2a. Figures 2b and 2c show the corresponding first and second derivatives, respectively. Clearly $\{f_m''\}$, the fundamental sequence for x^{-2}, is building up a large negative peak at $x = 0$, compensating for the large positive values of x^{-2} and permitting

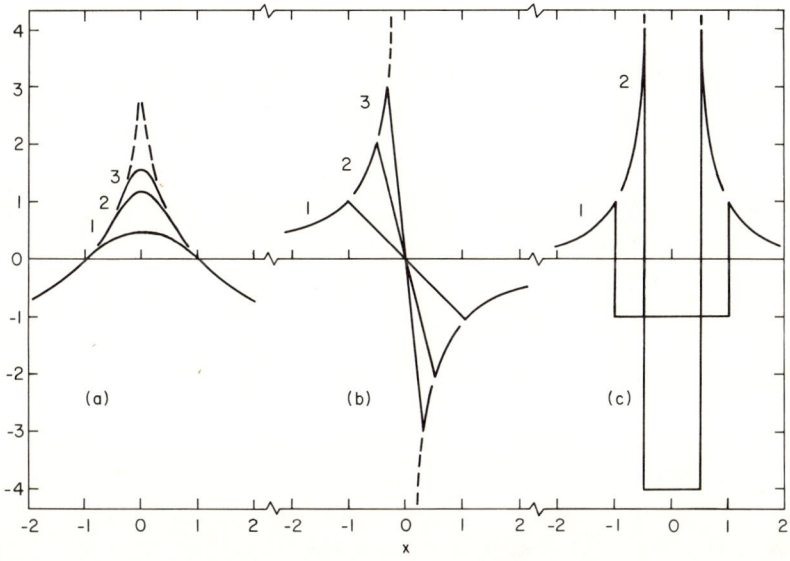

FIG. 2. Several members of the sequences (a) $\{f_m\}$, (b) $\{f_m'\}$, and (c) $\{f_m''\}$, where $\{f_m\}$ is defined in (5.42). Each curve is labeled by the corresponding value of m.

integrals across the singularity to converge. In fact, the peak is such that the total area under each of the f_m'' is exactly zero. A different choice in (5.42) would alter the appearance of Fig. 2 only in its details: The mechanism for convergence *must* occur at the origin, since f_m'' must tend to the ordinary function x^{-2} for $x \neq 0$, and therefore some sort of negative peak must occur near $x = 0$.

VI. Fourier Transforms

A. Fourier Transforms of Ordinary Functions

Before investigating the possibility of Fourier transforming generalized functions, we review the theory of Fourier transformation of ordinary functions. For most of this section we confine ourselves to the case of one variable.

The Fourier transform of $f(x)$, a complex-valued function of one real variable, is defined to be the function

$$\tilde{f}(q) \equiv \int_{-\infty}^{\infty} f(x)e^{iqx}\,dx \qquad (6.1)$$

provided that the integral exists. [We also use the notation $F_q[f(x)]$ for $\tilde{f}(q)$.] The existence of (6.1) is assured if

$$\int_{-\infty}^{\infty} |f(x)|\,dx \qquad (6.2)$$

converges, i.e., if f is absolutely integrable. The classical Riemann–Lebesgue lemma assures us that if (6.2) converges, then $\tilde{f}(q)$ must tend to zero as $|q| \to \infty$. If $f(x)$ is even or odd, respectively, (6.1) becomes a cosine transform

$$\tilde{f}(q) = 2\int_0^\infty f(x)\cos qx\,dx \qquad (6.3)$$

or a sine transform

$$\tilde{f}(q) = 2i\int_0^\infty f(x)\sin qx\,dx. \qquad (6.4)$$

Much of the usefulness of Fourier transforms lies in the fact that a large class of functions $f(x)$ can be reconstructed from their transforms $\tilde{f}(q)$ using the *Fourier inversion theorem*

$$f(x) = (2\pi)^{-1}\int_{-\infty}^\infty \tilde{f}(q)e^{-iqx}\,dq. \qquad (6.5)$$

This can be combined with the definition of $\tilde{f}(q)$ and written as

$$f(x) = (2\pi)^{-1}\int_{-\infty}^\infty e^{-iqx}\left[\int_{-\infty}^\infty f(x')\exp(iqx')\,dx'\right]dq$$
$$= (2\pi)^{-1}\int_{-\infty}^\infty \int_{-\infty}^\infty f(x')\exp(iq(x-x'))\,dx'\,dq, \qquad (6.6)$$

or more symbolically as

$$\tilde{\tilde{f}}(x) = F_x[F[f]] = 2\pi f(-x). \tag{6.7}$$

Except for the change in sign and the factor of 2π, the operator F would be its own inverse; f and $\tilde{f}$ are referred to as a *Fourier transform pair*. If $f(x)$ is an even function, (6.6) simplifies to

$$\begin{aligned} f(x) &= \pi^{-1} \int_0^\infty \int_{-\infty}^\infty f(x') \cos q(x' - x) \, dx' \, dq \\ &= 2\pi^{-1} \int_0^\infty \int_0^\infty f(x') \cos qx' \cos qx \, dx' \, dq. \end{aligned} \tag{6.8}$$

The question of necessary and sufficient conditions for the validity of (6.6) is quite involved. Several different sets of sufficient conditions are commonly discussed in textbooks on integral transforms (Titchmarsh, 1937; Sneddon, 1951) and advanced calculus (Apostol, 1957). The relationship among the various sets of conditions is not simple: Functions can be found which obey one but not another. For our present purposes it is sufficient to know that (6.6) is true if f is absolutely integrable and if f' exists at x.

The result (6.6) is easily made plausible. The right side can be written as

$$(2\pi)^{-1} \lim_{B\to\infty} \lim_{A\to\infty} \int_{-A}^A f(x') \left[\int_{-B}^B \exp(iq(x' - x)) \, dq \right] dx'$$

$$= \pi^{-1} \lim_{B\to\infty} \lim_{A\to\infty} \int_{-A}^A f(x') \frac{\sin B(x' - x)}{x' - x} \, dx'$$

$$= \pi^{-1} \lim_{B\to\infty} \lim_{A\to\infty} \int_{-BA-Bx}^{BA-Bx} f\left(x + \frac{t}{B}\right) \frac{\sin t}{t} \, dt. \tag{6.9}$$

Theorems on integration can now be used to justify the following heuristic evaluation of this integral. As $B \to \infty$, $f(x + t/B) \to f(x)$ for any finite t, and since only a region near $t = 0$ contributes substantially to the integral, the function f becomes a constant as far as the integration is concerned, giving

$$\pi^{-1} f(x) \int_{-\infty}^\infty t^{-1} \sin t \, dt = f(x) \tag{6.10}$$

as desired. For a proof based on complex variable theory, see Chapter 3, Section IX, by Silverstone.

TABLE I

TABLE OF FOURIER TRANSFORMS[a]

Function	Transform	
$f(x)$	$\tilde{f}(q) = F_q[f(x)] = \int_{-\infty}^{\infty} f(x) e^{iqx}\, dx$	(I. 1)
$f^{(m)}(x)$	$(-iq)^m \tilde{f}(q)$	(I. 2)
$x^m f(x)$	$(-i)^m d^m \tilde{f}(q)/dq^m$	(I. 3)
$f(x + a)$	$e^{-iqa}\tilde{f}(q)$	(I. 4)
$e^{-ibx} f(x)$	$\tilde{f}(q - b)$	(I. 5)
$f(ax)$	$\lvert a \rvert^{-1} \tilde{f}(q/a)$	(I. 6)
$(f * g)(x)$	$\tilde{f}(q)\tilde{g}(q)$	(I. 7)
$2\pi f(x) g(x)$	$(\tilde{f} * \tilde{g})(q)$	(I. 8)

[a] In this table, m is a positive integer, a and b are real constants.

Table I summarizes some Fourier transform results that will be useful in the next few pages. The required transforms are assumed to exist in each case. The proofs are not difficult; for example, (I.2) is obtained through integration by parts:

$$F_q[f'(x)] = \int_{-\infty}^{\infty} f'(x) e^{iqx}\, dx$$
$$= f(x) e^{iqx} \Big|_{-\infty}^{\infty} - \int_{-\infty}^{\infty} iq e^{iqx} f(x)\, dx = -iq\tilde{f}(q), \quad (6.11)$$

and the result

$$F_q[f^{(m)}] = (-iq)^m \tilde{f}(q) \quad (6.12)$$

follows by induction. Notice that (I.6) includes the useful result

$$F_q[f(-x)] = F_{-q}[f(x)].$$

B. THE SPACES S AND S'

1. Rapidly Decreasing Functions, S

By now the reader will have been conditioned to expect Fourier transforms of GFs to be defined in terms of transforms of the test functions. Indeed this is the route we shall follow, but one obstacle must first be overcome. D is not closed under Fourier transformation: If ϕ has a bounded support, $\tilde{\phi}$ cannot also have bounded support and hence is not in D. It is therefore necessary to enlarge our set of test functions and to define GFs on that set before proceeding further.

A set of test functions well suited for Fourier transformation is the set of *rapidly decreasing functions*, denoted by S (or by S_n to indicate n dimensions). For ϕ to be in S_n,

1. ϕ must be a real-valued function of n real variables having (partial) derivatives of all orders, and

2. ϕ and all its derivatives must decrease more rapidly than any power of the Cartesian coordinates x_i as $|\mathbf{r}| \to \infty$. In other words, for any two sets of n nonnegative integers $\{m_i\}$ and $\{k_i\}$, $i = 1, 2, \ldots, n$,

$$\left(\prod_{i=1}^{n} x_i^{m_i}\right) D^{\{k\}} \phi(\mathbf{r}) \to 0 \qquad \text{as} \quad |\mathbf{r}| \to \infty, \tag{6.13}$$

where $D^{\{k\}}$ is the differential operator

$$D^{\{k\}} = \frac{\partial^{\Sigma k_i}}{\partial x_1^{k_1} \cdots \partial x_n^{k_n}}. \tag{6.14}$$

Condition 2 relaxes the "bounded support" condition for functions in D while still restricting the behavior for large r. Clearly D is a proper subset of S. For instance, $\exp(-x^2)$ is easily shown to belong to S but certainly is not in D.

The introduction of S is justified by the following two results. First, all ϕ in S can be Fourier transformed, and the Fourier inversion theorem holds. Second, the real and imaginary parts of $\tilde{\phi}$ are in S_1 for any ϕ in S_1. This is shown as follows:

$$\frac{d^k}{dq^k} \tilde{\phi}(q) = \frac{d^k}{dq^k} \int_{-\infty}^{\infty} \phi(x) e^{iqx}\, dx = i^k \int_{-\infty}^{\infty} x^k \phi(x) e^{iqx}\, dx, \tag{6.15}$$

which can be integrated by parts m times to give

$$q^m \tilde{\phi}^{(k)}(q) = i^{k-m} \int_{-\infty}^{\infty} \frac{d^m}{dx^m}(x^k \phi(x)) e^{iqx}\, dx. \tag{6.16}$$

The transform on the right exists and, by the Riemann–Lebesgue lemma, approaches zero for large $|q|$. Since m and k were arbitrary nonnegative integers, condition 2 above is satisfied by both the real and imaginary parts of $\tilde{\phi}$. It is nearly trivial that they also satisfy 1, and therefore both the real and imaginary parts of $\tilde{\phi}$ are themselves test functions in S_1.

S and D have many properties in common. If ϕ_1 and ϕ_2 are in S_n, then ϕ_1', $\phi_1 + \phi_2$, and $\phi_1\phi_2$ are also in S_n. Also $\alpha\phi_1$ will be in S_n where α is any infinitely differentiable function of n real variables that does not disturb the large $|\mathbf{r}|$ behavior of ϕ_1 upon multiplication.

S is a linear space. Convergence of a sequence $\{\phi_j\}$ in S_n is defined as the uniform convergence of

$$\left(\prod_{i=1}^{n} x_i^{m_i}\right) D^{\{k\}} \phi_j(\mathbf{r}), \tag{6.17}$$

for any sets $\{m\}$ and $\{k\}$ as in (6.13), as $j \to \infty$. Although S contains D, D is dense in S; that is, any ϕ in S_n is the limit of some sequence $\{\phi_m\}$ in D_n.

2. Generalized Functions of Slow Growth, S'

We are now in a position to define a new set of GFs that are functionals on the space S. Let us begin by considering any real-valued function of n real variables, $f(\mathbf{r})$, that is dominated by powers of the x_i; i.e., there exists some set of positive integers $\{m_i\}$, $i = 1, 2, \ldots, n$, such that

$$\left(\prod_{i=1}^{n} x_i^{-m_i}\right) f(\mathbf{r}) \to 0 \quad \text{as} \quad |\mathbf{r}| \to \infty. \tag{6.18}$$

Such a function is said to be of *slow growth*, and, if locally integrable, will define on S_n the functional

$$f\langle\phi\rangle = \int f(\mathbf{r})\phi(\mathbf{r}) \, dV, \tag{6.19}$$

where ϕ is an arbitrary member of S_n. This functional is called a *regular GF of slow growth*. More generally, *any* real-valued continuous linear functional S_n is called an n-dimensional *GF of slow growth*,[†] and the space of these is denoted by S_n'. We shall refer to the S_n' collectively as S'. All these definitions are entirely parallel to those for D'.

S' is actually a proper subset of D'; for example, the function e^x defines a GF on D but not on S since e^x is not bounded by any polynomial in x. However, all GFs on D having bounded support can also be evaluated on S. All the operations we have defined for GFs in D' hold essentially without change for GFs in S'. S' is a linear space in its own right and is dense in D'. Convergence of sequences of GFs in S' is defined exactly as in D'.

[†] Many authors refer to this as a *tempered* GF.

3. Extension to Complex Values

Since Fourier transforms intrinsically involve complex-valued functions, it is helpful to extend S and S' accordingly. Therefore let us change condition 1 in Section VI.B.1 to include *complex*-valued functions of n real variables as test functions in S_n. This does not involve us in complex variable theory since the independent variables are still real. It simply means that we take all possible pairs (ϕ_1, ϕ_2) of the original real-valued test functions and combine them to form the complex test function

$$\phi = \phi_1 + i\phi_2 \qquad (6.20)$$

where $i^2 = -1$. A GF evaluated at ϕ is then just

$$f\langle\phi\rangle = f\langle\phi_1\rangle + if\langle\phi_2\rangle, \qquad (6.21)$$

so that GFs become mappings from the test functions onto the whole complex plane. This extension also has the effect that the transform of any member of S_1 (and *a fortiori* of D_1) is also in S_1: the operator F maps S_1 onto itself.

The GFs themselves are subject to this extension also. In defining regular GFs in S_n' we admit complex-valued locally integrable functions $f(x_1, x_2, \ldots, x_n)$ and form the functional[†]

$$f\langle\phi\rangle = \int f(\mathbf{r})\phi(\mathbf{r})\,dV. \qquad (6.22)$$

In general, we can combine *any* two real members of S_n', f_1 and f_2, giving the complex GF $f = f_1 + if_2$, whose evaluation at ϕ is

$$f\langle\phi\rangle = f_1\langle\phi_1\rangle - f_2\langle\phi_2\rangle + i[f_1\langle\phi_2\rangle + f_2\langle\phi_1\rangle]. \qquad (6.23)$$

The extension to complex values may be applied to D and D' as well as to S and S'. In both cases, the results we have previously obtained are left unchanged. The operation of complex conjugation, denoted by an asterisk, can now be defined for GFs:

$$f^*\langle\phi\rangle \equiv [f\langle\phi^*\rangle]^*. \qquad (6.24)$$

[†] In (6.22) one may replace the f in the integrand by f^*, the complex conjugate of f, in order to enhance the analogy between $f\langle\phi\rangle$ and the inner product in Hilbert space. However, a number of our results would then be slightly different, and actually a notation like $\langle f, \phi \rangle$ would be more appropriate to this development.

C. Fourier Transforms of Generalized Functions in S'

1. Definition

One could define Fourier transforms of GFs by starting with the classical definition (6.1) and the definition of definite integrals of GFs. However, the following route is faster and more general.

Consider a regular GF of slow growth, $\tilde{f}$, formed from a Fourier transformable function, $f(x)$. Now $\tilde{f}$ gives rise to the GF

$$\tilde{f}\langle\phi\rangle = \int_{-\infty}^{\infty} \tilde{f}(q)\phi(q)\,dq \tag{6.25}$$

which may be rewritten as[†]

$$\int_{-\infty}^{\infty}\left\{\int_{-\infty}^{\infty} f(x)e^{iqx}\,dx\right\}\phi(q)\,dq = \int f(x)\left\{\int \phi(q)e^{iqx}\,dq\right\}dx$$

$$= \int f(x)\tilde{\phi}(x)\,dx = f(x)\langle\tilde{\phi}(x)\rangle. \tag{6.26}$$

This result is a form of Parseval's theorem, and plays the same role for Fourier transformation as the integration by parts did for differentiation: We use

$$\tilde{f}\langle\phi\rangle \equiv f(x)\langle\tilde{\phi}(x)\rangle \tag{6.27}$$

to define $\tilde{f}$, the Fourier transform of any GF, $f\langle\phi\rangle$, in S_1'.

Since the Fourier transforms of all ϕ in S_1 are again in S_1, (6.27) is defined for all f in S_1'; that is, every member of S_1' has a Fourier transform and the latter is again in S_1'. For the special case of regular GFs, we have thus been led to a generalization of the ordinary notion of Fourier transform that does not require absolute integrability.

The GF counterpart of the Fourier inversion theorem is just as pleasant as the "existence" result in the last paragraph. Evaluating $\tilde{\tilde{f}}$, using (6.27), gives

$$\tilde{\tilde{f}}\langle\phi(x)\rangle = \tilde{f}(x)\langle\tilde{\phi}(x)\rangle = f(x)\langle\tilde{\tilde{\phi}}(x)\rangle = 2\pi f(x)\langle\phi(-x)\rangle, \tag{6.28}$$

[†] In this derivation and those that follow, it is often best to divorce the symbols x and q from their usual connotations of distance and wave number. The independent variable of a test function is ultimately a dummy variable, and both x and q should be regarded in this light.

so that the inversion theorem

$$\tilde{\tilde{f}}(x) = 2\pi f(-x) \tag{6.29}$$

is valid for *all* GFs in S_1'.

2. Properties of the Fourier Transform

Many properties of the Fourier transform of ordinary functions are also valid for GFs. In fact, *all* the entries in Table I hold for GFs, and can be verified using rather simple manipulations.[†] For example, to establish the validity of (I.2), we proceed as follows:

$$\begin{aligned}F_q[f^{(m)}]\langle\phi(q)\rangle &= f^{(m)}(q)\langle\tilde{\phi}(q)\rangle = (-1)^m f(q)\langle\tilde{\phi}^{(m)}(q)\rangle \\ &= (-1)^m f(q)\langle F_q[(ix)^m\phi(x)]\rangle \\ &= \tilde{f}(q)\langle(-iq)^m\phi(q)\rangle = (-iq)^m\tilde{f}(q)\langle\phi(q)\rangle. \end{aligned} \tag{6.30}$$

The proofs clearly hinge on the validity of Table I for the test functions themselves.

Given any f in S_1', $x^{-m}f(x)$ and $f^{(-m)}$, $m > 0$, are also well-defined members of S_1' and can therefore be Fourier transformed. This remark suggests that (I.2) and (I.3) in Table I may be extended to negative integer values of m when applied to GFs. This is in fact true. Letting $f^{(-m)} = g$, where m is a positive integer, we apply (I.2) to $g^{(m)}$, giving

$$\tilde{f}(q) = F_q[g^{(m)}] = (-iq)^m\tilde{g}(q) = (-iq)^m F_q[f^{(-m)}]. \tag{6.31}$$

To cast this in the form of (I.2), we could divide both sides by $(-iq)^m$, but not without adding m terms of the form $c_k\, \delta^{(k)}(q)$. The generalization of (I.3) for $m < 0$ follows a similar pattern.

It is simple to prove that Fourier transformation, just like differentiation, is a continuous, linear operation. It follows that any sequence or series of GFs convergent in S_1' can be Fourier transformed term by term, and the limit of the resulting sequence or series will be the transform of the original limit. Likewise, if a GF in S_1' is represented by a fundamental sequence, the transformed sequence, if it exists, will represent the transform of the GF.

[†] With the exception of (I.7) and (I.8); to prove these one needs the direct product, which we have not introduced here. Notice also the conditions required for the existence of the convolution product (Section III.C.2).

D. Particular Results

Having established some general properties of GF Fourier transforms, let us derive a few particular results. First we transform δ:

$$\tilde{\delta}\langle\phi\rangle = \delta\langle\tilde{\phi}\rangle = \tilde{\phi}(0) = \int_{-\infty}^{\infty} \phi(x)\,dx = 1\langle\phi\rangle, \tag{6.32}$$

and therefore, by the inversion theorem,

$$\tilde{1} = 2\pi\,\delta. \tag{6.33}$$

This justifies the equation

$$\int_{-\infty}^{\infty} e^{iqx}\,dx = 2\pi\,\delta(q), \tag{6.34}$$

which is often called an orthogonality relation for the exponential functions. Several useful results can be obtained by combining (6.33) with the entries in Table I. Using (I.3) gives

$$F_q[x^m]\langle\phi(q)\rangle = (-i)^m\left(\frac{d^m}{dq^m}\,\tilde{1}(q)\right)\langle\phi(q)\rangle = 2\pi(-i)^m\,\delta^{(m)}\langle\phi\rangle$$
$$= 2\pi i^m \phi^{(m)}(0), \qquad m \geq 0, \tag{6.35}$$

whose inverse is

$$F_q[\delta^{(m)}] = (-iq)^m, \qquad m \geq 0. \tag{6.36}$$

From (I.5) and (6.33) one obtains

$$F_q[e^{ibx}] = 2\pi\,\delta(q+b), \tag{6.37}$$

so that

$$F_q[\cos bx] = \tfrac{1}{2}F_q[e^{ibx} + e^{-ibx}] = \pi[\delta(q+b) + \delta(q-b)] \tag{6.38}$$

and

$$F_q[\sin bx] = -i\pi[\delta(q+b) - \delta(q-b)]. \tag{6.39}$$

Needless to say, these functions are not Fourier transformable in the ordinary sense.

2. Generalized Functions

Now let us try to transform the negative integer powers of x, beginning with x^{-1} (Section III.D.2):

$$F_q[x^{-1}]\langle\phi(q)\rangle = x^{-1}\langle\tilde{\phi}(x)\rangle = \int_0^\infty x^{-1}[\tilde{\phi}(x) - \tilde{\phi}(-x)]\,dx$$

$$= 2i \int_0^\infty x^{-1} \int_{-\infty}^\infty \phi(x') \sin xx'\,dx'\,dx. \qquad (6.40)$$

The two integrations may be interchanged, leaving a sign function, so that

$$F_q[x^{-1}] = \pi i \operatorname{sgn} q. \qquad (6.41)$$

From this result we immediately obtain

$$F_q[x^{-(m+1)}] = F_q\left[\frac{(-1)^m}{m!}(x^{-1})^{(m)}\right] = \frac{\pi i (iq)^m}{m!} \operatorname{sgn} q, \qquad m \geq 0. \qquad (6.42)$$

The inversion theorem can be applied to (6.42) to give

$$F_q[x_+^m] = i^{m+1} m! q^{-(m+1)} + \pi(-i)^m \delta^{(m)}, \qquad m \geq 0, \qquad (6.43)$$

with x_+^0 to be interpreted as $H(x)$. The results for x_-^m are similar, and will yield (6.35) when combined with (6.43).

Now let us turn to nonintegral powers of x. Since x_+^γ is the limit in S_1' of $e^{-\lambda x}x_+^\gamma$ as $\lambda \to 0_+$, we can evaluate the transform of the latter and then take the indicated limit. This procedure makes the problem tractable. In fact, for $\gamma > -1$, $e^{-\lambda x}x_+^\gamma$ is a regular GF and can be transformed classically. The result can be written in a closed form using the factorial function:[†]

$$\int_0^\infty x^\gamma e^{-\lambda x} e^{iqx}\,dx = \int_0^\infty x^\gamma \exp(ix(q+i\lambda))\,dx = i^{\gamma+1}\gamma!(q+i\lambda)^{-(\gamma+1)}. \qquad (6.44)$$

Taking the limit $\lambda \to 0_+$ gives

$$F_q[x_+^\gamma] = i\gamma![i^\gamma q_+^{-(\gamma+1)} - (-i)^\gamma q_-^{-(\gamma+1)}]. \qquad (6.45)$$

By induction it is then simple to show that (6.45) is actually valid for all γ (except negative integers, of course).

For the negative integer exponents, we can transform (5.20); for x_+^{-1} this will give

$$F_q[x_+^{-1}] = -C_E + \tfrac{1}{2}\pi i \operatorname{sgn} q - \ln|q|. \qquad (6.46)$$

[†] $\gamma! \equiv \Gamma(\gamma+1) \equiv \int_0^\infty x^\gamma e^{-x}\,dx, \qquad \gamma > -1; \qquad \gamma! = \gamma(\gamma-1)!.$

The Euler constant $C_E = 0.5772\ldots$ is the negative of the derivative of $\gamma!$ at $\gamma = 0$. From (6.46) and (5.23) we can obtain

$$F_q[x_+^{-(m+1)}] = \frac{(iq)^m}{m!}\left[-C_E + \sum_{k=1}^{m} k^{-1} + \tfrac{1}{2}\pi i \operatorname{sgn} q - \ln|q|\right]. \quad (6.47)$$

These relations can be combined and inverted to give a large variety of other results. For example, (6.46) gives

$$F_q[|x|^{-1}] = F_q[x_+^{-1} + x_-^{-1}] = -2(C_E + \ln|q|) \quad (6.48)$$

to which we can apply the inversion theorem to get

$$F_q[\ln|x|] = -\pi[|q|^{-1} + 2C_E \delta(q)]. \quad (6.49)$$

From (6.46),

$$\begin{aligned}F_q[\ln x_+] = F_q[(x_+^{-1})^{(-1)}] &= -iq^{-1}(C_E + \ln|q|) \\ &\quad - \tfrac{1}{2}\pi|q|^{-1} - C_E\pi\,\delta(q),\end{aligned} \quad (6.50)$$

where the arbitrary constant of integration is fixed by a comparison with (6.49).

The results of this section, together with Table I, should enable the reader to find Fourier transforms of many of the one-dimensional cases he will encounter.

As a sample application of these results, reference may be made to the analytic evaluation of certain integrals in molecular quantum mechanics (Silverstone and Todd, 1971). There one wishes to find the three-dimensional Fourier transform of a power of r times a spherical harmonic. This can be expressed in terms of the one-dimensional transforms of $|x|^m$ for various integer values of m. The answer can easily be assembled from the results obtained above. For positive powers, one gets

$$F_q[|x|^{2m}] = (-1)^m 2\pi\,\delta^{(2m)}(q), \qquad m \geq 0, \quad (6.51)$$

$$F_q[|x|^{2m-1}] = (-1)^m 2(2m-1)! q^{-2m}, \qquad m \geq 1; \quad (6.52)$$

for negative powers,

$$F_q[|x|^{-2m}] = (-1)^m \pi |q|^{2m-1}/(2m-1)!, \qquad m \geq 1, \quad (6.53)$$

$$F_q[|x|^{-(2m+1)}] = (-1)^m 2q^{2m}\left(-C_E + \sum_{k=1}^{2m} k^{-1} - \ln|q|\right)\!\bigg/(2m)!,$$

$$m \geq 0. \quad (6.54)$$

E. Transforms of Periodic Generalized Functions

1. Fourier Series

Many authors discuss periodic GFs and their Fourier analysis by introducing a new space containing periodic test functions, defining functionals on this space, and then developing an analog of the classical Fourier series. However, we shall obtain Fourier series more directly as a special case of the Fourier transform, as follows.

Consider f, a GF of slow growth in S_1', and let f be periodic:

$$f(x + a) = f(x), \tag{6.55}$$

where a is the period. Periodicity imposes the condition

$$(1 - e^{-iqa})\tilde{f}(q) = 0 \tag{6.56}$$

on the transform $\tilde{f}$, as may be seen by transforming both sides of (6.55). Clearly $\tilde{f}(q)$ equals zero in open intervals not containing the points $q_m = 2\pi m/a$, $m = 0, \pm 1, \pm 2, \ldots$. In fact, $1 - e^{-iqa} \sim ia(q - q_m)$ when $q \simeq q_m$, so that we have, locally, a problem of division by a factor linear in q. Hence (6.56) implies that $\tilde{f}(q)$ must be a sum of the form $\sum_m c_m \delta(q - q_m)$.

The following alternative argument may make this division problem clearer. Obviously (6.55) and (6.56) must hold for a replaced by $-a$. This is equivalent to insisting that both

$$(1 - \cos qa)\tilde{f}(q) = 0 \tag{6.57}$$

and

$$(\sin qa)\tilde{f}(q) = 0 \tag{6.58}$$

be satisfied. According to the identity

$$\sin x = x \prod_{m=1}^{\infty} (1 - x^2/\pi^2 m^2) = x \prod_{m=1}^{\infty} (1 - x/\pi m)(1 + x/\pi m), \tag{6.59}$$

$\sin qa$ can be regarded as a "polynomial of infinite order," and solving (6.58) reduces to division by the factors $1 \pm x/(\pi m)$. Assuming the infinity of factors to present no problem, the solution is

$$c_0 \delta(q) + \sum_{m=1}^{\infty} [c_m \delta(q - \pi m/a) + c_{-m} \delta(q + \pi m/a)] \tag{6.60}$$

where the c_m are arbitrary. In order that (6.57) be satisfied also, the odd-numbered c's must be zero. Thus periodicity imposes the form

$$\tilde{f}(q) = 2\pi \sum_{m=-\infty}^{\infty} C_m \delta(q - q_m), \qquad q_m = 2\pi m/a \qquad (6.61)$$

on $\tilde{f}$; the factor of 2π has been inserted for later convenience.

Applying the inversion theorem to (6.61) gives

$$f(-x) = \sum_m C_m F_x[\delta(q - q_m)] \qquad (6.62)$$

or

$$f(x) = \sum_{m=-\infty}^{\infty} C_m \exp(-iq_m x). \qquad (6.63)$$

A series of this form is called a *Fourier series* (written in complex form); it is the discrete analog of (6.5). We realize that we have just shown that *every* periodic GF in S_1' can be written as a Fourier series. Fourier series have already been mentioned in connection with the convergence of series (Section II.C.2).

To evaluate the C_m explicitly we can multiply both sides of (6.63) by $\exp(iq_{m'}x)$ and integrate over one period (whose endpoints are chosen to make our method of definite integration applicable). Equivalently, we could multiply (6.61) by $k^{-1} \sin k$ where $k = \frac{1}{2}a(q - q_{m'})$, to pick term m' out of the sum. In either case we get

$$C_m = a^{-1} \int_{\text{one period}} f(x) \exp(iq_m x)\, dx, \qquad (6.64)$$

which is exactly the usual formula for the Fourier coefficients.

2. Examples

As an example, consider the periodic member of S_1' defined by

$$f(x) = \sum_{k=-\infty}^{\infty} (x - k)^{-2}. \qquad (6.65)$$

According to (6.64) the Fourier coefficients are

$$C_m = \int_{-1/2}^{1/2} \sum_k (x - k)^{-2} e^{2\pi i m x}\, dx = \int_{-\infty}^{\infty} x^{-2} e^{2\pi i m x}\, dx = -2\pi^2 \mid m \mid, \qquad (6.66)$$

so that the Fourier series for f can be written as

$$f(x) = -4\pi^2 \sum_{m=1}^{\infty} m \cos 2\pi m x. \quad (6.67)$$

As a series of ordinary functions, (6.67) is obviously divergent for all x; yet as a series of GFs it converges in S'. In fact, a numerical check shows that if (6.67) is evaluated at the unit area test function

$$\phi(x) = 10^2 \pi^{-1/2} \exp[-10^4(x - \tfrac{1}{2})^2],$$

the result is $1.000494\ldots \pi^2$, and as more highly peaked test functions are used the result approaches

$$f(\tfrac{1}{2}) = 2 \sum_{k=1}^{\infty} (k - \tfrac{1}{2})^{-2} = 8 \sum_{k=1}^{\infty} (2k - 1)^{-2} = \pi^2. \quad (6.68)$$

Furthermore, since differentiation does not destroy the convergence of a series we immediately have results like

$$f'(x) = \sum_{k=-\infty}^{\infty} (-2)(x - k)^{-3} = 8\pi^3 \sum_{m=1}^{\infty} m^2 \sin 2\pi m x. \quad (6.69)$$

As an ordinary function, of course, (6.65) cannot be Fourier analyzed because of its singular behavior. As another example, the reader may wish to verify that the Fourier series for $\csc x = (\sin x)^{-1}$, defined as the derivative of the regular GF $\ln |\tan \tfrac{1}{2} x|$, is

$$\csc x = 2(\sin x + \sin 3x + \sin 5x + \cdots). \quad (6.70)$$

In dealing with classical Fourier transforms and series one soon finds empirically that a function with a discontinuous kth derivative (but with continuous lower order derivatives) exhibits a q^{-k-1} behavior in its transform (or m^{-k-1} in its Fourier coefficients). Notice that the series above fit this pattern, with k being negative. Lighthill (1958) makes these notions precise by giving several theorems on the asymptotic behavior of Fourier transforms and series.

3. Discrete Fourier Transforms

Equation (6.61) shows that the Fourier transform of a periodic function f is a row of equally spaced δ's, their amplitudes being determined by the properties of f. The inverse situation, (6.62), is a matter of some practical interest, since the row of δ's is a numerically convenient approximation to a continuous function, as discussed in Section II.C.3. From (6.62) we can immediately conclude that the Fourier transform of $g_\varDelta$ in (2.34) is a periodic GF having a period of $2\pi/\varDelta$. This transform is often referred to as a discrete Fourier transform of the function $f(x)$ being approximated in (2.34). Usually the numerical analyst must be careful to ensure that $|q| < \pi/\varDelta$ in such work, since the results for $|q| > \pi/\varDelta$ are spurious to the problem of obtaining a good approximation to the transform of $f(x)$ itself.

An amusing situation occurs when the amplitudes of the δ's [i.e., the $f(x_m)$] are themselves periodic with some period $M\varDelta$, so that $f(x_m) = f(x_{m+M})$. The Fourier transform of such a GF is also an equally spaced row of δ's with periodic amplitudes. The period is $2\pi/\varDelta$, as before, and the spacing of the δ's is $2\pi/(M\varDelta)$. A very special case is the "comb" of δ's, whose transform is another comb:

$$F_q\left[\sum_{m=-\infty}^{\infty} \delta(x - m\varDelta)\right] = 2\pi\varDelta^{-1} \sum_{m=-\infty}^{\infty} \delta(q - 2\pi m\varDelta^{-1}). \quad (6.71)$$

Written explicitly as a Fourier series, this is

$$\varDelta \sum_{m=-\infty}^{\infty} \delta(x - m\varDelta) = \sum_{m=-\infty}^{\infty} \exp(-2\pi i m x/\varDelta), \quad (6.72)$$

a familiar equation in solid-state theory.

F. SEVERAL DIMENSIONS

The extension of Fourier transformation to $n > 1$ dimensions presents no problems. Definition (6.1) for ordinary functions is changed to the n-dimensional volume integral

$$\tilde{f}(\mathbf{q}) = \int f(\mathbf{r}) e^{i\mathbf{q}\cdot\mathbf{r}} \, dV, \quad (6.73)$$

where the scalar product $\mathbf{q} \cdot \mathbf{r}$ has replaced the ordinary product qx. The inversion theorem becomes

$$\tilde{\tilde{f}}(\mathbf{r}) = (2\pi)^n f(-\mathbf{r}). \quad (6.74)$$

Even Table I changes very little upon extension to n dimensions; (I.2), for example, becomes

$$F_\mathbf{q}[\partial^m f/\partial x_k^m] = (-iq_k)^m \tilde{f}(\mathbf{q}), \quad (6.75)$$

and the changes in (I.3) are similar. In (I.4) and (I.5) the arguments of the exponentials simply become scalar products. The factor $|a|^{-1}$ in (I.6) becomes $|a|^{-n}$, (I.7) remains unchanged, and in (I.8) the factor (2π) becomes $(2\pi)^n$.

Fourier transforms of GFs of slow growth in n variables are also easily established. S_n and S_n' have already been defined, and the transforms of test functions in S_n are again in S_n. The transform $\tilde{f}\langle\phi\rangle$ is again defined to be $f\langle\tilde{\phi}\rangle$, and the modifications discussed for Table I hold for GFs, as expected.

It is well known that transform techniques are important tools in solving partial differential equations. This importance has been enhanced significantly by the introduction of GFs, since a much wider range of "functions" has thereby become Fourier transformable. Space does not permit us to display specific examples, but the reader is referred to Gel'fand et al. (1964) for such applications.

VII. Laplace Transforms

A. Laplace Transforms of Ordinary Functions

The object of Section VII is to show very briefly that the Laplace transform, which is closely related to the Fourier transform, can also be extended to GFs and retains many of its properties in the course of this extension. We begin by reviewing the definition and properties of Laplace transforms of ordinary functions. All of Section VII deals with one dimension.

Consider a function $f(x)$ that is identically zero for $x < 0$. (In spite of this restriction, such functions are of considerable interest since they are typical of the output of a passive system that receives no input prior to the "time" $x = 0$.) The function $f(x)$ may not be Fourier transformable, but let us suppose that there exists a real constant c such that $\exp(-s_1 x)f(x)$ is Fourier transformable whenever $s_1 > c$:

$$F_q[\exp(-s_1 x)f(x)] = \int_0^\infty \exp(-s_1 x)f(x)e^{iqx}\,dx. \qquad (7.1)$$

The result of this transformation can be regarded as a function of the complex variable $s = s_1 - iq$:

$$L_s[f] \equiv F_q[\exp(-s_1 x)f(x)] = \int_0^\infty e^{-sx}f(x)\,dx. \qquad (7.2)$$

$L_s[f]$ is called the *Laplace transform* of f. It exists everywhere in the half-plane $\text{Re}(s) = s_1 > c$ and is an analytic function of s in that half-plane. (Re stands for "real part of.")

As an example, we can transform the step function $H(x)$. In this case $c = 0$, and

$$L_s[H] = \int_0^\infty e^{-sx}\, dx = s^{-1}. \tag{7.3}$$

More generally, for the function[†] x_+^γ, $\gamma > -1$, we still have $c = 0$ and

$$L_s[x_+^\gamma] = \int_0^\infty e^{-sx} x^\gamma\, dx = \gamma!\, s^{-(\gamma+1)}, \qquad \gamma > -1. \tag{7.4}$$

Here γ is not restricted to integral values. Notice the role played by c in these examples: The transform becomes singular on the line $s_1 = c = 0$. In general, the half-plane $s_1 > c$ is the half-plane to the right of all singularities in the transform. Although the integral in (7.2) converges only for $s_1 > c$, the resulting formula [such as (7.3) or (7.4)] may be meaningful for $s_1 < c$ as well. In the present context this should be regarded as an analytic continuation of the transform into the left half-plane.

On any line parallel to the imaginary axis (i.e., $s_1 = $ constant $> c$) $L_s[f]$ is just the Fourier transform of $\exp(-s_1 x)f(x)$. Clearly we can apply the inversion theorem and recover $\exp(-s_1 x)f(x)$ from the transform:

$$\exp(-s_1 x)f(x) = (2\pi)^{-1} \int_{-\infty}^\infty L_{s_1-iq}[f] e^{-iqx}\, dq \tag{7.5}$$

or

$$f(x) = (2\pi)^{-1} \int_{-\infty}^\infty L_{s_1-iq}[f] e^{-iqx} \exp(s_1 x)\, dq$$

$$= (2\pi i)^{-1} \int_{s_1-i\infty}^{s_1+i\infty} L_s[f] e^{sx}\, ds, \qquad s_1 > c. \tag{7.6}$$

This is the Laplace inversion theorem; the contour of the integration is a line parallel to the imaginary axis. Of course the answer is independent of s_1 provided $s_1 > c$. As an application of the calculus of residues, the reader may wish to substitute (7.4) into (7.6) and verify that x_+^γ is indeed the result.

[†] It is understood throughout Section VII that Laplace transformable functions (and later GFs) have been set equal to zero for $x < 0$. Following our earlier notation, this is denoted by a subscript "plus" sign.

2. Generalized Functions

TABLE II

TABLE OF LAPLACE TRANSFORMS[a]

Function	Transform	
$f(x)$	$L_s[f] = \int_0^\infty e^{-sx} f(x)\, dx$	(II. 1)
$f^{(m)}(x)$	$s^m L_s[f] - \sum_{k=1}^m s^{k-1} f^{(m-k)}(0_+)$	(II. 2)
$f^{(-m)}(x)$	$s^{-m} L_s[f]$ (see footnote b)	(II. 3)
$x^m f(x)$	$(-1)^m (L_s[f])^{(m)}$	(II. 4)
$f(x + a)$	$e^{sa} L_s[f]$	(II. 5)
$e^{-bx} f(x)$	$L_{s+b}[f]$	(II. 6)
$f(ax)$	$a^{-1} F(s/a), \quad a > 0$	(II. 7)
$(f * g)(x)$	$L_s[f] L_s[g]$	(II. 8)

[a] In this table, m is a positive integer, a is a real constant, and b is a complex constant.
[b] The constants of integration are fixed by the requirement $f(x < 0) = 0$.

Table II is a Laplace transform analog of Table I. The derivations are quite straightforward; for example, (II.2) is obtained as follows:

$$L_s[f'] = \int_0^\infty e^{-sx} f'(x)\, dx = e^{-sx} f(x) \Big|_0^\infty + s \int_0^\infty e^{-sx} f(x)\, dx$$
$$= sL_s[f] - f(0_+). \tag{7.7}$$

Then (II.2) follows by induction. The result can be used to effect the well-known transformation of a linear differential equation with constant coefficients into an algebraic equation.

The other results in Table II are obtained in a similar fashion, the required transforms being assumed to exist for some c in each case. Of course the similarity to Table I is not surprising.

B. LAPLACE TRANSFORMS OF GENERALIZED FUNCTIONS

1. *Definition*

The Laplace transform, unlike the Fourier transform, does not map the text function space S_1 onto itself. Therefore we cannot use a definition analogous to (6.27). However, we can proceed by treating (7.2) as the evaluation of f at a test function behaving like e^{-sx}, but first we quickly lay some groundwork that will simplify the treatment of GFs equaling zero for $x < 0$.

First let us define two new spaces of test functions, D_+ and S_+. Both spaces contain infinitely differentiable complex-valued functions of one real variable and have no specially prescribed behavior for $x \to -\infty$. However, members of D_+ have their support bounded above, and members of S_+ are rapidly decreasing (Section VI.B.1) as $x \to +\infty$. Thus D_1 and S_1 are proper subsets of D_+ and S_+, respectively, and D_+ is a proper subset of S_+.

Now consider those GFs in D_1' and S_1' whose support is bounded below. These *right-sided* GFs form subspaces that we can denote by D_+' and S_+'. They can clearly be evaluated at the test functions in D_+ and S_+ respectively, and are ideally suited for a discussion of Laplace transforms. S_+' is of course a subset of D_+'.

Suppose f is some GF in D_+'. Let us form the product $e^{-bx}f(x)$ (for $b > 0$ this is valid even in S_+'), and assume that there is some real number c such that $e^{-bx}f(x)$ is actually in S_+' for $b > c$. The constant c may be positive, negative, or even $-\infty$. We have now reached our goal of defining the Laplace transform of f in terms of the evaluation of f at decaying exponentials, since the latter are members of S_+:

$$L_s[f] \equiv (e^{-bx}f(x))\langle \exp(-(s-b)x)\rangle, \qquad \mathrm{Re}(s) \geq b > c. \quad (7.8)$$

In the half-plane indicated, $L_s[f]$ is an ordinary analytic function of the complex variable s, and is actually independent of the choice of b. L_s is a continuous linear operator; for parametric GFs it commutes with differentiation and integration with respect to the parameter.

As an example, let us evaluate the Laplace transform of δ. Since $c = -\infty$ in this case, we expect the transform to be analytic for all s. The transform is

$$L_s[\delta] = (e^{-bx}\delta)\langle \exp(-(s-b)x)\rangle = 1. \quad (7.9)$$

$L_s[\delta']$ can also be found easily. From (3.14) it follows that $e^{-bx}\delta' = \delta' + b\,\delta$, so that

$$L_s[\delta'] = (\delta' + b\,\delta)\langle \exp(-(s-b)x)\rangle = (s-b) + b = s. \quad (7.10)$$

2. Properties of the Laplace Transform. Examples

The last example raises the question of extending Table II to Laplace transformable GFs in D_+'. Consider (II.2):

$$L_s[f'] = [(e^{-bx}f(x))' + be^{-bx}f(x)]\langle\exp(-(s-b)x)\rangle$$
$$= -e^{-bx}f(x)[(\exp(-(s-b)x))'] + bL_s[f]$$
$$= ((s-b)+b)L_s[f] = sL_s[f], \qquad (7.11)$$

and therefore

$$L_s[f^{(m)}] = s^m L_s[f]. \qquad (7.12)$$

Notice that this differs from the result (II.2) for ordinary functions in that it omits the summation terms. This discrepancy seems puzzling until one realizes that the summation terms result from the inability of ordinary functions to handle any discontinuous behavior of $f(x)$ at the origin. For instance, the ordinary derivative of $H(x)$ is zero (except at the origin, where it is undefined), and accordingly (II.2) gives the erroneous result

$$L_s[H'] = sL_s[H] - 1 = s \cdot s^{-1} - 1 = 0. \qquad (7.13)$$

By contrast, the GF derivative of H is δ, and the transform of δ is 1. This corresponds precisely to leaving off the summation in (II.2). Actually, similar terms would appear in the corresponding Fourier formula, (I.2), if one tried to handle the derivatives of discontinuous functions as ordinary functions.

The remainder of Table II extends to GFs without change and needs no further comment. However, we use the table to find the transforms of several singular GFs. First consider x_+^γ for $-m-1 < \gamma < -m$, where m is a positive integer. Combining (5.7), (7.4), and (II.2),

$$L_s[x_+^\gamma] = \prod_{j=1}^m (\gamma+j)^{-1} L_s[(x_+^{\gamma+m})^{(m)}]$$
$$= \prod_{j=1}^m (\gamma+j)^{-1} s^m (m+\gamma)! s^{-(\gamma+m+1)}$$
$$= \gamma! s^{-(\gamma+1)}, \qquad \gamma \neq -1, -2, -3, \ldots. \qquad (7.14)$$

Thus the form (7.4) is valid also for nonintegral negative γ, as one might have anticipated. We can differentiate (7.14) with respect to γ at $\gamma = 0$, giving

$$L_s[\ln x_+] = \frac{\partial}{\partial \gamma} [\gamma! s^{-(\gamma+1)}]_{\gamma=0} = -s^{-1}(C_E + \ln s), \qquad (7.15)$$

where $C_E = 0.5772\ldots$ is Euler's constant. Actually $\ln x_+$ is a regular GF in S_+', and (7.15) can be obtained by direct integration. From (7.15) and (5.21) we get

$$L_s[x_+^{-1}] = -(C_E + \ln s), \tag{7.16}$$

and $x_+^{-(m+1)}$ can now be transformed with the aid of (5.23):

$$L_s[x_+^{-(m+1)}] = \frac{(-s)^m}{m!} \left\{ \sum_{k=1}^{m} k^{-1} - \ln s - C_E \right\}. \tag{7.17}$$

We have been treating $L_s[f]$ as an ordinary function of the complex variable $s = s_1 + is_2$. Since this function is analytic in a right half-plane and hence locally integrable in the variable s_2, it can define a regular GF in S_1' simply by integration with a test function $\phi(s_2)$. Then one can obtain a GF inversion formula analogous to (7.6). In practice it is usually easier to consult Laplace transform tables. A table including some GF entries is given by Zemanian (1965) and by other references quoted therein. In this connection we also mention Liverman (1964) who discusses *direct* operational methods which achieve the same simplification as the usual transform techniques without the intermediate step of performing a transformation on the problem at hand.

VIII. Conclusion

In conclusion, we emphasize that all common operations with ordinary functions, with the exception of multiplication, have been extended satisfactorily to generalized functions. In the process we have achieved the various goals outlined in Section I, including the creation of a mathematical structure that encompasses the δ function. The simplicity of many of the results, particularly for differentiation and Fourier transformation, suggests that generalized functions, rather than ordinary functions, are the natural environment in which to apply these operations.

This introduction has only been able to scratch the surface of the developed body of theory related to generalized functions. We have not discussed operational calculus, GFs of complex variables, vector-valued GFs, and other spaces of GFs. We have merely hinted at the intimate connection between GFs and analytic functions of a complex variable, and have not dealt with the direct product and Schwartz's kernel theorem. For these subjects the reader is referred to the sources already quoted.

In spite of the introductory nature of this chapter, it is hoped that the reader has appreciated the beauty of the theory and seen some potential applications of it in his own work.

Acknowledgments

This chapter was written while I was at the Theoretical Physics Institute of the University of Alberta. I wish to thank the Institute for the facilities provided for this work and the National Research Council of Canada for its financial support. I am particularly grateful to Dr. John V. Olson for his careful reading of the manuscript and to Dr. John Stephenson for a number of helpful discussions. I also thank Drs. M. S. Wertheim, A. Z. Capri, and H. J. Silverstone for several helpful comments.

References

Apostol, T. M. (1957). "Mathematical Analysis." Addison-Wesley, Reading, Massachusetts.

Arsac, J. (1966). "Fourier Transforms and the Theory of Distributions." Prentice-Hall, Englewood Cliffs, New Jersey.

Baxter, R. J. (1968). Percus-Yevick equation for hard spheres with surface adhesion. *J. Chem. Phys.* **49**, 2770.

Beltrami, E. J., and Wohlers, M. R. (1966). "Distributions and the Boundary Values of Analytic Functions." Academic Press, New York.

Bouix, M. (1964). "Les Fonctions Généralisées ou Distributions." Masson, Paris.

Bremermann, H. (1965). "Distributions, Complex Variables, and Fourier Transforms." Addison-Wesley, Reading, Massachusetts.

Capri, A. Z. (1975). A forthcoming textbook on quantum mechanics (to be published).

Dirac P. A. M. (1926). The physical interpretation of quantum dynamics. *Proc. Roy. Soc.* (*London*) **A113**, 621.

Dirac, P. A. M. (1930). "The Principles of Quantum Mechanics." Oxford Univ. Press (Clarendon), London and New York.

Donoghue, W. F. Jr. (1969). "Distributions and Fourier Transforms." Academic Press, New York.

Edwards, R. W. (1965). "Functional Analysis—Theory and Applications." Holt, New York.

Erdélyi, A. (1961). From delta functions to distributions. *In* "Modern Mathematics for the Engineer" (E. F. Beckenbach, ed.), Chapter 1, 2nd ser. McGraw-Hill, New York.

Erdélyi, A. (1962). "Operational Calculus and Generalized Functions." Holt, New York.

Friedman, A. (1963). "Generalized Functions and Partial Differential Equations." Prentice-Hall, Englewood Cliffs, New Jersey.

Gårding, L., and Lions, J. L. (1959). Functional analysis. *Nuovo Cimento Suppl.* **14**, 9.

GEL'FAND, I. M., and SHILOV, G. E. (1964). "Generalized Functions." Vol. I, Academic Press, New York. See also the subsequent volumes: Vol. II (1968); Vol. III (1967); Gel'fand, I. M., and Vilenkin, N.Ya., Vol. IV (1964); Gel'fand, I. M., Graev, M. I., and Vilenkin, N.Ya, Vol. V (1966).

HADAMARD, J. (1923). "Lectures on Cauchy's Problem in Linear Partial Differential Equations." Yale Univ. Press, New Haven, Connecticut. (Also Dover, New York, 1952.)

HALPERIN, I. (1952). "Introduction to the Theory of Distributions." Univ. Toronto Press, Toronto.

HEAVISIDE, O. (1893). On operators in mathematical physics. *Proc. Roy. Soc. (London)* **52**, 504; (1894) **54**, 105.

JANTSCHER, L. (1971). "Distributionen." Walter de Gruyter, Berlin.

JAGER, E. M. DE (1964). "Applications of Distributions in Mathematical Physics." Mathematisch Centrum Amsterdam, Amsterdam.

JAGER, E. M. DE (1970). Theory of distributions. *In* "Mathematics Applied to Physics" (E. Roubine, ed.), Chapter II. Springer-Verlag, New York and UNESCO, Paris.

KOREVAAR, J. (1955). Distributions defined from the point of view of applied mathematics. *Nederl. Akad. Wetensch. Proc. Ser. A* **58**, 368–389, 483–503, 663–674.

LIGHTHILL, M. J. (1958). "Introduction to Fourier Analysis and Generalized Functions." Cambridge Univ. Press, London and New York.

LIVERMAN, T. P. G. (1964). "Generalized Functions and Direct Operational Methods," Vol. I. Prentice-Hall, Englewood Cliffs, New Jersey.

MARCHAND J.-P. (1962). "Distributions." North-Holland Publ., Amsterdam.

MESSIAH, A. (1966). "Quantum Mechanics." North-Holland Publ., Amsterdam.

MIKUSINSKI, J. G. (1959). "Operational Calculus." Pergamon, Oxford.

PAPOULIS, A. (1962). "The Fourier Integral and Its Applications." McGraw-Hill, New York.

SALTZER, C. (1958). The theory of distributions. *Advan. Appl. Mech.* **5**, 91.

SCHWARTZ, L. (1966). "Théorie des Distributions." Hermann, Paris (originally in two volumes, 1950, 1951).

SCHWEBER, S. S. (1961). "An Introduction to Relativistic Quantum Field Theory." Row, Evanston, Illinois.

SILVERSTONE, H. J., and TODD, H. D. (1971). Analytic evaluation of three-center one-electron integrals of $r^N Y_L^M(\theta, \phi)$ with Slater-type atomic orbitals. *Int. J. Quantum Chem.* **4**, 371.

SNEDDON, I. N. (1951). "Fourier Transforms." McGraw-Hill, New York.

STREATER, R. F., and WIGHTMAN, A. S. (1964). "PCT, Spin and Statistics, and all that." Benjamin, New York.

TEMPLE, G. (1955). "The Theory of Generalized Functions." *Proc. Roy. Soc. (London)* **A228**, 175.

TITCHMARSH, E. C. (1937). "Introduction to the Theory of Fourier Integrals." Oxford Univ. Press (Clarendon), London and New York.

TREVES, F. (1967). "Topological Vector Spaces, Distributions and Kernels." Academic Press, New York.

ZEMANIAN, A.H. (1965). "Distribution Theory and Transform Analysis." McGraw-Hill, New York.

Chapter 3

Complex Variable Theory

HARRIS J. SILVERSTONE

I. Introduction	152
II. Complex Numbers	153
A. The Complex Number Field	154
B. Notation and Conventions. Elementary Consequences	155
C. Geometric Representation	158
D. Powers and Roots	159
III. Analytic Functions of a Complex Variable	160
A. Functions of a Complex Variable	160
B. Limit, Continuity	161
C. Derivative	162
D. Cauchy–Riemann Equations	164
E. Analytic Functions	165
F. Conjugate Coordinates	166
G. Two-Dimensional Laplace Equation	166
IV. Complex Integration	167
A. Definition of the Complex Integral	168
B. Inequalities	171
C. Open Sets. Connectivity. Domains	172
D. The Fundamental Theorem of Calculus	172
E. Cauchy's Theorem	175
F. Cauchy's Integral Formula	179
G. Some Consequences of Cauchy's Integral Formula, Including the Fundamental Theorem of Algebra	181
H. Substitution Formula	183
I. Connection with Real Line Integrals	183
V. Power Series	184
A. Elementary Definitions and Results	184
B. Basic Theorem on the Convergence of Power Series	186
C. Integration and Differentiation of Series. Analyticity	187
D. Representation of Analytic Functions by Power Series	189
E. Laurent Series	190
F. Analytic Continuation	192

VI.	Elementary Functions.	193
	A. Singularities. Zeros. The Point at Infinity	193
	B. Algebraic Functions	194
	C. The Exponential and Related Functions	198
VII.	Evaluation of Real Definite Integrals	205
	A. Residue Theorem	205
	B. Five Classes of Integrals Amenable to the Residue Theorem	206
	C. Examples	211
	D. On Computing Residues	212
	E. Cauchy Principal Value	212
VIII.	Higher Transcendental Functions	214
	A. The Gamma Function	214
	B. The Beta Function	217
	C. The Hypergeometric Function	221
	D. Legendre Polynomials and Associated Legendre Functions as Examples of the Hypergeometric Function	226
	E. The Confluent Hypergeometric Functions	231
	F. Spherical Bessel Functions as Examples of Confluent Hypergeometric Functions	235
	G. Series for Exponential-Type Integral	237
IX.	On Fourier Transforms	238
X.	Quantum Chemistry Integrals	241
	A. Spherical Harmonics. Slater-Type Atomic Orbitals	241
	B. Outline of the Approach	242
	C. Fourier Transform of a Slater-Type Atomic Orbital	243
	D. Two-Center Overlap-Type Integral	243
	E. Fourier Transform of a Two-Center Product	246
	F. Evaluation of a (1-2)-Type Three-Center Integral	250
	G. Arbitrary Multicenter Integrals	255
XI.	A Formula of Lagrange and Nondegenerate Perturbation Theory	256
	A. Lagrange's Formula	256
	B. Nondegenerate Perturbation Theory	257
	References	259

I. Introduction

Complex variable theory is the differential and integral calculus of complex-valued functions of a complex-valued variable. A few elegant and powerful theorems, due mainly to Augustin Louis Cauchy (1787–1857) and to Karl Weierstrass (1815–1897) and Georg Friedrich Bernhard Riemann (1826–1866), contain the main results. Both the generality of the basic theorems and the remarkable interconnections among differentiation, integration, and power series expansion endow complex variable theory with a unique aesthetic unity.

Complex variable theory is a basic tool for many sciences. The classic application is to the solution of Laplace's equation in two dimensions, as in electrostatics, hydrodynamics, and temperature distribution. The very "language" of quantum mechanics is complex variable theory. In this chapter, two general applications are treated: the evaluation of real definite integrals, and the representation of so-called special functions by contour integrals. Specific applications are also given that are relevant to quantum chemistry.

The organization of the chapter is to proceed from the general to the specific. Sections II–V deal in turn with complex numbers, differentiation, integration, and power series. The major theorems occur in Sections III–V. Section VI introduces elementary functions and the notion of branch point. Major applications begin in Section VII with the evaluation of real definite integrals, and in Section VIII with the simpler higher transcendental functions—the gamma function, the beta function, the hypergeometric function, the confluent hypergeometric function, and special cases of these latter two. A brief excursion into Fourier transforms in Section IX leads into the evaluation of "Slater-type orbital" multicenter integrals in quantum chemistry in Section X. Finally, in Section XI "Lagrange's formula" is derived and applied to the formal solution of the nondegenerate Rayleigh–Schrödinger perturbation problem in quantum mechanics.

II. Complex Numbers

What are complex numbers? The reader is probably already able to reply: that a complex number z is composed of two real numbers, x and y, and a symbol i; that it is written

$$z = x + iy = x + yi; \qquad (2.1)$$

that i satisfies,

$$i^2 = -1; \qquad (2.2)$$

and that complex numbers otherwise obey the rules of ordinary algebra.

What is taken today for granted took hundreds of years to develop. The first *implicit* appearance of complex numbers is in sixteenth century formulas for the *real* roots of cubic and quartic polynomials, as, for instance, were published by Girolamo Cardano. That a polynomial could have *complex* roots, and that a polynomial of degree n has exactly n

roots, were recognized in the seventeenth century. But the birth of complex numbers was not complete until the doctoral thesis of Karl Friedrich Gauss and his proof of the fundamental theorem of algebra in 1799.

To characterize more precisely the complex number system, we first define a *field*. Loosely, a field is a set whose elements (numbers) can be combined by addition, subtraction, multiplication, and division. We then *construct* the complex field from the field of real numbers. The complex field is the smallest field that contains the real number field as a subfield and a solution of $z^2 = -1$.

A. The Complex Number Field

A *field* F is a set of elements closed under two binary operations, called addition and multiplication, which satisfy the following axioms (z, z_1, z_2, and z_3 denote any elements of F):

Associative Laws

$$(z_1 + z_2) + z_3 = z_1 + (z_2 + z_3) \tag{2.3}$$

$$(z_1 z_2) z_3 = z_1 (z_2 z_3). \tag{2.4}$$

Commutative Laws

$$z_1 + z_2 = z_2 + z_1 \tag{2.5}$$

$$z_1 z_2 = z_2 z_1. \tag{2.6}$$

Distributive Law

$$z_1(z_2 + z_3) = z_1 z_2 + z_1 z_3. \tag{2.7}$$

Additive Identity

There is an element 0 in F for which

$$z + 0 = z \quad \text{for all} \quad z \quad \text{in} \quad F. \tag{2.8}$$

Multiplicative Identity

There is an element 1 in F for which

$$z1 = z \quad \text{for all} \quad z \quad \text{in} \quad F. \tag{2.9}$$

3. Complex Variable Theory

Additive Inverse

For each z in F, there is an element in F, usually denoted $(-z)$, for which

$$z + (-z) = 0. \tag{2.10}$$

Multiplicative Inverse

For each nonzero z in F, there is an element in F, usually denoted z^{-1}, for which

$$zz^{-1} = 1. \tag{2.11}$$

The rational number system and the real number system are both examples of fields. We show how to construct the complex field from the real number field.

Let S denote the set of all ordered pairs (x, y) of real numbers x and y. Define equality, addition ($\oplus$), and multiplication ($\times$) on S by

$$(x_1, y_1) = (x_2, y_2) \Leftrightarrow \text{both } x_1 = x_2 \text{ and } y_1 = y_2 \tag{2.12}$$

$$(x_1, y_1) \oplus (x_2, y_2) = (x_1 + x_2, y_1 + y_2) \tag{2.13}$$

$$(x_1, y_1) \times (x_2, y_2) = (x_1 x_2 - y_1 y_2, x_1 y_2 + y_1 x_2). \tag{2.14}$$

The additive and multiplicative identities are $(0, 0)$ and $(1, 0)$. The additive and multiplicative inverses for (x, y) $[(x, y) \neq (0, 0)]$ are $(-x, -y)$ and $(x/(x^2 + y^2)^{1/2}, -y/(x^2 + y^2)^{1/2})$. One may quickly verify that under $\oplus$ and $\times$, S is a field; that under $\oplus$ and $\times$, the set of all elements of S of the form $(x, 0)$ is also a field, essentially the field of real numbers; that $(x, y) \times (x, y) = (-1, 0)$ has two solutions, $(0, 1)$ and $(0, -1)$,

$$(0, 1) \times (0, 1) = (0, -1) \times (0, -1) = (-1, 0); \tag{2.15}$$

and that any (x, y) can be written

$$(x, y) = (x, 0) \oplus (0, 1) \times (y, 0). \tag{2.16}$$

This field S is called the *complex field*, usually denoted by $\mathscr{C}$, and the (x, y) are called complex numbers.

B. Notation and Conventions. Elementary Consequences

The cumbersome (x, y) notation is seldom used, except in the context of the preceding construction. Instead, a "real number" $(x, 0)$ is simply written x, the *imaginary unit* $(0, 1)$ is written i, and an arbitrary (x, y) is written $x + iy$, following Eq. (2.16). A *pure imaginary* number is further shortened to iy from $0 + iy$, and 1 and 0 mean $(1, 0)$ and $(0, 0)$. The $\oplus$ and $\times$ are replaced by the usual $+$ and $\cdot$ (or nothing). Equations (2.13)–(2.15) are then

$$(x_1 + iy_1) + (x_2 + iy_2) = (x_1 + x_2) + i(y_1 + y_2) \tag{2.17}$$

$$(x_1 + iy_1)(x_2 + iy_2) = (x_1 x_2 - y_1 y_2) + i(x_1 y_2 + y_1 x_2) \tag{2.18}$$

$$i^2 = (-i)^2 = -1. \tag{2.19}$$

Subtraction and division are adaptations of addition and multiplication:

$$z_1 - z_2 = z_1 + (-z_2), \tag{2.20}$$

$$z_1/z_2 = z_1 z_2^{-1}, \quad z_2 \neq 0. \tag{2.21}$$

It is conventional to reserve z and w to denote complex numbers and x, y, u, and v to denote real numbers. The relations,

$$z = x + iy \tag{2.22}$$

$$w = u + iv \tag{2.23}$$

shall *invariably be understood*, if not explicitly stated.

Some conventional terminology, given $z = x + iy$, follows:

1. The *real part* of z, Re z, is x:

$$\operatorname{Re} z = x. \tag{2.24}$$

2. The *imaginary part* of z, Im z, is y (not iy):

$$\operatorname{Im} z = y. \tag{2.25}$$

3. The *absolute value* of z, $|z|$, also called modulus, magnitude, or length, is

$$|z| = (x^2 + y^2)^{1/2}. \tag{2.26}$$

4. The *argument* of z, arg z, is

$$\arg z = \arctan(y/x), \quad z \neq 0, \tag{2.27}$$

$$= \arccos(x/|z|) = \arcsin(y/|z|), \quad z \neq 0. \tag{2.28}$$

5. The *complex conjugate* of z, denoted both by $\bar{z}$ and by z^* (especially in the scientific literature), is

$$\bar{z} = x - iy. \tag{2.29}$$

Note that arg z is defined only up to integer multiples of 2π and that it is undefined for $z = 0$. A particular single-valued choice, denoted by Arg z and called the *principal value* of arg z, is fixed by

$$-\pi < \text{Arg } z \leq \pi. \tag{2.30}$$

Some elementary consequences of the definitions given above are

$$\text{Re } z = \tfrac{1}{2}(z + \bar{z}) \tag{2.31}$$

$$\text{Im } z = -\tfrac{1}{2}i(z - \bar{z}), \tag{2.32}$$

$$\overline{z_1 + z_2} = \bar{z}_1 + \bar{z}_2 \tag{2.33}$$

$$\overline{z_1 z_2} = \bar{z}_1 \bar{z}_2 \tag{2.34}$$

$$z\bar{z} = |z|^2 = |z^2|, \tag{2.35}$$

$$|z| \geq 0, \quad \text{with equality only when} \quad z = 0 \tag{2.36}$$

$$|\bar{z}| = |z| \tag{2.37}$$

$$|z_1 z_2| = |z_1||z_2| \tag{2.38}$$

$$|z_1/z_2| = |z_1|/|z_2|, \tag{2.39}$$

$$\arg(z_1 z_2) = \arg z_1 + \arg z_2 \quad (\text{modulo } 2\pi) \tag{2.40}$$

$$\arg(z_1/z_2) = \arg z_1 - \arg z_2 \quad (\text{modulo } 2\pi), \tag{2.41}$$

$$\text{Re } z \leq |z| \tag{2.42}$$

$$\text{Im}(z) \leq |z|, \tag{2.43}$$

$$|z_1 + z_2| \leq |z_1| + |z_2| \quad (\text{triangle inequality}) \tag{2.44}$$

$$|z_1 - z_2| \geq ||z_1| - |z_2|| \quad (\text{variant of triangle inequality}), \tag{2.45}$$

$$\left|\sum_{k=1}^{n} z_k w_k\right|^2 \leq \left(\sum_{k=1}^{n} |z_k|^2\right)\left(\sum_{k=1}^{n} |w_k|^2\right) \quad (\text{Cauchy's inequality}), \tag{2.46}$$

$$z = |z|(\cos \phi + i \sin \phi), \quad \text{where} \quad \phi = \arg z \quad (\text{polar form for } z). \tag{2.47}$$

To derive Eq. (2.46), note that $\sum_{k=1}^{n} |z_k - \lambda \bar{w}_k|^2 \geq 0$, and choose $\lambda = \sum_{k=1}^{n} z_k w_k / \sum_{k=1}^{n} |w_k|^2$. The derivations of the remaining equations are more straightforward and are left to the reader.

C. Geometric Representation

The ordered pair notation of Section II.A suggests and provides a one-to-one correspondence between complex numbers $z = x + iy$ and points (x, y) in a plane. Figure 1a shows the position vector corresponding to z. Note particularly the significance of $|z|$ and arg z [cf. Eq. (2.47)] and the location of the point corresponding to $\bar{z}$. The geometric picture of addition corresponds to vector addition (Fig. 1b), as is to be expected

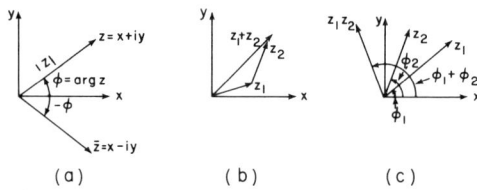

Fig. 1. Geometric representation of complex numbers by so-called Argand diagrams [after Jean Robert Argand (1806) but used already in 1797 by the Norwegian surveyor, Caspar Wessel]. (a) z and $\bar{z}$; (b) addition; (c) multiplication.

from Eq. (2.17). That multiplication has a geometric interpretation is a surprise! The position vector corresponding to $z_1 z_2$ has a length equal to the product of the lengths of z_1 and z_2 [Eq. (2.38)] and an angular coordinate equal to the sum of those for z_1 and z_2 [Eq. (2.40)], as illustrated in Fig. 1c:

$$\begin{aligned} z_1 z_2 &= |z_1| (\cos \phi_1 + i \sin \phi_1) |z_2| (\cos \phi_2 + i \sin \phi_2) \\ &= |z_1 z_2| [\cos(\phi_1 + \phi_2) + i \sin(\phi_1 + \phi_2)]. \end{aligned} \quad (2.48)$$

So straightforward and so conceptually useful is the geometric picture that the distinction between "complex number" and its corresponding "point" is customarily ignored. The terms complex number z and point z are used interchangeably, as are xy plane, complex plane, z plane, and complex z plane. The x axis is called the real axis, and the y axis the imaginary axis.

D. Powers and Roots

Computation of z^n and $z^{1/n}$ is an instructive yet elementary exercise. Let n be a positive integer. Inductive iteration of Eq. (2.48) gives

$$z^n = |z|^n(\cos\phi + i\sin\phi)^n \tag{2.49}$$
$$= |z|^n(\cos n\phi + i\sin n\phi), \qquad n \text{ an integer.} \tag{2.50}$$

Equation (2.50) is also valid when n is a negative integer (take reciprocals of both sides). The equation

$$(\cos\phi + i\sin\phi)^n = \cos n\phi + i\sin n\phi \tag{2.51}$$

is known as de Moivre's formula.

We now seek a solution w to the equation

$$w^n = z = |z|(\cos\phi + i\sin\phi), \qquad z \neq 0. \tag{2.52}$$

By Eq. (2.50), there are precisely n solutions,

$$w = |z|^{1/n}\{\cos[(\phi + 2\pi k)/n] + i\sin[(\phi + 2\pi k)/n]\},$$
$$k = 0, 1, \ldots, n-1, \tag{2.53}$$

where $|z|^{1/n}$ denotes the real, positive, nth root of $|z|$. There are thus n nth roots of z, denoted by the n-fold ambiguous symbol $z^{1/n}$.

The n numbers

$$(\omega_n)^k = \cos(2\pi k/n) + i\sin(2\pi k/n), \qquad k = 0, 1, \ldots, n-1, \tag{2.54}$$

are called the nth roots of unity. They lie on the unit circle at the vertices of a regular n-sided polygon. The case $n = 6$ is illustrated in Fig. 2.

Fig. 2. The six sixth roots of unity.

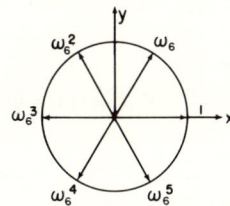

The rational powers of z, $z^{m/n}$ ($z \neq 0$, m and n both nonzero integers), can be defined by

$$z^{m/n} = (z^m)^{1/n} = |z|^{m/n}\{\cos[(m/n)(\phi + 2\pi k)] + i\sin[(m/n)(\phi + 2\pi k)]\}. \tag{2.55}$$

If m and n are relatively prime, then there are precisely $|n|$ such numbers.

III. Analytic Functions of a Complex Variable

The main thrust of this section is to define "analytic function" in terms of differentiation. The concepts of limit, continuity, and differentiation are developed in some detail. The main result is the connection of analyticity with the Cauchy–Riemann equations. Conjugate coordinates and the two-dimensional Laplace equation, both being derivative topics, are included at the end.

A. Functions of a Complex Variable

By a *function f of a complex variable z*, denoted by $f(z)$, is meant: (1) a "rule" or "correspondence" by which a definite complex number is associated with each value of the complex variable z; and (2) a set D of points z to which the "rule" applies. The set D is called the *domain of definition* of f.

As in real analysis, a small liberty is taken with the definition just given, and "function" is used in a second, distinct way: $f(z)$ may mean, as above, the "rule" itself, and implicitly the domain of definition, or $f(z)$ may mean the number associated with the number z by the "rule", as in "f takes on the value $f(z)$ at the point z."

The set of values taken on by $f(z)$ is called the *range* of f.

Inherent in part (1) of our definition of the word "function" is that $f(z)$ have only one value at z, i.e., that f should be *single valued*. It turns out that certain functions are so closely related to each other (in the sense of analytic continuation) that they are put under a common umbrella, *multiple-valued function*. An example is $z^{1/n}$, Eq. (2.53). This third usage of "function," considered in detail in Section VI, unfortunately is accompanied by a certain amount of semantic confusion. We refrain from using the redundant expression, "single-valued function," to mean "function," but use "multiple-valued function" as an inseparable term for that object.

Some simple functions are

$$f(z) = 1, \quad \text{with domain of definition} \quad = \mathscr{C} \qquad (3.1\text{a})$$

$$= z, \qquad = \mathscr{C} \qquad (3.1\text{b})$$

$$= z^2, \qquad = \mathscr{C} \qquad (3.1\text{c})$$

$$= z^n \quad (n \text{ a positive integer}), \quad = \mathscr{C} \qquad (3.1\text{d})$$

$$= \sum_{k=0}^{n} c_k z^k \quad (c_k \in \mathscr{C}, c_n \neq 0; \quad = \mathscr{C} \qquad (3.1\text{e})$$
$$\text{polynomial of degree } n),$$

$$= \bar{z}, \qquad = \mathscr{C} \qquad (3.1\text{f})$$

$$= |z|, \qquad = \mathscr{C} \qquad (3.1\text{g})$$

$$= z^{-n} \quad (n \text{ a positive integer}), \quad = \mathscr{C} - \{0\} \qquad (3.1\text{h})$$

$$= p_n(z)/q_m(z) \qquad = \mathscr{C} - \{\text{points at}$$
$$\text{(quotient of two polynomials,} \qquad \text{which } q_m = 0\} \qquad (3.1\text{i})$$
$$\text{called a rational function)},$$

$$= |z|^{1/n} \left(\cos \frac{\arg z}{n} + i \sin \frac{\arg z}{n} \right)$$
$$(-\pi < \arg z \le \pi), \qquad = \mathscr{C} \qquad (3.1\text{j})$$

$$= |z|^{1/n} \left(\cos \frac{\arg z}{n} + i \sin \frac{\arg z}{n} \right)$$
$$(0 \le \arg z < 2\pi) \qquad = \mathscr{C}. \qquad (3.1\text{k})$$

A complex function $f(z)$ corresponds to two real functions, u and v, of the two real variables x and y:

$$f(z) = u(x, y) + iv(x, y). \qquad (3.2)$$

Just as $z = x + iy$ is a standard notation, so is Eq. (3.2). Equation (3.2) permits results for functions of two real variables to be used for functions of a complex variable.

B. LIMIT, CONTINUITY

"Limit" and "continuity" hang on the notion of distance. The *distance* between z_1 and z_2 is defined to be the length of the vector connecting the two points in the complex plane: $|z_1 - z_2|$.

A *neighborhood* of z_0 is the interior of a circle centered on z_0; i.e., $\{z \mid |z - z_0| < r\}$. This symbol is read "the set of all points z satisfying $|z - z_0| < r$." The r is usually intended to be small.

We say that $f(z)$ *approaches a limit* f_0 as z approaches z_0, written,

$$f(z) \to f_0 \quad \text{as} \quad z \to z_0, \quad \text{or} \quad \lim_{z \to z_0} f(z) = f_0, \tag{3.3}$$

if and only if $f(z)$ is defined at all points of some neighborhood of z_0, except possibly at z_0 itself, and $|f(z) - f_0| \to 0$ as $|z - z_0| \to 0$. Although we have tied the definition of complex limit to real limit, it is important to realize that the limit process takes place in two dimensions. $f(z)$ must approach z_0 independently of the relative rates of $x \to x_0$ and $y \to y_0$. It is *explicitly* stated ("neighborhood") in the definition that z can approach z_0 from any direction. Of course, "$f(z)$ approaches f_0 as z approaches z_0 along the curve γ" has an obvious meaning, but such a qualified use of "limit" is the exception. Most of the general theorems refer to points that can be approached from any direction (i.e., interior points).

If $f(z) \to f_0$ at z_0, then $u(x, y) \to u_0$ and $v(x, y) \to v_0$ as both x and y approach x_0 and y_0, and conversely.

A number of elementary consequences follow from the definition of limit, the triangle inequality [Eq. (2.44)], and $|z_1 z_2| = |z_1||z_2|$:

$$\lim_{z \to z_0}\left(\sum_{k=1}^{N} f_k(z)\right) = \sum_{k=1}^{N} (\lim_{z \to z_0} f_k(z)) \tag{3.4}$$

$$\lim_{z \to z_0}\left(\prod_{k=1}^{N} f_k(z)\right) = \prod_{k=1}^{N} (\lim_{z \to z_0} f_k(z)) \tag{3.5}$$

$$\lim_{z \to z_0}(f_1(z)/f_2(z)) = (\lim_{z \to z_0} f_1(z))/(\lim_{z \to z_0} f_2(z)) \quad \text{if} \quad \lim f_2(z) \neq 0. \tag{3.6}$$

The function $f(z)$ is said to be *continuous* at z_0 if and only if $f(z)$ is defined at z_0 and

$$\lim_{z \to z_0} f(z) = f(z_0). \tag{3.7}$$

C. Derivative

The *derivative* of $f(z)$ at z_0, denoted variously by $(d/dz)f(z_0)$, $(d/dz)f$, df/dz, $f'(z_0)$, and f', is defined by

$$f'(z_0) = \lim_{h \to 0} \frac{f(z_0 + h) - f(z_0)}{h}, \tag{3.8}$$

provided the limit exists. This definition is formally identical with the real case, but the fundamental difference is that the limit must not depend on how h ($=\operatorname{Re} h + i \operatorname{Im} h$) approaches zero.

The rules for derivatives of sums, products, quotients, and compositions are the same as for the real case, being consequences of Eqs. (3.4)–(3.6) and (3.8):

$$(f+g)' = f' + g' \tag{3.9}$$

$$(fg)' = f'g + fg' \tag{3.10}$$

$$(f/g)' = (f'g - fg')/g^2 \tag{3.11}$$

$$\frac{d}{dz}f(g(z)) = \frac{df(g)}{dg}\frac{dg(z)}{dz} \quad \text{(chain rule)}. \tag{3.12}$$

The proofs are virtually identical with the real case and are not restated here.

Derivatives of polynomials and rational functions are easily obtained. An immediate consequence of the definition [Eq. (3.8)] is that

$$(d/dz)k = 0, \quad k \text{ a constant}, \tag{3.13}$$

$$(d/dz)z = 1. \tag{3.14}$$

The sum, product, and quotient rules, applied inductively, then give

$$(d/dz)z^n = nz^{n-1}, \quad n \text{ an integer}, \quad z \neq 0 \text{ when } n \text{ is negative}. \tag{3.15}$$

When the integer n is positive, Eq. (3.15) also follows from

$$(z_1^n - z_2^n)/(z_1 - z_2) = z_1^{n-1} + z_1^{n-2}z_2 + \cdots + z_1 z_2^{n-2} + z_2^{n-1}. \tag{3.16}$$

Thus, polynomials are differentiable everywhere, and rational functions are differentiable everywhere except where their denominators vanish, and their derivatives are obtained formally by the same rules as in real analysis.

The complex conjugate $\bar{z}$ of z does not have a derivative anywhere, since

$$(\overline{z+h} - \bar{z})/h = \bar{h}/h = [\cos(\arg h) - i \sin(\arg h)]^2 \tag{3.17}$$

approaches no limit as h approaches zero. (A "direction-dependent" limit is not enough.)

Similarly, $|z|$ has no derivative, and $|z|^2 = z\bar{z}$ has a derivative only at $z = 0$.

D. Cauchy–Riemann Equations

The Cauchy–Riemann equations are an immediate consequence of the independence of $f'(z_0)$ on how h approaches zero in Eq. (3.8). By alternatively computing f' with real h and pure imaginary h, one obtains

$$f'(z_0) = \lim_{h_x \to 0} \frac{f(z_0 + h_x) - f(z_0)}{h_x} = \frac{\partial f(z_0)}{\partial x} \tag{3.18}$$

$$= \lim_{h_y \to 0} \frac{f(z_0 + ih_y) - f(z_0)}{ih_y} = -i \frac{\partial f(z_0)}{\partial y} \tag{3.19}$$

or the Cauchy–Riemann equation in complex form,

$$(\partial/\partial x) f(z_0) = -i(\partial/\partial y) f(z_0). \tag{3.20}$$

In real form, the Cauchy–Riemann equations are

$$(\partial/\partial x) u(x_0, y_0) = (\partial/\partial y) v(x_0, y_0) \tag{3.21a}$$

$$(\partial/\partial y) u(x_0, y_0) = -(\partial/\partial x) v(x_0, y_0). \tag{3.21b}$$

Note: The existence of $f'(z_0)$ guarantees the continuity of f, u, and v at z_0 and the existence of the partial derivatives u_x, u_y, v_x, and v_y at (x_0, y_0).

Of more practical importance is the reverse: Given that u and v satisfy the Cauchy–Riemann equations at z_0, does $f = u + iv$ have a derivative at z_0? More information is needed. *If* the estimate is valid,

$$f(z_0 + h_x + ih_y) - f(z_0) = h_x(\partial/\partial x) f(z_0) + h_y(\partial/\partial y) f(z_0) + \varepsilon, \tag{3.22}$$

where $\varepsilon/|h|$ tends to zero as $|h|$ tends to zero, then the Cauchy–Riemann equation (3.20) in Eq. (3.8) gives

$$\lim_{h \to 0} \frac{f(z_0 + h_x + ih_y) - f(z_0)}{h_x + ih_y} = \lim_{h \to 0} \frac{(h_x + ih_y)(\partial/\partial x) f(z_0) + \varepsilon}{h_x + ih_y}$$

$$= \frac{\partial}{\partial x} f(z_0), \tag{3.23}$$

and the answer would be yes. The "if," however, cannot always be satisfied affirmatively, as is shown by the example

$$u(x, y) = v(x, y) = xy/(x^2 + y^2), \quad (x, y) \neq (0, 0) \tag{3.24}$$

$$= 0, \quad (x, y) = (0, 0) \tag{3.25}$$

$$u_x(0, 0) = u_y(0, 0) = v_x(0, 0) = v_y(0, 0) = 0 \tag{3.26}$$

The Cauchy–Riemann equations are trivially satisfied, but u and v are not even continuous at $(0, 0)$!

The problem is the same as the "differentiability" or existence of a "total differential" for a real function of two real variables. It is (more than) sufficient to assume additionally that u_x, u_y, v_x, and v_y are continuous at (x_0, y_0). Then by the mean value theorem,

$$u(x_0 + h_x, y_0 + h_y) - u(x_0, y_0)$$
$$= u(x_0 + h_x, y_0 + h_y) - u(x_0, y_0 + h_y) + u(x_0, y_0 + h_y) - u(x_0, y_0) \quad (3.27)$$

$$= u_x(x_0 + \theta_1 h_x, y_0 + h_y)h_x + u_y(x_0, y_0 + \theta_2 h_y)h_y,$$
$$0 \leq \theta_1 \leq 1, 0 \leq \theta_2 \leq 1, \quad (3.28)$$

and further, by continuity,

$$u(x_0 + h_x, y_0 + h_y) - u(x_0, y_0)$$
$$= u_x(x_0, y_0)h_x + u_y(x_0, y_0)h_y + \varepsilon_1 h_x + \varepsilon_2 h_y \quad (3.29)$$

where both ε_1 and ε_2 tend to zero as $|h|$ tends to zero. Equation (3.29) and a similar one for v clinch Eq. (3.22).

Thus, *if f has a derivative at z_0, the Cauchy–Riemann equations are satisfied. If the first partial derivatives satisfy the Cauchy–Riemann equations and are continuous at z_0, then f has a derivative at z_0.*

E. Analytic Functions

A function $f(z)$ is said to be *analytic at a point z_0* if $f'(z)$ exists at every point of some neighborhood of z_0. $f(z)$ is said to be *analytic in a region* (region used loosely for some portion of the complex plane) if $f(z)$ is analytic at every point of the region. Complex analysis is indeed the theory of analytic functions.

Polynomials are analytic in the entire z plane; rational functions are analytic for all z except where their denominators vanish (see Section III.C).

Analytic functions can be constructed from real functions with the aid of the Cauchy–Riemann equations. For instance, $f(z) \equiv e^x \cos y + i e^x \sin y$ satisfies the Cauchy–Riemann equation and has continuous partial derivatives everywhere. It is therefore analytic for all z.

If the derivative of an analytic function vanishes identically in some neighborhood of z_0, $f'(z) \equiv 0$ for $|z - z_0| < r$, then

$$u_x = u_y = v_x = v_y \equiv 0, \qquad |z - z_0| < r, \quad (3.30)$$

and $f(z)$ is a constant in that neighborhood. Moreover, if f is identically real or identically imaginary, then either f is not analytic or $f' \equiv 0$ and f is constant.

F. Conjugate Coordinates

This amusing, sometimes useful subject gives an interesting insight into analyticity. Note that $x = \frac{1}{2}(z + \bar{z})$, $y = -\frac{1}{2}i(z - \bar{z})$ have the same form as a coordinate transformation. If one regards an arbitrary complex function, $f = u(x, y) + iv(x, y)$, to be a function of the "independent" variables z and $\bar{z}$, one may compute formally

$$\frac{\partial f}{\partial z} = \frac{\partial x}{\partial z}\frac{\partial f}{\partial x} + \frac{\partial y}{\partial z}\frac{\partial f}{\partial y} = \frac{1}{2}\left(\frac{\partial f}{\partial x} - i\frac{\partial f}{\partial y}\right) \qquad (3.31)$$

$$\frac{\partial f}{\partial \bar{z}} = \frac{\partial x}{\partial \bar{z}}\frac{\partial f}{\partial x} + \frac{\partial y}{\partial \bar{z}}\frac{\partial f}{\partial y} = \frac{1}{2}\left(\frac{\partial f}{\partial x} + i\frac{\partial f}{\partial y}\right). \qquad (3.32)$$

The Cauchy–Riemann equation (3.20) is just $(\partial f/\partial \bar{z}) = 0$.

One can add rigor to these remarks and present some results of complex variable theory from this point of view. [See, for instance, Nehari (1968).] An old and interesting application related to spherical harmonics is found in Hobson (1955).

G. Two-Dimensional Laplace Equation

The basis for using complex variable theory to solve Laplace's equation in two dimensions is the Cauchy–Riemann equations. For an analytic function f, $(\partial/\partial x) = -i(\partial/\partial y)$, or $(\partial/\partial x)^2 = -(\partial/\partial y)^2$, so that f, u, and v are all solutions of the two-dimensional Laplace equation

$$\frac{\partial^2}{\partial x^2}f(z) + \frac{\partial^2}{\partial y^2}f(z) = 0, \qquad f(z) \text{ analytic} \qquad (3.33)$$

$$\frac{\partial^2}{\partial x^2}u(x, y) + \frac{\partial^2}{\partial y^2}u(x, y) = 0 \qquad (3.34)$$

$$\frac{\partial^2}{\partial x^2}v(x, y) + \frac{\partial^2}{\partial y^2}v(x, y) = 0. \qquad (3.35)$$

A solution of Laplace's equation is called a *harmonic function*. The real functions u and v (or v and $-u$) are said to be *conjugate harmonic functions* when $u + iv$ is an analytic function. Solving Laplace's equation is equivalent to finding an analytic function satisfying specified boundary conditions. This important application is discussed in many standard texts (e.g., Ahlfors, 1966; Churchill, 1960; Nehari, 1968) and by Henderson in Chapter 4, Section IX, of this volume.

We complete this section with an entertainingly useful exercise involving conjugate coordinates: how to find the conjugate harmonic function $v(x, y)$ and the analytic function $f(z)$ when given the harmonic function $u(x, y)$. Start with

$$f(z) = 2u(x, y) - \overline{f(z)}. \tag{3.36}$$

Since $f(z)$ is analytic, $(\partial/\partial \bar{z})f(z) = 0$, i.e., $\overline{(\partial/\partial z)f(z)} = 0$, so that $\overline{f(z)}$ is a function only of $\bar{z}$:

$$\overline{f(z)} = g(\bar{z}) = g(x - iy). \tag{3.37}$$

We use Eq. (3.37) to rewrite Eq. (3.36),

$$f(x + iy) = 2u(x, y) - g(x - iy), \tag{3.38}$$

into which we substitute formally

$$x = z/2, \quad y = z/2i \tag{3.39}$$

and find

$$f(z) = 2u(z/2, z/2i) - \bar{f}(0). \tag{3.40}$$

The usefulness of Eq. (3.40) hangs on the meaningfulness of $u(z/2, z/2i)$. The meaning is clear if $u(x, y)$ is a polynomial in x and y. Note that $v(x, y)$ is determined only up to an arbitrary additive constant.

IV. Complex Integration

The most important theorems in complex analysis, Cauchy's theorem and Cauchy's integral formula, are derived in this section. After defining the complex integral, we show that the usual rules of ordinary calculus are valid for the complex calculus. We derive Cauchy's theorem and integral formula, and then explore their immediate consequences.

A. Definition of the Complex Integral

In ordinary calculus, the definite integral of a function $f(x)$, piecewise continuous on the closed interval (a, b), is defined by

$$\int_a^b f(x)\,dx = \lim_{\substack{n\to\infty \\ \max\{\Delta x_k\}\to 0}} \sum_{k=1}^n f(\hat{x}_k)\,\Delta x_k, \qquad (4.1)$$

where

$$a = x_0 < x_1 < x_2 \cdots < x_n = b,$$
$$x_{k-1} \le \hat{x}_k \le x_k, \quad \text{and} \quad \Delta x_k = x_k - x_{k-1}. \qquad (4.2)$$

In the complex case there is the added freedom of specifying the path γ connecting the complex numbers a and b.

A path or curve γ in the complex plane is a geometric object. Its analytical description is called a parametric representation:

$$\gamma: z = z(t), \quad t_a \le t \le t_b, \quad z(t) \text{ continuous}, \qquad (4.3)$$
$$z(t_a) = a, \quad z(t_b) = b. \qquad (4.4)$$

γ can be conveniently subdivided by subdividing $[t_a, t_b]$:

$$t_a = t_0 < t_1 < t_2 \cdots < t_n = t_b, \quad t_{k-1} \le \hat{t}_k \le t_k, \qquad (4.5)$$
$$z_k = z(t_k), \quad \hat{z}_k = z(\hat{t}_k), \quad \Delta z_k = z_k - z_{k-1}. \qquad (4.6)$$

The integral of the complex function $f(z)$ over the path γ is then defined by

$$\int_\gamma f(z)\,dz = \lim_{\substack{n\to\infty \\ \max\{|\Delta z_k|\}\to 0}} \sum_{k=1}^n f(\hat{z}_k)\,\Delta z_k, \qquad (4.7)$$

provided that the limit exists.

1. Concerning Paths

A complex integral depends on the path γ, as well as on the endpoints and on $f(z)$. If γ is divided at a point into two paths, γ_1 and γ_2, then we write $\gamma = \gamma_1 + \gamma_2$, and

$$\int_{\gamma_1+\gamma_2} f(z)\,dz = \int_{\gamma_1} f(z)\,dz + \int_{\gamma_2} f(z)\,dz. \qquad (4.8)$$

The *length* of a path γ, denoted by L_γ, is defined by

$$L_\gamma = \lim_{\substack{n \to \infty \\ \max\{|\Delta z_k|\} \to 0}} \sum_{k=1}^{n} |\Delta z_k|, \tag{4.9}$$

provided that the limit exists.

Paths have a *sense* or *direction*. Some authors use $-\gamma$ to denote the path whose points are the same as for γ, but whose sense is from b to a. If γ is given by $z = z(t)$ ($t_a \leq t \leq t_b$), then a representation for $-\gamma$ is

$$-\gamma: z = z(t_a + t_b - t), \quad t_a \leq t \leq t_b. \tag{4.10}$$

Clearly,

$$\int_{-\gamma} f(z)\, dz = -\int_\gamma f(z)\, dz. \tag{4.11}$$

If $a = b$, γ is said to be *closed*. If γ is nonself-intersecting, i.e.,

$$z(t_1) = z(t_2) \Leftrightarrow t_1 = t_2, \tag{4.12}$$

then γ is said to be *simple*. If $a = b$ is the only exception to Eq. (4.12), then γ is called a *simple closed curve*. By convention, the *positive direction* on a simple closed curve is *counterclockwise*.

When $\int_\gamma f(z)\, dz$ depends only on the endpoints of γ, we may write without ambiguity,

$$\int_\gamma f(z)\, dz = \int_a^b f(z)\, dz. \tag{4.13}$$

When γ is a simple closed curve with counterclockwise sense, one may use the notation

$$\int_\gamma f(z)\, dz = \oint_\gamma f(z)\, dz. \tag{4.14}$$

The statements "$\int_\gamma f(z)\, dz$ depends only on the endpoints" and "$\int_\gamma f(z)\, dz = 0$ for all closed curves γ" are equivalent, for if γ_1 and γ_2 are any two paths connecting a to b, then $\gamma_1 - \gamma_2$ is a closed curve, and

$$\int_{\gamma_1 - \gamma_2} f(z)\, dz = \int_{\gamma_1} f(z)\, dz - \int_{\gamma_2} f(z)\, dz. \tag{4.15}$$

The importance of instantly associating these two statements together, when only one is read, cannot be overemphasized.

γ is said to be *piecewise differentiable* if

$$\frac{dz(t)}{dt} = \frac{dx(t)}{dt} + i\frac{dy(t)}{dt} \tag{4.16}$$

exists and is continuous for $t_a \leq t \leq t_b$, except for a finite number of jump discontinuities. We shall only consider piecewise differentiable γ. The terms path, curve, arc, and contour are used synonymously.

2. Existence of the Complex Integral

If it is assumed that $f(z)$ is piecewise continuous for z on γ, and that γ has finite length, then the existence of $\int_\gamma f(z)\,dz$ can be proved by the same method used to prove the existence of the real definite integral (4.1), *mutatis mutandis*. It is more instructive in a practical sense to assume additionally that γ is piecewise differentiable and to reduce the complex integral to two real integrals, whose existence is proved in elementary calculus. For piecewise differentiable γ, with an assist from the mean value theorem and Eqs. (4.1) and (4.16), and with due care to be taken at any finite discontinuities, we may recast Eq. (4.7) as

$$\int_\gamma f(z)\,dz = \lim_{\substack{n\to\infty \\ \max\{t_k - t_{k-1}\}\to 0}} \sum_{k=1}^n f(\hat{z}_k) \frac{z_k - z_{k-1}}{t_k - t_{k-1}}(t_k - t_{k-1}) \tag{4.17}$$

$$= \int_{t_a}^{t_b} \left[u(x(t), y(t))\frac{dx(t)}{dt} - v(x(t), y(t))\frac{dy(t)}{dt}\right] dt$$

$$+ i \int_{t_a}^{t_b} \left[u(x(t), y(t))\frac{dy(t)}{dt} + v(x(t), y(t))\frac{dx(t)}{dt}\right] dt, \tag{4.18}$$

or, more compactly and suggestively,

$$\int_\gamma f(z)\,dz = \int_{t_a}^{t_b} f(z(t))\frac{dz(t)}{dt}\,dt = \int_{t_a}^{t_b} f\frac{dz}{dt}\,dt. \tag{4.19}$$

It remains only to show that Eqs. (4.18) and (4.19) are independent of the parameterization of γ. Let $z = Z(s)$, $s_a \leq s \leq s_b$, be another piecewise differentiable representation of γ. Then $z(t(s)) = Z(s)$ implicitly defines a piecewise differentiable function

$$t = t(s), \quad s_a \leq s \leq s_b. \tag{4.20}$$

Starting with Eq. (4.19) [as shorthand for Eq. (4.18)] and using the usual rules for changing integration variables in ordinary calculus, we obtain

$$\int_\gamma f(z)\,dz = \int_{s_a}^{s_b} f(z(t(s))) \frac{dz(t)}{dt} \frac{dt}{ds}\,ds. \tag{4.21}$$

By the chain rule of ordinary calculus,

$$\frac{dz}{dt}\frac{dt}{ds} = \frac{dZ}{ds}, \tag{4.22}$$

so that

$$\int_\gamma f(z)\,dz = \int_{s_a}^{s_b} f(Z(s)) \frac{dZ(s)}{ds}\,ds, \tag{4.23}$$

which has the same form as Eq. (4.19).

3. When γ Runs to Infinity

If one or both of the endpoints a and b are at ∞, the integral is to be regarded as the limit of the integral for finite a and b, as a and/or b approach ∞ appropriately, provided that the limit exists.

B. Inequalities

The analog of

$$\left|\int_a^b f(x)\,dx\right| \le \int_a^b |f(x)|\,dx \le \max\{|f(x)|\} \cdot |b-a| \tag{4.24}$$

is enormously useful. First define, in the notation of Section IV.A.1,

$$\int_\gamma |f(z)|\,|dz| = \lim_{\substack{n\to\infty \\ \max\{|\Delta z_k|\}\to 0}} \sum_{k=1}^n |f(\hat{z}_k)|\,|\Delta z_k|, \tag{4.25}$$

or, equivalently for piecewise differentiable γ,

$$\int_\gamma |f(z)|\,|dz| = \int_{t_a}^{t_b} |f(z(t))|\,\left|\frac{dz(t)}{dt}\right|\,dt. \tag{4.26}$$

Then the triangle inequality (2.44) applied to the absolute value of Eq. (4.7) gives

$$\left|\int_\gamma f(z)\,dz\right| \le \int_\gamma |f(z)|\,|dz|. \tag{4.27}$$

If $|f(z)| \leq M$ on γ, then by Eqs. (4.25) and (4.9),

$$\left|\int_\gamma f(z)\, dz\right| \leq \int_\gamma |f(z)|\,|dz| \leq ML_\gamma. \tag{4.28}$$

C. Open Sets. Connectivity. Domains

The main theorems of complex analysis pertain to functions analytic in a kind of region in the complex plane called a *domain*. This use of "domain" is distinct from "domain of definition."

A set S of points in the complex plane is said to be *open* if, for each point z in S, there is a neighborhood of z contained entirely in S.

A set S of points in the complex plane is said to be *connected* if, for any two points z_1 and z_2 in S, there is a continuous path from z_1 to z_2 lying entirely in S.

A *domain* is an open, connected set.

A connected set is said to be *simply connected* if it has no holes. More precisely, simple connectivity means that any two paths joining the same endpoints can be continuously deformed, via paths in S, into each other. Otherwise the set is said to be *multiply connected*.

The open unit disk, $\{z\,|\,|z|<1\}$, is a simply connected domain. The punctured disk, $\{z\,|\,0<|z|<1\}$, is a *multiply connected domain*. The *closed* disk, $\{z\,|\,|z|\leq 1\}$, is neither open nor a domain.

The relevance of domain is best appreciated from the theorem on convergence of power series (Section V.B) and the theorem on the representation of analytic functions by power series (Section V.D).

D. The Fundamental Theorem of Calculus

We now prove the complex analog of the central result of ordinary calculus, the *fundamental theorem*,

$$\int_a^b \frac{df(x)}{dx}\, dx = f(b) - f(a). \tag{4.29}$$

1. Fundamental Theorem. Part 1

Theorem If $f(z)$ is analytic and $(d/dz)f(z)$ continuous in a domain D, and if γ is any piecewise differentiable path lying in D and having endpoints a and b, then

$$\int_\gamma \frac{df(z)}{dz}\, dz = f(b) - f(a). \tag{4.30}$$

Observe that the integral in Eq. (4.30) depends only on the endpoints a and b, and not on any other details of γ! The gist of the proof is to justify Eq. (4.32) in the sequence of equations:

$$\int_\gamma f(z)\,dz = \int_{t_a}^{t_b} f(z(t)) \frac{dz(t)}{dt}\,dt \qquad (4.31)$$

$$= \int_{t_a}^{t_b} \frac{d}{dt} f(z(t))\,dt \qquad (4.32)$$

$$= f(z(t))\Big|_{t_a}^{t_b} = f(b) - f(a), \qquad (4.33)$$

for Eq. (4.33) follows from Eq. (4.32) by the fundamental theorem of ordinary calculus applied separately to $\mathrm{Re}\,f$ and $\mathrm{Im}\,f$. To evaluate $(d/dt)f(z(t))$, use Eq. (3.22), which was justified for continuous (df/dz). To state the argument loosely, divide Eq. (3.22) by Δt and take the limit as $\Delta t \to 0$:

$$\frac{d}{dt} f(z(t)) = \frac{dx}{dt} \frac{\partial f(z)}{\partial x} + \frac{dy}{dt} \frac{\partial f(z)}{\partial y}. \qquad (4.34)$$

Then the Cauchy–Riemann equation (3.20) and Eq. (4.16) give

$$\frac{d}{dt} f(z(t)) = \frac{df(z)}{dz} \frac{dz}{dt}. \qquad (4.35)$$

An instructive exercise for the reader is to work through the equivalent derivation using the real functions u and v, and the Cauchy–Riemann equations (3.21).

2. *A Few Simple Integrals*

From Eq. (3.15),

$$\int_a^b z^n\,dz = (b^{n+1} - a^{n+1})/(n+1), \quad n \text{ an integer}, \quad n \neq -1. \qquad (4.36)$$

When n is negative, the integration path must avoid the origin.

The integration-by-parts formula remains valid. If in a domain D, $f(z)$ and $g(z)$ are analytic, and $f'(z)$ and $g'(z)$ are continuous, then by Eq. (3.10),

$$\int_\gamma f'(z)g(z)\,dz = f(b)g(b) - f(a)g(a) - \int_\gamma f(z)g'(z)\,dz. \qquad (4.37)$$

A nice counterexample is given by $\int_\gamma (1/z)\,dz$. We have not given a function whose derivative is $1/z$, so we cannot use the fundamental theorem. We can compute the integral directly, however, when γ is the circle $z = r\cos t + ir\sin t$ ($0 \leq t \leq 2\pi$, r real and positive). Note that $dz/dt = iz$.

$$\oint_{|z|=r} (1/z)\,dz = \int_0^{2\pi} (1/z)iz\,dt = \int_0^{2\pi} i\,dt = 2\pi i. \quad (4.38)$$

This *important result* is independent of r. If $1/z$ is the derivative of an analytic function, the domain of analyticity cannot contain the circle $|z| = r$. Otherwise $\oint_{|z|=r}(1/z)\,dz$ could not be nonzero.

3. Fundamental Theorem. Part 2

The real definite integral is a differentiable function of its upper limit. So is the complex integral, if it is otherwise independent of the path.

Theorem (Converse of the theorem of Section IV.D.1.) If in a domain D, $f(z)$ is continuous and $\int_\gamma f(z)\,dz$ depends only on the endpoints of γ, for all γ in D, then $f(z)$ is the derivative of a function $F(z)$ analytic in D.

Proof Fix z_0 and define $F(z)$ by

$$F(z) = \int_{z_0}^z f(\zeta)\,d\zeta. \quad (4.39)$$

As long as z_0, z, and γ are in D, the integral depends only on the endpoints. Two special paths facilitate verifying the Cauchy–Riemann equations and computing $F'(z)$. Let z_1 and z_2 ($z_1 = x_1 + iy$ and $z_2 = x + iy_2$) both belong to a neighborhood of $z = x + iy$, entirely contained in D. Then one can write

$$F(z) = \int_{z_0}^{z_1} f(\zeta)\,d\zeta + \int_{z_1}^z f(\zeta)\,d\zeta = \int_{z_0}^{z_2} f(\zeta)\,d\zeta + \int_{z_2}^z f(\zeta)\,d\zeta \quad (4.40)$$

$$= F(z_1) + \int_{x_1}^x [u(\xi,y) + iv(\xi,y)]\,d\xi \quad (4.41)$$

$$= F(z_2) + i\int_{y_1}^y [u(x,\eta) + iv(x,\eta)]\,d\eta. \quad (4.42)$$

The fundamental theorem of ordinary calculus, applied separately to the real and imaginary parts of Eq. (4.41), and then similarly to Eq. (4.42), yields

$$\partial F(z)/\partial x = f(z) \qquad (4.43)$$

$$\partial F(z)/\partial y = if(z), \qquad (4.44)$$

which verifies that the Cauchy–Riemann equations are satisfied. Since $f(z)$ was assumed continuous, by Section III.D and E, $F(z)$ is analytic with derivative $f(z)$, for all z in D.

E. Cauchy's Theorem

Independence of path is generally an inconvenient criterion. Cauchy's theorem makes the criterion the *analyticity* of $f(z)$.

1. *Cauchy's Theorem in a Simply Connected Domain. Statement*

Cauchy's Theorem Let $f(z)$ be analytic in a simply connected domain D, and let γ be any closed path in D. Then

$$\oint_\gamma f(z)\,dz = 0. \qquad (4.45)$$

If $f'(z)$ is also assumed to be continuous, then a simple proof follows from the theory of real line integrals (see Section IV.I). A proof not requiring the continuity of $f'(z)$ was given by E. Goursat, and Cauchy's theorem is often referred to as the Cauchy–Goursat theorem. We shall see in Section IV.F.2 that $f'(z)$ itself is analytic and *ipso facto* continuous whenever $f(z)$ is analytic.

2. *Cauchy's Theorem in a Simply Connected Domain. Proof*

The proof proceeds in stages. By first taking γ to be a rectangle, we focus on relating the local behavior of $f(z)$ at a point (analyticity) to its global behavior (integral about γ). The fundamental theorem provides the extension from rectangular to arbitrary γ when the domain is a disk. Finally, a simple procedure takes care of an arbitrarily shaped domain.

a. Rectangular γ. We start with γ as a rectangle and immediately cut it into four equal rectangles, $\gamma_1^{(1)}$, $\gamma_1^{(2)}$, $\gamma_1^{(3)}$, $\gamma_1^{(4)}$, as illustrated in Fig. 3.

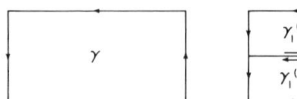 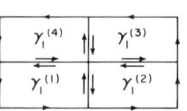

FIG. 3. Partition of γ into four equal rectangles.

In the sense of Eq. (4.8), both $\gamma = \gamma_1^{(1)} + \gamma_1^{(2)} + \gamma_1^{(3)} + \gamma_1^{(4)}$ and

$$\oint_\gamma f(z)\, dz = \sum_{k=1}^{4} \oint_{\gamma_1^{(k)}} f(z)\, dz. \qquad (4.46)$$

By the triangle inequality,

$$\left| \oint_\gamma f(z)\, dz \right| \le \sum_{k=1}^{4} \left| \oint_{\gamma_1^{(k)}} f(z)\, dz \right| \le 4 \max\left\{ \left| \oint_{\gamma_1^{(k)}} f(z)\, dz \right| \right\}. \qquad (4.47)$$

Denote by γ_1 the small rectangle for which $\left| \oint_{\gamma_1^{(k)}} f(z)\, dz \right|$ is maximum. Then

$$\left| \oint_\gamma f(z)\, dz \right| \le 4 \left| \oint_{\gamma_1} f(z)\, dz \right|. \qquad (4.48)$$

After carrying out this procedure n times, each time ending up with a rectangle γ_k that is one-quarter of the preceding γ_{k-1}, we obtain a nested sequence of rectangles,

$$\gamma \supset \gamma_1 \supset \gamma_2 \supset \cdots \supset \gamma_n. \qquad (4.49)$$

The sides of γ_n are 2^{-n} times the sides of γ, and

$$\left| \oint_\gamma f(z)\, dz \right| \le 4^n \left| \oint_{\gamma_n} f(z)\, dz \right|. \qquad (4.50)$$

As $n \to \infty$, the dimensions of γ_n tend to 0, and there is precisely one point z_0 contained in (or on) every rectangle of the sequence.

The analyticity of f at z_0 yields an estimate of $\left| \oint_{\gamma_n} f(z)\, dz \right|$ for sufficiently large n. Given $\varepsilon > 0$, there is a $\delta > 0$, such that

$$\left| \frac{f(z) - f(z_0)}{z - z_0} - f'(z_0) \right| < \varepsilon, \quad \text{when} \quad |z - z_0| < \delta. \qquad (4.51)$$

For all sufficiently large n, γ_n will be contained within the disk, $|z - z_0| < \delta$. By Eq. (4.36),

$$\oint_{\gamma_n} dz = \oint_{\gamma_n} (z - z_0)\, dz = 0 \qquad (4.52)$$

$$\oint_{\gamma_n} f(z)\, dz = \oint_{\gamma_n} [f(z) - f(z_0) - f'(z_0)(z - z_0)]\, dz. \qquad (4.53)$$

Then by the inequality (4.28),

$$\left| \oint_{\gamma_n} f(z)\, dz \right| < \varepsilon \max\{|z - z_0|\}_{z \text{ on } \gamma_n} L_{\gamma_n}. \tag{4.54}$$

Since $L_{\gamma_n} = 2^{-n} L_\gamma$, and $\max\{|z - z_0|\} \le \tfrac{1}{2} L_{\gamma_n} = 2^{-n-1} L_\gamma$,

$$\left| \oint_\gamma f(z)\, dz \right| \le 4^n \left| \oint_{\gamma_n} f(z)\, dz \right| \le \tfrac{1}{2} \varepsilon L_\gamma^{\,2}. \tag{4.55}$$

But ε is arbitrary, so $\oint_\gamma f(z)\, dz$ must vanish.

b. When D is a Disk. With the result just obtained, we construct $F(z)$ such that $F'(z) = f(z)$. Let z_0 be the center of the disk and z any point in the disk. The rectangle whose sides are parallel to the x and y axes, with opposite corners at z_0 and z, lies inside D, and

$$F(z) = i \int_{y_0}^{y} f(x_0 + i\eta)\, d\eta + \int_{x_0}^{x} f(\xi + iy)\, d\xi \tag{4.56}$$

$$= \int_{x_0}^{x} f(\xi + iy_0)\, d\xi + i \int_{y_0}^{y} f(x + i\eta)\, d\eta. \tag{4.57}$$

We compute $(\partial/\partial x) F(z)$ from Eq. (4.56) and $(\partial/\partial y) F(z)$ from Eq. (4.57) via the fundamental theorem of ordinary calculus:

$$(\partial/\partial x) F(z) = f(z) \tag{4.58}$$

$$(\partial/\partial y) F(z) = i f(z). \tag{4.59}$$

Since $f(z)$ is continuous and Eqs. (4.58) and (4.59) are the Cauchy–Riemann equations, $F(z)$ is analytic, and $F'(z) = f(z)$, for all z in D. By the complex fundamental theorem, $\oint_\gamma f(z)\, dz = 0$, for all closed γ in D.

c. When D is an Arbitrary Simply Connected Domain. Since any closed curve may be considered to be a sum of simple closed curves, we may without loss of generality consider a simple closed curve γ in D. Let d denote the shortest distance from γ to the boundary of D. Superimpose on D a grid of lines parallel to the coordinate axes and spaced $\tfrac{1}{2} d$ apart. We may write $\gamma = \sum_n \gamma^{(n)}$, where each $\gamma^{(n)}$ is a $\tfrac{1}{2} d \times \tfrac{1}{2} d$ square lying entirely inside γ or a partial square (part square inside γ, and part γ), and

$$\oint_\gamma f(z)\, dz = \sum_n \oint_{\gamma^{(n)}} f(z)\, dz. \tag{4.60}$$

The number of $\gamma^{(n)}$ is finite. Each $\gamma^{(n)}$ is contained in a disk of radius $\frac{1}{2}d$, lying entirely in D, so each term on the right-hand side of Eq. (4.60) vanishes.

3. Cauchy's Theorem in a Multiply Connected Domain

That $\oint_{|z|=r} z^{-1} dz = 2\pi i$, Eq. (4.38), shows the necessity of simple connectivity for Cauchy's theorem. There is, nevertheless, a very useful adaptation for multiply connected domains. Consider first the situation in which γ_1 and γ_2 are nonintersecting, simple closed curves, with γ_2 lying inside of γ_1, and both curves having counterclockwise sense. Let $f(z)$ be analytic both in the domain bounded on the outside by γ_1 and on the inside by γ_2, and at every point of γ_1 and γ_2. It is not necessary for $f(z)$ to be analytic everywhere inside of γ_2. Then

$$\oint_{\gamma_1} f(z) \, dz = \oint_{\gamma_2} f(z) \, dz. \tag{4.61}$$

The essence of the proof is illustrated in Fig. 4. Additional lines are introduced to form simple closed contours, Γ_1 and Γ_2. Each of Γ_1 and Γ_2 lies in simply connected domains in which $f(z)$ is analytic, and

$$\oint_{\gamma_1} f(z) \, dz - \oint_{\gamma_2} f(z) \, dz = \oint_{\Gamma_1} f(z) \, dz + \oint_{\Gamma_2} f(z) \, dz = 0. \tag{4.62}$$

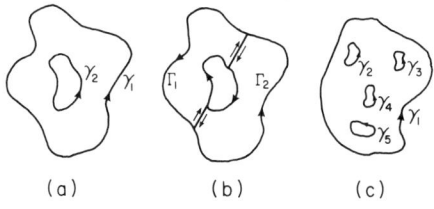

FIG. 4. Contours for multiply connected domains. (a) γ_1 and γ_2; (b) addition of two lines joining γ_1 and γ_2, so that $\Gamma_1 + \Gamma_2 = \gamma_1 - \gamma_2$; (c) a more general situation.

The general situation, illustrated in Fig. 4c, is that γ_1 surrounds the disjoint simple closed curves $\gamma_2, \gamma_3, \ldots, \gamma_n$, all taken counterclockwise, and $f(z)$ is analytic on each curve and in the multiply connected domain they bound. The adaptation of Cauchy's theorem is

$$\oint_{\gamma_1} f(z) \, dz = \sum_{k=2}^{n} \oint_{\gamma_k} f(z) \, dz. \tag{4.63}$$

A *major use* of Eqs. (4.61) and (4.63) in the practical evaluation of integrals is to replace a given contour γ_1 by a more tractable contour γ_2

(or $\gamma_2 + \cdots + \gamma_n$). Picturesque jargon is often employed, namely, "the contour γ_1 is *deformed* into the contour γ_2." As a simple example, let γ be a simple closed curve, and let z_0 and the circle $|z - z_0| = r$ lie entirely inside γ. We deform γ into $|z - z_0| = r$, then use Eq. (4.38):

$$\oint_\gamma (z - z_0)^{-1} dz = 2\pi i, \qquad z_0 \text{ inside } \gamma. \tag{4.64}$$

If z_0 were outside γ, the integral would vanish.

F. Cauchy's Integral Formula

Cauchy's integral formula, the single most important formula in complex variable theory, is the key to developing several important aspects of complex variable theory.

1. Cauchy's Integral Formula

Theorem Let $f(z)$ be analytic on and within a simple closed curve γ (i.e., in a simply connected domain that contains γ). Let z_0 lie inside (but not on) γ. Then

$$f(z_0) = \frac{1}{2\pi i} \oint_\gamma \frac{f(z)}{z - z_0} dz \qquad \text{(Cauchy's integral formula)}. \tag{4.65}$$

Note that if z_0 were outside γ, the integral would vanish.

Proof Use successively: Eq. (4.64); $\oint_\gamma dz = 0$; $\gamma = \gamma_1$ and $\gamma_2 =$ the circle, $|z - z_0| = \delta$, in Eq. (4.61); and a choice of ε and δ as in Eq. (4.51), to obtain

$$\frac{1}{2\pi i} \oint_\gamma \frac{f(z)}{z - z_0} dz - f(z_0) = \frac{1}{2\pi i} \oint_\gamma \frac{f(z) - f(z_0)}{z - z_0} dz \tag{4.66}$$

$$= \frac{1}{2\pi i} \oint_\gamma \left[\frac{f(z) - f(z_0)}{z - z_0} - f'(z_0) \right] dz \tag{4.67}$$

$$\left| \frac{1}{2\pi i} \oint_\gamma \frac{f(z)}{z - z_0} dz - f(z_0) \right| < \frac{1}{2\pi} \varepsilon \cdot 2\pi \delta = \varepsilon \delta. \tag{4.68}$$

Clearly, the left-hand side of Eq. (4.68) vanishes.

Cauchy's integral formula displays some of the remarkable tightness of complex variable theory. The values of an analytic function on the boundary of a region completely determine the value at every interior point!

2. Formula for $f^{(n)}(z)$

We now show, via Cauchy's integral formula, that the derivative of an analytic function is itself analytic, and that an analytic function possesses derivatives of all orders.

a. Formula for $f'(z)$. In Cauchy's integral formula, Eq. (4.65), put z for z_0 and ζ for z. Use also

$$\frac{1}{h}\left[\frac{1}{\zeta-z-h}-\frac{1}{\zeta-z}\right] = \left(\frac{1}{\zeta-z}\right)^2 + h\left(\frac{1}{\zeta-z}\right)^2\left(\frac{1}{\zeta-z-h}\right). \qquad (4.69)$$

Then

$$f'(z) = \lim_{h\to 0}\frac{1}{2\pi i}\oint_\gamma f(\zeta)\frac{1}{h}\left[\frac{1}{\zeta-z-h}-\frac{1}{\zeta-z}\right]d\zeta \qquad (4.70)$$

$$= \frac{1}{2\pi i}\oint_\gamma \frac{f(\zeta)}{(\zeta-z)^2}\,d\zeta + \lim_{h\to 0}\frac{h}{2\pi i}\oint_\gamma \frac{f(\zeta)}{(\zeta-z)^2(\zeta-z-h)}\,d\zeta. \qquad (4.71)$$

Since z is inside γ, $|\zeta-z|$ has a nonzero minimum for ζ on γ, and

$$\lim_{h\to 0}\left|\frac{h}{2\pi i}\oint\frac{f(\zeta)}{(\zeta-z)^2(\zeta-z-h)}\,d\zeta\right|$$
$$\leq \lim_{h\to 0}\frac{|h|}{2\pi}\frac{\max\{|f(\zeta)|\}L_\gamma}{\min\{|\zeta-z|^2\}[\min\{|\zeta-z|\}-|h|]} = 0. \qquad (4.72)$$

Thus, $f'(z)$ is given by the formula obtained by differentiating Cauchy's integral formula under the integral sign:

$$f'(z) = \frac{1}{2\pi i}\oint_\gamma \frac{f(\zeta)}{(\zeta-z)^2}\,d\zeta, \qquad z \text{ inside } \gamma. \qquad (4.73)$$

3. Complex Variable Theory

b. Formula for $f^{(2)}(z)$. Does $f'(z)$ have a derivative?

$$\lim_{h \to 0} \frac{f'(z+h) - f'(z)}{h}$$

$$= \lim_{h \to 0} \frac{1}{2\pi i} \oint_\gamma f(\zeta) \frac{1}{h} \left[\frac{1}{(\zeta - z - h)^2} - \frac{1}{(\zeta - z)^2} \right] d\zeta \quad (4.74)$$

$$= \frac{2}{2\pi i} \oint_\gamma \frac{f(\zeta)}{(\zeta - z)^3} d\zeta + \lim_{h \to 0} \frac{h}{2\pi i} \oint_\gamma f(\zeta)$$

$$\times \left[\frac{2}{(\zeta - z - h)(\zeta - z)^3} + \frac{1}{(\zeta - z - h)^2(\zeta - z)^2} \right] d\zeta. \quad (4.75)$$

As in Eqs. (4.71) and (4.72), the second integral in Eq. (4.75) vanishes, and

$$f^{(2)}(z) = \frac{2}{2\pi i} \oint_\gamma \frac{f(\zeta)}{(\zeta - z)^3} d\zeta, \quad z \text{ inside } \gamma. \quad (4.76)$$

c. Formula for $f^{(n)}(z)$. By induction, $f^{(n)}(z)$ exists and is analytic wherever $f(z)$ is analytic. Further,

$$f^{(n)}(z) = \frac{n!}{2\pi i} \oint_\gamma \frac{f(\zeta)}{(\zeta - z)^{n+1}} d\zeta, \quad z \text{ inside } \gamma, \quad (4.77)$$

as is obtained by integrating $(2\pi i)^{-1} \oint_\gamma f^{(n)}(\zeta)(\zeta - z)^{-1} d\zeta$ by parts, n times. [This argument was not used to obtain Eq. (4.76), because the analyticity of both $f'(z)$ and $f^{(2)}(z)$ was not yet shown.]

The integral formula (4.77) displays the remarkable interconnection between complex integration and differentiation. It is also the formula obtained by differentiating Cauchy's integral formula n times and interchanging the order of integration and differentiation.

G. SOME CONSEQUENCES OF CAUCHY'S INTEGRAL FORMULA, INCLUDING THE FUNDAMENTAL THEOREM OF ALGEBRA

One consequence is that analytic functions are infinitely differentiable. A few more quick consequences follow.

1. Bound for $|f^{(n)}(z)|$

Let $f(z)$ be analytic on and within a circle of radius r, centered on z, and let $|f(\zeta)| \leq M$, for ζ on the circle. By Eqs. (4.77) and (4.28),

$$|f^{(n)}(z)| \leq Mn!r^{-n} \quad \text{(Cauchy's estimate).} \quad (4.78)$$

2. Liouville's Theorem

If $f(z)$ is analytic and bounded everywhere in the complex plane, then $f(z)$ is constant.

Proof Take $n = 1$, and let $r \to \infty$ in Cauchy's estimate.

A main use of Liouville's theorem is to prove the fundamental theorem of algebra.

3. Fundamental Theorem of Algebra

A polynomial $p_n(z)$, of degree $n > 0$, has at least one complex root.

Proof If $p_n(z)$ were never zero, $1/p_n(z)$ and $p_n(z)$ itself would be constant.

4. Morera's Theorem

If $f(z)$ is continuous in a domain D, and if $\int_\gamma f(z)\, dz = 0$, for all closed γ in D, then $f(z)$ is analytic in D.

Section IV.D.3 implies that the integral of $f(z)$ is analytic, and $f(z)$ is its derivative. But the derivative of an analytic function is analytic.

5. Theorem

Let $g(\zeta)$ be piecewise continuous on a simple closed curve γ. Then

$$f(z) = \frac{1}{2\pi i} \int_\gamma \frac{g(\zeta)}{\zeta - z}\, d\zeta, \qquad z \text{ inside } \gamma, \tag{4.79}$$

defines a function analytic in the simply connected domain bounded by γ.

Proof The integral clearly exists, and the derivation of Eq. (4.73), *mutatis mutandis*, provides a formula for $f'(z)$. Note that Eq. (4.79) may not be used for z on γ, and that $f(z)$ does not necessarily approach $g(\zeta)$ as z approaches ζ on γ.

6. Maximum Modulus Theorem

Let $f(z)$ be analytic in a domain D. Then $|f(z)|$ has no maximum in D.

Proof

$$|f(z_0)| \leq |(2\pi i)^{-1} \oint_{|z-z_0|=\varepsilon} f(z)(z-z_0)^{-1} dz| \leq \max\{|f(z)|\}_{|z-z_0|=\varepsilon},$$

for all z_0 in D, and for all sufficiently small ε.

H. SUBSTITUTION FORMULA

The complex version of the substitution formula is a consequence of both Cauchy's theorem and the fundamental theorem, part 2.

Let $z = z(\zeta)$, considered as a mapping, map the curve γ_ζ one to one onto the curve γ_z. Let $f(z)$ and $z(\zeta)$ be analytic functions of z and ζ in appropriate domains. Then

$$\int_{\gamma_z} f(z)\, dz = \int_{\gamma_\zeta} f(z(\zeta)) \frac{dz(\zeta)}{d\zeta}\, d\zeta. \tag{4.80}$$

Proof Both $\int^{z(\zeta)} f(\hat{z})\, d\hat{z}$ and $\int^{\zeta} f(z(\hat{\zeta}))(dz(\hat{\zeta})/d\hat{\zeta})\, d\hat{\zeta}$ have the same derivative with respect to ζ.

I. CONNECTION WITH REAL LINE INTEGRALS

The complex integral is composed of two real line integrals. In the notation of Section IV.A, a line integral in two dimensions over the piecewise differentiable path γ is given by

$$\int_{t_a}^{t_b} \left[p(x(t), y(t)) \frac{dx(t)}{dt} + q(x(t), y(t)) \frac{dy(t)}{dt} \right] dt$$

$$= \int_{\gamma} [p(x, y)\, dx + q(x, y)\, dy], \tag{4.81}$$

where p and q are continuous on γ. Via Eq. (4.18) in line integral notation, namely,

$$\int_{\gamma} f(z)\, dz = \int_{\gamma} [u(x, y)\, dx - v(x, y)\, dy]$$
$$+ i \int_{\gamma} [v(x, y)\, dx + u(x, y)\, dy], \tag{4.82}$$

results for real line integrals can be applied to complex integrals.

The following are relevant results for real line integrals: If p, q, p_y, and q_x are continuous functions of x and y in a domain D, then

1. the line integral $\int_\gamma (p\,dx + q\,dy)$ depends only on the endpoints of γ, if and only if there is a function $\phi(x,y)$, differentiable in D, for which $p = \phi_x$ and $q = \phi_y$;

2. further, if D is simply connected, and γ is any simple closed curve in D, then $\oint_\gamma (p\,dx + q\,dy) = 0$, if and only if $p_y = q_x$ everywhere in D;

3. still further (with D simply connected),

$$\oint_\gamma (p\,dx + q\,dy) = \iint_{\text{area enclosed by }\gamma} (q_x - p_y)\,dx\,dy \qquad (4.83)$$

(Green's theorem or Gauss's theorem, restricted to two dimensions).

Result 1 is essentially the same result as the fundamental theorem, part 2 (Section IV.D.3). Either result 2 or 3 provides an immediate proof of Cauchy's theorem, if $f'(z)$ is assumed to be continuous. In either case, one first takes $p = u$ and $q = -v$, then $p = v$ and $q = u$, and obtains results separately for each term in Eq. (4.82).

V. Power Series

A main result of this section is that convergent power series and analytic function are synonymous. Other important results concern the convergence, differentiation, and integration of power series, the existence of Laurent series, and the concept of analytic continuation.

A. Elementary Definitions and Results

The definitions of infinite series and convergence are virtually the same as in real analysis.

The expressions $\sum_{k=0}^\infty c_k$, $\sum_{k=0}^\infty f_k(z)$, where $c_k \in \mathscr{C}$ and $f_k(z) \in \mathscr{C}$, are called *infinite series*.

The expressions $\sum_{k=0}^\infty c_k z^k$, $\sum_{k=0}^\infty c_k(z - z_0)^k$, where $c_k \in \mathscr{C}$, are called *power series* in z and in $z - z_0$.

The nth *partial sum* of an infinite series is the sum of the terms with indices less than or equal to n, e.g.,

$$s_n(z) = \sum_{k=0}^n c_k z^k. \qquad (5.1)$$

3. Complex Variable Theory

The series is said to *converge* (pointwise) to $s(z)$ if, given any $\varepsilon > 0$, there exists an N such that

$$|s_n(z) - s(z)| < \varepsilon \qquad \text{for all} \quad n \geq N. \tag{5.2}$$

A series, e.g., $\sum_k c_k z^k$, is said to *converge uniformly* for z in some region R if the N in Eq. (5.2) can be chosen independently of z, and to *converge absolutely* if $\sum_k |c_k z^k|$ converges. The series is said to satisfy the *Cauchy test for convergence*, if, given any $\varepsilon > 0$, there exists an N such that

$$|s_n(z) - s_m(z)| < \varepsilon \qquad \text{for all} \quad n \geq N \quad \text{and} \quad m \geq N. \tag{5.3}$$

In the case of a power series,

$$|s_n(z) - s_m(z)| = |c_{m+1}z^{m+1} + c_{m+2}z^{m+2} + \cdots + c_n z^n|, \qquad n > m. \tag{5.4}$$

If Cauchy's test is satisfied, then the series converges, and conversely. The converse follows from

$$|s_n - s_m| < |s_n - s| + |s_m - s|. \tag{5.5}$$

Proof That Cauchy's Test Implies Convergence For fixed m, there are infinitely many s_n lying within a radius ε of s_m. These s_n possess a point of accumulation s. There can be only one such point, because only a finite number of s_n can be further apart than any fixed distance.

An absolutely convergent series converges, since, e.g.,

$$|c_{m+1}z^{m+1} + c_{m+2}z^{m+2} + \cdots + c_n z^n| \leq |c_{m+1}z^{m+1}| + |c_{m+2}z^{m+2}|$$
$$+ \cdots + |c_n z^n|. \tag{5.6}$$

An extremely useful convergence test is the *Weierstrass M test*. Let $\sum b_k$ converge absolutely, let $\sum f_k$ be an infinite series, and let there be an $M > 0$ for which

$$|f_k| \leq M |b_k|, \qquad k = 0, 1, 2, \ldots. \tag{5.7}$$

Then $\sum f_k$ converges absolutely.

Proof Use Cauchy's test and

$$|f_{m+1}| + \cdots + |f_n| \leq M(|b_{m+1}| + \cdots + |b_n|). \tag{5.8}$$

The *geometric series*, $\sum_{k=0}^{\infty} z^k = 1 + z + z^2 + \cdots$, is frequently used as a comparison series. When $|z| < 1$, the geometric series converges to $(1-z)^{-1}$:

$$s_n(z) - (1-z)^{-1} = -z^{n+1}(1-z)^{-1} \to 0 \quad \text{as} \quad n \to \infty, \quad |z| < 1. \tag{5.9}$$

When $|z| \geq 1$, the series diverges.

B. Basic Theorem on the Convergence of Power Series

The convergence properties of complex power series are reasonably neat. A power series converges in the interior of a circle, called the *circle of convergence*, and it diverges everywhere in the exterior. It may or may not converge at points on the circle. A simple but useful convergence theorem is given first, followed by a sharper result.

Theorem If $\sum_{k=0}^{\infty} c_k \zeta^k$ converges, then $\sum_{k=0}^{\infty} c_k z^k$ converges absolutely for all $|z| < |\zeta|$. For fixed $r < |\zeta|$, convergence in $|z| \leq r$ is uniform.

Proof $\sum c_k \zeta^k$ converges $\Rightarrow c_k \zeta^k \to 0$ as $k \to \infty \Rightarrow$ there is an M such that $|c_k \zeta^k| \leq M$, for all k. Then, since

$$|c_k z^k| = |c_k \zeta^k| |z/\zeta|^k \leq M |z/\zeta|^k, \tag{5.10}$$

the $\sum c_k z^k$ converges absolutely when $|z/\zeta| < 1$, by the Weierstrass M test. Uniformity follows from

$$|c_{m+1} z^{m+1} + \cdots + c_n z^n| \leq |c_{m+1} r^{m+1}| + \cdots + |c_n r^n|, \quad |z| \leq r. \tag{5.11}$$

Simple reasoning permits a sharper formulation. Every power series has a *radius of convergence* R ($0 \leq R \leq \infty$). For $|z| < R$, convergence is absolute. For fixed $r < R$, convergence is uniform in $|z| \leq r$. For $|z| > R$, the series diverges.

Theorem R is given by Hadamard's formula

$$\frac{1}{R} = \limsup_{n \to \infty} |c_n|^{1/n}. \tag{5.12}$$

Proof Define R by Eq. (5.12). We show convergence for $|z| < R$ and divergence for $|z| > R$. Let $\varepsilon: 0 < \varepsilon < 1$. By definition, lim sup means

$$|c_n|^{1/n} > 1/[R(1+\varepsilon)], \qquad \text{for infinitely many } n, \qquad (5.13)$$

$$|c_n|^{1/n} > 1/[R(1-\varepsilon)], \qquad \text{for at most finitely many } n. \qquad (5.14)$$

First let $|z| > R$, and pick ε such that $|z| > R(1+\varepsilon)$. Then

$$|c_n z^n| > |c_n R^n (1+\varepsilon)^n| > 1, \qquad \text{for infinitely many } n, \qquad (5.15)$$

and the series cannot converge. Second, let $|z| < R$, and pick ε such that $|z| < R(1-\varepsilon)$. By Eq. (5.14), there is an M such that $|c_n|R^n \times (1-\varepsilon)^n \le M$, for all n, and

$$|c_n z^n| = |c_n R^n (1-\varepsilon)^n| \, |z/[R(1-\varepsilon)]|^n \le M \, |z/[R(1-\varepsilon)]|^n. \qquad (5.16)$$

Convergence follows from the Weierstrass M test.

Simple Examples The geometric series $\sum z^k$ has $R = 1$. The series $\sum k! z^k$ has $R = 0$, while $\sum (k!)^{-1} z^k$ has $R = \infty$. The series $\sum z^{k+2}/[(k+1)(k+2)]$ and the series $\sum (k+1)(k+2) z^k$ both have $R = 1$, but the first converges for all $|z| = 1$, whereas the second diverges for all $|z| = 1$.

The complex plane is the natural milieu for power series. The conventional illustration of the complex insight is the power series for $(1 + x^2)^{-1}$. This function is infinitely differentiable for all real x, but its power series, $\sum (-x^2)^k$, converges only for $x^2 < 1$. As a function of a complex variable, the divergence of the power series for $(1 + z^2)^{-1}$ when $|z| \ge 1$ is transparently understandable from the convergence theorem and the behavior of $(1 + z^2)^{-1}$ at $z = \pm i$.

C. Integration and Differentiation of Series. Analyticity

We turn now to the analysis of power series. The essential result is that, within their circle of convergence, power series are analytic functions, and they may be integrated and differentiated term by term.

Theorem Let $\sum c_k z^k$ have radius of convergence $R > 0$; let $|z| < R$; let γ lie in $|z| < R$; let L_γ be finite; and let $g(z)$ be continuous for z on γ. Then

$$\sum_{k=0}^{\infty} c_k z^k \text{ is continuous,} \qquad |z| < R \tag{5.17}$$

$$\int_\gamma g(z) \sum_{k=0}^{\infty} c_k z^k \, dz = \sum_{k=0}^{\infty} c_k \int_\gamma g(z) z^k \, dz \tag{5.18}$$

$$\sum_{k=0}^{\infty} c_k z^k \text{ is analytic,} \qquad |z| < R \tag{5.19}$$

$$\int_0^z \sum_{k=0}^{\infty} c_k z^k \, dz = \sum_{k=0}^{\infty} c_k z^{k+1}/(k+1) \tag{5.20}$$

$$(d/dz) \sum_{k=0}^{\infty} c_k z^k = \sum_{k=0}^{\infty} k c_k z^{k-1}, \tag{5.21}$$

and the radius of convergence of the integrated and derived series [Eqs. (5.20) and (5.21)] is also R.

Proof Continuity and term by term integrability are direct consequences of uniform convergence on $|z| \leq r < R$. Assume Eqs. (5.17) and (5.18) to be true. Then for any simple closed curve γ in $|z| < R$,

$$\oint_\gamma \sum_{k=0}^{\infty} c_k z^k \, dz = \sum_{k=0}^{\infty} c_k \oint_\gamma z^k \, dz = 0, \tag{5.22}$$

so that $\sum c_k z^k$ is analytic, by Morera's theorem. Equation (5.20) is a trivial consequence of Eq. (5.18), and Eq. (5.21) follows from Eq. (5.18) by choosing $g(z)$ to be $(z - z_0)^{-2}$ and γ to be a simple closed curve enclosing z_0. The statement on radius of convergence is clinched by $\lim_{n\to\infty} n^{1/n} = 1$ and Hadamard's formula.

To prove Eq. (5.17), let $|z - z_0| < \delta$, where $\delta > 0$ will be fixed later, let r satisfy $|z| < r < R$, $|z_0| < r < R$, and let s_N denote the Nth partial sum. Consider

$$\left| \sum_{k=0}^{\infty} c_k z^k - \sum_{k=0}^{\infty} c_k z_0^k \right| \leq \left| \sum_{k=0}^{\infty} c_k z^k - s_N(z) \right| + |s_N(z) - s_N(z_0)|$$
$$+ \left| \sum_{k=0}^{\infty} c_k z_0^k - s_N(z_0) \right|. \tag{5.23}$$

Pick $\varepsilon > 0$. By uniformity of convergence, N can be chosen so that

$$\left| \sum_{k=0}^{\infty} c_k \zeta^k - s_N(\zeta) \right| < \varepsilon/3 \qquad \text{for all} \quad |\zeta| \leq r. \tag{5.24}$$

3. Complex Variable Theory

By continuity of the polynomial $s_N(z)$ at z, δ can be chosen so that

$$|s_N(z) - s_N(z_0)| < \varepsilon/3 \quad \text{whenever} \quad |z - z_0| < \delta. \tag{5.25}$$

The right-hand side of Eq. (5.23) is thus less than ε, when $|z - z_0| < \delta$, and $\sum c_k z^k$ is continuous.

To prove Eq. (5.18), start with

$$\left| \int_\gamma g(z) \sum_{k=0}^\infty c_k z^k \, dz - \sum_{k=0}^n c_k \int_\gamma g(z) z^k \, dz \right|$$
$$= \left| \int_\gamma g(z) \left[\sum_{k=0}^\infty c_k z^k - s_n(z) \right] dz \right|. \tag{5.26}$$

Fix $\varepsilon > 0$. By uniform convergence, N can be chosen so that

$$\left| \sum_{k=0}^\infty c_k z^k - s_n(z) \right| < \varepsilon,$$

for all $n \geq N$ and all $|z| \leq r$. Then

$$\left| \int_\gamma g(z) \sum_{k=0}^\infty c_k z^k \, dz - \sum_{k=0}^n c_k \int_\gamma g(z) z^k \, dz \right| < \max\{|g|\} \varepsilon L_\gamma, \tag{5.27}$$

which clinches Eq. (5.18).

The preceding theorem depended almost entirely on the *uniform convergence* of the infinite series. The reader may readily verify a more general result (to be used in the next section).

Theorem on Continuity and Integrability of Uniformly Convergent Series of Continuous Functions Let $f_k(z)$ be continuous for z in a domain D, let γ be any curve in D of finite length, and let $\sum_{k=0}^\infty f_k(z)$ converge uniformly for z in D. Then $\sum_{k=0}^\infty f_k(z)$ is continuous in D, and the series may be integrated term by term. If each $f_k(z)$ is analytic in D, then so is $\sum_{k=0}^\infty f_k(z)$, and the series may be differentiated term by term.

D. Representation of Analytic Functions by Power Series

A convergent power series is an analytic function. The reverse is also true.

Theorem If $f(z)$ is analytic in $|z - z_0| < r$, then $f(z)$ can be represented by a *Taylor* series at z_0,

$$f(z) = \sum_{k=0}^{\infty} \frac{f^{(k)}(z_0)}{k!} (z - z_0)^k, \qquad |z - z_0| < r. \qquad (5.28)$$

Proof Given z, choose r_1 such that $|z - z_0| < r_1 < r$. Then substitute into Cauchy's integral formula

$$(\zeta - z)^{-1} = [(\zeta - z_0) - (z - z_0)]^{-1} = \sum_{k=0}^{\infty} (z - z_0)^k (\zeta - z_0)^{-k-1}, \qquad (5.29)$$

which converges uniformly with respect to ζ, for ζ on the circle $|\zeta - z_0| = r_1$. Term by term integration gives Eq. (5.28).

The radius of convergence of the Taylor series for $f(z)$ about z_0 must be at least as large as the distance to the nearest point at which $f(z)$ is not analytic.

Taylor series are unique: Power series can be differentiated term by term, so that the coefficient of $(z - z_0)^k$ is always $f^{(k)}(z_0)/k!$.

E. Laurent Series

Consider the simple function $f(z) = (z - 1)^{-1}(z - 2)^{-1}$. The Taylor series about $z = 0$,

$$(z-1)^{-1}(z-2)^{-1} = -(z-1)^{-1} + (z-2)^{-1} \qquad (5.30)$$

$$= \sum_{k=0}^{\infty} (1 - 2^{-k-1}) z^k, \qquad |z| < 1, \qquad (5.31)$$

diverges when $|z| > 1$. In the annulus $1 < |z| < 2$, $(z-1)^{-1}$ can be expanded in powers of z^{-1}, and $(z-2)^{-1}$ in powers of $z/2$ (both series being just the geometric series):

$$(z-1)^{-1}(z-2)^{-1} = (z-2)^{-1} - z^{-1}(1 - z^{-1})^{-1} \qquad (5.32)$$

$$= -\sum_{k=0}^{\infty} z^k/2^{k+1} - \sum_{k=1}^{\infty} z^{-k-1}, \qquad 1 < |z| < 2. \qquad (5.33)$$

The net result is a series in positive and negative powers of z that converges in the annulus $1 < |z| < 2$, but that diverges when $|z| < 1$ or $|z| > 2$. This behavior is general.

Theorem Let $f(z)$ be analytic in the annulus $R_2 < |z - z_0| < R_1$. Then $f(z)$ possesses the *Laurent series* about z_0,

$$f(z) = \sum_{k=0}^{\infty} c_k(z - z_0)^k + \sum_{k=1}^{\infty} c_{-k}(z - z_0)^{-k}, \qquad (5.34)$$

where

$$c_k = \frac{1}{2\pi i} \oint_{|\zeta - z_0| = r_1} \frac{f(\zeta)}{(\zeta - z_0)^{k+1}} \, d\zeta, \qquad k = 0, 1, 2, \ldots, \qquad (5.35)$$

$$c_{-k} = \frac{1}{2\pi i} \oint_{|\zeta - z_0| = r_2} f(\zeta)(\zeta - z_0)^{k-1} \, d\zeta, \qquad k = 1, 2, 3, \ldots, \qquad (5.36)$$

and $R_2 < r_2 < |z - z_0| < r_1 < R_1$. The series $\sum c_k(z - z_0)^k$ converges absolutely for $|z - z_0| < R_1$ and uniformly for $|z - z_0| \leq r_1$, and the series $\sum c_{-k}(z - z_0)^{-k-1}$ converges absolutely for $|z - z_0| > R_2$ and uniformly for $|z - z_0| \geq r_2$.

Proof Use Cauchy's theorem as adapted for multiply connected domains, Eq. (4.63), with γ_1 the circle $|\zeta - z_0| = r_1$, γ_2 the circle $|\zeta - z_0| = r_2$, γ_3 the circle $|\zeta - z| = \varepsilon$ (ε being small enough not to cause any difficulties), and with ζ for z and $f(\zeta)/(\zeta - z)$ for $f(z)$, to obtain

$$f(z) = \frac{1}{2\pi i} \oint_{|\zeta - z| = \varepsilon} \frac{f(\zeta)}{\zeta - z} \, d\zeta$$

$$= \oint_{|\zeta - z_0| = r_1} \frac{f(\zeta)}{\zeta - z} \, d\zeta - \oint_{|\zeta - z_0| = r_2} \frac{f(\zeta)}{\zeta - z} \, d\zeta. \qquad (5.37)$$

On $|\zeta - z_0| = r_1$, use $(\zeta - z)^{-1} = \sum_{k=0}^{\infty} (z - z_0)^k (\zeta - z_0)^{-k-1}$, and on $|\zeta - z_0| = r_2$, use $(\zeta - z)^{-1} = -\sum_{k=1}^{\infty} (\zeta - z_0)^{k-1}(z - z_0)^{-k}$. Term by term integration yields Eqs. (5.34)–(5.36), while the convergence statement is a consequence of the convergence theorem and term by term integrability theorem of Sections V.B and V.C.

Remarks (1) If $f(z)$ is analytic for $|z| < R_1$, then all the c_{-k} vanish.

(2) Laurent series are unique [substitute the Laurent series about z_0 for $f(\zeta)$ into Eqs. (5.35) and (5.36), and integrate term by term].

(3) In practice, Laurent series are seldom obtained by evaluating the integrals in Eqs. (5.35) and (5.36). The use of *ad hoc* methods, as, for example, in Eq. (5.33), is justified by the uniqueness of the result.

F. Analytic Continuation

We conclude the power series section with a discussion of analytic continuation.

In real analysis, a differentiable function defined on a line segment (a, b) can be extended to an adjacent line segment (b, c) in a virtually arbitrary manner. Not so in complex analysis. Analytic extensions are essentially unique. ("Essentially" is explained below.)

Uniqueness Theorem Let $f_1(z)$ and $f_2(z)$ both be analytic in the same domain D. Further, let $f_1(z) = f_2(z)$ at all points of an arc γ in D. Then $f_1(z) = f_2(z)$ everywhere in D.

Proof The proof is a constructive use of power series. Let z_0 be on γ. Derivatives at z_0 can be computed using only points on γ, so that $f_1^{(k)}(z_0) = f_2^{(k)}(z_0)$, for $k = 0, 1, 2, \ldots$. Thus the Taylor series at z_0 for $f_1(z)$ and $f_2(z)$ are identical, and $f_1(z) \equiv f_2(z)$ in the largest circle about z_0 that fits in D. Consider the point z_1 in D, not inside this circle. Let z_0 be connected to z_1 by a path γ_{01} of finite length L lying entirely in D. Let d be the shortest distance from γ_{01} to the boundary of D. Consider the sequence of open circles of radius d, centered at points spaced $d/2$ apart measured along γ_{01} from z_0. All these circles are contained in D. The center of the nth circle is contained in the $(n-1)$st circle on the segment of γ_{01} falling in the $(n-1)$st circle. The point z_1 is contained in at least the last circle of this finite sequence (the number of circles is approximately $2L/d$). By induction, $f_1(z) \equiv f_2(z)$ in each circle. Consequently $f_1(z_1) = f_2(z_1)$.

In the next section certain real functions defined on the real axis will be extended to analytic functions defined in domains containing the real axis. By the uniqueness theorem, if such an extension exists, it is unique.

Suppose that $f_1(z)$ is analytic in a domain D_1, that $f_2(z)$ is analytic in a domain D_2, that the intersection of D_1 with D_2 is a (nonempty) domain D, and that $f_1(z) \equiv f_2(z)$ everywhere in D. Then $f_2(z)$ is said to be an *analytic continuation* of $f_1(z)$ from D_1 into D_2, and vice versa. (See Fig. 5a.) The construction used in proving the uniqueness theorem clearly can be used to continue analytically a function from one domain into another, or from a smaller domain into a larger domain, to the extent permitted by the function itself.

Let $f_1(z)$ be analytic in a domain D_1, and let the domains D_2 and D_3 intersect D_1 and each other as illustrated in Fig. 5b. Let z_0 be in both

FIG. 5. Intersecting domains for analytic continuation.

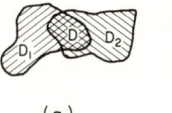
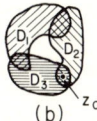
(a) (b)

D_2 and D_3 but not in D_1. Let there be analytic continuations $f_2(z)$ and $f_3(z)$ of $f_1(z)$ from D_1 into D_2 and D_3, respectively. Both $f_2(z)$ and $f_3(z)$ are uniquely determined, but it is not necessarily true that $f_2(z_0) = f_3(z_0)$. Stated alternatively, $f_2(z)$ is not necessarily a direct analytic continuation of $f_3(z)$ from D_2 into D_3, even though both are direct analytic continuations of $f_1(z)$. Specific examples in the next section illustrate this situation.

Finally, we contrast a property of power series with a property of Cauchy's integral formula: The former permits values of $f(z)$ to be determined in an outer region from values in an inner region, whereas the latter permits values of $f(z)$ to be determined in an inner region from values in an outer region.

VI. Elementary Functions

We have already encountered the simplest elementary functions: polynomials, rational functions, and nth roots. In this section, properties of these and other elementary functions, including the exponential and logarithm, are developed in some detail.

A. SINGULARITIES. ZEROS. THE POINT AT INFINITY

Some terminology facilitating the description of functions is given first.

A function analytic in the entire z plane is called an *entire function*. Examples: polynomials; $\sum_{k=0}^{\infty} z^k/k!$.

A point z_0 at which $f(z)$ is not analytic is called a *singularity* of f. Example: $(z - z_0)^{-1}$.

A point z_0 is called an *isolated singularity* of $f(z)$ if $f(z)$ is analytic everywhere in some neighborhood of z_0, except at the point z_0 itself. Example: same as preceding example.

Let z_0 be an isolated singularity of $f(z)$, and let $f(z)$ have the Laurent series

$$f(z) = \sum_{k=0}^{\infty} c_k(z-z_0)^k + \sum_{k=1}^{\infty} c_{-k}(z-z_0)^{-k}, \quad 0 < |z-z_0| < \varepsilon. \quad (6.1)$$

1. If $c_{-k} = 0$ for all $k \geq 1$, then f is said to have a *removable singularity* at z_0 [removed by defining $f(z_0)$ to be c_0]. A removable singularity is an unnatural, correctable singularity.

2. If $c_{-n} \neq 0$, $c_{-m} = 0$ for all $m > n$, then f is said to have a *pole of order n* at z_0.

3. If $c_{-n} \neq 0$ for infinitely many $n > 0$, then f is said to have an *essential singularity* at z_0. Example: $\sum_{k=0}^{\infty} (z - z_0)^{-k}/k!$.

4. The $\sum_{k=1}^{\infty} c_{-k}(z - z_0)^{-k}$ in Eq. (6.1) where z_0 is an isolated singularity of $f(z)$, is called the *principal part* of f at z_0.

Note: If f has a pole of order n at z_0, then

$$g(z) = (z - z_0)^n f(z), \qquad z \neq z_0 \qquad (6.2)$$

$$= c_{-n}, \qquad z = z_0 \qquad (6.3)$$

$$= \sum_{k=0}^{\infty} c_{k-n}(z - z_0)^k \qquad (6.4)$$

is analytic and nonvanishing at z_0.

If the first n terms of the Taylor series for $f(z)$ at z_0 vanish,

$$f(z) = \sum_{k=n}^{\infty} c_k(z - z_0)^k, \qquad n \geq 1, c_n \neq 0, \qquad (6.5)$$

then f is said to have a *zero of order n* at z_0.

Simple pole and *simple zero* are synonymous with first-order pole and first-order zero.

Note: $f(z)$ has a pole of order n at z_0 if and only if $1/f(z)$ has a zero of order n at z_0.

If $f(1/\zeta)$ has a pole or zero at $\zeta = 0$, we say that $f(z)$ has a pole or zero at $z = \infty$.

By the behavior of $f(z)$ at the *point at infinity*, we mean the behavior of $f(1/\zeta)$ at the point $\zeta = 0$.

By *extended complex plane*, we mean the complex plane plus the point at infinity.

B. Algebraic Functions

Let $w = w(z)$ and let $p(w, z)$ be a polynomial in w and z for which $p(w(z), z) \equiv 0$. Then $w(z)$ is said to be an *algebraic function* of z. Polynomials, rational functions, and nth roots are examples of algebraic functions that we discuss below. For a discussion of more general algebraic functions, the reader is referred to other texts (e.g., Ahlfors, 1966).

1. Polynomials and Rational Functions

Polynomials are entire functions. We add here one more fact: If $p_n(z)$ is a polynomial of degree n,

$$p_n(z) = \sum_{k=0}^{n} c_k z^k, \qquad c_n \neq 0, \tag{6.6}$$

then by the fundamental theorem of algebra and the division algorithm [in this case, $p_n(z) = (z - a)p_{n-1}(z) + b$ where p_{n-1} is a polynomial of degree $n - 1$, and b is a constant], $p_n(z)$ has precisely n (not necessarily distinct) roots, $z_1, z_2, \ldots, z_n$, and the unique factorization

$$p_n(z) = c_n(z - z_1)(z - z_2) \cdots (z - z_n). \tag{6.7}$$

The additional features of rational functions are essentially those of the reciprocal of a polynomial. From Eq. (6.7), the singularities of the reciprocal of a polynomial are poles at the zeros of the polynomial. A useful fact is that the reciprocal of a polynomial can be resolved into *partial fractions*. Let $z_1, z_2, \cdots, z_m$ all be distinct. Then

$$(z - z_1)^{-n_1}(z - z_2)^{-n_2} \cdots (z - z_m)^{-n_m} = \sum_{j=1}^{m} \sum_{k=0}^{n_j-1} c_{jk}(z - z_j)^{-n_j+k} \tag{6.8}$$

where the constants c_{lk} are given explicitly by

$$c_{lk} = (1/k!)(d/dz_l)^k \prod_{\substack{j=1 \\ (j \neq l)}}^{m} (z_l - z_j)^{-n_j}. \tag{6.9}$$

That Eq. (6.8) exists in principle follows inductively from

$$(z - z_1)^{-1}(z - z_2)^{-1} = [(z - z_1)^{-1} - (z - z_2)^{-1}](z_1 - z_2)^{-1}. \tag{6.10}$$

The explicit formula for c_{lk} is derived in Section VII.D.

In many texts, the special rational function

$$w(z) = c[(z - z_1)/(z - z_2)], \qquad z_1 \neq z_2, \tag{6.11}$$

is discussed in detail. Variously known as a *linear transformation, bilinear transformation, linear fractional transformation,* and *Möbius transformation,* its importance is that it maps the extended z plane one to one onto the extended w plane, and that it maps circles into circles (with straight lines regarded as limiting cases of circles).

2. nth Roots. Multiple-Valued Functions

The singularities of the nth roots of z, called *branch points* and *branch cuts*, are not isolated; nth roots lead naturally to the concepts *multiple-valued function* and *Riemann surface*.

We proceed heuristically. Equations (2.52)–(2.54) define n nth roots of z, which we denote by $w_k(z)$,

$$w_k(z) = |z|^{1/n}\left(\cos\frac{\arg z}{n} + i\sin\frac{\arg z}{n}\right),$$

$$(2k-1)\pi < \arg z \leq (2k+1)\pi, \qquad k = 0, 1, 2, \ldots, n-1,$$
(6.12)

$$= \omega_n{}^k w_0(z). \tag{6.13}$$

Each $w_k(z)$ satisfies the Cauchy–Riemann equations in $(2k-1)\pi < \arg z < (2k+1)\pi$, and has the derivative

$$\frac{dw_k(z)}{dz} = \frac{1}{n}\frac{1}{z}w_k(z), \quad \text{i.e.,} \quad \frac{1}{n}z^{(1/n)-1}. \tag{6.14}$$

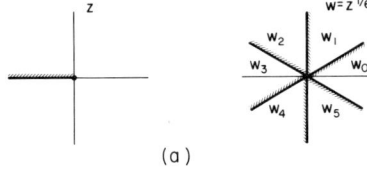

(a)

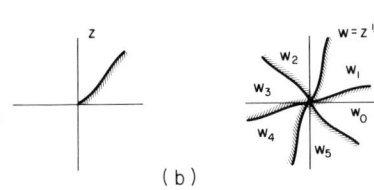

(b)

FIG. 6. Branch cuts, domains, and ranges for the six sixth roots of z. (The shading is only to facilitate visualization.) (a) Branch cut along $(-x)$ axis; (b) another branch cut.

Each nth root is analytic in the domain consisting of the entire complex plane with the origin and negative real axis removed. Figure 6 illustrates the domain and range of each w_k for $n = 6$.

We examine in detail the nature of the singularities of $w_k(z)$.

   a. dw_k/dz does not exist at $z = 0$. There is a singularity at the origin.
   b. The origin is not an isolated singularity. For $x > 0$, $\varepsilon > 0$, then

$$\lim_{\varepsilon \to 0}[\text{Arg}(-x + i\varepsilon) - \text{Arg}(-x - i\varepsilon)] = 2\pi,$$

and

$$\lim_{\varepsilon \to 0} [w_0(-x + i\varepsilon) - w_0(-x - i\varepsilon)] = (1 - \omega_n^{-1})w_0(-x) \neq 0.$$

Thus the domain of analyticity can contain no circle or other simple closed curve that encloses the origin.

c. Any simple curve running from the origin to infinity defines a domain in which arg z can be both single valued and continuous and in which n analytic nth roots can be defined (see Fig. 6).

d. The "domain" is as much a part of the definition of a function as the "rule." Thus there can be as many sets of analytic nth root functions as there are simple curves drawn from the origin to infinity. Nevertheless, the totality of values taken on remains the same.

e. The simple curve that runs from 0 to ∞ is called a *branch cut*. The two endpoints, 0 and ∞, are called *branch points*. The functions $w_k(z)$ are called *branches* of $z^{1/n}$. For definiteness,

$$|z|^{1/n}\left(\cos\frac{\text{Arg } z}{n} + i\sin\frac{\text{Arg } z}{n}\right), \quad -\pi < \text{Arg } z \leq \pi, \quad (6.15)$$

with domain of analyticity $|\text{Arg } z| < \pi$, is called the *principal value* or *principal branch* of $z^{1/n}$.

The branches of $z^{1/n}$ are even more intimately connected than by the simple relation, Eq. (6.13), or by the fact that they all satisfy $w^n = z$. Let D_1 denote the second quadrant and D_2 the left half-plane. Let $\pi/2 < \arg z < \pi$ in D_1, and $\pi/2 < \arg z < 3\pi/2$ in D_2. In the sense of Section V.F, $w_1(z)$ is the *analytic continuation* of $w_0(z)$ across the negative real axis from the second quadrant into the third quadrant. Indeed, examination of Eq. (6.12) and Fig. 6 shows that each $w_k(z)$ is the analytic *continuation* of $w_{k-1}(z)$ from the second quadrant across the branch cut to the third, with w_{n-1} leading to w_0.

This situation is unified by the concept of *multiple-valued function*. One says that $z^{1/n}$ is a multiple-valued function. When a suitable branch cut is specified, the resulting $w_k(z)$ are said to be the *single-valued branches* of the *multiple-valued function* $z^{1/n}$. It is to be emphasized that "multiple-valued function" means not a function in the sense of Section III.A, but a "*collection* of functions related by analytic continuation."

More precisely, we say that $f_1(z)$ and $f_2(z)$, both analytic in the same simply connected domain D, are *branches of the same multiple-valued analytic function* $f(z)$ if there is a finite sequence of functions, $g_1(z), g_2(z), \ldots, g_n(z)$, analytic in the simply connected domains $D_1, D_2, \ldots, D_n$, such that $D \cap D_1$, $D_1 \cap D_2$, ..., $D_n \cap D$ are all nonempty simply connected domains, and such that g_1 is the analytic continuation of f_1, g_2 is the analytic continuation of g_1, ..., g_n is the analytic continuation of g_{n-1}, and f_2 is the analytic continuation of g_n.

3. Riemann Surface

Further insight into multiple-valued functions is provided by the notion *Riemann surface*, which we discuss for $z^{1/n}$.

Imagine taking n sheets of paper, infinite in extent, and slicing each from 0 to $-\infty$ on the real axis. Number the sheets from 0 to $n-1$. Join the second quadrant of each sheet to the third quadrant of the next sheet at the cut. Imagine also connecting sheet $n-1$ back to sheet 0. Assign $w_k(z)$ to sheet k. Follow a path that winds around the origin n times, from sheet 0 onto sheet 1, from sheet 1 onto sheet 2, ..., from sheet $n-2$ to sheet $n-1$, from sheet $n-1$ to sheet 0, back to the starting point. The value of $z^{1/n}$ varies continuously along this path. The hypothetical construction is called the *Riemann surface* for $z^{1/n}$. The individual cut planes are called *Riemann sheets*. One crosses from one Riemann sheet to the next when crossing a branch cut. We may regard $z^{1/n}$ as a single-valued analytic function of z, where z varies continuously on the Riemann surface for $z^{1/n}$.

C. The Exponential and Related Functions

Nonalgebraic functions are called transcendental. The exponential, logarithmic, trigonometric, and hyperbolic functions are the elementary transcendental functions of real analysis. In complex analysis, these different functions all turn out to be aspects of a single analytic function, the exponential.

1. Definition and Properties of the Exponential Function

Four equations are basic to the exponential function, denoted by $\exp(z)$ or just e^z:

$$e^z = \sum_{k=0}^{\infty} z^k/k! \qquad (6.16)$$

$$(d/dz)e^z = e^z, \qquad e^0 = 1 \qquad (6.17)$$

$$\exp(z_1)\exp(z_2) = \exp(z_1 + z_2) \qquad (6.18)$$

$$e^{x+iy} = e^x(\cos y + i \sin y). \qquad (6.19)$$

Any one of Eqs. (6.16) to (6.19) can be chosen as fundamental. The other three then follow. (Here e^x means the known real exponential function of a real variable x.) We adopt the power series (6.16) as the definition of e^z. Then the following comments pertain:

(a) The exponential function is an entire function of z.
(b) Since $\sum_{k=0}^{\infty} x^k/k!$ is the known power series for the real exponential, e^z is the unique analytic extension of e^x.
(c) Differentiation of Eq. (6.16) term by term gives Eq. (6.17).
(d) Equation (6.18) follows from Eq. (6.17) and from

$$(d/dz)(e^z e^{a-z}) = 0.$$

(e) Equation (6.19) follows from Eqs. (6.18) and (6.16) and from the known power series for the real functions $\cos y$ and $\sin y$.
(f) If $|z| = r$, and $\arg z = \theta$, then

$$z = |z| [\cos(\arg z) + i \sin(\arg z)] = |z| \exp(i \arg z) = re^{i\theta}. \quad (6.20)$$

(g) Modulus of e^z:

$$|e^z| = e^x; \quad |e^{i\theta}| = 1, \quad \theta \text{ real}. \quad (6.21)$$

(h) e^z is never zero or infinity (z finite).
(i) e^z is periodic with period $2\pi i$. The only period for e^z is $2\pi i$:

$$\exp(z + 2\pi i) = e^z \quad (6.22)$$

$$e^{2\pi i} = 1 \quad (6.23)$$

$$e^{\pi i} = -1. \quad (6.24)$$

(j) Complex conjugate:

$$\overline{e^z} = e^{\bar{z}}. \quad (6.25)$$

(k) In general, if $f(z)$ is analytic and $f(x)$ is real, then $\overline{f(z)} = f(\bar{z})$. This is easy to see if $f(z)$ is analytic in a circle centered on Re z with radius greater than Im z.
(l) $w = e^z$ maps lines of constant x into circles $|w| = e^x$, and lines of constant y into rays $\arg w = y$. The domain $\{x < 0, |y| < \pi\}$ is mapped into $\{|w| < 1\}$, and $\{x > 0, |y| < \pi\}$ is mapped into $\{|w| > 1\}$, as in Fig. 7.

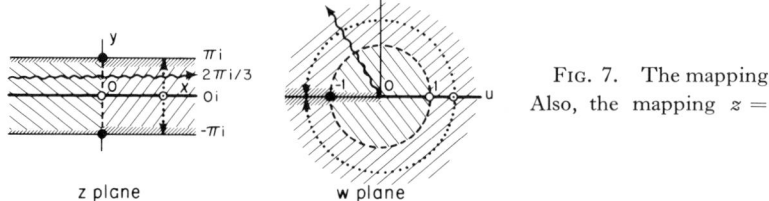

Fig. 7. The mapping $w = e^z$. Also, the mapping $z = \text{Log } w$.

2. The Trigonometric and Hyperbolic Functions

For complex z, the functions $\sin z$, $\cos z$, $\sinh z$, and $\cosh z$ are defined by

$$\sin z = (e^{iz} - e^{-iz})/2i \qquad (6.26)$$

$$\cos z = (e^{iz} + e^{-iz})/2 \qquad (6.27)$$

$$\sinh z = (e^z - e^{-z})/2 = -i \sin iz \qquad (6.28)$$

$$\cosh z = (e^z + e^{-z})/2 = \cos iz. \qquad (6.29)$$

The trigonometric and hyperbolic sines and cosines are entire functions. They satisfy a number of simple relations that are immediate consequences of their definitions via exponentials:

$$\sin z = \sum_{k=0}^{\infty} (-1)^k z^{2k+1}/(2k+1)! \qquad (6.30)$$

$$\cos z = \sum_{k=0}^{\infty} (-1)^k z^{2k}/(2k)! \qquad (6.31)$$

$$\sinh z = \sum_{k=0}^{\infty} z^{2k+1}/(2k+1)! \qquad (6.32)$$

$$\cosh z = \sum_{k=0}^{\infty} z^{2k}/(2k)!, \qquad (6.33)$$

$$e^{iz} = \cos z + i \sin z, \quad z \text{ complex}, \qquad (6.34)$$

$$(d/dz) \sin z = \cos z, \quad (d/dz) \sinh z = \cosh z \qquad (6.35)$$

$$(d/dz) \cos z = -\sin z, \quad (d/dz) \cosh z = \sinh z, \qquad (6.36)$$

$$\sin^2 z + \cos^2 z = 1 = \cosh^2 z - \sinh^2 z, \qquad (6.37)$$

$$\overline{\sin z} = \sin \bar{z}, \quad \overline{\cos z} = \cos \bar{z} \qquad (6.38)$$

$$\overline{\sinh z} = \sinh \bar{z}, \quad \overline{\cosh z} = \cosh \bar{z}, \qquad (6.39)$$

$$2\,|\sin z\,|^2 = \cosh 2y - \cos 2x \tag{6.40}$$
$$2\,|\sinh z\,|^2 = \cosh 2x - \cos 2y \tag{6.41}$$
$$2\,|\cos z\,|^2 = \cosh 2y + \cos 2x \tag{6.42}$$
$$2\,|\cosh z\,|^2 = \cosh 2x + \cos 2y, \tag{6.43}$$

$$\sin(z_1 + z_2) = \sin z_1 \cos z_2 + \cos z_1 \sin z_2 \tag{6.44}$$
$$\sinh(z_1 + z_2) = \sinh z_1 \cosh z_2 + \cosh z_1 \sinh z_2 \tag{6.45}$$
$$\cos(z_1 + z_2) = \cos z_1 \cos z_2 - \sin z_1 \sin z_2 \tag{6.46}$$
$$\cosh(z_1 + z_2) = \cosh z_1 \cosh z_2 + \sinh z_1 \sinh z_2, \tag{6.47}$$

$$\sin(z + z_0) = \sin z \quad \text{(all } z\text{)} \Leftrightarrow z_0 = \pm 2\pi n, \quad n = 0, 1, 2, \ldots \tag{6.48}$$
$$\cos(z + z_0) = \cos z \quad \text{(all } z\text{)} \Leftrightarrow z_0 = \pm 2\pi n, \quad n = 0, 1, 2, \ldots, \tag{6.49}$$

$$\sin z = 0 \Leftrightarrow z = \pm \pi n, \quad n = 0, 1, 2, \ldots \tag{6.50}$$
$$\cos z = 0 \Leftrightarrow z = (\pm n + \tfrac{1}{2})\pi, \quad n = 0, 1, 2, \ldots, \tag{6.51}$$

$$\sin z = \sin(\pi - z) = -\sin(\pi + z) = -\sin(-z) \tag{6.52}$$
$$\cos z = -\cos(\pi - z) = -\cos(z + \pi) = \cos(-z), \tag{6.53}$$

$$\sin(\tfrac{1}{2}\pi - z) = \cos z, \qquad \cos(\tfrac{1}{2}\pi - z) = \sin z. \tag{6.54}$$

By Eqs. (6.30)–(6.33) and the known power series for the real functions, $\sin z$, $\cos z$, $\sinh z$, and $\cosh z$ are the unique analytic extensions of the corresponding real functions. All the identities for the real functions carry over to the complex functions. Note also Eq. (6.34) for e^{iz}. Note especially the location of the zeros of $\sin z$ and $\cos z$. The zeros are all simple zeros.

The derivations of most of Eqs. (6.30)–(6.54) are fairly simple. We single out only Eq. (6.50):

$$\sin z = 0, \quad \Rightarrow e^{iz} - e^{-iz} = 0, \quad \Rightarrow e^{2iz} = 1, \quad \Rightarrow 2iz = \pm 2\pi i n. \tag{6.55}$$

The other trigonometric and hyperbolic functions are defined in terms of $\sin z$, $\cos z$, $\sinh z$, and $\cosh z$, as in the real case.

$$\tan z = \sin z/\cos z, \qquad \tanh z = \sinh z/\cosh z, \tag{6.56}$$
$$\cot z = 1/\tan z, \qquad \coth z = 1/\tanh z, \tag{6.57}$$
$$\sec z = 1/\cos z, \qquad \operatorname{sech} z = 1/\cosh z, \tag{6.58}$$
$$\csc z = 1/\sin z, \qquad \operatorname{csch} z = 1/\sinh z. \tag{6.59}$$

The usual addition and differentiation formulas hold. For example,

$$(d/dz) \tan z = 1/\cos^2 z. \tag{6.60}$$

Note that $\tan z$ has period π, simple poles at $(n + \tfrac{1}{2})\pi$ ($n = 0, \pm 1, \pm 2, \ldots$), and simple zeros at $n\pi$ ($n = 0, \pm 1, \pm 2, \ldots$).

3. *The Logarithm. Principal Branch*, Log z

The logarithm is the inverse of the exponential; $\log z$ is a multiple-valued function. We first discuss the *principal value* or *principal branch* of $\log z$, denoted Log z, whose domain of analyticity is the plane cut from 0 to $-\infty$ along the negative real axis. The equations basic to the logarithm are

$$\text{Log } z = \int_1^z \zeta^{-1}\, d\zeta, \quad z \neq 0 \neq \zeta, \quad -\pi < \text{Arg } z \leq \pi$$
$$-\pi < \text{Arg } \zeta \leq \pi \tag{6.61}$$

$$\text{Log } z = \text{Log } |z| + i \text{ Arg } z, \quad -\pi < \text{Arg } z \leq \pi, \tag{6.62}$$

$$(d/dz) \text{Log } z = z^{-1}, \quad \text{Log } 1 = 0, \quad z \neq 0, \quad |\text{Arg } z| < \pi \tag{6.63}$$

$$e^{\text{Log } z} = z, \quad z \neq 0; \quad \text{Log } e^z = z, \quad -\pi < \text{Im } z \leq \pi, \tag{6.64}$$

$$\text{Log } z = -\sum_{k=0}^{\infty} (1-z)^{k+1}/(k+1), \quad |1-z| < 1 \tag{6.65}$$

$$\text{Log}(1-z) = -\sum_{k=0}^{\infty} z^{k+1}/(k+1), \quad |z| < 1, \tag{6.66}$$

$$\text{Log}(z_1 z_2) = \text{Log } z_1 + \text{Log } z_2, \quad -\pi < \text{Arg } z_1 + \text{Arg } z_2 \leq \pi \tag{6.67}$$

$$= \text{Log } z_1 + \text{Log } z_2 - 2\pi i, \quad \pi < \text{Arg } z_1 + \text{Arg } z_2 \leq 2\pi \tag{6.68}$$

$$= \text{Log } z_1 + \text{Log } z_2 + 2\pi i, \quad -2\pi < \text{Arg } z_1 + \text{Arg } z_2 \leq -\pi. \tag{6.69}$$

One could start with any one of Eqs. (6.61) to (6.69) and obtain the others as consequences. We choose Eq. (6.61) to define Log z. Then the following remarks are pertinent:

(a) Since z^{-1} is analytic in the cut plane $|\operatorname{Arg} z| < \pi$, and the cut plane is a simply connected domain, by Cauchy's theorem and the converse of the fundamental theorem (Section IV.D.3), $\int_1^z \zeta^{-1} d\zeta$ defines a function, analytic in the cut plane, with derivative given by Eq. (6.63). The values of Log z on Arg $z = \pi$ are obtained by continuity from above.

(b) Log $|z| + i$ Arg z [Eq. (6.62)] corresponds to the path, $1 \leq \zeta \leq |z|$, then $|\zeta| = |z|$. When z is real and positive, Log z is Log $|z|$, the real, natural logarithm, so that Log z is the unique analytic extension of Log x.

(c) Log(0) is not defined.

(d) The range of Log z is the infinite strip, $-\pi < \operatorname{Im} z \leq \pi$. (See Fig. 7.)

(e) Equations (6.64) may be regarded as the consequence either of the formula for the real logarithm or of (d/dz) Log $e^z = 1$ (use chain rule).

(f) The power series is the term by term integral of the geometric series.

(g) The addition formula follows from Eq. (6.64).

(h) Note that

$$\operatorname{Log}(1/z) = -\operatorname{Log} z, \qquad |\operatorname{Arg} z| < \pi. \tag{6.70}$$

4. *Logarithmic Branch Point.* log z

By remark (a) just given, any cut from 0 to ∞ by a simple curve makes $\int_1^z \zeta^{-1} d\zeta$ an analytic logarithm. The situation is analogous to the $z^{1/n}$ discussed in Sections VI.B.2 and VI.B.3. Log z is one branch of a multiple-valued function, log z. The values of log z are

$$\log z = \operatorname{Log} z + 2\pi i n, \qquad n = 0, \pm 1, \pm 2, \ldots . \tag{6.71}$$

One uses log z to denote both the multiple-valued function and any particular single-valued branch. The Riemann surface for log z is analogous to that for $z^{1/n}$, except that the sheets form an infinite spiral. There is no first sheet or last sheet. Equations (6.61)–(6.69) may be rewritten with log z, provided one qualifies each by the phrase, "for a suitable branch of log z." Equations (6.74) and (6.75) below, however, hold for all branches.

$$\log z = \int_1^z \zeta^{-1}\,d\zeta, \qquad z \neq 0 \neq \zeta \tag{6.72}$$

$$\log z = \text{Log}\,|z| + i\arg z, \tag{6.73}$$

$$(d/dz)\log z = z^{-1} \qquad \text{(all branches)} \tag{6.74}$$

$$e^{\log z} = z \qquad \text{(all branches)} \tag{6.75}$$

$$\log e^z = z, \tag{6.76}$$

$$\log z = 2\pi n\,i - \sum_{k=0}^{\infty}(1-z)^{k+1}/(k+1), \qquad |1-z| < 1 \tag{6.77}$$

$$\log(1-z) = 2\pi n\,i - \sum_{k=0}^{\infty} z^{k+1}/(k+1), \qquad |z| < 1 \tag{6.78}$$

$$\log(z_1 z_2) = \log z_1 + \log z_2. \tag{6.79}$$

5. Inverse Trigonometric and Hyperbolic Functions

Since the trigonometric and hyperbolic functions are periodic, the inverse functions are multiple valued. Indeed, they can be explicitly expressed in terms of the logarithm. We discuss only arc sin z, which is typical. In every equation immediately below containing arc sin and log, the words "for a suitable branch" are understood.

$$z = \sin w = (e^{iw} - e^{-iw})/2i \tag{6.80}$$

can be inverted to give

$$w = \arcsin z = -i\log(iz + (1-z^2)^{1/2}). \tag{6.81}$$

The multiple valuedness arises from both the log and the $(1-z^2)^{1/2}$. From the log, one set of branches is characterized by values differing by $2\pi n$, whereas the second set reflects the identity $\sin(\pi - w) = \sin w$, in that

$$-i\log(iz + (1-z^2)^{1/2}) - i\log(iz - (1-z^2)^{1/2}) = -i\log(-1)$$
$$= (2n+1)\pi. \tag{6.82}$$

One can easily compute

$$(d/dz)\arcsin z = (1-z^2)^{-1/2}. \tag{6.83}$$

The other inverse functions behave similarly, with identities and differentiation formulas being the same as their real counterparts.

6. The Functions z^α and α^z

Define z^α and α^z by

$$z^\alpha = \exp(\alpha \log z) \tag{6.84}$$

$$\alpha^z = \exp(z \log \alpha). \tag{6.85}$$

Some remarks follow:

(a) When α is not an integer, z^α is multiple valued, with a branch cut from 0 to ∞ required to specify single-valued branches. On any branch

$$(d/dz)z^\alpha = \alpha z^{\alpha-1}. \tag{6.86}$$

When α is a negative integer, there is a pole at $z = 0$.

(b) The values of z^α on two contiguous branches differ by a *factor* $e^{2\pi i \alpha}$. The number of branches is infinite, unless α is rational.

(c) When α is rational, the definition above agrees with Eq. (2.55) $[z^{m/n} = (z^m)^{1/n}]$.

(d) Given α, there are an infinite number of such functions α^z, differing by factors $e^{2\pi i n z}$. When a particular value of $\log \alpha$ is fixed, α^z is an entire function, and

$$(d/dz)\alpha^z = \alpha^z \log \alpha. \tag{6.87}$$

(e) The exponential e^z agrees with Eq. (6.85) in the sense $\alpha = e$, provided that one uses $\log e = 1$ in Eq. (6.85), and not $1 + 2\pi i n$ ($n \neq 0$).

VII. Evaluation of Real Definite Integrals

Some classes of real definite integrals can be recast as integrals around simple closed curves in the complex plane, $\oint_\gamma f(z)\, dz$. If $f(z)$ is analytic on and within γ, the integral vanishes. If $f(z)$ has a single isolated singularity inside γ at z_0, then the integral is $2\pi i$ times the coefficient of $(z - z_0)^{-1}$ in the Laurent series for $f(z)$ about z_0. This seemingly trivial observation is the basis for a powerful technique for evaluating integrals and is stated more precisely below.

A. Residue Theorem

Let $f(z)$ have an isolated singularity at z_0. Then the *residue* of $f(z)$ at z_0 is defined as the coefficient of $(z - z_0)^{-1}$ in the Laurent series for $f(z)$ about z_0.

Residue Theorem Let $f(z)$ be analytic on and within a simple closed curve γ, except at a finite number of points $z_1, z_2, \ldots, z_n$, which do not lie on γ. Then

$$\oint_\gamma f(z)\, dz = 2\pi i \sum_{k=1}^{n} \{\text{residue of } f(z) \text{ at } z_k\}. \tag{7.1}$$

Equation (7.1) is an immediate consequence of Cauchy's theorem [Eq. (4.63)] and Eq. (5.36).

B. Five Classes of Integrals Amenable to the Residue Theorem

As examples of applications of the residue theorem, we evaluate integrals of five different classes. In what follows, $p_m(z)$ and $q_n(z)$ denote polynomials of degrees m and n, $r(\cos\theta, \sin\theta)$ denotes a rational function of $\cos\theta$ and $\sin\theta$ with no poles on $0 \leq \theta \leq 2\pi$, and α and k are real numbers.

First the integrals, and the results:

$$\int_{-\infty}^{\infty} \frac{p_m(x)}{q_n(x)}\, dx$$
$$= 2\pi i \times \sum \left\{\text{residues of } \frac{p_m}{q_n} \text{ in upper half plane}\right\}$$
$$q_n(x) \neq 0, \quad -\infty < x < \infty; \quad n \geq m+2, \tag{7.2}$$

$$\int_{-\infty}^{\infty} e^{ikx} \frac{p_m(x)}{q_n(x)}\, dx$$
$$= 2\pi i \times \sum \left\{\text{residues of } \left(e^{ikz}\frac{p_m}{q_n}\right) \text{ in upper half plane}\right\},$$
$$q_n(x) \neq 0, \quad -\infty < x < \infty; \quad n \geq m+1; \quad k > 0, \tag{7.3}$$

$$= -2\pi i \times \sum \left\{\text{residues of } \left(e^{ikz}\frac{p_m}{q_n}\right) \text{ in lower half plane}\right\},$$
$$q_n(x) \neq 0, \quad -\infty < x < \infty; \quad n \geq m+1; \quad k < 0, \tag{7.4}$$

$$\int_0^{\infty} x^\alpha \frac{p_m(x)}{q_n(x)}\, dx$$
$$= \frac{2\pi i}{1 - e^{2\pi i \alpha}} \sum \left\{\text{residues of } \left(z^\alpha \frac{p_m}{q_n}\right) \text{ in cut plane}\right\},$$
$$q_n(x) \neq 0, \quad 0 \leq x < \infty; \quad -1 < \alpha < 1; \quad \alpha \neq 0;$$
$$n > m + \alpha + 1; \quad 0 < \arg z < 2\pi, \tag{7.5}$$

$$\int_0^\infty \frac{p_m(x)}{q_n(x)} dx$$
$$= -\sum \left\{ \text{residues of } \left(\log z \, \frac{p_m}{q_n} \right) \text{ in cut plane} \right\},$$
$$q_n(x) \neq 0, \quad 0 \leq x < \infty; \quad n \geq m+2; \quad 0 < \arg z < 2\pi, \quad (7.6)$$

$$\int_0^{2\pi} r(\cos\theta, \sin\theta) \, d\theta$$
$$= 2\pi \sum \left\{ \text{residues of } \left[z^{-1} r\left(\frac{z+z^{-1}}{2}, \frac{z-z^{-1}}{2i} \right) \right] \text{ inside } |z|=1 \right\}. \quad (7.7)$$

The Derivations Equation (7.7) is obtained by a transformation to an integral around the unit circle. Substitute $z = e^{i\theta}$, $\cos\theta = (z+z^{-1})/2$, $\sin\theta = (z-z^{-1})/2i$, and $d\theta = -iz^{-1} dz$. Then

$$\int_0^{2\pi} r(\cos\theta, \sin\theta) \, d\theta = -i \oint_{|z|=1} z^{-1} r\left(\frac{z+z^{-1}}{2}, \frac{z-z^{-1}}{2i} \right) dz, \quad (7.8)$$

and Eq. (7.7) follows by the residue theorem.

Equation (7.2) is *in essence* derived by shortening the integration path to $[-R, R]$ and then by closing the integration path by a large semicircle. Since p_m/q_n vanishes like $R^{-(n-m)} \leq R^{-2}$ for large $|z| = R$, the integral around the semicircle vanishes at least as fast as R^{-1}:

$$\lim_{R \to \infty} \int_0^\pi \frac{p_m(Re^{i\theta})}{q_n(Re^{i\theta})} iRe^{i\theta} \, d\theta = 0, \quad n \geq m+2. \quad (7.9)$$

(The semicircle $|z| = R$, $0 \leq \arg z \leq \pi$, has the parametric representation $z = Re^{i\theta}$, $0 \leq \theta \leq \pi$.) Then

$$\int_{-\infty}^\infty \frac{p_m(x)}{q_n(x)} dx = \lim_{R \to \infty} \int_{-R}^R \frac{p_m(x)}{q_n(x)} dx \quad (7.10)$$

$$= \lim_{R \to \infty} \left\{ \int_{-R}^R \frac{p_m(x)}{q_n(x)} dx + \int_0^\pi \frac{p_m(Re^{i\theta})}{q_n(Re^{i\theta})} iRe^{i\theta} \, d\theta \right\} \quad (7.11)$$

$$= \lim_{R \to \infty} \oint_{\gamma_R} \frac{p_m(z)}{q_n(z)} dz. \quad (7.12)$$

In Eq. (7.12), γ_R denotes the path from $-R$ to R along the real axis, then from R to $-R$ along the semicircle $z = Re^{i\theta}$, $0 \leq \theta \leq \pi$. Figure 8a shows γ_R. Equation (7.2) follows by the residue theorem.

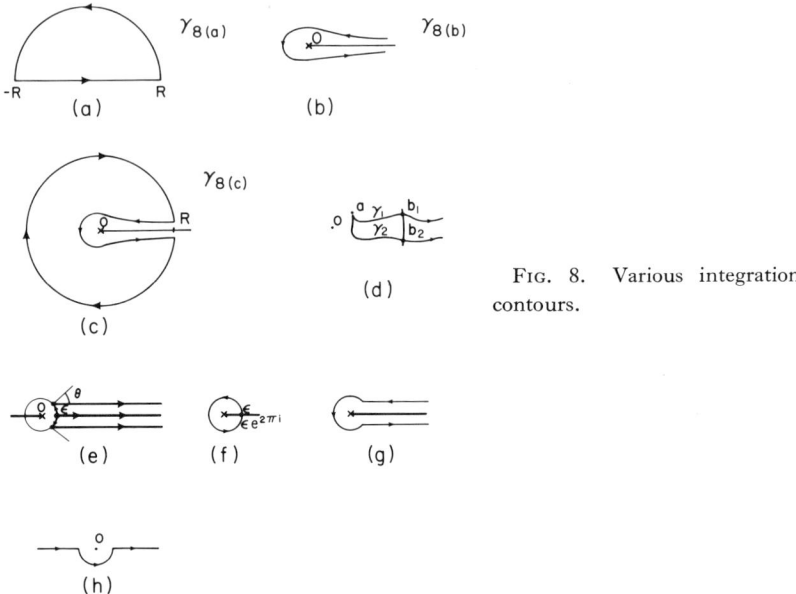

Fig. 8. Various integration contours.

A picturesque description of the procedure just given is that *the contour is closed at infinity by a semicircle in the upper half-plane, where the integrand vanishes sufficiently fast, and then the integral is evaluated by the residue theorem.*

For Eq. (7.3), note that $|e^{ikz}| = e^{-ky}$. When $k > 0$, e^{ikz} is bounded in the upper half-plane, and when $k < 0$, in the lower half-plane. When $n \geq m + 2$ and $k > 0$, the contour can be closed at infinity in the upper half-plane, as for Eq. (7.2), and Eq. (7.3) results. When $k < 0$, the same argument is valid for the lower half-plane, and Eq. (7.4) results. The minus sign results from the clockwise sense of the closed contour.

That Eqs. (7.3) and (7.4) hold when $n = m + 1$ is known as *Jordan's lemma*. The exponential factor helps make the integral over the semicircle vanish as $R \to \infty$. Care is required, however, because $|e^{ikz}| = 1$ when z is on the real axis. We use the estimates

$$\sin \theta \geq 2\theta/\pi, \qquad 0 \leq \theta \leq \tfrac{1}{2}\pi \qquad (7.13)$$

$$\sin \theta \geq 2 - 2\theta/\pi, \qquad \tfrac{1}{2}\pi \leq \theta \leq \pi, \qquad (7.14)$$

and that zp_m/q_{m+1} is bounded as $z \to \infty$,

$$\left| \frac{zp_m(z)}{q_{m+1}(z)} \right| \leq M \qquad \text{as} \qquad |z| \to \infty. \qquad (7.15)$$

Then Jordan's lemma is clinched by (when $k > 0$)

$$\left| \int_{\substack{|z|=R \\ (\mathrm{Im}\,z \geq 0)}} e^{ikz} \frac{p_m(z)}{q_n(z)}\, dz \right| \leq M \int_{\substack{|z|=R \\ (\mathrm{Im}\,z \geq 0)}} |e^{ikz}|\,|z^{-1}|\,|dz| \quad \text{as } R \to \infty \tag{7.16}$$

$$= M \int_0^\pi \exp(-kR \sin\theta)\, d\theta \tag{7.17}$$

$$\leq M \left(\int_0^{\pi/2} \exp(-kR 2\theta/\pi)\, d\theta + \int_{\pi/2}^\pi \exp(-kR(2 - 2\theta/\pi))\, d\theta \right) \tag{7.18}$$

$$= M \frac{\pi}{kR} (1 - e^{-kR}). \tag{7.19}$$

A more elaborate manipulation of contours leads to Eqs. (7.5) and (7.6). In preview, the path from 0 to ∞ is first replaced by the contour $\gamma_{8(b)}$ of Fig. 8b, which is essentially ∞ to 0, then around the origin so that z^α becomes $e^{2\pi i \alpha} x^\alpha$, followed by 0 to ∞. Then $\gamma_{8(b)}$ is "closed at ∞" by a large circle (Fig. 8c). The residue theorem applies to the closed contour.

The first detail of the derivation is a slight *extension of Cauchy's theorem* when one endpoint is at ∞. Let D_1 denote the domain $\{|z| > r, \theta_1 < \arg z < \theta_2\}$; let $f(z)$ be analytic in a simply connected domain D that contains D_1; and let $|f(z)| \leq M |z|^{-1-\sigma}$ ($\sigma > 0$) for all z in D_1. Let both γ_1 and γ_2 extend to ∞ in D_1 from the same initial point a in D, and let all points on γ_1 and γ_2 with $|z| > r$ lie in D_1. Then

$$\int_{\gamma_1} f(z)\, dz = \int_{\gamma_2} f(z)\, dz = \int_a^\infty f(z)\, dz. \tag{7.20}$$

The proof is obtained by cutting both γ_1 and γ_2 by the circle $|z| = R > r$ at the points b_1 and b_2 (Fig. 8d). Then, using Cauchy's theorem,

$$\int_{\gamma_1} f(z)\, dz = \lim_{b_1 \to \infty \text{ on } \gamma_1} \int_a^{b_1} f(z)\, dz = \lim_{R \to \infty} \int_a^{b_1} f(z)\, dz \tag{7.21}$$

$$= \lim_{R \to \infty} \int_a^{b_2} f(z)\, dz + \lim_{R \to \infty} \int_{b_2}^{b_1} f(z)\, dz \tag{7.22}$$

$$= \int_{\gamma_2} f(z)\, dz + \lim_{R \to \infty} \int_{b_2}^{b_1} f(z)\, dz. \tag{7.23}$$

But, $\left| \int_{b_1}^{b_2} f(z)\, dz \right| \leq 2\pi M R^{-\sigma}$.

The second detail is that for $\varepsilon > 0$,

$$\int_\varepsilon^\infty x^\alpha \frac{p_m(x)}{q_n(x)} dx = \left(\int_\varepsilon^{\varepsilon e^{i\theta}} + \int_{\varepsilon e^{i\theta}}^{\infty+\varepsilon e^{i\theta}}\right) z^\alpha \frac{p_m(z)}{q_n(z)} dz \quad (7.24)$$

$$= \left(\int_\varepsilon^{\varepsilon e^{-i\theta}} + \int_{\varepsilon e^{-i\theta}}^{\infty+\varepsilon e^{-i\theta}}\right) z^\alpha \frac{p_m(z)}{q_n(z)} dz. \quad (7.25)$$

See Fig. 8e. If in Eq. (7.25), Arg $z + 2\pi$ is used for arg z, one obtains

$$\int_\varepsilon^\infty x^\alpha \frac{p_m(x)}{q_n(x)} dx = e^{-2\pi i \alpha}\left(\int_{\varepsilon e^{2\pi i}}^{\varepsilon e^{(2\pi-\theta)i}} + \int_{\varepsilon e^{(2\pi-\theta)i}}^{\infty e^{2\pi i}+\varepsilon e^{(2\pi-\theta)i}}\right) z^\alpha \frac{p_m(z)}{q_n(z)} dz. \quad (7.26)$$

The third detail is to add $-\int_{\varepsilon e^{i\theta}}^{\infty+\varepsilon e^{i\theta}}$, $\int_{\varepsilon e^{(2\pi-\theta)i}}^{\infty e^{2\pi i}+\varepsilon e^{(2\pi-\theta)i}}$, $\int_\varepsilon^{\varepsilon e^{2\pi i}}$ (Fig. 8f) to obtain an integral over $\gamma_{8(g)}$ (Fig. 8g),

$$(e^{2\pi i \alpha} - 1)\int_\varepsilon^\infty x^\alpha \frac{p_m(x)}{q_n(x)} dx + \int_\varepsilon^{\varepsilon e^{2\pi i}} z^\alpha \frac{p_m(z)}{q_n(z)} dz$$

$$= \left(\int_{\infty+\varepsilon e^{i\theta}}^{\varepsilon e^{i\theta}} + \int_{\varepsilon e^{i\theta}}^{\varepsilon e^{(2\pi-\theta)i}} + \int_{\varepsilon e^{(2\pi-\theta)i}}^{\infty e^{2\pi i}+\varepsilon e^{(2\pi-\theta)i}}\right) z^\alpha \frac{p_m(z)}{q_n(z)} dz \quad (7.27)$$

$$= \int_{\gamma_{8(g)}} z^\alpha \frac{p_m(z)}{q_n(z)} dz. \quad (7.28)$$

Note the location of the branch cut in Fig. 8g. Note also that by the extension of Cauchy's theorem applied separately to each endpoint, $\int_{\gamma_{8(g)}}$ is independent of ε and θ—indeed, $\int_{\gamma_{8(g)}} = \int_{\gamma_{8(b)}}$—provided that $q_n(z)$ has no zeros on or "inside" $\gamma_{8(g)}$ or $\gamma_{8(b)}$.

The fourth detail is to evaluate the left-hand side of Eq. (7.27) at a convenient value of ε: $\varepsilon = 0$. Since

$$\left|\int_\varepsilon^{\varepsilon e^{2\pi i}} z^\alpha \frac{p_m(z)}{q_n(z)} dz\right| \lesssim 2\pi |\varepsilon|^{\alpha+1} p_m(0)/q_n(0), \quad (7.29)$$

and since $(\alpha + 1) > 0$,

$$\lim_{\varepsilon \to 0}\left[(e^{2\pi i \alpha} - 1)\int_\varepsilon^\infty \cdots dx + \int_\varepsilon^{\varepsilon e^{2\pi i}} \cdots dz\right] = (e^{2\pi i \alpha} - 1)\int_0^\infty \cdots dx.$$

By assumption, $e^{2\pi i \alpha} \neq 1$, so that

$$\int_0^\infty x^\alpha \frac{p_m(x)}{q_n(x)} dx = (e^{2\pi i \alpha} - 1)^{-1} \int_{\gamma_{8(b)}} z^\alpha \frac{p_m(z)}{q_n(z)} dz. \quad (7.30)$$

The fifth detail is to "close the contour at ∞" (Fig. 8c). The justification is virtually the same as in the derivation of Eq. (7.2). The residue theorem then applies to the integral over the closed contour $\gamma_{8(c)}$.

3. Complex Variable Theory

The derivation of Eq. (7.6) is based on the relation

$$\log(xe^{2\pi i}) = \log x + 2\pi i, \tag{7.31}$$

rather than $(xe^{2\pi i})^\alpha = e^{2\pi i \alpha} x^\alpha$. The analog of Eq. (7.26) is

$$\left(\int_{\varepsilon e^{2\pi i}}^{\varepsilon e^{(2\pi-\theta)i}} + \int_{\varepsilon e^{(2\pi-\theta)i}}^{\infty e^{2\pi i}+\varepsilon e^{(2\pi-\theta)i}} \right) \log z \, \frac{p_m(z)}{q_n(z)} \, dz$$

$$= \int_{\varepsilon e^{2\pi i}}^{\infty e^{2\pi i}} \log z \, \frac{p_m(z)}{q_n(z)} \, dz \tag{7.32}$$

$$= \int_\varepsilon^\infty \log x \, \frac{p_m(x)}{q_n(x)} \, dx + 2\pi i \int_\varepsilon^\infty \frac{p_m(x)}{q_n(x)} \, dx. \tag{7.33}$$

Otherwise the considerations are very similar to those leading to Eq. (7.5). As an exercise, the reader might obtain Eq. (7.6) from Eq. (7.5) by letting $\alpha \to 0$.

Notation Contours similar to $\gamma_{8(b)}$ appear so frequently that it is convenient to have a more suggestive notation. The symbols $\int_\infty^{(0+)}$ and $\int_{\gamma_{8(b)}}$ are equivalent. $\int_{+\infty}^{(0+)}$ is read, "the integral about a path that starts at the 'point' $+\infty$, circles the origin counterclockwise, and returns to $+\infty$."

C. Examples

Some examples of Eqs. (7.2)–(7.6) are

$$\int_{-\infty}^\infty (1 + x^2)^{-1} \, dx = \pi \tag{7.34}$$

$$\int_{-\infty}^\infty \frac{\cos kx}{1 + x^2} \, dx = \int_{-\infty}^\infty \frac{e^{ikx}}{1 + x^2} \, dx = \pi e^{-|k|}, \quad k \text{ real} \tag{7.35}$$

$$\int_0^\infty x^\alpha (1 + x^2)^{-1} \, dx = \pi \sin(\tfrac{1}{2}\pi\alpha)/\sin(\pi\alpha) \tag{7.36}$$

$$\int_0^\infty (x + 1)^{-2}(x + 2)^{-1} \, dx = 1 - \log 2 \tag{7.37}$$

$$\int_0^{2\pi} (a + \cos \theta)^{-1} \, d\theta = -2i \oint_{|z|=1} (z^2 + 2az + 1)^{-1} \, dz$$

$$= 2\pi(a^2 - 1)^{-1/2}, \quad a > 1. \tag{7.38}$$

Another example, $\int_{-\infty}^\infty x^{-1} \sin x \, dx$, is slightly outside the framework of Eq. (7.3), although not the spirit, since $x^{-1} e^{\pm ix}$ has a pole at $x = 0$.

Since $z^{-1} \sin z$ is analytic at $z = 0$ ($z^{-1} \sin z = 1$ at $z = 0$), the contour can be deformed to avoid the origin, as in Fig. 8h. Then

$$\int_{-\infty}^{\infty} x^{-1} \sin x \, dx = \int_{\gamma_{8(h)}} z^{-1} \sin z \, dz$$

$$= (2i)^{-1} \int_{\gamma_{8(h)}} z^{-1} e^{iz} \, dz - (2i)^{-1} \int_{\gamma_{8(h)}} z^{-1} e^{-iz} \, dz. \quad (7.39)$$

The contour for the e^{iz} term can be closed at ∞ in the upper half-plane, enclosing the pole at $z = 0$; the contour for the e^{-iz} can be closed at ∞ in the lower half-plane, in which there is no singularity. The result is

$$\int_{-\infty}^{\infty} x^{-1} \sin x \, dx = \pi. \quad (7.40)$$

D. On Computing Residues

There is one standard situation whereby computing a residue is equivalent to taking a derivative. Let $f(z)$ be analytic at z_0, so that

$$f(z) = \sum_{k=0}^{\infty} f^{(k)}(z_0)(z - z_0)^k / k! \quad (7.41)$$

converges in a neighborhood of z_0. The residue of $(z - z_0)^{-n} f(z)$ at z_0 is $f^{(n-1)}(z_0)/(n-1)!$:

$$\text{residue at } z_0 \text{ of } \left\{ \frac{f(z)}{(z - z_0)^n} \right\} = \frac{1}{(n-1)!} \left(\frac{d}{dz_0} \right)^{n-1} f(z_0). \quad (7.42)$$

As an example, the coefficient c_{lk} of Eq. (6.8) is $(2\pi i)^{-1} \oint_{|z-z_l|=\varepsilon} dz$ of $(z - z_l)^{n_l - k - 1} \times$ the right side of Eq. (6.8). Equation (6.9) is just the residue at z_l, by Eq. (7.42).

E. Cauchy Principal Value

Let $f(z)$ be analytic at $z = 0$, and let $f(0) \neq 0$. In applications one sometimes encounters expressions like $\int_{-a}^{b} x^{-1} f(x) \, dx$, where $a > 0$ and $b > 0$, which are superficially meaningless. More properly, one encounters limits of related expression which are discussed here.

Since $x^{-1}[f(x) - f(0)]$ is analytic at the origin [the value at zero being $f'(0)$], it is sufficient to consider integrals of z^{-1}.

3. Complex Variable Theory

Consider

$$\int_{-a}^{-\varepsilon} x^{-1} \, dx = \text{Log}(\varepsilon/a) \tag{7.43}$$

$$\int_{\varepsilon}^{b} x^{-1} \, dx = \text{Log}(b/\varepsilon). \tag{7.44}$$

The *symmetric limit*, called the *Cauchy principal value* and indicated by $\mathscr{P}$, exists and is given by

$$\mathscr{P}\int_{-a}^{b} x^{-1} \, dx = \lim_{\varepsilon \to 0} \left(\int_{-a}^{-\varepsilon} + \int_{\varepsilon}^{b} \right) x^{-1} \, dx = \text{Log}(b/a) \tag{7.45}$$

$$\mathscr{P}\int_{-a}^{b} x^{-1} f(x) \, dx = \lim_{\varepsilon \to 0} \left(\int_{-a}^{-\varepsilon} + \int_{\varepsilon}^{b} \right) x^{-1} f(x) \, dx \tag{7.46}$$

$$= \int_{-a}^{b} x^{-1} [f(x) - f(0)] \, dx + f(0) \, \text{Log}(b/a). \tag{7.47}$$

Another situation is represented by

$$\lim_{\varepsilon \to 0} \int_{-a}^{b} (x + i\varepsilon)^{-1} f(x) \, dx, \qquad \varepsilon > 0. \tag{7.48}$$

For nonzero ε, there is no singularity on the real axis. Indeed,

$$\int_{-a}^{b} (x + i\varepsilon)^{-1} f(x) \, dx = \int_{-a}^{b} (x + i\varepsilon)^{-1} [f(x) - f(-i\varepsilon)] \, dx$$

$$+ \int_{-a}^{b} (x + i\varepsilon)^{-1} f(-i\varepsilon) \, dx, \tag{7.49}$$

$$= \int_{-a}^{b} (x + i\varepsilon)^{-1} [f(x) - f(-i\varepsilon)] \, dx$$

$$+ f(-i\varepsilon) \left(\text{Log} \frac{b + i\varepsilon}{a - i\varepsilon} - i\pi \right). \tag{7.50}$$

Taking the limit $\varepsilon \to 0$, and using Eq. (7.47), one obtains

$$\lim_{\varepsilon \to 0} \int_{-a}^{b} (x + i\varepsilon)^{-1} f(x) \, dx = \mathscr{P} \int_{-a}^{b} x^{-1} f(x) \, dx - i\pi f(0). \tag{7.51}$$

Similarly,

$$\lim_{\varepsilon \to 0} \int_{-a}^{b} (x - i\varepsilon)^{-1} f(x) \, dx = \mathscr{P} \int_{-a}^{b} x^{-1} f(x) \, dx + i\pi f(0). \tag{7.52}$$

Equations (7.51) and (7.52) are sometimes referred to by the notation

$$(x \pm i\varepsilon)^{-1} = \mathscr{P} x^{-1} \mp i\pi \, \delta(x) \tag{7.53}$$

where $\delta(x)$ denotes the Dirac delta function [Chapter 2 of this volume, or Lighthill (1958)].

An alternative but equivalent situation to $\overline{\lim}(x \pm i\varepsilon)^{-1}$ is where the *contour is indented to avoid the singularity*. Thus (cf. Fig. 8h),

$$\left(\int_{-a}^{-\varepsilon} + \int_{\varepsilon e^{\pi i}}^{\varepsilon e^{2\pi i}} + \int_{\varepsilon}^{b}\right) z^{-1} f(z)\, dz = \lim_{\varepsilon \to 0} \int_{-a}^{b} (x - i\varepsilon)^{-1} f(x)\, dx. \quad (7.54)$$
$$\text{(Re} z < 0)$$

Note that in the sense of Eqs. (7.51)–(7.53)

$$\mathscr{P} x^{-1} = \tfrac{1}{2}[(x + i\varepsilon)^{-1} + (x - i\varepsilon)^{-1}], \quad (7.55)$$

i.e., $\mathscr{P} \int_{-a}^{b} x^{-1} f(x)\, dx$ is the average of avoiding $x = 0$ by indenting both above and below the origin. For example,

$$\mathscr{P} \int_{-\infty}^{\infty} x^{-1} e^{ix}\, dx = \pi i. \quad (7.56)$$

VIII. Higher Transcendental Functions

Functions that are not algebraic are called transcendental. Several transcendental functions appear frequently in applications: The gamma function, the hypergeometric function, and the confluent hypergeometric function are the most important. These functions can be defined by contour integrals, which provide an excellent vehicle for both developing the properties of these functions and for developing further techniques in contour integration.

A. THE GAMMA FUNCTION

1. *Euler's Integral*

The gamma function is defined, for Re $z > 0$, by

$$\Gamma(z) = \int_0^\infty t^{z-1} e^{-t}\, dt, \quad \text{Re } z > 0. \quad (8.1)$$

$\Gamma(z)$ is a natural generalization of the factorial

$$\Gamma(n+1) = n! \quad (8.2)$$

and it coincides with $\int_{-\infty}^{\infty} \tau^{2n} e^{-\tau^2}\, d\tau$, when z is $n + \tfrac{1}{2}$ (substitute $t = \tau^2$).

For example,
$$\Gamma(\tfrac{1}{2}) = \pi^{1/2}. \tag{8.3}$$

Since $|t^z| = t^x$ $(t \geq 0)$, the condition for convergence of Euler's integral is $x > 0$.

2. Recurrence and Reflection Formulas

Integration by parts yields the recurrence formula. The reflection formula is derived in Section VIII.B.3.

$$z\Gamma(z) = \Gamma(z+1) \quad \text{(recurrence formula)} \tag{8.4}$$

$$\Gamma(z) = [z(z+1) \cdots (z+n-1)]^{-1}\Gamma(z+n)$$
$$\text{(iterated recurrence formula)} \tag{8.5}$$

$$\Gamma(z)\Gamma(1-z) = \pi/\sin \pi z \quad \text{(reflection formula)}. \tag{8.6}$$

3. Analyticity

Differentiating Euler's integral under the integral sign, one obtains heuristically

$$(d/dz)\Gamma(z) = \int_0^\infty t^{z-1} e^{-t} \log t \, dt, \quad \text{Re } z > 0, \tag{8.7}$$

which also converges in Re $z > 0$. More rigorously,

$$\left| \frac{\Gamma(z+h) - \Gamma(z)}{h} - \int_0^\infty t^{z-1} e^{-t} \log t \, dt \right|$$
$$= \left| \int_0^\infty t^{z-1} e^{-t} \left(\frac{t^h - 1}{h} - \log t \right) dt \right| \tag{8.8}$$

$$\leq \left| \int_0^1 t^{z-1} e^{-t} \left(\frac{t^h - 1}{h} - \log t \right) dt \right|$$
$$+ \left| \int_1^\infty t^{z-1} e^{-t} \left(\frac{t^h - 1}{h} - \log t \right) dt \right| \tag{8.9}$$

$$\leq |h| \left| \int_0^1 t^{h-1} \left[\int_0^t (1/\tau') \left(\int_0^{\tau'} \tau^{z-1} e^{-\tau} d\tau \right) d\tau' \right] dt \right|$$
$$+ |h| \left| \int_1^\infty t^x e^{-t} dt \right| \tag{8.10}$$

$$\leq |h| [x^{-2}(\text{Re } h + x)^{-1} + e^{-1}]. \tag{8.11}$$

For the first terms in Eqs. (8.10) and (8.11), one integrates by parts twice and uses

$$\left| \int_0^1 t^{h-1} \left[\int_0^t (1/\tau') \left(\int_0^{\tau'} \tau^{z-1} e^{-\tau} \, d\tau \right) d\tau' \right] dt \right|$$

$$\leq \int_0^1 t^{\operatorname{Re} h - 1} \left[\int_0^t (1/\tau') \left(\int_0^{\tau'} \tau^{x-1} \, d\tau \right) d\tau' \right] dt = x^{-2}(\operatorname{Re} h + x)^{-1}. \quad (8.12)$$

For the second term, one uses

$$\left| \frac{t^h - 1}{h} - \log t \right| = \left| \sum_{k=2}^{\infty} \frac{h^{k-1}(\log t)^k}{k!} \right| \leq |h| \sum_{k=0}^{\infty} \frac{|\log t|^k}{k!} = |h| t,$$

$$|h| \leq 1, \quad t > 1 \quad (8.13)$$

$$\left| \int_1^{\infty} t^z e^{-t} \, dt \right| \leq \int_1^{\infty} t^x e^{-t} \, dt \leq \int_1^{\infty} e^{-t} \, dt = e^{-1}. \quad (8.14)$$

Euler's formula defines an analytic function of z in the half-plane $\operatorname{Re} z > 0$.

The right side of Eq. (8.5) is analytic for $\operatorname{Re} z > -n$, $z \neq 0, -1, -2, \ldots, -n+1$, and it coincides with $\Gamma(z)$ when $\operatorname{Re} z > 0$. It thus provides the analytic continuation of $\Gamma(z)$ to $\operatorname{Re} z \leq 0$.

The only singularities of $\Gamma(z)$ in the finite plane are simple poles at $z = 0, -1, -2, \ldots$.

4. Hankel's Formula

In the derivation of Eq. (7.5) of Section VII.B, the $\int_0^{\infty}$ was replaced by $\int_{\infty}^{(0+)}$. The motivation was to be able to close the contour at ∞. The same manipulation is invoked here for Euler's formula. The detailed justification is virtually the same and is not repeated, but the motivation is different: to obtain a formula, analytic for a wider range of z. Since the integration contour no longer passes through the singularity of the integrand at $z = 0$, only the equivalent of Eq. (8.13) for complex t is needed to clinch the existence of $\Gamma'(z)$. The resulting formulas are all referred to as Hankel's formula. In the three versions given below, the first two reflect alternate ways of indicating $\arg t$, while the third is a combination of the first with the reflection formula, followed by $z \to 1 - z$.

$$\Gamma(z) = (e^{2\pi i z} - 1)^{-1} \int_\infty^{(0+)} t^{z-1} e^{-t} \, dt,$$

$\arg t = 0$ at "beginning", $\quad z \neq$ integer $\quad$ (8.15)

$$= -(2i \sin \pi z)^{-1} \int_\infty^{(0+)} (-t)^{z-1} e^{-t} \, dt,$$

$\arg(-t) = -\pi i$ at "beginning", $\quad z \neq$ integer $\quad$ (8.16)

$$1/\Gamma(z) = -e^{\pi i z}(2\pi i)^{-1} \int_\infty^{(0+)} t^{-z} e^{-t} \, dt,$$

$\arg t = 0$ at "beginning", $\quad$ valid for all z. $\quad$ (8.17)

Hankel's formula is equivalent to Euler's integral when $\mathrm{Re}\, z > 0$, but it represents an analytic function in a much larger domain. It therefore represents the analytic continuation of $\Gamma(z)$ to the larger domain.

5. Zeros of $\Gamma(z)$

A direct consequence of the reflection formula (8.6) is that $\Gamma(z)$ has no zeros for finite z.

B. THE BETA FUNCTION

The integral

$$B(p, q) = \int_0^1 t^{p-1}(1-t)^{q-1} \, dt, \quad \mathrm{Re}\, p > 0, \quad \mathrm{Re}\, q > 0, \quad (8.18)$$

defines the beta function when $\mathrm{Re}\, p > 0$ and $\mathrm{Re}\, q > 0$. The beta function is closely related to the gamma function, but more important, the extensions of the integral representation involve manipulations that are among the most intricate encountered in elementary complex analysis.

1. Relation to Gamma Function

The beta function is computed from the gamma function by

$$B(p, q) = \Gamma(p)\Gamma(q)/\Gamma(p + q), \tag{8.19}$$

an equation that also provides an analytic continuation with respect to both p and q to $\operatorname{Re} p \leq 0$ and $\operatorname{Re} q \leq 0$, and shows that

$$B(p, q) = B(q, p). \tag{8.20}$$

Equation (8.19) is derived as follows:

$$\Gamma(p)\Gamma(q) = \int_0^\infty s^{p-1} e^{-s}\, ds \int_0^\infty t^{q-1} e^{-t}\, dt \tag{8.21}$$

$$= 4 \int_0^\infty \int_0^\infty x^{2p-1} y^{2q-1} \exp[-(x^2 + y^2)]\, dx\, dy,$$
$$s = x^2, \quad t = y^2 \tag{8.22}$$

$$= \left(2 \int_0^\infty r^{2p+2q-1} \exp(-r^2)\, dr\right)\left(2 \int_0^{\pi/2} \cos^{2p-1} \theta \sin^{2q-1} \theta\, d\theta\right)$$
$$\text{(polar coordinates)} \tag{8.23}$$

$$= \int_0^\infty \varrho^{p+q-1} e^{-\varrho}\, d\varrho \int_0^1 \tau^{p-1}(1 - \tau)^{q-1}\, d\tau,$$
$$\varrho = r^2, \quad \tau = \cos^2 \theta \tag{8.24}$$

$$= \Gamma(p + q)B(p, q). \tag{8.25}$$

2. Other Integral Representations

As for $\Gamma(z)$, the multiple-valuedness of t^{p-1} and $(1 - t)^{q-1}$ can be exploited to obtain integral representations that both avoid the branch points at $t = 0$ and $t = 1$, and are valid when $\operatorname{Re} p \leq 0$ and $\operatorname{Re} q \leq 0$.

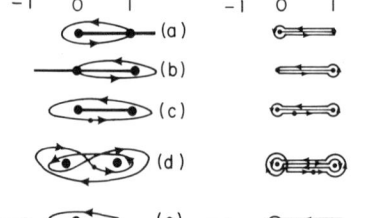

FIG. 9. Contours for Eqs. (8.26)–(8.32): (a) Eq. (8.26); (b) Eq. (8.27); (c) Eqs. (8.28) and (8.29); (d) Pochhammer's contour, for Eq. (8.30); (e) Eq. (8.32).

The contours are illustrated in Fig. 9. The integrals are

$$B(p, q) = e^{-\pi i p}(2i \sin \pi p)^{-1} \int_1^{(0+)} t^{p-1}(1-t)^{q-1}\,dt,$$
$$\text{Re } q > 0, \qquad p \neq \text{integer} \quad (8.26)$$

$$= -e^{-\pi i q}(2i \sin \pi q)^{-1} \int_0^{(1+)} t^{p-1}(1-t)^{q-1}\,dt,$$
$$\text{Re } p > 0, \qquad q \neq \text{integer} \quad (8.27)$$

$$B(p, 1-p) = e^{\pi i p}(2i \sin \pi p)^{-1} \int_x^{(1+, 0+)} t^{p-1}(1-t)^{-p}\,dt,$$
$$0 < x < 1, \quad \arg x = \arg(1-x) = 0; \quad p \neq \text{integer} \quad (8.28)$$

$$= (2i \sin \pi p)^{-1} \oint_{|t|=r>1} t^{p-1}(t-1)^{-p}\,dt,$$
$$\arg t = 0 \quad \text{when} \quad t = r; \quad p \neq \text{integer}. \quad (8.29)$$

$$B(p, q) = -\exp(-\pi i (p+q))(4 \sin \pi p \sin \pi p)^{-1}$$
$$\times \int_{\substack{x \\ (0<x<1)}}^{(1+, 0+, 1-, 0-)} t^{p-1}(1-t)^{q-1}\,dt,$$
$$p \neq \text{integer}, \quad q \neq \text{integer} \quad (8.30)$$

$$B(p, q) = \int_1^\infty s^{-p-q}(s-1)^{q-1}\,ds,$$
$$\text{Re } q > 0, \quad \text{Re } p > 0 \quad (t = 1/s) \quad (8.31)$$

$$= e^{-\pi i q}(2i \sin \pi q)^{-1} \int_\infty^{(0+)} (1+t)^{-p-q} t^{q-1}\,dt,$$
$$\text{Re } p > 0, \quad q \neq \text{integer}. \quad (8.32)$$

Apart from Eq. (8.31), the detailed derivation of each parallels the first part of the derivation of Eq. (7.5) in which $\int_0^\infty$ is replaced by $(e^{2\pi i \alpha} - 1)^{-1} \int_\infty^{(0+)}$, except that the upper limit in Eq. (8.18) is 1, not ∞. By Cauchy's theorem, the contour for each case can be deformed to consist of straight line segments along the real axis and infinitesimal circles about $z = 0$ and $z = 1$, as indicated by the right-hand member of each pair in Fig. 9. When $\text{Re } p > 0$ and $\text{Re } q > 0$, the infinitesimal circles contribute infinitesimally, and the straight line paths are essentially the original integration path. The various integrals provide analytic continuations of $B(p, q)$ to $\text{Re } p \leq 0$, or $\text{Re } q \leq 0$, or both, depending on which singularity of the integrand has been avoided.

The integrand of Eqs. (8.28) and (8.29) is essentially different from the others. The function $t^{p-1}(1-t)^{-p}$ changes by a factor $e^{2\pi i p}$ or $e^{-2\pi i p}$

when arg t or arg$(1 - t)$ is increased by 2π. If *both* increase by 2π, the function is unchanged. Thus, a satisfactory branch cut for $t^{p-1}(1-t)^{-p}$ runs from $t = 0$ to $t = 1$, as shown in Fig. 9c.

The $\int_x^{(1+,0+,1-,0-)} dt$ ($0 < x < 1$) in Eq. (8.30) is read: the integral from $t = x$ (on the real axis between 0 and 1), counterclockwise around $t = 1$, counterclockwise around $t = 0$, then clockwise around $t = 1$, clockwise around $t = 0$, and back to $t = x$, as illustrated in Fig. 9d. This contour is called *Pochhammer's contour*. So far we have only considered integrands that have been analytic, and *ipso facto* single valued, in the domain in which the integration path lay (except perhaps for the endpoints). If necessary, branch cuts specified single-valued branches of multiple-valued functions, and integration contours were kept in the corresponding domains of analyticity. In Eq. (8.30) the integration contour does *not* stay in a domain in which a single-valued branch of the integrand can be defined. Moreover, the integrand is taken to vary continuously on the contour (if branch cuts were drawn, then a change of $\pm 2\pi$ in arg t or arg$(1 - t)$ would be indicated when the path crossed a branch cut). Since the branch points at $t = 0$ and $t = 1$ are each circled twice, once counterclockwise and once clockwise, both arg t and arg$(1 - t)$ return to the same value (0) at the endpoint $t = x$, as at the initial point $t = x$, so that $t^{p-1}(1 - t)^{q-1}$ is continuous at all points of the contour.

To what extent can Pochhammer's contour be deformed without changing the value of the integral? Cauchy's theorem can be applied to any *segment* that lies in a domain in which $t^{p-1}(1 - t)^{q-1}$ is single valued. Consequently any deformation in the domain $\{z \mid z \neq 0,\ z \neq 1\}$, that is, any deformation that does not change how the contour loops around $z = 0$ and $z = 1$, does not change the value of the integral.

We follow the phase of the integrand; i.e.,

$$t^{p-1}(1 - t)^{q-1}/[\,|t|^{p-1}\,|\,1 - t\,|^{q-1}\,]$$

on the right-hand contour of Fig. 9d. On the first horizontal segment, it is 1; on the second, $e^{2\pi i q}$; on the third, $\exp(2\pi i(p + q))$; and on the fourth, $e^{\pi i p}$. When direction is taken into account,

$$\int_x^{(1+,0+,1-,0-)} t^{p-1}(1 - t)^{q-1}\,dt$$
$$= (1 - e^{2\pi i q} + e^{2\pi i(p+q)} - e^{2\pi i p})\int_0^1 t^{p-1}(1 - t)^{q-1}\,dt,$$
$$\operatorname{Re} p > 0, \quad \operatorname{Re} q > 0 \quad (8.33)$$
$$= -4\,e^{\pi i(p+q)} \sin \pi p \sin \pi q\, B(p, q). \quad (8.34)$$

3. Gamma Function Reflection Formula

Equations (8.29) and (8.25), with a Laurent series expansion, give an immediate proof of Eq. (8.6):

$$\Gamma(z)\Gamma(1-z) = B(z, 1-z) \tag{8.35}$$

$$= (2i \sin \pi z)^{-1} \oint_{|t|=r>1} t^{z-1}(t-1)^{-z}\, dt \tag{8.36}$$

$$= (2i \sin \pi z)^{-1} \oint_{|t|=r>1} t^{-1}(1-t^{-1})^{-z}\, dt \tag{8.37}$$

$$= (2i \sin \pi z)^{-1}$$
$$\times \oint_{|t|=r>1} \sum_{k=0}^{\infty} t^{-k-1} \frac{1 \cdot z(z+1) \cdots (z+k-1)}{k!}\, dt \tag{8.38}$$

$$= \pi/\sin \pi z, \tag{8.39}$$

since only the $k=0$ term has a nonzero integral. We note in passing that the Taylor series for $(1+z)^n$ is

$$(1+z)^n = \sum_{k=0}^{\infty} \frac{\Gamma(n+1)}{\Gamma(n+1-k)k!} z^k, \qquad |z| < 1, \tag{8.40}$$

and that when n is a positive integer, the summation is finite (binomial theorem). The expansion holds when n is a negative integer via

$$\frac{\Gamma(n+1)}{\Gamma(n+1-k)} = \frac{\sin(n+1-k)\pi}{\sin(n+1)\pi} \frac{\Gamma(k-n)}{\Gamma(-n)} = \frac{(-1)^k \Gamma(k-n)}{\Gamma(-n)}. \tag{8.41}$$

C. THE HYPERGEOMETRIC FUNCTION

1. Definition

The hypergeometric function is defined by

$$F(a,b;c;z) = \frac{\Gamma(c)}{\Gamma(b)\Gamma(c-b)} \int_0^1 t^{b-1}(1-t)^{c-b-1}(1-zt)^{-a}\, dt,$$
$$\text{Re } c > \text{Re } b > 0, \qquad z \text{ not real and } \geq 1. \tag{8.42}$$

Several important "special functions" are special cases of $F(a, b; c; z)$, a notable example being the Legendre polynomials. The intention of this section is to show how the integral representation leads naturally to some of the important properties of $F(a, b; c; z)$. More complete treatments of the hypergeometric function can be found in Abramowitz and Stegun (1964), Erdélyi (1953), and Whittaker and Watson (1927).

2. Analyticity

The locations of the branch points, branch cuts, and integration path for Eq. (8.42) are illustrated in Fig. 10. From Eq. (8.42) and from the discussion of the beta function, it is apparent that $F(a, b; c; z)$ is an analytic function of each parameter a, b and c, and it is an analytic function of z with a branch cut running from $z = 1$ to ∞. For specific values of the parameters, the branch point at $z = 1$ may become a pole or even no singularity at all. For example, if a is a negative integer, $F(a, b; c; z)$ is a polynomial of degree $-a$ in z.

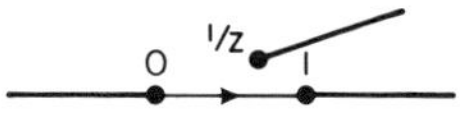

FIG. 10. Branch points, branch cuts, and integration path for $F(a, b; c; z)$, Eq. (8.42).

More general versions of Eq. (8.42) can be obtained with any of the contours of Fig. 9 [cf. Eqs. (8.26)–(8.32)]. With Pochhammer's contour,

$$F(a, b; c; z) = \frac{\Gamma(c)}{\Gamma(b)\Gamma(c-b)} \frac{-e^{-\pi i c}}{4 \sin \pi b \sin \pi(c-b)}$$

$$\times \int_{\substack{t_0 \\ (0 < t_0 < 1)}}^{(1+, 0+, 1-, 0-)} t^{b-1}(1-t)^{c-b-1}(1-zt)^{-a}\, dt \qquad (8.43)$$

$$= -\frac{\Gamma(1-b)\Gamma(1+b-c)\Gamma(c)e^{-\pi i c}}{4\pi^2}$$

$$\times \int_{t_0}^{(1+, 0+, 1-, 0-)} t^{b-1}(1-t)^{c-b-1}(1-zt)^{-a}\, dt, \qquad (8.44)$$

valid when b is not a positive integer, $c - b$ is not a positive integer, and the contour does not intersect the branch cut from $t = 1/z$ to ∞.

3. Important Formulas

We discuss the derivation of the following formulas from Eqs. (8.42)–(8.44):

$$F(a, b; c; z) = 1 + \frac{ab}{c} z + \frac{a(a+1)b(b+1)}{c(c+1)} \frac{z^2}{2!} + \cdots \quad (8.45)$$

$$= \sum_{k=0}^{\infty} \frac{(a)_k (b)_k}{(c)_k} \frac{z^k}{k!}$$

$(|z| < 1;$ Gauss's hypergeometric series) (8.46)

$(a)_k = \Gamma(a+k)/\Gamma(a)$ (Pochhammer's symbol) (8.47)

$F(a, b; c; z) = F(b, a; c; z)$ (symmetry of a and b) (8.48)

$F(a, c; c; z) = (1 - z)^{-a}$ (geometric series when $a = 1$) (8.49)

$(d/dz)F(a, b; c; z) = (ab/c)F(a+1, b+1; c+1; z) \quad (8.50)$

$F(a, b; c; 0) = 1 \quad (8.51)$

$$F(a, b; c; 1) = \frac{\Gamma(c)\Gamma(c-b-a)}{\Gamma(c-a)\Gamma(c-b)} \quad (8.52)$$

$$\lim_{c \to -n} \frac{1}{\Gamma(c)} F(a, b; c; z) = \frac{(a)_{n+1}(b)_{n+1}}{(n+1)!} z^{n+1}$$
$$\times F(a+n+1, b+n+1; n+2; z) \quad (8.53)$$

$$F(-n, b; -n-l; z) = \sum_{k=0}^{n} \frac{n!}{(n-k)!} \frac{(n+l-k)!}{(n+l)!} (b)_k \frac{z^k}{k!} \quad (8.54)$$

$(c - b - 1)F(a, b; c; z) + bF(a, b+1; c; z)$
$- (c - 1)F(a, b; c - 1; z) = 0 \quad (8.55)$

$$F(a, b; c; z) = (1 - z)^{-a} F\left(a, c - b; c; \frac{z}{z-1}\right) \quad (8.56)$$

$$= (1 - z)^{-b} F\left(c - a, b; c; \frac{z}{z-1}\right) \quad (8.57)$$

$$= (1 - z)^{c-a-b} F(c - a, c - b; c; z) \quad (8.58)$$

$$= (-z)^{-a} \frac{\Gamma(c)\Gamma(b-a)}{\Gamma(b)\Gamma(c-a)} F(1 + a - c, a; 1 + a - b; 1/z)$$

$$+ (-z)^{-b} \frac{\Gamma(c)\Gamma(a-b)}{\Gamma(a)\Gamma(c-b)} F(1 + b - c, b; 1 - a + b; 1/z),$$
$$|\arg(-z)| < \pi \quad (8.59)$$

$$\left\{ z(1-z) \frac{d^2}{dz^2} + [c - (a+b+1)z] \frac{d}{dz} - ab \right\} F(a, b; c; z) = 0. \quad (8.60)$$

Gauss's hypergeometric series, Eqs. (8.45) and (8.46), results from the Taylor series for $(1 - zt)^{-a}$ [Eq. (8.40)] put into Eqs. (8.42)–(8.44), and then from Eqs. (8.18), (8.30), (8.19), and (8.6). The radius of convergence follows from $\lim_{k \to \infty} \{(a)_k(b)_k/[(c)_k k!]\}^{1/k} = 1$.

Equations (8.48)–(8.50) follow easily from the series expansion. Note that the geometric series is a special case. The value at $z = 1$ follows directly from the integral representation of $B(b - a, c - b)$.

When c is a negative integer, $F(a, b; c; z)$ is undefined, but the limit in Eq. (8.53) follows from, say, Eq. (8.43) and the observation

$$\frac{\Gamma(-a+1)(-z)^k}{\Gamma(-a+1-k)k!} B(b+k, -n-b) = \frac{(a)_k z^k}{k!} \frac{\Gamma(b+k)\Gamma(-n-b)}{\Gamma(-n+k)} = 0,$$
$$k \leq n \quad (8.61)$$

$$= \frac{(a)_k z^k}{k!} \frac{\Gamma(b+k)\Gamma(-n-b)}{(k-n-1)!},$$
$$k \geq n+1. \quad (8.62)$$

If, however, first a becomes a negative integer, $-n$, so that $F(-n, b; c; z)$ is a polynomial, and then c becomes an integer at least as negative as a, then Eq. (8.54) results.

Note: $(-n)_k/(-n-l)_k = (n+1-k)_k/(n+l+1-k)_k$.

The six hypergeometric functions for which one of a, b, and c is increased or decreased by 1 are said to be *contiguous* to $F(a, b; c; z)$. Any contiguous function can be expressed as a linear combination of $F(a, b; c; z)$ and any other contiguous function, where the coefficients are rational functions in a, b, c, z. Equation (8.55) is typical. It follows from

$$t^{b-1}(1-t)^{c-b-1}(1-zt)^{-a}$$
$$= t^{b-1}(1-t)^{c-b-2}(1-zt)^{-a} - t^b(1-t)^{c-b-2}(1-zt)^{-a}. \quad (8.63)$$

Equations (8.56)–(8.59) are examples of *linear transformation formulas*. Equation (8.56) results from the substitution $t \to 1 - t$. Equation (8.57) is (8.56) with a and b interchanged, while Eq. (8.58) is (8.56) applied to (8.57). Equation (8.59) is a little more subtle. Refer to Fig. 10, and note that $\int_0^1 dt = \int_0^{1/z} dt + \int_{1/z}^1 dt$. By straightforward changes of variable,

$$\int_0^{1/z} t^{b-1}(1-t)^{c-b-1}(1-zt)^{-a} dt$$
$$= \frac{\Gamma(b)\Gamma(1-a)}{\Gamma(1-a+b)} z^{-b} F(1+b-c, b; 1-a+b; 1/z) \tag{8.64}$$

$$\int_{1/z}^1 t^{b-1}(1-t)^{c-b-1}(1-zt)^{-a} dt$$
$$= \frac{\Gamma(c-b)\Gamma(1-a)}{\Gamma(1-a-b+c)} (-1)^a z^{1-a-b}(1-1/z)^{c-a-b}$$
$$\times F(1-a, 1-b; 1-a-b+c; 1-z), \tag{8.65}$$

so that

$$F(a, b; c; z) = \frac{\Gamma(c)\Gamma(1-a)}{\Gamma(c-b)\Gamma(1-a+b)} z^{-b} F(1+b-c, b; 1-a+b, 1/z)$$
$$+ \frac{\Gamma(c)\Gamma(1-a)}{\Gamma(b)\Gamma(1-a-b+c)} (-1)^a z^{1-a-b}(1-1/z)^{c-a-b}$$
$$\times F(1-a, 1-b; 1-a-b+c; 1-z). \tag{8.66}$$

Here, $(-1)^a$ is $\exp(\pm \pi i a)$, where the sign is taken to be the same as for arg z. By switching a and b, a similar equation results. Elimination of $F(1-a, 1-b; 1-a-b+c; 1-z)$ from the two gives Eq. (8.59.)

The proof that $F(a, b; c; z)$ is a solution of the hypergeometric differential equation (8.60) is typical for integral representations of solutions to linear differential equations.

The gist is that:

1. Eq. (8.60) is of the form

$$\hat{L}_z \int_\gamma f(t, z) dt = 0, \tag{8.67}$$

where $\hat{L}_z$ is the differential operator in braces, and $f(t, z)$ is $t^{b-1}(1-t)^{c-b-1} \times (1-zt)^{-a}$;

2. $\hat{L}_z f(t, z) = (\partial/\partial t) g(t, z)$, where $g(t, z) = -at^b(1-t)^{c-b}(1-zt)^{-a-1}$;

and

3. $g(t, z)$ has the same value at both endpoints of γ. Thus,

$$\hat{L}_z \int_\gamma f(t, z) dt = g(t, z) \Big|_{t_{\text{initial}}}^{t_{\text{final}}} = 0 \tag{8.68}$$

and $\int_\gamma f(t, z) dt$ is the desired solution.

D. LEGENDRE POLYNOMIALS AND ASSOCIATED LEGENDRE FUNCTIONS AS EXAMPLES OF THE HYPERGEOMETRIC FUNCTION

Important and somewhat typical examples of the hypergeometric function are the Legendre polynomials and associated Legendre functions. We show how some of their properties are special cases of those of the hypergeometric function, and how other properties can be derived using complex variable theory. These functions are met again in Chapter 4, Section VII.

1. *Legendre Polynomials*

   a. *Definition*

$$P_n(z) = F(-n, n+1; 1; (1-z)/2). \tag{8.69}$$

   b. *Important Formulas*

$$P_n(z) = (2^{n+1}\pi i)^{-1} \oint^{(z+)} (t-z)^{-n-1}(t^2-1)^n \, dt \quad \text{(Schläfli)} \tag{8.70}$$

$$= 2^n(\pi i)^{-1} \oint^{(1+)}_{(|t-1|<2)} (t-z)^n (t^2-1)^{-n-1} \, dt \tag{8.71}$$

$$= (2^n n!)^{-1} (d/dz)^n (z^2-1)^n \quad \text{(Rodrigues)} \tag{8.72}$$

$$= \pi^{-1} \int_0^\pi (z + (z^2-1)^{1/2} \cos \phi)^n \, d\phi \quad \text{(Laplace)} \tag{8.73}$$

$$= \pi^{-1} \int_0^\pi (z + (z^2-1)^{1/2} \cos \phi)^{-n-1} \, d\phi$$
$$\text{(Jacobi)}, \quad |\arg z| < \pi/2, \tag{8.74}$$

$$\int_{-1}^1 P_n(z) P_m(z) \, dz = \delta_{nm} 2/(2n+1) \quad \text{(orthogonality)}, \tag{8.75}$$

$$(n+1)P_{n+1}(z) - (2n+1)zP_n(z) + nP_{n-1}(z) = 0 \quad \text{(recurrence)} \tag{8.76}$$

$$\{(1-z^2)(d/dz)^2 - 2z(d/dz) + n(n+1)\}P_n(z) = 0, \tag{8.77}$$

$$(1 + h^2 - 2hz)^{-1/2} = \sum_{k=0}^\infty h^k P_k(z), \quad |h|^2 + 2|hz| < 1$$
$$\text{(generating function)}. \tag{8.78}$$

   c. *On the Derivation of Eqs.* (8.70)–(8.78). These equations are, except for Eqs. (8.75) and (8.78), special cases of general relations for $F(a, b; c; z)$. Equations (8.70) and (8.71) are essentially the integral representations corresponding to the contour of Fig. 9c [cf. Eq. (8.29)]:

$$F\left(-n, n+1; 1; \frac{1-z}{2}\right)$$
$$= \frac{1}{2\pi i} \oint_{|\tau|=r>1} \tau^n (\tau-1)^{-n-1} \left(1 - \frac{1-z}{2}\tau\right)^n d\tau \quad (8.79)$$

$$F\left(n+1, -n; 1; \frac{1-z}{2}\right)$$
$$= \frac{1}{2\pi i} \oint_{(|\tau|<2/|1-z|)}^{(0+)} \tau^{-n-1}(\tau-1)^n \left(1 - \frac{1-z}{2}\tau\right)^{-n-1} d\tau. \quad (8.80)$$

[It is necessary to use $\Gamma(-n) \sin \pi n = -\pi/\Gamma(n+1)$.] The substitution $\tau = (1-t)/(1-z)$ yields Eqs. (8.70) and (8.71). The residue theorem applied to Schläfli's formula yields Rodrigues's formula. The Laplace integrals (the one found by Jacobi is called Laplace's integral of the second kind) are Eqs. (8.70) and (8.71) with the substitution $t = z + (z^2 - 1)^{1/2} e^{i\phi}$. The orthogonality is obtained by using Rodrigues's formula and integrating by parts. Note that $\int_{-1}^{1} (1-t^2)^n dt = 2^{2n+1} \times B(n+1, n+1)$. The recurrence formula (8.76) is related to Eq. (8.55); it is a special case of such a relation among $F(a+1, b-1; c; z)$, $F(a, b; c; z)$, and $F(a-1, b+1; c; z)$. Equation (8.76) is also a reflection of the relation

$$[(1 + h^2 - 2hz)(d/dh) + h - z](1 + h^2 - 2hz)^{-1/2} = 0. \quad (8.81)$$

The differential equation (8.77) is Eq. (8.60) with appropriate substitutions.

The generating function can be obtained by directly summing the right-hand side of Eq. (8.78) via Schläfli's integral:

$$\sum_{k=0}^{\infty} h^k P_k(z) = \frac{1}{2\pi i} \sum_{k=0}^{\infty} \oint_{(|t-z|=1)}^{(z+)} (t-z)^{-1} \left(\frac{h}{2} \frac{t^2-1}{t-z}\right)^n dt \quad (8.82)$$
$$= \frac{1}{2\pi i} \oint_{(|t-z|=1)} [t - z - \tfrac{1}{2}h(t^2-1)]^{-1} dt, \quad (8.83)$$

which converges for small enough h; the circle $|t - z| = 1$ is chosen for convenience. The quadratic is factorable:

$$t - z - \tfrac{1}{2}h(t^2-1) = -\tfrac{1}{2}h\left(t - \frac{1 + (1 + h^2 - 2hz)^{1/2}}{h}\right)$$
$$\times \left(t - \frac{1 - (1 + h^2 - 2hz)^{1/2}}{h}\right). \quad (8.84)$$

The root $(1 - (1 + h^2 - 2hz)^{1/2})/h \approx z$ lies inside the contour, while $(1 + (1 + h^2 - 2hz)^{1/2})/h \approx 2/h$ lies outside. The residue at
$$(1 - (1 + h^2 - 2hz)^{1/2})/h \quad \text{is} \quad 1/(1 + h^2 - 2hz)^{1/2}.$$

2. Associated Legendre Functions

a. Definition, Important Equations. The analogs of Eqs. (8.69)–(8.78) are similarly derived. In physical applications, z is usually real, and $|z| \leq 1$. Where $(1-z^2)^{1/2}$ appears below, the principal branch and $|z^2| \leq 1$ are assumed. The definition and consequences, stated without further comment, are

$$P_n^m(z) = (1-z^2)^{m/2}(d/dz)^m F(-n; n+1; 1; (1-z)/2) \tag{8.85a}$$

$$= (1-z^2)^{m/2} 2^{-m} \frac{(n+m)!}{(n-m)!} \frac{1}{m!}$$
$$\times F\left(-n+m, n+1+m; 1+m; \frac{1-z}{2}\right),$$
$$m = 0, 1, 2, \ldots, n \tag{8.85b}$$

$$= (1-z^2)^{m/2}(1+z)^{-m} \frac{(n+m)!}{(n-m)!} \frac{1}{m!}$$
$$\times F\left(-n, n+1; 1+m; \frac{1-z}{2}\right) \quad \text{[cf. Eq. (8.58)]} \tag{8.86}$$

$$= (1-z^2)^{m/2} 2^{-n-1}(\pi i)^{-1}(n+m)!(n!)^{-1}$$
$$\times \oint^{(z+)} (t-z)^{-n-m-1}(t^2-1)^n \, dt \tag{8.87a}$$

$$= (1-z^2)^{-m/2}(-1)^m 2^{-n-1}(\pi i)^{-1}(n+m)!(n!)^{-1}$$
$$\times \oint^{(z+)} (t-z)^{-n+m-1}(t^2-1)^n \, dt \tag{8.87b}$$

$$= (1-z^2)^{m/2}(-1)^m 2^n (\pi i)^{-1} n! [(n-m)!]^{-1}$$
$$\times \oint_{|t-1|<2}^{(1+)} (t-z)^{n-m}(t^2-1)^{-n-1} \, dt \tag{8.88}$$

$$= 2^{-n}(n!)^{-1}(1-z^2)^{m/2}(d/dz)^{n+m}(z^2-1)^n \tag{8.89a}$$

$$= 2^{-n}(n!)^{-1}[(-1)^m(n+m)!/(n-m)!](1-z^2)^{-m/2}$$
$$\times (d/dz)^{n-m}(z^2-1)^n \tag{8.89b}$$

$$= (1-z^2)^{m/2}(d/dz)^m P_n(z) \tag{8.90}$$

$$= i^{-m}(n+m)!(2\pi n!)^{-1} \int_0^{2\pi} e^{im\phi}(z+i(1-z^2)^{1/2}\cos\phi)^n \, d\phi \tag{8.91}$$

$$= i^m n! [2\pi(n-m)!]^{-1} \int_0^{2\pi} e^{im\phi}(z+i(1-z^2)^{1/2}\cos\phi)^{-n-1} \, d\phi,$$
$$\text{Re } z > 0, \tag{8.92}$$

$$\int_{-1}^{1} P_{n'}^{m}(z)P_{n}^{-m}(z)\,dz = \delta_{nn'}(-1)^{m}2/(2n+1) \tag{8.93}$$

$$\int_{-1}^{1} P_{n'}^{m}(z)P_{n}^{m}(z)\,dz = \delta_{nn'}\frac{(n+m)!}{(n-m)!}\frac{2}{2n+1}, \tag{8.94}$$

$$(n+1-m)P_{n+1}^{m}(z) - (2n+1)zP_{n}^{m}(z) + (n+m)P_{n-1}^{m}(z) = 0 \tag{8.95}$$

$$(1-z^2)^{1/2}P_{n}^{m+1}(z) - 2mzP_{n}^{m}(z) + (n+m)(n-m+1)(1-z^2)^{1/2}P_{n}^{m-1}(z) = 0 \tag{8.96}$$

$$\{(1-z^2)(d/dz)^2 - 2z(d/dz) + n(n+1) - m^2/(1-z^2)\}P_{n}^{m}(z) = 0, \tag{8.97}$$

$$(1+h^2-2hz)^{-m-1/2}h^m(1-z^2)^{m/2}(2m-1)!! = \sum_{k=0}^{\infty} h^{k+m}P_{k+m}^{m}(z), \tag{8.98}$$

where

$$(2m-1)!! = (2m)!/(2^m m!) = 1 \cdot 3 \cdot 5 \cdot \cdots \cdot (2m-1). \tag{8.99}$$

b. Negative Order, $P_n^{-m}(z)$. Equation (8.85b) defines $P_n^m(z)$ when m is a negative integer, if Eq. (8.53) is appropriately invoked:

$$P_n^{-M}(z) \equiv \lim_{c \to 1-M} (1-z^2)^{m/2} 2^{-m} \frac{(n+m)!}{(n-m)!} \frac{1}{\Gamma(c)}$$

$$\times F\left(-n+m; n+1+m; c; \frac{1-z}{2}\right),$$

$$M \geq 0, \quad m = -M \tag{8.100}$$

$$= (-1)^M (1-z^2)^{-M/2}(1-z)^M \frac{1}{M!}$$

$$\times F\left(-n, n+1; 1+M; \frac{1-z}{2}\right) \tag{8.101}$$

$$= (-1)^M \frac{(n-M)!}{(n+M)!} P_n^M(z) \tag{8.102}$$

$$P_n^m(z) = (-1)^m \frac{(n+m)!}{(n-m)!} P_n^{-m}(z),$$

$$m = n, n-1, \ldots, 0, \ldots, -n+1, -n. \tag{8.103}$$

Thus P_n^{-m} is a multiple of P_n^m, and vice versa. Equations (8.85b)–(8.89) and (8.91)–(8.97) are valid for positive and negative m; Eqs. (8.90) and (8.98) are valid only for positive m.

c. Addition Formula. An extremely useful equation in applications is the addition formula:

$$P_n(\cos\Theta) = \sum_{m=-n}^{n} (-1)^m P_n{}^m(\cos\theta_1) P_n^{-m}(\cos\theta_2) \exp[im(\phi_1 - \phi_2)], \quad (8.104)$$

$$\cos\Theta = \cos\theta_1 \cos\theta_2 + \sin\theta_1 \sin\theta_2 \cos(\phi_1 - \phi_2). \quad (8.105)$$

By the addition formula, a function of the relative angle between two vectors is reduced to products in which the angular coordinates of the two vectors appear in separate factors. A derivation follows:

$$\sum_{m=-n}^{n} (-1)^m P_n{}^m(z_1) P_n^{-m}(z_2) e^{im\psi}$$

$$= \frac{e^{in\psi}}{\pi i} \left(\frac{1-z_1^2}{1-z_2^2}\right)^{n/2} \int_{|t_1-1|<2}^{(1+)} dt_1 \, (t_1^2-1)^{-n-1}$$

$$\times \left\{ \sum_{m=-n}^{n} \left[(t_1-z_1) e^{-i\psi}\left(\frac{1-z_2^2}{1-z_1^2}\right)^{1/2}\right]^{n-m} \right.$$

$$\left. \times \frac{1}{2\pi i} \int^{(z_2^+)} (t_2-z_2)^{-(n-m)-1}(t_2^2-1)^n \, dt_2 \right\} \quad (8.106)$$

$$= \frac{e^{in\psi}}{\pi i} \left(\frac{1-z_1^2}{1-z_2^2}\right)^{n/2} \oint_{|t-1|<2}^{(1+)} (t^2-1)^{-n-1}$$

$$\times \left\{ \left[z_2 + (t-z_1) e^{-i\psi}\left(\frac{1-z_2^2}{1-z_1^2}\right)^{1/2}\right]^2 - 1 \right\}^n dt, \quad (8.107)$$

$$\sum_{h=0}^{\infty} h^n \sum_{m=-n}^{n} (-1)^m P_n{}^m(z_1) P_n^{-m}(z_2) e^{im\psi}$$

$$= \frac{1}{\pi i} \oint^{(1+)} \left[t^2 - 1 - h e^{i\psi}\left(\frac{1-z_1^2}{1-z_2^2}\right)^{1/2} \right.$$

$$\left. \times \left\{ \left[z_2 + (t-z_1) e^{-i\psi}\left(\frac{1-z_2^2}{1-z_1^2}\right)^{1/2}\right]^2 - 1 \right\} \right]^{-1} dt \quad (8.108)$$

$$= \frac{1}{\pi i} \oint^{(1+)} \left\{ t^2\left[1 - h e^{-i\psi}\left(\frac{1-z_2^2}{1-z_1^2}\right)^{1/2}\right] \right.$$

$$- 2ht\left[z_2 - z_1 e^{-i\psi}\left(\frac{1-z_2^2}{1-z_1^2}\right)^{1/2}\right]$$

$$- \left[1 - 2hz_1 z_2 - h e^{i\psi}(1-z_1^2)^{1/2}(1-z_2^2)^{1/2}\right.$$

$$\left.\left. + h z_1^2 e^{-i\psi}\left(\frac{1-z_2^2}{1-z_1^2}\right)^{1/2}\right]\right\}^{-1} dt \quad (8.109)$$

$$= \{1 + h^2 - 2h[z_1 z_2 + (1-z_1^2)^{1/2}(1-z_2^2)^{1/2}\cos\psi]\}^{-1/2}. \quad (8.110)$$

Equation (8.106) uses both representations (8.87) and (8.88). Equation (8.107) follows from Eqs. (4.77), (5.28), and the observation that the first $(2n + 1)$ terms of the Taylor series for a polynomial of degree $2n$ *are* the Taylor series. The right-hand side of Eq. (8.108) is the sum of the geometric series, which converges for sufficiently small $|h|$. The denominator of Eq. (8.109) is just a quadratic polynomial in t. The two roots are at

$$t = t_{\pm} = \left[1 - he^{-i\psi}\left(\frac{1-z_2^2}{1-z_1^2}\right)^{1/2}\right]^{-1}\left\{h\left[z_2 - z_1 e^{-i\psi}\left(\frac{1-z_2^2}{1-z_1^2}\right)^{1/2}\right]\right.$$
$$\left. \pm [1 + h^2 - 2h(z_1 z_2 + (1-z_1^2)^{1/2}(1-z_2^2)^{1/2}\cos\psi)]^{1/2}\right\} \quad (8.111)$$

$$\approx \pm 1 + O(h). \quad (8.112)$$

The root t_+ (near 1 for small h) lies inside the contour, the root t_- lies outside, and the result is $2\pi i \times$ residue at t_+ [Eq. (8.110)]. Finally, (8.110) is the generating function for the $P_n(\cos\Theta)$, by Eqs. (8.105) and (8.78).

An alternative derivation of the addition theorem is given by Henderson in Chapter 4.

E. THE CONFLUENT HYPERGEOMETRIC FUNCTIONS

1. *Definition of* $M(a, b, z)$

The *confluent hypergeometric function* of the first kind, Kummer's function, is defined by

$$M(a, b, z) = \frac{\Gamma(b)}{\Gamma(a)\Gamma(b-a)} \int_0^1 t^{a-1}(1-t)^{b-a-1} e^{zt}\, dt,$$
$$\text{Re } b > \text{Re } a > 0. \quad (8.113)$$

It may be obtained as a limit of a hypergeometric function:

$$\lim_{n\to\infty} F(-n, a; b; -z/n)$$
$$= \lim_{n\to\infty} \frac{\Gamma(b)}{\Gamma(a)\Gamma(b-a)} \int_0^1 t^{a-1}(1-t)^{b-a-1}(1 + zt/n)^n\, dt \quad (8.114)$$

$$= \frac{\Gamma(b)}{\Gamma(a)\Gamma(b-a)} \int_0^1 t^{a-1}(1-t)^{b-a-1} e^{zt}\, dt. \quad (8.115)$$

The nomenclature reflects the formation of an irregular singular point at ∞ by the *confluence* of two regular singular points ($1/z$ and ∞) of the hypergeometric differential equation.

As with $F(a, b; c; z)$, $M(a, b, z)$ is an analytic function of the parameters a, b, and the variable z. Any of the contours of Fig. 9 introduced for $B(a, b)$ can be used, when appropriate, for $M(a, b\ z)$. For example, the equation

$$M(a, b, z) = -\tfrac{1}{4}\pi^{-2}e^{-\pi ib}\Gamma(b)\Gamma(1-a)\Gamma(1+a-b)$$
$$\times \int_{\substack{t_0 \\ (0<t_0<1)}}^{(1+,0+,1-,0-)} t^{a-1}(1-t)^{b-a-1}e^{zt}\,dt \qquad (8.116)$$

is valid when neither b, $1-a$, nor $1+a-b$ is a negative integer. $M(a, b, z)$ is an entire function of z [cf. Eq. (8.123)].

Comprehensive treatments of confluent hypergeometric functions include those by Slater (1960), Erdélyi (1953), and Whittaker and Watson (1927). Here we sketch how some of the important results follow from the integral representation.

2. *Differential Equation*

$M(a, b, z)$ satisfies Kummer's differential equation:

$$\{z(d/dz)^2 + (b-z)(d/dz) - a\}M(a, b, z) = 0. \qquad (8.117)$$

In the notation of Section VIII.C.3, Eq. (8.67), take $f(t, z) = t^{a-1} \times (1-t)^{b-a-1}e^{zt}$ and $g(t, z) = -t^a(1-t)^{b-a}e^{zt}$.

3. *Definition of $U(a, b, z)$*

The confluent hypergeometric function of the second kind, $U(a, b, z)$, is also a solution of Kummer's equation:

$$U(a, b, z) = [\Gamma(a)]^{-1}\int_0^\infty t^{a-1}(1+t)^{b-a-1}e^{-zt}\,dt, \qquad \text{Re } a > 0. \qquad (8.118)$$

[Take $f(t, z) = t^{a-1}(1+t)^{b-a-1}e^{-zt}$ and $g(t, z) = -t^a(1+t)^{b-a}e^{-zt}$.]
$U(a, b, z)$ is multiple valued [cf. Eq. (8.125)], with branch point at $z = 0$.

A more general integral representation for $U(a, b, z)$ [cf. Eq. (8.15)] is

$$U(a, b, z) = (2\pi i)^{-1}e^{-\pi ia}\Gamma(1-a)\int_\infty^{(0+)} t^{a-1}(1+t)^{b-a-1}e^{-zt}\,dt. \qquad (8.119)$$

4. Recurrence Formulas

The identity $(1 \pm t)^{b-a-1} = (1 \pm t)^{b-a-2} \pm t(1 \pm t)^{b-a-2}$ yields

$$(1 + a - b)M(a, b, z) - aM(a + 1, b, z) + (b - 1)M(a, b - 1, z) = 0 \tag{8.120}$$

$$U(a, b, z) - aU(a + 1, b, z) - U(a, b - 1, z) = 0 \tag{8.121}$$

which are analogous to Eq. (8.55) and typical of a number of such equations in the references cited above.

5. Series Expansion

Expansion of e^{zt} and term-by-term integration of Eq. (8.115) [cf. Eqs. (8.45) and (8.46)] give

$$M(a, b, z) = 1 + \frac{a}{b} z + \frac{a(a+1)}{b(b+1)} \frac{z^2}{2!} + \cdots \tag{8.122}$$

$$= \sum_{k=0}^{\infty} \frac{(a)_k}{(b)_k} \frac{z^k}{k!}, \tag{8.123}$$

which clearly has an infinite radius of convergence. $F(a, b; c; z)$ and $M(a, b, z)$ are special cases, $_2F_1(a, b; c; z)$ and $_1F_1(a; b; z)$, respectively, of the *generalized hypergeometric function*

$$_kF_l(a_1, a_2, \ldots, a_k; b_1, b_2, \ldots, b_l; z) = \sum_{n=0}^{\infty} \frac{(a_1)_n(a_2)_n \cdots (a_k)_n}{(b_1)_n(b_2)_n \cdots (b_l)_n} \frac{z^n}{n!}. \tag{8.124}$$

A series for $U(a, b, z)$ is implicit in the formula

$$U(a, b, z) = \frac{\pi}{\sin \pi b} \left[\frac{M(a, b, z)}{\Gamma(1 + a - b)\Gamma(b)} - \frac{z^{1-b} M(1 + a - b, 2 - b, z)}{\Gamma(a)\Gamma(2 - b)} \right], \tag{8.125}$$

which clearly displays the nature of the multiple valuedness of $U(a, b, z)$. The derivation is quite instructive. The manipulation of the contour is essentially $\int_0^\infty = \int_0^{-1} + \int_{-1}^\infty$, and involves drawing both branch cuts to $+\infty$ along the real axis. We treat the case $\operatorname{Re} a > 0$ for simplicity.

Then
$$U(a, b, z) = (e^{2\pi i b} - 1)^{-1}[\Gamma(a)]^{-1}$$
$$\times \left\{ \int_{\infty}^{(-1+)} - \int_{0}^{(-1+)} \right\} t^{a-1}(1 + t)^{b-a-1} e^{-zt} \, dt. \quad (8.126)$$

For the first integral, deform the contour so that $|t| > 1$ always. Then
$$(e^{2\pi i b} - 1)^{-1}[\Gamma(a)]^{-1} \int_{\substack{\infty \\ |t|>1}}^{(-1+)} t^{b-2}(1 + t^{-1})^{b-a-1} e^{-zt} \, dt$$
$$= (e^{2\pi i b} - 1)^{-1}[\Gamma(a)]^{-1}$$
$$\times \int_{\infty}^{(0+)} \sum_{k=0}^{\infty} \frac{\Gamma(1 + a - b + k)(-1)^k}{\Gamma(1 + a - b)k!} t^{b-2-k} e^{-zt} \, dt \quad (8.127)$$
$$= \sum_{k=0}^{\infty} (1 + a - b)_k \frac{\Gamma(b - 1 - k)(-1)^k}{\Gamma(a)k!}$$
$$\times z^{1-b+k} \left[\frac{\pi}{\sin \pi(b - 1 - k)\Gamma(b - 1 - k)\Gamma(2 - b + k)} \right] \quad (8.128)$$
$$= -\frac{\pi}{\sin \pi b} z^{1-b} \frac{M(1 + a - b, 2 - b, z)}{\Gamma(a)\Gamma(2 - b)}. \quad (8.129)$$

The second term is amenable to the substitution $\tau = -t$:
$$-(e^{2\pi i b} - 1)^{-1}[\Gamma(a)]^{-1} e^{\pi i(a-1)} \int_{0}^{(-1+)} (-t)^{a-1}(1 + t)^{b-a-1} e^{-zt} \, dt$$
$$= -(e^{2\pi i b} - 1)^{-1}[\Gamma(a)]^{-1} e^{\pi i a} \int_{0}^{(1+)} \tau^{a-1}(1 - \tau)^{b-a-1} e^{z\tau} \, d\tau \quad (8.130)$$
$$= -(e^{2\pi i b} - 1)^{-1} \frac{\Gamma(b - a)}{\Gamma(b)} e^{\pi i a}[1 - e^{2\pi i(b-a)}] M(a, b, z) \quad (8.131)$$
$$= \frac{\pi}{\sin \pi b} \frac{M(a, b, z)}{\Gamma(b)\Gamma(1 + a - b)}. \quad (8.132)$$

6. Kummer's Transformations

These are the analogs of the linear transformation formulas (8.56)–(8.59). Set $\tau = 1 - t$ in Eq. (8.113). Then
$$M(a, b, z) = e^z M(b - a, b, -z). \quad (8.133)$$

The second Kummer transformation is a direct consequence of Eq. (8.125):
$$U(a, b, z) = z^{1-b} U(1 + a - b, 2 - b, z). \quad (8.134)$$

F. Spherical Bessel Functions as Examples of Confluent Hypergeometric Functions

The confluent hypergeometric family includes the Bessel function family, the Airy function, the Hermite and Laguerre polynomials, and the exponential integral, to mention a few. The spherical Bessel function is important in many applications—scattering in quantum mechanics being a good example. We show how some of its properties are special cases of those of the confluent hypergeometric function, and how others can be derived by complex variable theory. Bessel functions are encountered again in Chapter 4, Section VI.

1. Definition of $j_n(z)$. The Double Factorial

$$j_n(z) = [(3/2)_n]^{-1}(\tfrac{1}{2}z)^n e^{-iz} M(n+1, 2n+2, 2iz) \quad (8.135)$$

$$= [(2n+1)!!]^{-1} z^n e^{-iz} M(n+1, 2n+2, 2iz). \quad (8.136)$$

The spherical Bessel function is a special case of Bessel's function:

$$j_n(z) = [\pi/(2z)]^{1/2} J_{n+1/2}(z). \quad (8.137)$$

The double factorial, defined in Eq. (8.99) for add integers, appears repeatedly in formulas related to $j_n(z)$. It is completely characterized by

$$(2n)!! = 2^n n! = 2^n \Gamma(n+1) = 2^n (1)_n, \quad n = 0, 1, 2, \ldots \quad (8.138)$$

$$(2n-1)!! = (2n)!/(2n)!! = 2^n \Gamma(n+\tfrac{1}{2})/\Gamma(\tfrac{1}{2}) = 2^n (\tfrac{1}{2})_n,$$
$$n = 0, 1, 2, \ldots \quad (8.139)$$

$$(-2n-1)!! = 2^{-n} \Gamma(-n+\tfrac{1}{2})/\Gamma(\tfrac{1}{2}) = (-1)^n/(2n-1)!!,$$
$$n = 0, \pm 1, \pm 2, \ldots \quad (8.140)$$

$$1/(2n)!! = 0, \quad n = -1, -2, \ldots . \quad (8.141)$$

2. Differential Equation and Integral Representation

Directly from Eqs. (8.136), (8.117), and (8.113), one has

$$\{(d/dz)^2 + 2z^{-1}(d/dz) + 1 - n(n+1)z^{-2}\} j_n(z) = 0, \quad (8.142)$$

$$j_n(z) = z^n e^{-iz} 2^n (n!)^{-1} \int_0^1 t^n (1-t)^n e^{2izt} \, dt \quad (8.143)$$

$$= z^n (2^{n+1} n!)^{-1} \int_{-1}^1 (1-s^2)^n e^{izs} \, ds, \quad t = (s+1)/2. \quad (8.144)$$

3. Rayleigh's Formula, Power Series, Recurrence Formulas

Integration by parts of Eq. (8.144) and $ise^{izs} = (d/dz)e^{izs}$ gives

$$j_n(z) = (-z)^n (z^{-1} d/dz)^n z^{-1} \sin z \quad \text{(Rayleigh's formula)}. \quad (8.145)$$

In principle, a power series for $j_n(z)$ could be obtained from Eqs. (8.136), (8.123), and the power series for e^{-iz}. However, Rayleigh's formula and the power series for $\sin z$ give directly

$$j_n(z) = \sum_{k=0}^{\infty} (-1)^k z^{n+2k} / [(2n + 2k + 1)!!(2k)!!]. \quad (8.146)$$

Substitution in the appropriate recurrence formulas for M would yield recurrence formulas for $j_n(z)$. In this short exposition, an *ad hoc* approach is more efficient. The result,

$$(d/dz)z^{-n}j_n(z) = z^{-n}[(d/dz) - n/z]j_n(z) = -z^{-n}j_{n+1}(z), \quad (8.147)$$

follows from Rayleigh's formula, while

$$(d/dz)z^{n+1}j_n(z) = z^{n+1}[(d/dz) + (n+1)/z]j_n(z) = z^{n+1}j_{n-1}(z) \quad (8.148)$$

uses Rayleigh's formula and the identity

$$(d/dz)z^{2n+1}(z^{-1} d/dz)^n z^{-1} = z^{2n}(z^{-1} d/dz)^{n-1} z^{-1} (d/dz)^2. \quad (8.149)$$

Together, Eqs. (8.147) and (8.148) yield

$$nj_{n-1}(z) - (n+1)j_{n+1}(z) = (2n+1)(d/dz)j_n(z) \quad (8.150)$$

$$j_{n-1}(z) + j_{n+1}(z) = (2n+1)z^{-1}j_n(z). \quad (8.151)$$

4. Connection with Legendre Polynomials. Plane Wave Expansion

Equations (8.72) and (8.144) give immediately

$$j_n(z) = \tfrac{1}{2} i^{-n} \int_{-1}^{1} P_n(s) e^{izs} \, ds. \quad (8.152)$$

We derive additionally that

$$e^{izs} = \sum_{n=0}^{\infty} (2n+1) i^n j_n(z) P_n(s). \quad (8.153)$$

[Note that if we knew that the Legendre polynomials were complete, then Eq. (8.153) follows from Eq. (8.152) and the orthogonality of the Legendre polynomials, Eq. (8.75).] We substitute

$$s^n/n! = \sum_{\substack{k \\ (n-k \text{ even}, k \leq n)}} (2k+1)P_k(s)/[(n+k+1)!!(n-k)!!] \qquad (8.154)$$

[which is easily computed via Eqs. (8.152) and (8.75)] into the power series for e^{izs} and rearrange the terms. [The double series converges absolutely; an estimate of $|P_n(z)|$, $|P_n(s)| \leq (|s| + |s^2 - 1|^{1/2})^n$ follows from Laplace's integral, Eq. (8.73).] The result is

$$e^{izs} = \sum_{n=0}^{\infty} (iz)^n s^n/n! \qquad (8.155)$$

$$= \sum_{n=0}^{\infty} \left[\sum_k \frac{(iz)^n}{(n+k+1)!!(n-k)!!} (2k+1)P_k(s) \right] \qquad (8.156)$$

$$= \sum_{k=0}^{\infty} \left(\sum_{\substack{m=0 \\ (n=k+2m)}}^{\infty} \frac{(iz)^{k+2m}}{(2k+2m+1)!!(2m)!!} \right)(2k+1)P_k(s) \qquad (8.157)$$

$$= \sum_{k=0}^{\infty} (2k+1)i^k j_k(z) P_k(s). \qquad (8.158)$$

Equation (8.158) often occurs combined with the addition formula (8.104), in a form called the "plane wave expansion":

$$e^{i\mathbf{k}\cdot\mathbf{r}} = \sum_{l=0}^{\infty} (2l+1)i^l j_l(kr)(-1)^m P_l^m(\cos\theta) P_l^{-m}(\cos\theta_k) \exp(im(\phi - \phi_k)), \qquad (8.159)$$

where (r, θ, ϕ) and (k, θ_k, ϕ_k) are the spherical polar coordinates of $\mathbf{r}$ and $\mathbf{k}$.

G. Series for Exponential-Type Integral

The exponential-type integral $E_n(z)$ (Abramowitz and Stegun, 1964), which is defined by

$$E_n(z) = \int_1^{\infty} t^{-n} e^{-zt} \, dt, \qquad (8.160)$$

is related to the confluent hypergeometric function U by

$$E_n(z) = e^{-z} U(1, -n+2, z). \qquad (8.161)$$

In Section X.E a series expansion is needed for $E_n(z)$, for n a positive integer. The expansion is derived by first letting n be nonintegral, by using Eqs. (8.125), (8.133), $[M(n, n, z) = e^z]$, and (8.123), then letting $n \to N =$ positive integer. With $\psi(n) = [(d/dn)\Gamma(n)]/\Gamma(n)$, one obtains

$$E_N(z) = \lim_{n \to N} e^{-z} \frac{\pi}{\sin \pi(2-n)} \left[\frac{M(1, 2-n, z)}{\Gamma(n)\Gamma(2-n)} - z^{n-1} \frac{M(n, n, z)}{\Gamma(1)\Gamma(n)} \right] \quad (8.162)$$

$$= \lim_{n \to N} \frac{\pi}{\sin \pi(2-n)} \left[\frac{M(1-n, 2-n, -z)}{\Gamma(n)\Gamma(2-n)} - \frac{z^{n-1}}{\Gamma(n)} \right] \quad (8.163)$$

$$= \lim_{n \to N} -\sum_{k=0}^{\infty} \frac{1}{(k-n+1)} \frac{(-z)^k}{k!} + \frac{\pi(-1)^{N-1}}{\sin \pi(N-n)} \frac{z^{n-1}}{\Gamma(n)} \quad (8.164)$$

$$= -\sum_{\substack{k=0 \\ (k \neq N-1)}}^{\infty} \frac{1}{(k-N+1)} \frac{(-z)^k}{k!}$$

$$- \frac{(-z)^{N-1}}{(N-1)!} [\log z - \psi(N)]. \quad (8.165)$$

IX. On Fourier Transforms

The Fourier transform of $f(x)$ is defined by

$$\text{FT}\{f(x)\} = \int_{-\infty}^{\infty} f(x) e^{ikx} \, dx. \quad (9.1)$$

The central result, which is established below, is the *Fourier transform inversion formula*

$$(2\pi)^{-1} \int_{-\infty}^{\infty} e^{-ikx} \, \text{FT}\{f(x)\} \, dk = f(x); \quad (9.2)$$

that is,

$$(2\pi)^{-1} \int_{-\infty}^{\infty} e^{-ikx} \left(\int_{-\infty}^{\infty} e^{ikx'} f(x') \, dx' \right) dk = f(x). \quad (9.3)$$

Many computations are easier to carry out through the intermediacy of Fourier transforms.

3. Complex Variable Theory

In this section, the basic results on Fourier transforms are established for functions $f(x)$ that, together with their Fourier transforms, are analytic in a strip that includes the entire real axis, and that vanish strongly enough at ∞ both for the integrals below to exist, and for the integrated term in each integral by parts formula to vanish.

Given Eqs. (9.1)–(9.3), one may readily establish (interchange differentiation and integration)

$$\text{FT}\{(d/dx)^n f(x)\} = (-ik)^n \, \text{FT}\{f(x)\} \qquad \text{(integrate by parts)} \qquad (9.4)$$

$$\text{FT}\{x^n f(x)\} = (-i d/dk)^n \, \text{FT}\{f(x)\} \qquad (9.5)$$

$$\text{FT}\{f(x - x_0)\} = \exp(ikx_0) \, \text{FT}\{f(x)\} \qquad (9.6)$$

$$\int_{-\infty}^{\infty} f_1(x) f_2(x) \, dx = (2\pi)^{-1} \int_{-\infty}^{\infty} \text{FT}\{f_1(x)\}_{(-k)} \, \text{FT}\{f_2(x)\} \, dk \qquad (9.7)$$

$$\int_{-\infty}^{\infty} f_1^*(x) f_2(x) \, dx = (2\pi)^{-1} \int_{-\infty}^{\infty} \text{FT}^*\{f_1(x)\} \, \text{FT}\{f_2(x)\} \, dk. \qquad (9.8)$$

Equations (9.7) and (9.8) are two versions of the *Fourier transform convolution theorem*. The asterisk (*) denotes complex conjugate, and $\text{FT}\{f_1(x)\}_{(-k)}$ denotes

$$\text{FT}\{f_1(x)\}_{(-k)} = \int_{-\infty}^{\infty} e^{-ikx} f_1(x) \, dx. \qquad (9.9)$$

To derive the inversion formula under the assumptions given above, we use the extension of Cauchy's theorem given in Section VII.B and the fact that the order of integration in a double integral can be interchanged when the integrand vanishes uniformly at ∞ in the two variables. Thus for real x,

$$(2\pi)^{-1} \int_{-\infty}^{\infty} dk \int_{-\infty}^{\infty} dx' \, \exp(ik(x' - x)) f(x')$$

$$= \left(\int_{-\infty}^{0} dk \int_{-\infty}^{\infty} dx' + \int_{0}^{\infty} dk \int_{-\infty}^{\infty} dx' \right) \exp(ik(x' - x)) f(x') \qquad (9.10)$$

$$= \int_{-\infty}^{0} dk \int_{-\infty - i\varepsilon}^{\infty - i\varepsilon} dz' \, \exp(ik(z' - x)) f(z')$$

$$+ \int_{0}^{\infty} dk \int_{-\infty + i\varepsilon}^{\infty + i\varepsilon} dz' \, \exp(ik(z' - x)) f(z'), \qquad \varepsilon > 0. \qquad (9.11)$$

Since $\text{Re}[ik(z' - x)]$ is less than 0 for *both* terms, the $\int dk$ can be evaluated first. Then

$$(2\pi)^{-1} \int_{-\infty}^{\infty} dk \int_{-\infty}^{\infty} dx' \exp[ik(x' - x)] f(x')$$

$$= (2\pi i)^{-1} \int_{-\infty - i\varepsilon}^{\infty - i\varepsilon} (z' - x)^{-1} f(z') \, dz'$$

$$- (2\pi i)^{-1} \int_{-\infty + i\varepsilon}^{\infty + i\varepsilon} (z' - x)^{-1} f(z') \, dz' \qquad (9.12)$$

$$= (2\pi i)^{-1} \oint^{(x+)} (z' - x)^{-1} f(z') \, dz' = f(x). \qquad (9.13)$$

Fourier Transforms in Three Dimensions The three-dimensional Fourier transform of $f(\mathbf{r})$ $[\mathbf{r} = (x, y, z)]$ is defined in the natural way by

$$\text{FT}\{f(\mathbf{r})\} = \int_{-\infty}^{\infty} dx \int_{-\infty}^{\infty} dy \int_{-\infty}^{\infty} dz \exp(i(k_x x + k_y y + k_z z)) f(\mathbf{r}) \quad (9.14)$$

$$= \int e^{i\mathbf{k} \cdot \mathbf{r}} f(\mathbf{r}) \, dV, \qquad dV = dx \, dy \, dz. \qquad (9.15)$$

An essential point is that if $f(\mathbf{r})$ vanishes rapidly and uniformly at ∞, then the order of integration is immaterial. Indeed, in the next section we find it convenient to use spherical polar coordinates.

A Special Form of the Convolution Theorem A situation to which Eq. (9.8) is applicable, perhaps unexpectedly, is

$$\int_{-\infty}^{\infty} dx_1 \int_{-\infty}^{\infty} dx_2 f_1^*(x_1) f_{12}(x_1 - x_2) f_2(x_2)$$

$$= (2\pi)^{-1} \int_{-\infty}^{\infty} \text{FT}^*\{f_1(x)\} \, \text{FT}\{f_{12}(x)\} \, \text{FT}\{f_2(x)\} \, dk, \quad (9.16)$$

since

$$\text{FT}\left\{\int_{-\infty}^{\infty} dx_2 f_{12}(x - x_2) f_2(x_2)\right\} = \text{FT}\{f_{12}(x)\} \, \text{FT}\{f_2(x)\}. \quad (9.17)$$

The three-dimensional version is

$$\int dV_1 \int dV_2 f_1^*(\mathbf{r}_1) f_{12}(\mathbf{r}_1 - \mathbf{r}_2) f_2(\mathbf{r}_2)$$
$$= (2\pi)^{-3} \int \mathrm{FT}^*\{f_1(\mathbf{r})\}\, \mathrm{FT}\{f_{12}(\mathbf{r})\}\, \mathrm{FT}\{f_2(\mathbf{r})\}\, d^3\mathbf{k}, \quad (9.18)$$

where $d^3\mathbf{k}$ denotes $dk_x\, dk_y\, dk_z$.

X. Quantum Chemistry Integrals

An important problem in quantum chemistry is the calculation of the energy levels of molecules. Frequently, it is necessary to evaluate integrals of the form

$$I_{ab}(\mathbf{R}) = \int \psi_a^*(\mathbf{r}) \hat{A} \psi_b(\mathbf{r} - \mathbf{R})\, dV \quad (10.1)$$

$$I_{cd;ab}(\mathscr{R}_1, \mathscr{R}_2, \mathbf{R}) = \iint [\psi_a^*(\mathbf{r}_2) \psi_b(\mathbf{r}_2 - \mathscr{R}_2)]^*$$
$$\times r_{12}^{-1} [\psi_c^*(\mathbf{r}_1 - \mathbf{R}) \psi_d(\mathbf{r}_1 - \mathbf{R} - \mathscr{R}_1)]\, dV_1\, dV_2, \quad (10.2)$$

where $\hat{A}$ denotes $1, -\tfrac{1}{2}\nabla^2, r^{-1}$, or $|\mathbf{r} - \mathscr{R}|^{-1}$, and the functions $\psi(\mathbf{r})$ are so-called atomic orbitals.

Much has been written on the evaluation of these integrals using various approaches. The purpose here is to indicate how complex variable theory is applicable by working a few specific examples. The application is taken from Silverstone (1966, 1967, 1968a,b), Silverstone and Kay (1968), Silverstone and Todd (1971), Kay and Silverstone (1969a,b, 1970), Kay, Todd, and Silverstone (1969a,b), and Todd, Kay, and Silverstone (1970), to which the reader is referred for the general cases and more detailed discussions.

A. Spherical Harmonics. Slater-Type Atomic Orbitals

A commonly used atomic orbital is the so-called Slater-type atomic orbital:

$$\psi(\mathbf{r}) = \psi_{nlm\zeta}(\mathbf{r}) = r^{n-1} e^{-\zeta r} Y_l^m(\theta, \phi) \quad (10.3)$$

where ζ is a constant, called the *orbital exponent*, and n, l, m are integers satisfying $n - 1 \geq l \geq |m|$. The $Y_l^m(\theta, \phi)$ are called *spherical harmonics* and are defined by

$$Y_l^m(\theta, \phi) = (-1)^m \left[\frac{2l+1}{4\pi} \frac{(l-m)!}{(l+m)!} \right]^{1/2} P_l^m(\cos\theta) e^{im\phi} \quad (10.4)$$

$$= (-1)^m Y_l^{-m*}(\theta, \phi). \quad (10.5)$$

By Eq. (8.94) and by $\int_0^{2\pi} \exp(i(m'-m)\phi) \, d\phi = 2\pi \, \delta_{mm'}$, the Y_l^m are orthogonal and normalized to unity:

$$\int_0^{2\pi} d\phi \int_0^{\pi} d\theta \sin\theta \, Y_l^{m*}(\theta, \phi) Y_{l'}^{m'}(\theta, \phi) = \delta_{ll'} \, \delta_{mm'}. \quad (10.6)$$

In the derivations below, we repeatedly use both the orthonormality of the Y_l^m [Eq. (10.6)], the relation between Y_l^{-m} and Y_l^{m*} [Eq. (10.5)], and the plane wave expansion [Eq. (8.159)] expressed in terms of the Y_l^m,

$$e^{i\mathbf{k}\cdot\mathbf{r}} = \sum_{l=0}^{\infty} 4\pi i^l j_l(kr) \sum_{m=-l}^{l} (-1)^m Y_l^m(\theta, \phi) Y_l^{-m}(\theta_k, \phi_k). \quad (10.7)$$

Also, the spherical polar coordinates for the vectors $\mathbf{r}$, $\mathbf{k}$, $\mathbf{R}$, $\mathscr{R}_1$, and $\mathscr{R}_2$ are written (r, θ, ϕ), (k, θ_k, ϕ_k), (R, θ_R, ϕ_R), $(\mathscr{R}_1, \theta_{\mathscr{R}_1}, \phi_{\mathscr{R}_1})$, and $(\mathscr{R}_2, \theta_{\mathscr{R}_2}, \phi_{\mathscr{R}_2})$.

B. Outline of the Approach

The use of complex variable theory to evaluate integrals of the type $I_{ab}(\mathbf{R})$ and $I_{cd;ab}(\mathscr{R}_1, \mathscr{R}_2, \mathbf{R})$ is based on six equations: the Fourier transform inversion and convolution theorems, Eqs. (9.2) and (9.18); Eq. (9.6); the plane wave expansion, Eq. (10.7); the orthogonality of the Y_l^m, Eq. (10.6); and the residue theorem, Eq. (7.1). In each case the fundamental steps are

1. transformation to Fourier transform variable via convolution theorem;
2. introduction of plane wave expansion to bring all angular dependence into spherical harmonics;
3. evaluation of angular integrations by orthogonality formula;
4. manipulation of final radial integration $\int_0^{\infty} dk$ so as to be able to use the residue theorem.

C. Fourier Transform of a Slater-Type Atomic Orbital

Basic to this application is the $\text{FT}\{\psi_{nlm\zeta}(\mathbf{r})\}$:

$$\text{FT}\{\psi_{nlm\zeta}(\mathbf{r})\} = \int e^{i\mathbf{k}\cdot\mathbf{r}}\,\psi_{nlm\zeta}(\mathbf{r})\,dV \tag{10.8}$$

$$= \int \sum_{\lambda,\mu} 4\pi i^{\lambda} j_{\lambda}(kr) Y_{\lambda}^{\mu}(\theta_k, \phi_k) Y_{\lambda}^{\mu*}(\theta, \phi) r^{n-1} e^{-\zeta r} Y_l^m(\theta, \phi)\,dV \tag{10.9}$$

$$= f_{nl\zeta}(k) Y_l^m(\theta_k, \phi_k), \tag{10.10}$$

$$f_{nl\zeta}(k) = 4\pi i^l \int_0^\infty j_l(kr) r^{n-1} e^{-\zeta r} r^2\,dr \tag{10.11}$$

$$= 4\pi i^l \int_0^\infty (-k)^l (k^{-1}\,d/dk)^l k^{-1} r^{n-l} \sin kr\, e^{-\zeta r}\,dr \tag{10.12}$$

$$= 2\pi i^{l-1}(-k)^l (k^{-1}\,d/dk)^l k^{-1}$$
$$\times \int_0^\infty r^{n-l}[\exp(-(\zeta-ik)r) - \exp(-(\zeta+ik)r)]\,dr \tag{10.13}$$

$$= 2\pi i^{l-1}(n-l)!(-k)^l(k^{-1}\,d/dk)^l k^{-1}$$
$$\times [(\zeta-ik)^{l-n-1} - (\zeta+ik)^{l-n-1}]. \tag{10.14}$$

At this stage it is unnecessary to carry out the differentiations in Eq. (10.14). Rayleigh's formula for $j_l(kr)$ and the definition of $\Gamma(n-l+1)$ were used in Eqs. (10.12) and (10.14). Note that the order of integration and differentiation was interchanged in Eqs. (10.12) and (10.13). Here and subsequently, we leave the justification to the reader.

D. Two-Center Overlap-Type Integral

The integrals are named in part by the number of distinct atomic centers. The one-center integrals, characterized by $\mathscr{R}_1 = \mathscr{R}_2 = \mathbf{R} = 0$, present no problem and are not discussed here. The simplest two-center integral is Eq. (10.1) with $\hat{A} = 1$, and ψ_a and ψ_b both with $n=1$, $l=m=0$ [Eq. (10.3)]. We compute the "overlap integral"

$$I_{ab} = \int \psi_{100\zeta_a}^*(\mathbf{r})\psi_{100\zeta_b}(\mathbf{r}-\mathbf{R})\,dV. \tag{10.15}$$

In the course of computing I_{ab}, we are led to consider four special functions of the confluent hypergeometric family.

Via the convolution theorem, Eq. (9.8), and Eqs. (9.6), (10.10), and (10.14), one obtains

$$I_{ab} = (2\pi)^{-3} \int \text{FT}^*\{\psi_{100\zeta_a}(\mathbf{r})\} \, \text{FT}\{\psi_{100\zeta_b}(\mathbf{r} - \mathbf{R})\} \, d^3\mathbf{k} \tag{10.16}$$

$$= (2\pi)^{-3} \int e^{i\mathbf{k}\cdot\mathbf{R}} \, \text{FT}^*\{\psi_{100\zeta_a}(\mathbf{r})\} \, \text{FT}\{\psi_{100\zeta_b}(\mathbf{r})\} \, d^3\mathbf{k} \tag{10.17}$$

$$= (2\pi)^{-3} \int e^{i\mathbf{k}\cdot\mathbf{R}} f^*_{10\zeta_a}(k) f_{10\zeta_b}(k) (Y_0^{\,0})^2 \, d^3\mathbf{k}. \tag{10.18}$$

We use the plane wave expansion, the orthonormality of the Y_l^m, and that $Y_0^{\,0} = (4\pi)^{-1/2}$, to obtain

$$I_{ab} = (2\pi)^{-3} \int_0^\infty j_0(kR)\{2\pi i^{-1}k^{-1}[(\zeta_a - ik)^{-2} - (\zeta_a + ik)^{-2}]\}^*$$
$$\times \{2\pi i^{-1}k^{-1}[(\zeta_b - ik)^{-2} - (\zeta_b + ik)^{-2}]\}k^2 \, dk \tag{10.19}$$

$$= -(2\pi)^{-1} \int_0^\infty \frac{\sin(kR)}{kR}$$
$$\times [(\zeta_a - ik)^{-2} - (\zeta_a + ik)^{-2}][(\zeta_b - ik)^{-2} - (\zeta_b + ik)^{-2}] \, dk. \tag{10.20}$$

Since the integrand is an even function of k, Eq. (10.20) can be written as a special case of Eq. (7.3) (note that the integrand is not singular at $k = 0$):

$$I_{ab} = -\frac{1}{4\pi i} \int_{-\infty}^\infty e^{ikR}(kR)^{-1}[(\zeta_a - ik)^{-2} - (\zeta_a + ik)^{-2}]$$
$$\times [(\zeta_b - ik)^{-2} - (\zeta_b + ik)^{-2}] \, dk \tag{10.21}$$

$$= -\tfrac{1}{2} \sum \{\text{residues at } k = i\zeta_a, i\zeta_b\} \tag{10.22}$$

$$= \tfrac{1}{2}(\partial/\partial\zeta_a) \exp(-\zeta_a R)(\zeta_a R)^{-1}[(\zeta_b + \zeta_a)^{-2} - (\zeta_b - \zeta_a)^{-2}]$$
$$+ \tfrac{1}{2}(\partial/\partial\zeta_b) \exp(-\zeta_b R)(\zeta_b R)^{-1}[(\zeta_a + \zeta_b)^{-2} - (\zeta_a - \zeta_b)^{-2}],$$
$$\zeta_a \neq \zeta_b \tag{10.23}$$

$$= -\tfrac{1}{12}(d/d\zeta_a)^3 \exp(-\zeta_a R)(\zeta_a R)^{-1}$$
$$+ (d/d\zeta_a) \exp(-\zeta_a R)(\zeta_a R)^{-1}(\zeta_a + \zeta_b)^{-2}, \quad \zeta_a = \zeta_b. \tag{10.24}$$

The cases for more general values of the atomic orbital parameters work out similarly (Silverstone, 1966).

Despite the simplicity of the derivation of Eqs. (10.23) and (10.24), these equations do not exhibit clearly the behavior of $I_{ab}(R)$ when either

$R \sim 0$ or $\zeta_a \sim \zeta_b$. An alternative approach, involving the exponential-type integral functions α_n and $\hat{\alpha}_n$, gives a more transparent formula (Todd, Kay, and Silverstone, 1970).

1. *The Functions $\alpha_n(z)$ and $\hat{\alpha}_n(z)$*

Define $\alpha_n(z)$ (Abramowitz and Stegun, 1964) and $\hat{\alpha}_n(z)$ (Silverstone, 1968a) by

$$\alpha_n(z) = \int_1^\infty t^n e^{-zt}\, dt = \sum_{k=0}^{n} \frac{n!}{(n-k)!} z^{-k-1} e^{-z}, \qquad n = 0, 1, 2, \ldots \tag{10.25}$$

$$\hat{\alpha}_n(z) = -\int_0^1 t^n e^{-zt}\, dt = -\sum_{k=0}^{\infty} \frac{(-z)^k}{k!(n+k+1)}, \qquad n = 0, 1, 2, \ldots \tag{10.26}$$

$$= \alpha_n(z) - n!/z^{n+1}. \tag{10.27}$$

$\hat{\alpha}_n(z)$ is an entire function of z [Eq. (10.26)], while $\alpha_n(z)$ has a pole of order $n+1$ at the origin [Eq. (10.27)]. As $z \sim \infty$, $\alpha_n(z) \sim z^{-1}e^{-z}$ [Eq. (10.25)].

2. *Alternative Formula*

Returning to Eq. (10.20), we use the evenness of the integrand and Eq. (10.27) to write

$$I_{ab} = (2\pi)^{-1} \int_{-\infty}^{\infty} (kR)^{-1} \sin kR\, (\zeta_a + ik)^{-2}[(\zeta_b - ik)^{-2} - (\zeta_b + ik)^{-2}]\, dk \tag{10.28}$$

$$= (2\pi)^{-1} \int_{-\infty}^{\infty} k^{-1}R \sin kR\, \{\alpha_1[(\zeta_a + ik)R] - \hat{\alpha}_1[(\zeta_a + ik)R]\}$$
$$\times [(\zeta_b - ik)^{-2} - (\zeta_b + ik)^{-2}]\, dk \tag{10.29}$$

$$= (2\pi)^{-1} \int_{-\infty}^{\infty} k^{-1}R \sin kR\, \alpha_1[(\zeta_a+ik)R][(\zeta_b-ik)^{-2} - (\zeta_b+ik)^{-2}]\, dk$$

$$- (2\pi)^{-1} \int_{-\infty}^{\infty} k^{-1}R(-2i)^{-1}e^{-ikR}\hat{\alpha}_1[(\zeta_a+ik)R]$$
$$\times [(\zeta_b - ik)^{-2} - (\zeta_b + ik)^{-2}]\, dk$$

$$- (2\pi)^{-1} \int_{-\infty}^{\infty} k^{-1}R(2i)^{-1}e^{ikR}\hat{\alpha}_1[(\zeta_a+ik)R]$$
$$\times [(\zeta_b - ik)^{-2} - (\zeta_b + ik)^{-2}]\, dk. \tag{10.30}$$

The integrands of the first two terms are bounded in the lower half-plane, and the third integrand is bounded in the upper half-plane. Closing the contour at ∞ in the appropriate half-plane, we evaluate each via the residue theorem. The poles inside the contour are either at $k = i\zeta_b$ or $k = -i\zeta_b$. The result is

$$\begin{aligned}
I_{ab} &= (d/d\zeta_b)R\zeta_b^{-1} i \sin(i\zeta_b R)\alpha_1[(\zeta_a + \zeta_b)R] \\
&\quad - \tfrac{1}{2}(d/d\zeta_b)R\zeta_b^{-1} \exp(-\zeta_b R)\hat{a}_1[(\zeta_a + \zeta_b)R] \\
&\quad + \tfrac{1}{2}(d/d\zeta_b)R\zeta_b^{-1} \exp(-\zeta_b R)\hat{a}_1[(\zeta_a - \zeta_b)R], \quad (10.31) \\
&= -(d/d\zeta_b)R\zeta_b^{-1} \sinh(\zeta_b R)\alpha_1[(\zeta_a + \zeta_b)R] \\
&\quad - (d/d\zeta_b)R\zeta_b^{-1} \exp(-\zeta_b R)\tfrac{1}{2}\{\hat{a}_1[(\zeta_a + \zeta_b)R] - \hat{a}_1[(\zeta_a - \zeta_b)R]\}. \quad (10.32)
\end{aligned}$$

Before taking the $(d/d\zeta_b)$ explicitly, we introduce two more functions.

3. Modified Spherical Bessel Functions

Still more concise formulas are obtained by introducing the modified spherical Bessel functions, $\mathcal{K}_l$ and $\mathcal{I}_l$:

$$\mathcal{K}_l(z) = (-z)^l (z^{-1} d/dz)^l z^{-1} e^{-z} \quad (10.33)$$

$$= \sum_{m=0}^{l} ((l+m)!/(l-m)!(2m)!!) z^{-m-1} e^{-z} \quad (10.34)$$

$$\mathcal{I}_l(z) = z^l (z^{-1} d/dz)^l z^{-1} \sinh z = \sum_{m=0}^{\infty} z^{l+2m}/[(2l+2m+1)!!(2m)!!] \quad (10.35)$$

$$= i^{-l} j_l(iz) \quad (10.36)$$

$$= -\tfrac{1}{2}[\mathcal{K}_l(-z) + (-1)^l \mathcal{K}_l(z)]. \quad (10.37)$$

With $\mathcal{K}_l$ and $\mathcal{I}_l$, I_{ab} becomes

$$\begin{aligned}
I_{ab} &= -R^2 (d/d\zeta_b)(\mathcal{I}_0(\zeta_b R)\alpha_1[(\zeta_a + \zeta_b)R] \\
&\quad + \mathcal{K}_0(\zeta_b R)\tfrac{1}{2}\{\hat{a}_1[(\zeta_a + \zeta_b)R] - \hat{a}_1[(\zeta_a - \zeta_b)R]\}) \quad (10.38) \\
&= R^3 \mathcal{I}_0(\zeta_b R)\alpha_2[(\zeta_a + \zeta_b)R] - R^3 \mathcal{I}_1(\zeta_b R)\alpha_1[(\zeta_a + \zeta_b)R] \\
&\quad + R^3 \mathcal{K}_0(\zeta_b R)\tfrac{1}{2}\{\hat{a}_2[(\zeta_a + \zeta_b)R] + \hat{a}_2[(\zeta_a - \zeta_b)R]\} \\
&\quad + R^3 \mathcal{K}_1(\zeta_b R)\tfrac{1}{2}\{\hat{a}_1[(\zeta_a + \zeta_b)R] - \hat{a}_1[(\zeta_a - \zeta_b)R]\}. \quad (10.39)
\end{aligned}$$

E. Fourier Transform of a Two-Center Product

The more complicated integrals represented by Eqs. (10.1) and (10.2) involve the Fourier transform

$$G_{ab}(\mathbf{k}, \mathcal{R}) = \text{FT}\{\psi_a^*(\mathbf{r})\psi_b(\mathbf{r} - \mathcal{R})\} \quad (10.40)$$

$$= \int e^{i\mathbf{k}\cdot\mathbf{r}} \psi_a^*(\mathbf{r})\psi_b(\mathbf{r} - \mathcal{R}) \, dV. \quad (10.41)$$

We evaluate a simple case that illustrates the general considerations involved: We take $n_a = n_b = 1$, $l_a = m_a = l_b = m_b = 0$. Then

$$G_{ab}(\mathbf{k}, \mathscr{R}) = \int e^{i\mathbf{k}\cdot\mathbf{r}} \psi^*_{100\zeta_a}(r)\psi_{100\zeta_b}(\mathbf{r} - \mathscr{R})\, dV \tag{10.42}$$

$$= (4\pi)^{-1} \int e^{i\mathbf{k}\cdot\mathbf{r}} \exp(-\zeta_a |\mathbf{r}|) \exp(-\zeta_b |\mathbf{r} - \mathscr{R}|)\, dV. \tag{10.43}$$

The functional dependence on $|\mathbf{r}|$ and $|\mathbf{r} - \mathscr{R}|$ is somewhat inconvenient. A more convenient form is obtained at the cost of an infinite series. The $\exp(-\zeta_b |\mathbf{r} - \mathscr{R}|)$ is expanded in $Y_l^m(\theta, \phi)$ and functions of r:

$$\exp(-\zeta_b |\mathbf{r} - \mathscr{R}|) = \sum_{l=0}^{\infty} v_l(r, \mathscr{R}) \sum_{m=-l}^{l} (-1)^m Y_l^{-m}(\theta_\mathscr{R}, \phi_\mathscr{R}) Y_l^m(\theta, \phi). \tag{10.44}$$

$$v_l(r, \mathscr{R}) = 4\pi(-d/d\zeta_b)\zeta_b \mathscr{I}_l(\zeta_b r)\mathscr{K}_l(\zeta_b \mathscr{R}), \qquad r < \mathscr{R} \tag{10.45}$$

$$= 4\pi(-d/d\zeta_b)\zeta_b \mathscr{I}_l(\zeta_b \mathscr{R})\mathscr{K}_l(\zeta_b r), \qquad \mathscr{R} < r. \tag{10.46}$$

A most straightforward way to derive this and similar expansions is via the inverse Fourier transform (Silverstone, 1967; Kay, Todd, and Silverstone, 1969a,b)

$$\exp(-\zeta_b |\mathbf{r} - \mathscr{R}|) = (2\pi)^{-3} \int \exp(-i\mathbf{k}\cdot(\mathbf{r} - \mathscr{R}))\, \mathrm{FT}\{\exp(-\zeta_b r)\}\, d^3\mathbf{k} \tag{10.47}$$

$$= (2\pi)^{-3} \int \{\sum_{l_1, m_1} 4\pi i^{-l_1} j_{l_1}(kr) Y_{l_1}^{m_1}(\theta, \phi) Y_{l_1}^{m_1*}(\theta_k, \phi_k)\}$$
$$\{\sum_{l_2, m_2} 4\pi i^{l_2} j_{l_2}(k\mathscr{R})(-1)^{m_2} Y_{l_2}^{-m_2}(\theta_\mathscr{R}, \phi_\mathscr{R})$$
$$\times Y_{l_2}^{m_2}(\theta_k, \phi_k)\} f_{10\zeta_b}(k)\, d^3\mathbf{k} \tag{10.48}$$

$$= \sum_{l=0}^{\infty} \sum_{m=-l}^{l} (-1)^m Y_l^{-m}(\theta_\mathscr{R}, \phi_\mathscr{R}) Y_l^m(\theta, \phi) v_l(r, \mathscr{R}) \tag{10.49}$$

$$v_l(r, \mathscr{R}) = (2/\pi) \int_0^\infty j_l(kr) j_l(k\mathscr{R}) 2\pi i^{-1} k^{-1}$$
$$\times [(\zeta_b - ik)^{-2} - (\zeta_b + ik)^{-2}] k^2\, dk. \tag{10.50}$$

Equation (10.50) is of a type similar to Eq. (7.3). When $\mathscr{R} > r$, we write [via Eq. (8.145) and the evenness of the integrand of Eq. (10.50)]

$$\nu_l(r, \mathscr{R}) = -2 \int_{-\infty}^{\infty} (-\mathscr{R})^l (\mathscr{R}^{-1} d/d\mathscr{R})^l \mathscr{R}^{-1} e^{ik\mathscr{R}} k^{-l} j_l(kr)$$
$$\times [(\zeta_b - ik)^{-2} - (\zeta_b + ik)^{-2}] \, dk. \tag{10.51}$$

We close the contour at ∞ in the upper half-plane and take $2\pi i \times$ residue at $k = i\zeta_b$:

$$\nu_l(r, \mathscr{R}) = -4\pi(-\mathscr{R})^l (\mathscr{R}^{-1} d/d\mathscr{R})^l \mathscr{R}^{-1} (d/d\zeta_b) \exp(-\zeta_b \mathscr{R})(i\zeta_b)^{-l}$$
$$\times j_l(i\zeta_b r) \tag{10.52}$$
$$= 4\pi(-d/d\zeta_b)\zeta_b \mathscr{K}_l(\zeta_b \mathscr{R}) \mathscr{I}_l(\zeta_b r), \quad \mathscr{R} > r. \tag{10.53}$$

When $\mathscr{R} < r$, Eq. (10.46) results.

With the expansion (10.44), the Fourier transform of Eq. (10.43) reduces to a sum of one-dimensional integrals

$$G_{ab}(\mathbf{k}, \mathscr{R}) = (4\pi)^{-1} \int e^{i\mathbf{k}\cdot\mathbf{r}} \exp(-\zeta_a r) \sum_{l,m} \nu_l(r, \mathscr{R})(-1)^m$$
$$\times Y_l^{-m}(\theta_\mathscr{R}, \phi_\mathscr{R}) Y_l^m(\theta, \phi) \, dV \tag{10.54}$$
$$= \sum_{l=0}^{\infty} \sum_{m=-l}^{l} (-1)^m Y_l^{-m}(\theta_\mathscr{R}, \phi_\mathscr{R}) Y_l^m(\theta_k, \phi_k) G_l(k, \mathscr{R}) \tag{10.55}$$

$$G_l(k, \mathscr{R}) = i^l \int_0^{\infty} j_l(kr) \nu_l(r, \mathscr{R}) \exp(-\zeta_a r) r^2 \, dr \tag{10.56}$$
$$= 4\pi i^l (-d/d\zeta_b) \zeta_b \mathscr{K}_l(\zeta_b \mathscr{R}) \int_0^{\mathscr{R}} j_l(kr) \mathscr{I}_l(\zeta_b r) \exp(-\zeta_a r) r^2 \, dr$$
$$+ 4\pi i^l (-d/d\zeta_b) \zeta_b \mathscr{I}_l(\zeta_b \mathscr{R}) \int_{\mathscr{R}}^{\infty} j_l(kr) \mathscr{K}_l(\zeta_b r)$$
$$\times \exp(-\zeta_a r) r^2 \, dr. \tag{10.57}$$

The second integral is readily evaluated with the aid of Rayleigh's formula and the exponential integral (Section VIII.G):

$$\int_{\mathscr{R}}^{\infty} j_l(kr) \mathscr{K}_l(\zeta_b r) \exp(-\zeta_a r) r^2 \, dr$$
$$= (-k)^l (k^{-1} d/dk)^l k^{-1} (-\zeta_b)^l (\zeta_b^{-1} d/d\zeta_b)^l \zeta_b^{-1}$$
$$\times \int_{\mathscr{R}}^{\infty} r^{-2l} \exp(-(\zeta_a + \zeta_b) r) \sin kr \, dr \tag{10.58}$$
$$= (-k)^l (k^{-1} d/dk)^l k^{-1} (-\zeta_b)^l (\zeta_b^{-1} d/d\zeta_b)^l \zeta_b^{-1} \mathscr{R}^{1-2l} (2i)^{-1}$$
$$\times \{E_{2l}[(\zeta_a + \zeta_b - ik)\mathscr{R}] - E_{2l}[(\zeta_a + \zeta_b + ik)\mathscr{R}]\}. \tag{10.59}$$

3. Complex Variable Theory

The first integral, if we are careful about the lower limit, can be similarly treated:

$$\int_0^{\mathscr{R}} j_l(kr)\mathscr{I}_l(\zeta_b r) \exp(-\zeta_a r) r^2\, dr$$

$$= \lim_{\varepsilon \to 0} \int_\varepsilon^{\mathscr{R}} j_l(kr)\mathscr{I}_l(\zeta_b r) \exp(-\zeta_a r) r^2\, dr \tag{10.60}$$

$$= \lim_{\varepsilon \to 0} (-k)^l (k^{-1} d/dk)^l k^{-1} \zeta_b{}^l (\zeta_b^{-1} d/d\zeta_b)^l \zeta_b^{-1} (4i)^{-1}$$
$$\times r^{1-2l} \{E_{2l}[(\zeta_a + \zeta_b - ik)r] - E_{2l}[(\zeta_a - \zeta_b - ik)r]$$
$$- E_{2l}[(\zeta_a + \zeta_b + ik)r] + E_{2l}[(\zeta_a - \zeta_b + ik)r]\}\Big|_\varepsilon^{\mathscr{R}}. \tag{10.61}$$

Before taking the limit, we define

$$\tilde{E}_n(z) = E_n(z) + \frac{(-z)^{n-1}}{(n-1)!}[\log z - \psi(n)]. \tag{10.62}$$

By Eq. (8.165), $\tilde{E}_n(z)$ is an entire function of z. Moreover,

$$r^{1-2l} E_{2l}[(\zeta_a + \zeta_a - ik)r]\Big|_\varepsilon^{\mathscr{R}}$$
$$= r^{1-2l} \tilde{E}_{2l}[(\zeta_a + \zeta_b + ik)r]\Big|_\varepsilon^{\mathscr{R}} - \frac{[-(\zeta_a + \zeta_b - ik)]^{2l-1}}{(2l-1)!}$$
$$\times [\log \mathscr{R} - \log \varepsilon]; \tag{10.63}$$

that is, the terms in $\log(\zeta_a + \zeta_b - ik)$ cancel. Since

$$p_{2l-1}(\zeta_b) = (\zeta_a + \zeta_b - ik)^{2l-1} - (\zeta_a - \zeta_b - ik)^{2l-1}$$
$$- (\zeta_a + \zeta_b + ik)^{2l-1} + (\zeta_a - \zeta_b + ik)^{2l-1} \tag{10.64}$$

is both a polynomial of degree $(2l-1)$ in ζ_b and an odd function of ζ_b, one finds that $(\zeta_b^{-1} d/d\zeta_b)^l \zeta_b^{-1} p_{2l-1}(\zeta_b) = 0$, and that the E's in Eq. (10.61) can be replaced by $\tilde{E}$'s. Finally, since the first $2l$ terms in the series expansion of the four $\tilde{E}$'s vanish because of the $(\zeta_b d/d\zeta_b)^l \zeta_b^{-1}$, the limit $\varepsilon \to 0$ yields zero for the lower limit. The result for $G_l(k, \mathscr{R})$ is

$$G_l(k, \mathscr{R}) = 4\pi i^{l-1}(-d/d\zeta_b)\zeta_b \mathscr{I}_l(\zeta_b \mathscr{R})(-\zeta_b)^l(\zeta_b^{-1} d/d\zeta_b)^l \zeta_b^{-1}$$
$$\times (-k)^l(k^{-1} d/dk)^l k^{-1}\mathscr{R}^{1-2l}$$
$$\times \tfrac{1}{2}\{E_{2l}[(\zeta_a + \zeta_b - ik)\mathscr{R}] - E_{2l}[(\zeta_a + \zeta_b + ik)\mathscr{R}]\}$$
$$+ 4\pi i^{l-1}(-d/d\zeta_b)\zeta_b \mathscr{K}_l(\zeta_b \mathscr{R})\zeta_b{}^l(\zeta_b^{-1} d/d\zeta_b)^l \zeta_b^{-1}$$
$$\times (-k)^l(k^{-1} d/dk)^l k^{-1}\mathscr{R}^{1-2l}$$
$$\times \tfrac{1}{4}\{\tilde{E}_{2l}[(\zeta_a + \zeta_b - ik)\mathscr{R}] - \tilde{E}_{2l}[(\zeta_a - \zeta_b - ik)\mathscr{R}]$$
$$- \tilde{E}_{2l}[(\zeta_a + \zeta_b + ik)\mathscr{R}] + \tilde{E}_{2l}[(\zeta_a - \zeta_b + ik)\mathscr{R}]\}. \tag{10.65}$$

A convenient *shorthand notation* is

$$g^{(2)}(x \pm y) = g(x+y) - g(x-y) \tag{10.66}$$

$$g^{(4)}(x \pm y \pm z) = g(x+y+z) - g(x-y+z) - g(x+y-z)$$
$$+ g(x-y-z), \tag{10.67}$$

$$[\cdots \mathscr{I}_l(\zeta_b \mathscr{R}) \cdots] = (-d/d\zeta_b)\zeta_b \mathscr{I}_l(\zeta_b \mathscr{R})\zeta_b{}^l(\zeta_b^{-1} d/d\zeta_b)^l \zeta_b^{-1} \mathscr{R}^{1-2l} \tag{10.68}$$

$$[\cdots \mathscr{K}_l(\zeta_b \mathscr{R}) \cdots] = (-d/d\zeta_b)\zeta_b \mathscr{K}_l(\zeta_b \mathscr{R})\zeta_b{}^l(\zeta_b^{-1} d/d\zeta_b)^l \zeta_b^{-1} \mathscr{R}^{1-2l}. \tag{10.69}$$

Then

$$\begin{aligned}G_l(k, \mathscr{R}) &= 4\pi i^{l-1}[\cdots \mathscr{I}_l(\zeta_b \mathscr{R}) \cdots] \\ &\quad \times k^l(k^{-1} d/dk)^l k^{-1} \tfrac{1}{2} E_{2l}^{(2)}[(\zeta_a + \zeta_b \mp ik)\mathscr{R}] \\ &\quad + 4\pi i^{l-1}(-1)^l[\cdots \mathscr{K}_l(\zeta_b \mathscr{R}) \cdots] \\ &\quad \times k^l(k^{-1} d/dk)^l k^{-1} \tfrac{1}{4} \tilde{E}_{2l}^{(4)}[(\zeta_a \pm \zeta_b \mp ik)\mathscr{R}]. \end{aligned} \tag{10.70}$$

F. Evaluation of a (1-2)-Type Three-Center Integral

The final integral we evaluate in this section is typical of the three- and four-center integrals. Its evaluation involves more details than in the preceding examples, but the basic manipulations are essentially similar.

Consider

$$I_{c,ab}(\mathbf{R}, \mathscr{R}) = \iint \psi_{100\zeta_c}^*(\mathbf{r}_1) r_{12}^{-1} \psi_{100\zeta_a}^*(\mathbf{r}_2 - \mathbf{R}) \psi_{100\zeta_b}(\mathbf{r}_2 - \mathbf{R} - \mathscr{R}) \, dV_1 \, dV_2 \tag{10.71}$$

By Eqs. (9.18) and (9.6),

$$\begin{aligned}I_{c,ab}(\mathbf{R}, \mathscr{R}) &= (2\pi)^{-3} \int \mathrm{FT}^*\{\psi_{100\zeta_c}(\mathbf{r})\} \, \mathrm{FT}\{r^{-1}\} \\ &\quad \times e^{i\mathbf{k}\cdot\mathbf{R}} \, \mathrm{FT}\{\psi_{100\zeta_a}^*(\mathbf{r})\psi_{100\zeta_b}(\mathbf{r} - \mathscr{R})\} \, d^3\mathbf{k}. \end{aligned} \tag{10.72}$$

Two of the three FTs are given by Eqs. (10.10), (10.14), (10.55), and (10.70); the $\mathrm{FT}\{r^{-1}\}$ requires additional comment.

3. Complex Variable Theory

Fourier Transform of $1/r$. Straightforwardly one calculates

$$\text{FT}\{r^{-1}\} = \int e^{i\mathbf{k}\cdot\mathbf{r}} r^{-1}\, dV = 4\pi \int_0^\infty j_0(kr)\, r\, dr \tag{10.73}$$

$$= 4\pi k^{-1} \int_0^\infty \sin kr\, dr. \tag{10.74}$$

Clearly, the usual definition of $\text{FT}\{f(\mathbf{r})\}$ fails. On the other hand, from Eq. (10.14),

$$\lim_{\zeta \to 0} \text{FT}\{r^{-1} e^{-\zeta r}\} = \lim_{\zeta \to 0} 2\pi i^{-1} k^{-1}[(\zeta - ik)^{-1} - (\zeta + ik)^{-1}] \tag{10.75}$$

$$= 4\pi k^{-2}. \tag{10.76}$$

Thus we define $\text{FT}\{r^{-1}\}$ to be

$$\text{FT}\{r^{-1}\} = \lim_{\zeta \to 0} \text{FT}\{r^{-1} e^{-\zeta r}\} = 4\pi k^{-2}. \tag{10.77}$$

Note that the inverse Fourier transform of $4\pi k^{-2}$ is r^{-1}. A more rigorous justification of Eq. (10.77) can be given in the theory of generalized functions [Chapter 2 of this volume; Lighthill (1958)].

Reduction to a Radial Integration Equation (10.72) becomes

$$I_{c,ab}(\mathbf{R}, \mathscr{R}) = (2\pi)^{-3} \int \{f_{10\zeta_c}(k)(4\pi)^{-1/2}\}\{4\pi k^{-2}\}$$
$$\times \Big\{ \sum_{l_1,m_1} 4\pi i^{l_1} j_{l_1}(kR) Y_{l_1}^{m_1}(\theta_R, \phi_R) Y_{l_1}^{m_1*}(\theta_k, \phi_k) \Big\}$$
$$\times \Big\{ \sum_{l_2,m_2} (-1)^{m_2} Y_{l_2}^{-m_2}(\theta_{\mathscr{R}}, \phi_{\mathscr{R}}) Y_{l_2}^{m_2}(\theta_k, \phi_k) G_l(k, \mathscr{R}) \Big\} d^3\mathbf{k} \tag{10.78}$$

$$= \sum_{l=0}^\infty \sum_{m=-l}^l \pi^{3/2}(-1)^m Y_l^m(\theta_R, \phi_R) Y_l^{-m}(\theta_{\mathscr{R}}, \phi_{\mathscr{R}}) I_{c,ab}^l \tag{10.79}$$

$$I_{c,ab}^l = \pi^{-3} i^l \int_0^\infty f_{10\zeta_c}(k) j_l(kR) G_l(k, \mathscr{R})\, dk$$

$$= \tfrac{1}{2}\pi^{-3} i^l \int_{-\infty}^\infty f_{10\zeta_c}(k) j_l(kR) G_l(k, \mathscr{R})\, dk. \tag{10.80}$$

This last integral is like Eq. (7.3), but more complicated. It can readily be evaluated by the residue theorem after closing the contour appropriately at ∞. Note that the various terms in the integrand behave at ∞ like

$$\{k^{-3}\}\{e^{\pm ikR} k^{-1}\}\{e^{\pm ik\mathscr{R}} k^{-1}\} \tag{10.81}$$

and

$$\{k^{-3}\}\{e^{\pm ikR}k^{-1}\}\{(\zeta_b^{-1}\,d/d\zeta_b)^l\zeta_b^{-1}k^l(k^{-1}\,d/dk)^l k^{-1}(\zeta_a \pm \zeta_b \mp ik)^{2l-1}$$
$$\times \log[(\zeta_a \pm \zeta_b \mp ik)\mathscr{R}]\}, \tag{10.82}$$

so that the appropriate half-plane will depend on whether $\mathscr{R} > R$ or $\mathscr{R} < R$. The only singularities of the integrand are second-order poles at $k = \pm i\zeta_c$, logarithmic branch points at $k = \pm i(\zeta_a + \zeta_b)$, and a nascent simple pole at the origin. Avoiding the logarithmic branch cuts requires some dexterity.

Via Eqs. (10.14) and (10.70), we put Eq. (10.80) in the form

$$I_{c,ab}^l = I^{l,1} + I^{l,2} \tag{10.83}$$

$$I^{l,1} = (-1)^{l+1}4\pi^{-1}\int_{-\infty}^{\infty} k^{-1}[(\zeta_c - ik)^{-2} - (\zeta_c + ik)^{-2}]j_l(kR)$$
$$\times \{[\cdots \mathscr{T}_l(\zeta_b\mathscr{R}) \cdots]k^l(k^{-1}\,d/dk)^l k^{-1}\tfrac{1}{2}E_{2l}^{(2)}[(\zeta_a + \zeta_b \mp ik)\mathscr{R}]\}\,dk$$
$$\tag{10.84}$$

$$I^{l,2} = -4\pi^{-1}\int_{-\infty}^{\infty} k^{-1}[(\zeta_c - ik)^{-2} - (\zeta_c + ik)^{-2}]j_l(kR)$$
$$+ \{[\cdots \mathscr{K}_l(\zeta_b\mathscr{R}) \cdots]k^l(k^{-1}\,d/dk)^l k^{-1}\tfrac{1}{4}\tilde{E}_{2l}^{(4)}[(\zeta_a \pm \zeta_b \mp ik)\mathscr{R}]\}\,dk.$$
$$\tag{10.85}$$

Evaluation of $I^{l,1}$ The integrand of $I^{l,1}$ is an even function of k, so that

$$I^{l,1} = (-1)^{l+1}4\pi^{-1}\mathscr{P}\int_{-\infty}^{\infty} k^{-1}[(\zeta_c - ik)^{-2} - (\zeta_c + ik)^{-2}]j_l(kR)$$
$$\times \{[\cdots \mathscr{T}_l(\zeta_b\mathscr{R}) \cdots]k^l(k^{-1}\,d/dk)^l k^{-1}E_{2l}[(\zeta_a + \zeta_b - ik)\mathscr{R}]\}\,dk.$$
$$\tag{10.86}$$

The $\mathscr{P}$ arises as in Eqs. (7.55) and (7.56). The integrand of Eq. (10.86) has a simple pole at $k = 0$. When $\mathscr{R} > R$, $E_{2l}[(\zeta_a + \zeta_b - ik)\mathscr{R}]$ dominates the behavior at ∞, the contour can be closed at ∞ in the upper half-plane, and the residue theorem gives

$$I^{l,1} = 2\pi i \times \{(\text{residue at } k = i\zeta_c) + \tfrac{1}{2}(\text{residue at } k = 0)\} \tag{10.87}$$
$$= [\cdots \mathscr{T}_l(\zeta_b\mathscr{R}) \cdots]\{8(-1)^l(d/d\zeta_c)\mathscr{T}_l(\zeta_c R)\zeta_c^{-1+l}$$
$$\times (\zeta_c^{-1}\,d/d\zeta_c)^l\zeta_c^{-1}E_{2l}[(\zeta_a + \zeta_b + \zeta_c)\mathscr{R}]$$
$$+ 16\zeta_c^{-3}R^l(2l+1)^{-1}E_{2l}[(\zeta_a + \zeta_b)\mathscr{R}]\}, \quad \mathscr{R} > R. \tag{10.88}$$

3. Complex Variable Theory

For $I^{l,1}$ when $R > \mathscr{R}$, first add and subtract a term and manipulate as follows:

$$I^{l,1} = (-1)^{l+1}4\pi^{-1}\int_{-\infty}^{\infty} k^{-1}[(\zeta_c - ik)^{-2} - (\zeta_c + ik)^{-2}]j_l(kR)$$
$$\times [\cdots \mathscr{T}_l(\zeta_b\mathscr{R}) \cdots]k^l(k^{-1}\,d/dk)^l k^{-1}\tfrac{1}{2}\{E_{2l}^{(2)}[(\zeta_a + \zeta_b \mp ik)\mathscr{R}]$$
$$- (\mathscr{R}/R)^{2l-1}E_{2l}^{(2)}[(\zeta_a + \zeta_b \mp ik)R]$$
$$+ (\mathscr{R}/R)^{2l-1}E_{2l}^{(2)}[(\zeta_a + \zeta_b \mp ik)R]\}\,dk \tag{10.89}$$

$$= i^l 4\pi^{-1}\mathscr{P}\int_{-\infty}^{\infty} k^{-1}[(\zeta_c - ik)^{-2} - (\zeta_c + ik)^{-2}]\mathscr{K}_l(-ikR)$$
$$\times [\cdots \mathscr{T}_l(\zeta_b\mathscr{R}) \cdots]k^l(k^{-1}\,d/dk)^l k^{-1}\tfrac{1}{2}\{E_{2l}^{(2)}[(\zeta_a + \zeta_b \mp ik)\mathscr{R}]$$
$$- (\mathscr{R}/R)^{2l-1}E_{2l}^{(2)}[(\zeta_a + \zeta_b \mp ik)R]\} + (-1)^{l+1}4\pi^{-1}$$
$$\times \mathscr{P}\int_{-\infty}^{\infty} k^{-1}[(\zeta_c - ik)^{-2} - (\zeta_c + ik)^{-2}]j_l(kR)[\cdots \mathscr{T}_l(\zeta_b\mathscr{R}) \cdots]$$
$$\times k^l(k^{-1}\,d/dk)^l k^{-1}(\mathscr{R}/R)^{2l-1}E_{2l}[(\zeta_a + \zeta_b - ik)R]. \tag{10.90}$$

The behavior at ∞ is dominated by $e^{\pm ikR}$. The first integral of Eq. (10.90) can be closed at ∞ in the upper half-plane. The subtracted term has been chosen to cancel out the logarithmic branch point at $k = i(\zeta_a + \zeta_b)$, and the residue theorem applies. Indeed, from Eq. (8.165) [cf. Eqs. (10.61)–(10.65)],

$$(k^{-1}\,d/dk)^l k^{-1}\{E_{2l}^{(2)}[(\zeta_a + \zeta_b \mp ik)\mathscr{R}] - (\mathscr{R}/R)^{2l-1}E_{2l}^{(2)}[(\zeta_a + \zeta_b \mp ik)R]\}$$
$$= (k^{-1}\,d/dk)^l k^{-1}\{\tilde{E}_{2l}[(\zeta_a + \zeta_b \mp ik)\mathscr{R}] - (\mathscr{R}/R)^{2l-1}$$
$$\times \tilde{E}_{2l}[(\zeta_a + \zeta_b \mp ik)R]\}. \tag{10.91}$$

The second integral vanishes uniformly at ∞ and can be treated like $I^{l,1}$ in the $\mathscr{R} > R$ case. The result is

$$I^{l,1} = [\cdots \mathscr{T}_l(\zeta_b\mathscr{R}) \cdots](-8(d/d\zeta_c)\mathscr{K}_l(\zeta_c R)\zeta_c^{-1+l}$$
$$\times (\zeta_c^{-1}\,d/d\zeta_c)^l\zeta_c^{-1}\tfrac{1}{2}\{\tilde{E}_{2l}^{(2)}[(\zeta_a + \zeta_b \pm \zeta_c)\mathscr{R}]$$
$$- (\mathscr{R}/R)^{2l-1}\tilde{E}_{2l}^{(2)}[(\zeta_a + \zeta_b \pm \zeta_c)R]$$
$$+ 8(-1)^l(d/d\zeta_c)\mathscr{T}_l(\zeta_c R)\zeta_c^{-1+l}(\zeta_c^{-1}\,d/d\zeta_c)^l\zeta_c^{-1}(\mathscr{R}/R)^{2l-1}$$
$$\times E_{2l}[(\zeta_a + \zeta_b + \zeta_c)R]$$
$$+ 16\zeta_c^{-3}(2l+1)^{-1}R^{-l-1}\{\mathscr{R}^{2l+1}\hat{a}_1[(\zeta_a + \zeta_b)\mathscr{R}]$$
$$- \mathscr{R}^{2l-1}R^2\hat{a}_1[(\zeta_a + \zeta_b)R]\}$$
$$+ 16\zeta_c^{-3}(2l+1)^{-1}R^l(\mathscr{R}/R)^{2l-1}E_2[(\zeta_a + \zeta_b)R]), \quad R > \mathscr{R}. \tag{10.92}$$

We have used the fact [cf. Eqs. (10.62), (8.160), and (10.25)–(10.27)]

$$(-d/dz)^{n+m}\tilde{E}_n(z) = \hat{a}_m(z). \tag{10.93}$$

Evaluation of $I^{l,2}$ When $R > \mathscr{R}$, the $e^{\pm ikR}$ dominates the integrand of $I^{l,2}$, Eq. (10.85). The sequence of steps to evaluate $I^{l,2}$ is $j_l(kR) \to -\mathscr{P}i^{-l}\mathscr{K}_l(-ikR) \to 2\pi i \times \{(\text{residue at } k = i\zeta_c) + \frac{1}{2}(\text{residue at } k = 0)\}$. The result is

$$I^{l,2} = [\cdots \mathscr{K}_l(\zeta_b\mathscr{R})\cdots]\{-8(-1)^l(d/d\zeta_c)\mathscr{K}_l(\zeta_c R)\zeta_c^{-1+l} \\ \times (\zeta_c^{-1} d/d\zeta_c)^l \zeta_c^{-1} \tfrac{1}{4}\tilde{E}_{2l}^{(4)}[(\zeta_a \pm \zeta_b \pm \zeta_c)\mathscr{R}] \\ + 16(-1)^l \zeta_c^{-3}(2l+1)^{-1}R^{-l-1}\mathscr{R}^{2l+1}\tfrac{1}{2}\hat{a}_1^{(2)}[(\zeta_a \pm \zeta_b)\mathscr{R}]\}. \tag{10.94}$$

When $\mathscr{R} > R$, $I^{l,2}$ is analogous to $I^{l,1}$ with $R > \mathscr{R}$. The "E part" of the $\tilde{E}$ factors in Eq. (10.85) dominates the $j_l(kR)$ at ∞, but the "log part" is dominated by the $j_l(kR)$. We add and subtract a term with $\tilde{E}_{2l}^{(4)}[(\zeta_a \pm \zeta_b \mp ik)\mathscr{R}]$ replaced by $(\mathscr{R}/R)^{2l-1}\tilde{E}_{2l}^{(4)}[(\zeta_a \pm \zeta_b \mp ik)R]$, and note that

$$(k^{-1} d/dk)^l k^{-1}\{\tilde{E}_{2l}^{(4)}[(\zeta_a \pm \zeta_b \mp ik)\mathscr{R}] - (\mathscr{R}/R)^{2l-1} \\ \times \tilde{E}_{2l}^{(4)}[(\zeta_a \pm \zeta_b \mp ik)R]\} \\ = (k^{-1} d/dk)^l k^{-1}\{E_{2l}^{(4)}[(\zeta_a \pm \zeta_b \mp ik)\mathscr{R}] - (\mathscr{R}/R)^{2l-1} \\ \times E_{2l}^{(4)}[(\zeta_a \pm \zeta_b \mp ik)R]\}. \tag{10.95}$$

For the moment, we assume that $\zeta_a > \zeta_b$, so that the logarithmic branch cuts from $\pm i(\zeta_a - \zeta_b)$ to $\pm i\infty$ do not cross the real axis (the integration path). Then $I^{l,2}$ becomes

$$I^{l,2} = 4\pi^{-1} i^{-l} \mathscr{P} \int_{-\infty}^{\infty} k^{-1}[(\zeta_c - ik)^{-2} - (\zeta_c + ik)^{-2}]\mathscr{K}_l(-ikR) \\ \times [\cdots \mathscr{K}_l(\zeta_b\mathscr{R})\cdots]k^l(k^{-1} d/dk)^l k^{-1}(\mathscr{R}/R)^{2l-1} \\ \times \tfrac{1}{4}\tilde{E}_{2l}^{(4)}[(\zeta_a \pm \zeta_b \mp ik)R]\, dk \\ - 4\pi^{-1} \mathscr{P} \int_{\infty}^{\infty} k^{-1}[(\zeta_c - ik)^{-2} - (\zeta_c + ik)^{-2}]j_l(kR) \\ \times [\cdots \mathscr{K}_l(\zeta_b\mathscr{R})\cdots]k^l(k^{-1} d/dk)^l k^{-1}\tfrac{1}{2}\{E_{2l}^{(2)}[(\zeta_a \pm \zeta_b - ik)\mathscr{R}] \\ - (\mathscr{R}/R)^{2l-1}E_{2l}^{(2)}[(\zeta_a \pm \zeta_b - ik)R]\}\, dk. \tag{10.96}$$

Both contours can be closed at ∞ in the upper half-plane, and the result

3. Complex Variable Theory

is $2\pi i \times \{(\text{residue at } k = i\zeta_c) + \tfrac{1}{2}(\text{residue at } k = 0)\}$:

$$\begin{aligned}
I^{l,2} = [\cdots \mathscr{K}_l(\zeta_b \mathscr{R}) \cdots]&(-8(-1)^l (d/d\zeta_c) \mathscr{K}_l(\zeta_c R) \zeta_c^{-1+l} \\
&\times (\zeta_c^{-1} d/d\zeta_c)^l \zeta_c^{-1} \tfrac{1}{4} (\mathscr{R}/R)^{2l-1} \tilde{E}_{2l}^{(4)}[(\zeta_a \pm \zeta_b \pm \zeta_c) R] \\
&+ 8(d/d\zeta_c) \mathscr{I}_l(\zeta_c R) \zeta_c^{-1+l} (\zeta_c^{-1} d/d\zeta_c)^l \zeta_c^{-1} \tfrac{1}{2} \\
&\times \{\tilde{E}_{2l}^{(2)}[(\zeta_a \pm \zeta_b + \zeta_c) \mathscr{R}] - (\mathscr{R}/R)^{2l-1} \tilde{E}_{2l}^{(2)}[(\zeta_a \pm \zeta_b + \zeta_c) R]\} \\
&+ 16(-1)^l \zeta_c^{-3}(2l+1)^{-1} R^{-l+1} \mathscr{R}^{2l-1} \tfrac{1}{2} \tilde{a}_1^{(2)}[(\zeta_a \pm \zeta_b) R] \\
&+ 16(-1)^l \zeta_c^{-3}(2l+1)^{-1} R^{l} \tfrac{1}{2} \{\tilde{E}_{2l}^{(2)}[(\zeta_a \pm \zeta_b) \mathscr{R}] - (\mathscr{R}/R)^{2l-1} \\
&\times \tilde{E}_{2l}^{(2)}[(\zeta_a \pm \zeta_b) R]\}), \quad \mathscr{R} > R, \quad\quad (10.97)
\end{aligned}$$

where

$$\begin{aligned}
(\zeta_b^{-1} d/d\zeta_b)^l \zeta_b^{-1} \\
\times \{E_{2l}^{(2)}[(\zeta_a \pm \zeta_b + \zeta_c)\mathscr{R}] &- (\mathscr{R}/R)^{2l-1} E_{2l}^{(2)}[(\zeta_a \pm \zeta_b + \zeta_c) R]\} \\
= (\zeta_b^{-1} d/d\zeta_b)^l \zeta_b^{-1} \\
\times \{\tilde{E}_{2l}^{(2)}[(\zeta_a \pm \zeta_b + \zeta_c)\mathscr{R}] &- (\mathscr{R}/R)^{2l-1} \tilde{E}_{2l}^{(2)}[(\zeta_a \pm \zeta_b + \zeta_c) R]\}, \\
& \quad\quad (10.98)
\end{aligned}$$

and similar equations have been used to replace the $E^{(2)}$'s by corresponding $\tilde{E}^{(2)}$'s in Eq. (10.97). Note that the right side of Eq. (10.97) is analytic at $(\zeta_a - \zeta_b) = 0$. One may infer that the restriction $(\zeta_a > \zeta_b)$ can be removed.

This completes the evaluation of $I_{c;ab}$, Eq. (10.71).

G. Arbitrary Multicenter Integrals

The specific examples worked above illustrate how complex variable theory can be used in general to evaluate multicenter integrals in quantum chemistry. The results given contain derivatives that must be dealt with systematically, but this belongs to the realm of ordinary differential calculus, not to complex analysis. The evaluation of the more general cases have been discussed in the references cited at the beginning of this section. The details are indeed more intricate, but the basic principles of the method are the same.

XI. A Formula of Lagrange and Nondegenerate Perturbation Theory

In a mémoire read in 1770, Lagrange discussed finding roots of equations, and functions of these roots. The solution he gave is easily derived by complex variable theory. The formula itself gives a quick formal solution of Rayleigh–Schrödinger perturbation theory in quantum mechanics, an application discovered by Sack (1969) and by Silverstone and Holloway (1970).

A. Lagrange's Formula

Let C be a simple closed curve, and let $f(z)$ and $\phi(z)$ both be analytic on and within C. Let a be inside C, and let $\zeta = \zeta(t)$ be the only root inside C of the equation

$$\zeta = a + t\phi(\zeta), \qquad 0 \leq |t| < R. \tag{11.1}$$

Lagrange's formula is the expansion

$$f(\zeta) = f(a) + \sum_{n=1}^{\infty} t^n (n!)^{-1} (d/da)^{n-1} \{ f'(a)[\phi(a)]^n \}. \tag{11.2}$$

Derivation By assumption $z - a - t\phi(z)$ has a simple zero at $z = \zeta$, so that $(1 - t\phi'(z))/(z - a - t\phi(z))$ has a simple pole at $z = \zeta$ (cf. Section VI.A). The Laurent series about $z = \zeta$ is

$$\frac{1 - t\phi'(z)}{z - a - t\phi(z)} = \frac{1}{z - \zeta} + \left[-\frac{1}{2} \frac{t\phi''(\zeta)}{1 - t\phi'(\zeta)} \right] + \cdots. \tag{11.3}$$

$[1 - t\phi'(\zeta) \neq 0$, since ζ is a simple zero.] By Cauchy's theorem, the residue theorem, and integration by parts,

$$f(\zeta) - f(a) = (2\pi i)^{-1} \oint_C f(z) \left[\frac{1 - t\phi'(z)}{z - a - t\phi(z)} - \frac{1}{z - a} \right] dz \tag{11.4}$$

$$= -(2\pi i)^{-1} \oint_C f'(z) \log\left[1 - t\frac{\phi(z)}{z - a} \right] dz. \tag{11.5}$$

For small enough t,

$$f(\zeta) - f(a) = (2\pi i)^{-1} \oint_C \sum_{n=1}^{\infty} (1/n) t^n [\phi(z)/(z - a)]^n f'(z) \, dz \tag{11.6}$$

and
$$f(\zeta) = f(a) + \sum_{n=1}^{\infty} t^n (n!)^{-1} (d/da)^{n-1} \{f'(a)[\phi(a)]^n\}, \tag{11.7}$$

by the theorem of Section V.C on the integration of uniformly convergent series and by the residue theorem.

B. NONDEGENERATE PERTURBATION THEORY

This final section is intended only for the reader familiar with the formalism of quantum mechanics.

A standard quantum mechanical problem is to solve the eigenvalue equation
$$H\,|\,\psi\rangle = E\,|\,\psi\rangle, \tag{11.8}$$

where H is a Hermitian operator, E an eigenvalue, and $|\,\psi\rangle$ the corresponding eigenvector. Perturbation theory refers to methods treating H as a perturbation of a Hermitian operator whose eigenvalue and eigenvector are known:
$$H = H^{(0)} + \lambda V \tag{11.9}$$
$$(H^{(0)} - E^{(0)})\,|\,\psi^{(0)}\rangle = 0 \tag{11.10}$$
$$\lim_{\lambda \to 0} |\,\psi\rangle = |\,\psi^{(0)}\rangle. \tag{11.11}$$

We denote by Q the projection operator
$$Q = 1 - |\,\psi^{(0)}\rangle\langle\psi^{(0)}\,|. \tag{11.12}$$

We further assume that $E^{(0)}$ is a nondegenerate eigenvalue of $H^{(0)}$, and that E is a nondegenerate eigenvalue of H. In particular, the operator $Q/(E^{(0)} - H^{(0)})$,
$$\frac{Q}{E^{(0)} - H^{(0)}} = \lim_{\varepsilon \to 0} Q\,\frac{1}{E^{(0)} + \varepsilon - H^{(0)}}\,Q, \tag{11.13}$$

exists. In the equations that follow, all quantities are assumed to exist, and all series to converge.

The Brillouin–Wigner series for E and ψ are
$$E = E^{(0)} + \sum_{k=0}^{\infty} \langle\psi^{(0)}\,|\,\lambda V \left[\frac{Q}{E - H^{(0)}}\,\lambda V\right]^k |\,\psi^{(0)}\rangle \tag{11.14}$$
$$\psi = \sum_{k=0}^{\infty} \left[\frac{Q}{E - H^{(0)}}\,\lambda V\right]^k |\,\psi^{(0)}\rangle \tag{11.15}$$

where Eq. (11.14) must first be solved for E, then its solution used in Eq. (11.15). Note that Eq. (11.14) has the same form as Eq. (11.1) with $z = E$, $a = E^{(0)}$, and

$$t\phi(z) = \sum_{k=0}^{\infty} \langle \psi^{(0)} | \lambda V \left[\frac{Q}{E - H^{(0)}} \lambda V \right]^k | \psi^{(0)} \rangle.$$

The Rayleigh–Schrödinger series for E and ψ are the power series expansions in λ:

$$E = \sum_{M=0}^{\infty} \lambda^n E^{(n)} \qquad (11.16)$$

$$|\psi\rangle = \sum_{n=0}^{\infty} \lambda^n |\psi^{(n)}\rangle. \qquad (11.17)$$

The explicit form of the $E^{(n)}$ and $|\psi^{(n)}\rangle$ can be obtained from the Brillouin–Wigner series by first applying Lagrange's formula

$$E = E^{(0)} + \sum_{n=1}^{\infty} (n!)^{-1} \left(\frac{d}{dE^{(0)}} \right)^{n-1}$$
$$\times \left\{ \sum_{k=0}^{\infty} \langle \psi^{(0)} | \lambda V \left[\frac{Q}{E^{(0)} - H^{(0)}} \lambda V \right]^k | \psi^{(0)} \rangle \right\}^n \qquad (11.18)$$

$$|\psi\rangle = \sum_{k=0}^{\infty} \left[\frac{Q}{E^{(0)} - H^{(0)}} \lambda V \right]^k | \psi^{(0)} \rangle$$
$$+ \sum_{n=1}^{\infty} (n!)^{-1} \left(\frac{d}{dE^{(0)}} \right)^{n-1} \left\{ \left[\frac{d}{dE^{(0)}} \sum_{k=0}^{\infty} \left[\frac{Q}{E^{(0)} - H^{(0)}} \lambda V \right]^k | \psi^{(0)} \rangle \right] \right.$$
$$\times \left. \left[\sum_{k=0}^{\infty} \langle \psi^{(0)} | \lambda V \left[\frac{Q}{E^{(0)} - H^{(0)}} \lambda V \right]^k | \psi^{(0)} \rangle \right]^n \right\}. \qquad (11.19)$$

Then use the multinomial expansion

$$(x_1 + x_2 + \cdots)^n = \sum_{\sigma_1 + \sigma_2 + \cdots = n} \frac{n!}{\sigma_1! \sigma_2! \cdots} x_1^{\sigma_1} x_2^{\sigma_2} \cdots, \qquad (11.20)$$

and collect terms having the same power of λ. We use the abbreviations

$$R^{-1} = \frac{Q}{E^{(0)} - H^{(0)}} \qquad (11.21)$$

$$\langle V \rangle = \langle \psi^{(0)} | V | \psi^{(0)} \rangle \qquad (11.22)$$

$$\langle \cdots \rangle = \langle \psi^{(0)} | \cdots | \psi^{(0)} \rangle. \qquad (11.23)$$

Note that

$$\left(-\frac{d}{dE^{(0)}}\right)^n R^{-1} = n! \frac{Q}{(E^{(0)} - H^{(0)})^{n+1}} = n! R^{-n-1}. \quad (11.24)$$

The rearranged series are then

$$E = E^{(0)} + \sum_{N=1}^{\infty} \lambda^N \sum_{\sigma_1+2\sigma_2+3\sigma_3+\cdots=N} \frac{(d/dE^{(0)})^{-1+\sigma_1+\sigma_2+\sigma_3+\cdots}}{\sigma_1!\sigma_2!\sigma_3\cdots}$$
$$\times \langle V \rangle^{\sigma_1} \langle VR^{-1}V \rangle^{\sigma_2} \langle VR^{-1}VR^{-1}V \rangle^{\sigma_3} \cdots \quad (11.25)$$

$$= E^{(0)} + \lambda \langle V \rangle + \lambda^2 \langle VR^{-2}V \rangle$$
$$+ \lambda^3 (\langle VR^{-1}VR^{-1}V \rangle - \langle V \rangle \langle VR^{-2}V \rangle) + \cdots \quad (11.26)$$

$$|\psi\rangle = |\psi^{(0)}\rangle + \sum_{N=1}^{\infty} \lambda^N \{(R^{-1}V)^N | \psi^{(0)} \rangle$$
$$+ \sum_{j=1}^{N-1} \sum_{\sigma_1+2\sigma_2+3\sigma_3+\cdots=N-j} \frac{(d/dE^{(0)})^{-1+\sigma_1+\sigma_2+\sigma_3+\cdots}}{\sigma_1!\sigma_2!\sigma_3!\cdots}$$
$$\times [\langle V \rangle^{\sigma_1} \langle VR^{-1}V \rangle^{\sigma_2} \langle VR^{-1}VR^{-1}V \rangle^{\sigma_3} \cdots$$
$$\times (d/dE^{(0)})(R^{-1}V)^j | \psi^{(0)} \rangle]\} \quad (11.27)$$

$$= |\psi^{(0)}\rangle + \lambda R^{-1}V | \psi^{(0)} \rangle$$
$$+ \lambda^2 [R^{-1}VR^{-1}V | \psi^{(0)} \rangle - \langle V \rangle R^{-2}V | \psi^{(0)} \rangle] + \cdots. \quad (11.28)$$

Acknowledgment

This work was supported in part by the Alfred P. Sloan Foundation.

Special References

Abramowtiz, M., and Stegun, I. A. (eds.) (1964). Handbook of Mathematical Functions. *Nat. Bur. Std. Appl. Math. Ser.* No. 55, Washington, D. C.

Ahlfors, L. V. (1966). "Complex Analysis," 2nd ed. McGraw-Hill, New York.

Churchill, R. V. (1960). "Complex Variables and Applications," 2nd ed. McGraw-Hill, New York.

Erdélyi, A. (ed.) (1953). "Higher Transcendental Functions," Vols. 1–3. McGraw-Hill, New York.

Hobson, E. W. (1955). "The Theory of Spherical and Ellipsoidal Harmonics," pp. 133–135. Chelsea, New York.

Kay, K. G., and Silverstone, H. J. (1969a). *J. Chem. Phys.* **51**, 956.

Kay, K. G., and Silverstone, H. J. (1969b). *J. Chem. Phys.* **51**, 4287.

Kay, K. G., and Silverstone, H. J. (1970). *J. Chem. Phys.* **53**, 4269.

KAY, K. G., TODD, H. D., and SILVERSTONE, H. J. (1969a). *J. Chem. Phys.* **51**, 2359.
KAY, K. G., TODD, H. D., and SILVERSTONE, H. J. (1969b). *J. Chem. Phys.* **51**, 2363.
LIGHTHILL, M. J. (1958). "Introduction to Fourier Analysis and Generalised Functions." Cambridge Univ. Press, London and New York.
NEHARI, Z. (1968). "Introduction to Complex Analysis." Allyn and Bacon, Boston, Massachusetts.
SACK, R. A. (1969). Private communication.
SILVERSTONE, H. J. (1966). *J. Chem. Phys.* **45**, 4337.
SILVERSTONE, H. J. (1967). *J. Chem. Phys.* **47**, 537.
SILVERSTONE, H. J. (1968a). *J. Chem. Phys.* **48**, 4098.
SILVERSTONE, H. J. (1968b). *J. Chem. Phys.* **48**, 4106.
SILVERSTONE, H. J., and HOLLOWAY, T. T. (1970). *J. Chem. Phys.* **52**, 1472.
SILVERSTONE, H. J., and KAY, K. G. (1968). *J. Chem. Phys.* **48**, 4108.
SILVERSTONE, H. J., and TODD, H. D. (1971). *Int. J. Quantum Chem.* **4**, 371.
SLATER, L. J. (1960). "Confluent Hypergeometric Functions." Cambridge Univ. Press, London and New York.
TODD, H. D., KAY, K. G., and SILVERSTONE, H. J. (1970). *J. Chem. Phys.* **53**, 3951.
WHITTAKER, E. T., and WATSON, G. N. (1927). "A Course of Modern Analysis," 4th ed. Cambridge Univ. Press, London and New York.

Chapter 4

Boundary-Value Problems

DOUGLAS HENDERSON

I.	Introduction	262
II.	Some Typical Boundary-Value Problems	263
	A. Elastic String	263
	B. Heat Equation	264
	C. Potential Problems	264
	D. Schroedinger's Equation	265
	E. Some Further Comments	265
III.	The D'Alembert Solution of the Wave Equation	268
IV.	Separation of Variables	269
	A. Plucked String	269
	B. Comments Regarding the Separation of Variables	273
	C. Three Heat Conduction Problems	275
	D. Relation to the D'Alembert Solution	279
V.	Eigenvalues, Eigenfunctions, and Expansion Problems	280
	A. Orthogonal Sets of Functions	281
	B. Abstract Vector Space	284
	C. Sturm–Liouville Problem	285
	D. Another Heat Problem	290
VI.	Boundary-Value Problems in Cylindrical Coordinates	291
	A. Laplace's Equation in Cylindrical Coordinates	292
	B. Bessel Functions	292
	C. Modified Bessel Functions	295
	D. Some Relations for Bessel Functions	296
	E. Orthogonality of the Bessel Functions	298
	F. Three Examples	300
VII.	Boundary-Value Problems in Spherical Coordinates	304
	A. Laplace's Equation in Spherical Coordinates	304
	B. Legendre Polynomials	306
	C. Associated Legendre Functions	307
	D. Orthogonality of the Legendre Functions	309
	E. The Addition Theorem for Spherical Harmonics	310
	F. Three Examples	312

VIII. Green's Functions 316
    A. Introductory Example 316
    B. Green's Function for the Sturm–Liouville Operator 319
    C. Solution of Potential Problems by Green's Functions 324
    D. Green's Function for a Sphere 326
IX. Laplace Transform Methods 327
    A. A Heat Problem in a Semiinfinite Rod 328
    B. A Heat Problem in a Finite Rod 329
X. Conformal Mapping 331

    References 335

I. Introduction

The solution of a differential equation subject to some conditions on the boundary of the domain of the system (so-called *boundary conditions* or, in the case of time, *initial conditions*, which are merely a type of boundary condition) is, perhaps, the most common problem in applied mathematics. All such problems are referred to as *boundary-value problems*.

It is assumed that the reader is already familiar with the solution of such problems when there is only one independent variable and only ordinary differential equations arise. The concern of this chapter is with boundary-value problems that have two or more independent variables where we must deal with partial differential equations.

The level of the treatment is quite elementary and only a few topics can be considered. The emphasis is on the use of the techniques described in this chapter; little consideration is given to such important mathematical considerations as existence, convergence, differentiability, or integrability of solutions. Nor do we let the question of the uniqueness of the solution of a boundary-value problem detain us. Our point of view is that nature is reproducible. By that it is meant that if identical experiments are performed, identical results are obtained. Thus, if we have formulated a given problem correctly, any solution satisfying *both* the partial differential equation *and* the boundary and initial conditions is the general solution of that problem, and that all such solutions, even if not at first apparent, are equivalent. However, these considerations should not be construed as implying that existence theorems are not important. For example, even assuming that nature is reproducible, it is only after the proof of the existence and uniqueness of the solution that we can be confident that we have formulated the problem correctly.

The literature concerning boundary-value problems is vast. It is hoped that the reader will find this chapter a simple and useful introduction to this literature. Further reading may be found in the books listed in the References at the end of this chapter.

II. Some Typical Boundary-Value Problems

In this section a few boundary-value problems are discussed briefly. A better physical feeling would be obtained if the equations could be derived. However, space limitations prohibit this. Derivations may be found in the books by Churchill (1963) and Wylie (1966) as well as others.

A. Elastic String

Consider an elastic string of length l which is stretched under a tension T between two points on the x axis. If ϱ is the mass per unit length of the string and if the displacement of the string is small and is entirely in the xy plane, then the deflection $y(x, t)$ of the string, where t is the time, satisfies the equation

$$c^2\, \partial^2 y/\partial x^2 = \partial^2 y/\partial t^2 + g, \tag{2.1}$$

where $c^2 = T/\varrho$ has the dimensions of velocity. The meaning of c is discussed in Section III. The second term on the right-hand side of (2.1) gives the effect of the gravitational force. If ϱg is small compared to T, then (2.1) becomes the *wave equation*:

$$c^2\, \partial^2 y/\partial x^2 = \partial^2 y/\partial t^2. \tag{2.2}$$

The boundary conditions appropriate to the elastic string with fixed endpoints are

$$y(0, t) = y(l, t) = 0 \tag{2.3}$$

and appropriate initial conditions would be the initial displacement and velocity. It is to be observed that we need as many different boundary conditions for a certain independent variable as the maximum order of the derivatives with respect to that variable.

Quite obviously, we can make the vibrating string problem as complex as we please. For example, we could allow ϱ to be a function of x or we could consider $y(0, t)$ to be some function of t. However, we do not consider such complications here as they involve few new concepts.

Closely related to the vibrating string is the vibrating membrane. For this system the wave equation becomes

$$c^2\left(\frac{\partial^2 z}{\partial x^2} + \frac{\partial^2 z}{\partial y^2}\right) = \frac{\partial^2 z}{\partial t^2}, \tag{2.4}$$

or more simply

$$c^2 \nabla^2 z = \partial^2 z/\partial t^2, \tag{2.5}$$

where ∇ is the vector differential operator introduced in Chapter 1.

B. Heat Equation

The temperature $u(x, y, z, t)$ of a body, measured in any units, is governed by the equation

$$k\nabla^2 u = \partial u/\partial t, \tag{2.6}$$

where k is called the thermal diffusivity or more simply the *diffusivity* and t is the time. Equation (2.6) is the *heat equation*, also called the *diffusion equation*. One possible boundary condition would be that u has some constant value at some boundary. Another possibility would be that some boundary is insulated. For the latter case, at that boundary,

$$\partial u/\partial n = 0, \tag{2.7}$$

where n is the direction perpendicular to the boundary. An appropriate initial condition would be the initial temperature. Note that only one initial condition is needed because (2.6) is first order in the time.

In many cases, we are interested in the steady-state temperature distribution where $\partial u/\partial t = 0$. Thus, (2.6) reduces to *Laplace's equation*

$$\nabla^2 u = 0. \tag{2.8}$$

Again we could make this problem more complex by allowing k to vary or by allowing for heat sources or sinks in the body. However, such complications are beyond the scope of this chapter.

C. Potential Problems

As an example of a potential problem consider an electrostatic field. It is well known that the electric field vector $\mathbf{E}$ satisfies the equation

$$\nabla \cdot \mathbf{E} = 4\pi\varrho/\varepsilon, \tag{2.9}$$

where $\varrho(x, y, z)$ is the electric charge density and ε is the dielectric constant. Further, **E** is the gradient of a scalar field. In particular, if

$$\mathbf{E} = -\nabla\phi, \tag{2.10}$$

where ϕ is the electrostatic potential and ε is a constant, then ϕ satisfies *Poisson's equation*

$$\nabla^2\phi = -4\pi\varrho/\varepsilon. \tag{2.11}$$

When $\varrho = 0$, (2.11) reduces to Laplace's equation.

If the boundary is formed by a conductor, then the appropriate boundary condition is

$$\phi = \text{constant} \tag{2.12}$$

on the boundary. On the other hand, if there is a change of dielectric constant at some boundary, then the boundary conditions are

$$\phi_1 = \phi_2 \tag{2.13}$$

and

$$\varepsilon_1 \, \partial\phi_1/\partial n = \varepsilon_2 \, \partial\phi_2/\partial n. \tag{2.14}$$

Equation (2.14) is valid only if there is no surface charge at the interface.

D. Schroedinger's Equation

One of the most important classes of boundary-value problems in chemistry is that involving Schroedinger's equation:

$$-\frac{\hbar^2}{2m} \nabla^2\psi + V\psi = i\hbar \frac{\partial\psi}{\partial t}, \tag{2.15}$$

where ψ is the wave function, V the potential, and t the time.

In as much as Volumes III, IV, and V of this treatise include discussions of the solutions of (2.15), we will concentrate our attention on (2.5), (2.6), and (2.11).

E. Some Further Comments

There are many more boundary-value problems that arise in physics and chemistry but the examples just given suffice to illustrate the techniques. There are a number of similarities among the boundary-value problems which we have considered. For example, they involve the operator ∇^2 and, in many cases, the equation to be solved is Laplace's equation.

More important, each of the systems is *linear*. That is, none of the terms in the equation and the boundary conditions contains powers of the dependent variable higher than first. The system is *homogeneous* if all of the terms in the equation and the boundary conditions contain the dependent variable to the same degree. If the system is both linear and homogeneous, then any linear combination of solutions of the system is a solution of the system. This is the *principle of superposition* of solutions.

If the system is linear but nonhomogeneous, we must modify the principle of superposition slightly. The solution of the corresponding homogeneous system is called the *complementary function*. For a linear nonhomogeneous system, the general solution is the sum of the complementary function and any *particular integral* which satisfies the system. These principles have already been met in the theory of linear ordinary differential equations.

These concepts form the basis of the techniques that are used to solve boundary-value problems. However, before developing these techniques, it is of interest to introduce a few additional definitions.

Dirichlet boundary conditions fix the value of the dependent variable on the boundary. On the other hand, *Neumann boundary conditions* fix the value of the normal gradient there while *intermediate or mixed boundary conditions* fix some linear combination of the value and the normal gradient there. *Cauchy boundary conditions* fix *both* the value and the normal gradient on the boundary. Boundary conditions of the form of (2.13) and (2.14) are *matching boundary conditions*. If the dependent variable is periodic in some independent variable, then we have *periodic boundary conditions*.

It is frequently convenient to distinguish between three types of equations. For simplicity, let us restrict ourselves to equations with two independent variables. The most general form of a second-order partial differential equation is

$$A \frac{\partial^2 u}{\partial x^2} + 2B \frac{\partial^2 u}{\partial x \, \partial y} + C \frac{\partial^2 u}{\partial y^2} = F\left(x, y, u, \frac{\partial u}{\partial x}, \frac{\partial u}{\partial y}\right), \qquad (2.16)$$

where, if the equation is linear in u, F has the form

$$F = D \frac{\partial u}{\partial x} + E \frac{\partial u}{\partial y} + Gu + H \qquad (2.17)$$

and $A, B, \ldots, H$ are functions of x and y. We may distinguish between the following cases:

1. $AC - B^2 > 0$, *elliptic equation*
2. $AC - B^2 < 0$, *hyperbolic equation*
3. $AC - B^2 = 0$, *parabolic equation*.

Thus, Laplace's equation, the wave equation, and the heat equation are simple examples of elliptic, hyperbolic, and parabolic equations, respectively.

With this classification of boundary conditions and equations it is possible to specify under which conditions solutions to a boundary-value problem exist. These results are summarized in Table I. At first sight it would appear that one must worry a great deal about the existence of solutions. However, in practice this is not so. For example, because time is one of the variables in the wave equation and the heat equation, we will not want a solution of these equations for closed boundaries. The other aspects of Table I arise quite naturally as we solve individual boundary-value problems. For example, because the heat equation has only a first-order derivative with respect to time, it turns out that the time

TABLE I

Existence of Solutions to Boundary-Value Problems

Conditions	Boundary	Equations		
		Elliptic	Hyperbolic	Parabolic
Dirichlet or Neumann	Open	Insufficient	Insufficient	Unique, stable solution (unstable in negative direction)
	Closed	Unique, stable solution	Solution not unique	Solution overspecified
Cauchy	Open	Solution unstable	Unique, stable solution	Solution overspecified
	Closed	Solution overspecified	Solution overspecified	Solution overspecified

dependence is exponential. Thus, the solution can be stable only with respect to the positive time direction. For a detailed discussion the reader is referred to Chapter 6 of Morse and Feshbach (1953).

III. The D'Alembert Solution of the Wave Equation

The boundary-value problems we have considered above can be solved by means of the method of separation of variables. However, before discussing this general method it is of interest to consider an elegant special method due to D'Alembert for solving the elastic string problem.

We consider the wave equation

$$c^2 \partial^2 y/\partial x^2 = \partial^2 y/\partial t^2. \tag{3.1}$$

Let

$$\xi = x - ct \tag{3.2}$$

$$\eta = x + ct. \tag{3.3}$$

The wave equation becomes

$$\frac{\partial^2 y}{\partial \xi \, \partial \eta} = 0 \tag{3.4}$$

which, when integrated, gives

$$y = f(\eta) + g(\xi). \tag{3.5}$$

Thus the general solution of the wave equation is

$$y(x, t) = f(x + ct) + g(x - ct). \tag{3.6}$$

Hence, the solution consists of two waves traveling in opposite directions with velocities $+c$ and $-c$. So far f and g are arbitrary functions. Let us choose f and g so that the initial conditions and boundary conditions are satisfied. It we take as the initial conditions

$$y(x, 0) = D(x) \tag{3.7}$$

and

$$\partial y(x, 0)/\partial t = V(x), \tag{3.8}$$

then

$$y(x, t) = \tfrac{1}{2}[D(x + ct) + D(x - ct)] + \frac{1}{2c} \int_{x-ct}^{x+ct} V(s) \, ds. \tag{3.9}$$

If the string is infinite, then this is the solution. If the string is finite, then we must show that we can satisfy the boundary conditions by defining $D(x)$ and $V(x)$ outside the interval in a suitable way. The boundary conditions are

$$y(0, t) = y(l, t) = 0, \qquad (3.10)$$

where l is the length of the string. Since the D and V parts of the solution are independent, they must separately satisfy the boundary conditions. Therefore,

$$D(ct) + D(-ct) = 0 \qquad (3.11)$$
$$D(l + ct) + D(l - ct) = 0 \qquad (3.12)$$

for all t. Hence, D must be an odd function of period $2l$. Similarly, V must be an odd function of period $2l$.

Thus for the finite string, we may use (3.9) to obtain $y(x, t)$ for $0 \leq x \leq l$ if we extend D and V by means of the rules given above.

IV. Separation of Variables

The first method of solving a boundary-value problem which is considered is the *method of separation of variables*. In applying this method it is usually convenient to follow the following procedure.

1. Set aside some conditions. These are usually the initial conditions, but in steady-state problems it would be one or more boundary conditions.
2. If the remaining system is nonhomogeneous, find a particular integral, ψ.
3. Find the complementary function φ for the corresponding homogeneous system by means of separation of variables and the principle of superposition.
4. Choose the arbitrary constants in the complementary function so that the conditions set aside in step 1 are satisfied.

If the domain of the system is infinite, it may be better to use integral transforms. We consider this situation in Sections IV.C and IX.

A. Plucked String

Consider the problem of a string of length l in a gravitational field which, at time $t = 0$, is plucked at a distance a from one end and displaced a distance b. Thus we must solve

$$c^2 \, \partial^2 y/\partial x^2 = \partial^2 y/\partial t^2 + g, \tag{4.1}$$

subject to the boundary conditions

$$y(0, t) = y(l, t) = 0 \tag{4.2}$$

and the initial conditions

$$\begin{aligned} y(x, 0) &= bx/a & 0 \le x \le a \\ &= b(l - x)/(l - a), & a \le x \le l \end{aligned} \tag{4.3}$$

and

$$\partial y(x, 0)/\partial t = 0, \qquad 0 \le x \le l. \tag{4.4}$$

We now follow the procedure outlined above.

First Step Set aside the initial conditions and consider the remaining system.

Second Step Find a particular integral. Look for one which is a function only of x. Thus we must solve

$$d^2 \psi/dx^2 = g/c^2 \tag{4.5}$$

subject to

$$\psi(0) = \psi(l) = 0. \tag{4.6}$$

Thus

$$\psi(x) = (g/2c^2)x(x - l). \tag{4.7}$$

Third Step Find the complementary function for the homogeneous system

$$c^2 \, \partial^2 \varphi/\partial x^2 = \partial^2 \varphi/\partial t^2 \tag{4.8}$$

$$\varphi(0, t) = \varphi(l, t). \tag{4.9}$$

The principle of superposition can now be used. We look for solutions of the form

$$\varphi(x, t) = X(x)T(t). \tag{4.10}$$

In seeking solutions of the form (4.10) it appears that we are making such a drastic approximation that we have no hope of finding the general

4. Boundary-Value Problems

solution. This is not the case. If we can find a solution that satisfies (4.1)–(4.4), then this must be the general solution. In any case (4.10) is not so drastic an assumption as it would seem at first. We will find that our solution is a superposition of terms whose variables are separated. However, the solution does *not* have separated variables.

Substitution of (4.10) into (4.8) gives

$$c^2(X''/X) = T''/T. \qquad (4.11)$$

Since x and t are independent variables, Eq. (4.11) can be valid only if

$$c^2(X''/X) = -\lambda^2 \qquad (4.12)$$

and

$$T''/T = -\lambda^2, \qquad (4.13)$$

where λ is a constant which, at this point, is not necessarily real. Thus

$$X = A \sin(\lambda x/c) + B \cos(\lambda x/c) \qquad (4.14)$$

and

$$T = C \sin \lambda t + D \cos \lambda t. \qquad (4.15)$$

The boundary condition at $x = 0$ requires that $B = 0$, and the boundary condition at $x = l$ requires that

$$\sin(\lambda l/c) = 0, \qquad (4.16)$$

which means that

$$\lambda = n\pi c/l \qquad (4.17)$$

for $n = 1, 2, 3, \ldots$. We now see that we were justified in our choice of sign in Eqs. (4.12) and (4.13). If we had chosen a positive sign or, equivalently, used (4.12) and (4.13) with an imaginary value for λ, we would have had exponential functions (or hyperbolic functions) appearing in our expression for X. Such functions could not be made to vanish at both $x = 0$ and $x = l$.

Thus, using the principle of superposition, the characteristic function is

$$\varphi(x, t) = \sum_{n=1}^{\infty} \{A_n \sin(n\pi ct/l) + B_n \cos(n\pi ct/l)\} \sin(n\pi x/l). \qquad (4.18)$$

Hence, our solution is expressed in terms of a *Fourier series*.

Fourth Step We now choose A_n and B_n so that our general solution,

$$y(x, t) = (g/2c^2)x(x - l)
+ \sum_{n=1}^{\infty} \{A_n \sin(n\pi ct/l) + B_n \cos(n\pi ct/l)\} \sin(n\pi x/l), \quad (4.19)$$

satisfies the initial conditions. The second initial condition requires that $A_n = 0$. Thus we must choose the B_n to satisfy

$$y(x, 0) = (g/2c^2)x(x - l) + \sum_{n=1}^{\infty} B_n \sin(n\pi x/l). \quad (4.20)$$

We multiply (4.20) by $\sin(m\pi x/l)$, integrate from 0 to l, and make use of the *orthogonality condition*

$$\int_0^l \sin(n\pi x/l) \sin(m\pi x/l)\, dx = (l/2)\, \delta_{mn}, \quad (4.21)$$

where δ_{mn} is the *Kronecker delta* which is zero when $m \neq n$ and unity when $m = n$. Thus

$$B_n = (2/l) \int_0^l y(x, 0) \sin(n\pi x/l)\, dx - (g/lc^2) \int_0^l x(x - l) \sin(n\pi x/l)\, dx. \quad (4.22)$$

Substitution of (4.3) into (4.22) and integration yield the required solution to the problem:

$$y(x, t) = (g/2c^2)x(x - l) + \sum_{n=1}^{\infty} B_n \cos(n\pi ct/l) \sin(n\pi x/l), \quad (4.23)$$

where

$$B_n[2b/a(l - a)](l/n\pi)^2 + (2g/lc^2)(l/n\pi)^3[1 - (-1)^n]. \quad (4.24)$$

The function

$$\varphi_n(x, t) = \{A_n \sin(n\pi ct/l) + B_n \cos(n\pi ct/l)\} \sin(n\pi x/l) \quad (4.25)$$

is called the *nth normal mode of vibration*. The *natural frequency* of this mode is

$$\omega_n = n\pi c/l \quad \text{radians/unit time} \quad (4.26)$$

or

$$f_n = nc/2l \quad \text{cycles/unit time.} \quad (4.27)$$

4. Boundary-Value Problems

We note that the natural frequency of the nth harmonic is n times the natural frequency of the fundamental note.

B. Comments Regarding the Separation of Variables

We see that the key to our solution of the vibrating string problem given above was the separation of the variables. We briefly discuss conditions under which a separation of variables is possible. In addition, our general considerations allow us to separate variables more quickly than was done in the preceding example.

First, we can separate the variables in the partial differential equation

$$Ly = 0, \qquad (4.28)$$

if the operator L can be written

$$L = L_1 + L', \qquad (4.29)$$

where L_1 is a differential operator depending only on x_1 and L' depends only on $x_2, \ldots, x_n$. Thus we look for solutions of the form

$$y = f_1(x_1)g(x_2, \ldots, x_n). \qquad (4.30)$$

Hence, substituting (4.29) and (4.30) into (4.28), we obtain

$$gL_1 f_1 + f_1 L'g = 0. \qquad (4.31)$$

Therefore,

$$(L_1 f_1/f_1) + (L'g/g) = 0. \qquad (4.32)$$

The variables $x_1, \ldots, x_n$ are independent. Thus

$$L_1 f_1/f_1 = \lambda_1 \qquad (4.33)$$

and

$$L'g/g = -\lambda_1. \qquad (4.34)$$

Suppose that the solutions to (4.33) are $f_1(x_1, \lambda_1)$ and that the solutions to (4.34) are $g(x_2, \ldots, x_n, \lambda_1)$. Then

$$y(x_1, \ldots, x_n) = \sum_{\lambda_1} f_1(x_1, \lambda_1) g(x_2, \ldots, x_n, \lambda_1) \qquad (4.35)$$

is the general solution of (4.28).

If the possible values of λ_1 are not discrete, then the sum in (4.35) becomes an integral. We meet such problems shortly.

Let us consider (4.34). It may turn out that

$$L' = L_2 + L'' \tag{4.36}$$

$$L'' = L_3 + L''', \tag{4.37}$$

and so on, where L_j depends only on x_j. Hence, we can separate each of the variables in turn. Thus, the final form of the solution is

$$y(x_1, \ldots, x_n) = \sum_{\lambda_1 \cdots \lambda_{n-1}} f_1(x_1, \lambda_1) \cdots f_n(x_n, \lambda_n), \tag{4.38}$$

where

$$-\lambda_n = \lambda_1 + \cdots + \lambda_{n-1}. \tag{4.39}$$

Second, we can separate a variable which is *cyclic*. An independent variable is cyclic only when it appears in derivatives with respect to that variable. For example, consider

$$\frac{\partial^2 y}{\partial x_1^2} + \frac{x_2}{x_2 + x_3} \frac{\partial y}{\partial x_1} + x_3^2 \frac{\partial^2 y}{\partial x_2 \, \partial x_3} = 0. \tag{4.40}$$

We see that x_1 is a cyclic variable, whereas x_2 and x_3 are not. Try as a solution:

$$y = \exp(\alpha x_1) g(x_2, x_3, \alpha). \tag{4.41}$$

We obtain

$$\alpha^2 \exp(\alpha x_1) g + \frac{x_2}{x_2 + x_3} \alpha \exp(\alpha x_1) g + x_3^2 \exp(\alpha x_1) \frac{\partial^2 g}{\partial x_2 \, \partial x_3} = 0. \tag{4.42}$$

The exponentials can be divided out and the remaining equation does not contain x_1. Thus

$$y(x_1, x_2, x_3) = \sum_\alpha \exp(\alpha x_1) g(x_2, x_3, \alpha) \tag{4.43}$$

is the general solution.

It may happen that all variables are cyclic. For example, consider Laplace's equation:

$$\frac{\partial^2 u}{\partial x^2} + \frac{\partial^2 u}{\partial y^2} + \frac{\partial^2 u}{\partial z^2} = 0. \tag{4.44}$$

All three variables are cyclic. Thus, the solution is

$$u(x, y, z) = \sum_{\alpha \beta} A_{\alpha \beta} e^{\alpha x} e^{\beta y} e^{\gamma z}, \tag{4.45}$$

where
$$\alpha^2 + \beta^2 + \gamma^2 = 0. \tag{4.46}$$

Obviously, other than for the trivial case $\alpha = \beta = \gamma = 0$, one or more of α, β, and γ must be imaginary. However, this is no problem. It only means that trigonometric functions, rather than exponential functions, appear in part of (4.45).

If only even-order derivatives of the cyclic variable x_1 appear, then if $\exp(\alpha x_1)g(x_1, \ldots, x_n)$ is a solution, so is $\exp(-\alpha x_1)g(x_1, \ldots, x_n)$ also. Thus we will write the solution as

$$y(x_1, \ldots, x_n) = \sum_\alpha \exp(\pm \alpha x_1) g(x_1, \ldots, x_n) \tag{4.47}$$

where we interpret $e^{\pm \alpha x}$ as meaning any linear combination of $e^{\alpha x}$ and $e^{-\alpha x}$. The values of the coefficients in the linear combination are determined by the boundary conditions.

Finally, we note that it is necessary that both the equation and the boundary conditions be separable in the *same* coordinate system. To separate the variables in the boundary condition it is necessary that we use the coordinate system natural to the problem. Thus the boundary conditions determine the coordinate system and the form of the solution. This is why we refer to these problems as boundary-value problems.

C. Three Heat Conduction Problems

1. A very thin sheet of metal is in the xy plane and has vertices at the points $(0, 0)$, $(a, 0)$, (a, b), and $(0, b)$. The upper and lower faces of the sheet are perfectly insulated, so that the conduction is purely two-dimensional. Initially the temperature distribution in the sheet is $u(x, y, 0) = f(x, y)$. There are no heat sources or sinks in the sheet. Let us find the temperature, given that the edges from $(0, 0)$ to $(a, 0)$ and $(0, b)$ are maintained at temperature T and the other two edges are insulated. Thus

$$k\left[\frac{\partial^2 u}{\partial x^2} + \frac{\partial^2 u}{\partial y^2}\right] = \frac{\partial u}{\partial t} \tag{4.48}$$

$$u(0, y, t) = u(x, 0, t) = T \tag{4.49}$$

$$\frac{\partial u(a, y, t)}{\partial x} = \frac{\partial u(x, b, t)}{\partial y} = 0 \tag{4.50}$$

and
$$u(x, y, 0) = f(x, y). \tag{4.51}$$

First, we set aside (4.51); second, we take $\psi = T$; and third, we consider the homogeneous system formed by replacing (4.49) by

$$\varphi(0, y, t) = \varphi(x, 0, t) = 0. \tag{4.52}$$

Let us try

$$\varphi = e^{\pm i\alpha x}e^{\pm i\beta y} e^{-\gamma t}. \tag{4.53}$$

We choose this form because the boundary conditions suggest trigonometric functions for the x and y coordinates. We choose a negative sign for the exponential because the exponential of a positive quantity would diverge as $t \to \infty$. However, we have not committed ourselves by using (4.53). For example, if we had chosen an incorrect form, we would merely find that, for example, α or β were imaginary or that γ was negative.

Substituting (4.53) into (4.48) gives

$$k(\alpha^2 + \beta^2) = \gamma. \tag{4.54}$$

The boundary conditions at $x = y = 0$ require that only $\sin \alpha x$ and $\sin \beta y$ appear. The boundary conditions and $x = a$ and $y = b$ require that

$$\cos \alpha a = \cos \beta b = 0. \tag{4.55}$$

Thus, $\alpha = n\pi/2a$ and $\beta = m\pi/2b$, where n and m are odd integers. Hence

$$u(x, y, t) = T + \sum_{n,m=1,3}^{\infty} A_{nm} \exp\{-[(n/a)^2 + (m/b)^2](k\pi^2 t/4)\}$$
$$\times \sin(n\pi x/2a) \sin(m\pi y/2a). \tag{4.56}$$

Finally, we fit the initial condition

$$f(x, y) = T + \sum_{n,m=1,3}^{\infty} A_{nm} \sin(n\pi x/2a) \sin(m\pi y/2a). \tag{4.57}$$

Using the orthogonality of the trigonometric functions, we obtain

$$A_{nm} = (4/ab) \int_0^a \int_0^b [f(x, y) - T] \sin(n\pi x/2a) \sin(m\pi y/2b) \, dx \, dy. \tag{4.58}$$

2. Let us consider the sheet described above and calculate the steady-state temperature distribution when the $y = 0$ edge is maintained at a

temperature T and the other edges are maintained at $0°$. Thus

$$\frac{\partial^2 u}{\partial x^2} + \frac{\partial^2 u}{\partial y^2} = 0 \tag{4.59}$$

$$u(0, y, t) = u(a, y, t) = 0 \tag{4.60}$$

$$u(x, 0, t) = T \tag{4.61}$$

$$u(x, b, t) = 0. \tag{4.62}$$

First, we set aside (4.61) and second, we note that $\psi = 0$. Let us try

$$\varphi = e^{\pm i\alpha x} e^{\pm \beta y}. \tag{4.63}$$

Hence

$$\alpha^2 = \beta^2. \tag{4.64}$$

The boundary conditions at $x = 0$ and $y = b$ require that only $\sin \alpha x$ and $\sinh \beta(b - y)$ appear. The boundary condition at $x = a$ requires that

$$\alpha = n\pi/a, \tag{4.65}$$

for $n = 1, 2, 3, \ldots$. Thus

$$u(x, y) = \sum_{n=1}^{\infty} A_n \sin(n\pi x/a) \sinh[n\pi(b - y)/a]. \tag{4.66}$$

Finally, the remaining condition, Eq. (4.61), can be fit in the usual way. The result is

$$A_n = (2T/n\pi)[\sinh(n\pi b/a)]^{-1}[1 - (-1)^n]. \tag{4.67}$$

3. A slender rod whose curved surface is perfectly insulated stretches from $x = 0$ to $x = \infty$. Let us find the temperature of the rod if the $x = 0$ end of the rod is maintained at $0°$ and if initially the temperature along the rod is given by $u(x, 0) = f(x)$. Thus

$$k \frac{\partial^2 u}{\partial x^2} = \frac{\partial u}{\partial t} \tag{4.68}$$

$$u(0, t) = 0 \tag{4.69}$$

$$u(x, 0) = f(x). \tag{4.70}$$

First, we set aside (4.70) and second, we note that $\psi = 0$. Let us try

$$\varphi = e^{-\alpha t} e^{\pm i\beta x}. \tag{4.71}$$

In (4.71) α must be positive because otherwise the solution would diverge as t became large. Thus

$$\alpha = k\beta^2. \tag{4.72}$$

The boundary condition (4.69) means that only $\sin \beta x$ is present. Since there are no other boundary conditions, any value of β is acceptable. Thus

$$u(x, t) = \int_0^\infty A(\beta) \exp(-k\beta^2 t) \sin \beta x \, d\beta. \tag{4.73}$$

Finally we fit the initial condition

$$f(x) = \int_0^\infty A(\beta) \sin \beta x \, d\beta. \tag{4.74}$$

The obtain an expression for the function $A(\beta)$ we make use of the Fourier integral theorem (see Chapter 2, Section VI.A) which states that if (4.74) is valid, then

$$A(\beta) = (2/\pi) \int_0^\infty f(s) \sin \beta s \, ds. \tag{4.75}$$

Hence, our formal result is

$$u(x, t) = (2/\pi) \int_0^\infty \exp(-k\beta^2 t) \sin \beta x \left[\int_0^\infty f(s) \sin \beta s \, ds \right] d\beta. \tag{4.76}$$

We see that, in this case, the solution is expressed in terms of a *Fourier integral* rather than a Fourier series.

We can simplify (4.76) by first writing it in the form,

$$u(x, t) = (1/\pi) \int_0^\infty \int_0^\infty \exp(-k\beta^2 t) f(s) [\cos \beta(x-s) - \cos \beta(x+s)] \, ds \, d\beta, \tag{4.77}$$

and then making use of the result

$$\int_0^\infty \exp(-\lambda y^2) \cos \mu y \, dy = \tfrac{1}{2}(\pi/\lambda)^{1/2} \exp(-\mu^2/4\lambda), \tag{4.78}$$

to obtain

$$u(x, t) = \frac{1}{2(\pi k t)^{1/2}} \int_0^\infty f(s) \left\{ \exp\left[-\frac{(x-s)^2}{4kt} \right] - \exp\left[-\frac{(x+s)^2}{4kt} \right] \right\} ds. \tag{4.79}$$

On changing variables, we obtain

$$u(x, t) = (1/\sqrt{\pi}) \int_{-x/2(kt)^{1/2}}^{\infty} \exp(-\eta^2) f[x + 2\eta(kt)^{1/2}] \, d\eta$$
$$- (1/\sqrt{\pi}) \int_{x/2(kt)^{1/2}}^{\infty} \exp(-\eta^2) f[-x + 2\eta(kt)^{1/2}] \, d\eta. \tag{4.80}$$

In the particular case $f(x) = T_0$, we obtain

$$u(x, t) = T_0 \, \text{erf}[x/2(kt)^{1/2}], \tag{4.81}$$

where the *error function*, erf(y), is defined by

$$\text{erf}(y) = (2/\sqrt{\pi}) \int_0^y \exp(-\eta^2) \, d\eta. \tag{4.82}$$

D. Relation to the D'Alembert Solution

Let us return to the infinite elastic string. Thus we must solve

$$c^2 \, \partial^2 y/\partial x^2 = \partial^2 y/\partial t^2, \tag{4.83}$$

with initial conditions

$$y(x, 0) = f(x) \tag{4.84}$$

and

$$\partial y(x, 0)/\partial t = 0. \tag{4.85}$$

We try

$$u = e^{\pm i\alpha x} e^{\pm i\beta t}. \tag{4.86}$$

Equation (4.85) requires that only $\cos \beta t$ appear. Thus, since there are no boundary conditions to restrict the acceptable value of α, we have

$$y(x, t) = \int_{-\infty}^{\infty} A(\alpha) e^{i\alpha x} \cos c\alpha t \, d\alpha. \tag{4.87}$$

The initial condition (4.84) requires that

$$f(x) = \int_{-\infty}^{\infty} A(\alpha) e^{i\alpha x} \, d\alpha. \tag{4.88}$$

The Fourier integral theorem gives

$$A(\alpha) = (1/2\pi) \int_{-\infty}^{\infty} f(s) e^{-i\alpha s} \, ds. \tag{4.89}$$

Thus,

$$y(x, t) = (1/2\pi) \int_{-\infty}^{\infty} \int_{-\infty}^{\infty} f(s) \exp(i\alpha(x - s)) \cos c\alpha t \, ds \, d\alpha$$

$$= (1/\pi) \int_{0}^{\infty} \int_{-\infty}^{\infty} f(s) \cos \alpha(x - s) \cos c\alpha t \, ds \, d\alpha. \qquad (4.90)$$

Equation (4.90) can be written in the form

$$y(x, t) = (1/2\pi) \int_{0}^{\infty} \int_{-\infty}^{\infty} f(s)[\cos \alpha(x-s+ct) + \cos \alpha(x-s-ct)] \, ds \, d\alpha. \qquad (4.91)$$

Using the Fourier integral theorem we obtain

$$y(x, t) = \tfrac{1}{2}[f(x + ct) + f(x - ct)], \qquad (4.92)$$

which is the result obtained from the D'Alembert solution.

V. Eigenvalues, Eigenfunctions, and Expansion Problems

In the problems that we have considered thus far, we have found that in the process of separating variables we were usually led to ordinary differential equations of the form

$$Ly = \lambda y, \qquad (5.1)$$

where L is a linear differential operator, and some homogeneous boundary conditions. Often λ is a separation parameter, but it may arise in other ways. For example, it may be the energy in a quantum mechanical problem.

Usually the boundary conditions restrict the possible values of λ to some discrete set, $\lambda_1, \lambda_2, \ldots$. This discrete set of values of λ is called the set of *eigenvalues*. This word is a linguistic disaster resulting from an incomplete translation of the German word, *Eigenwert*, meaning *characteristic value*. The corresponding solutions $y_1, y_2, \ldots$, are called *eigenfunctions*. An equation of the form (5.1) is called an *eigenvalue problem*.

If there are two or more eigenfunctions corresponding to an eigenvalue, then the eigenvalue is said to be *degenerate*.

Quite obviously, $y = 0$ satisfies (5.1) for any L and for any λ. We specifically exclude the identically zero function.

4. Boundary-Value Problems

There was no difficulty in satisfying the initial conditions of the problems that we have considered thus far because the eigenfunctions with which we dealt were the trigonometric functions which arise in the Fourier series. Thus we are able to use the orthogonality conditions for the Fourier series to satisfy the initial conditions.

We must be careful not to obtain an exaggerated picture of the role of Fourier series in boundary-value problems. In most problems, the set of eigenfunctions will *not* form a Fourier series. Clearly, we must generalize the concept of the Fourier series.

A. Orthogonal Sets of Functions

It was the orthogonality property of the Fourier series which permitted us to fit the initial conditions of our problems. Fourier series are not the only of set functions with such a property. In fact, it is only one of infinitely many sets of functions with such a property. Also it is no coincidence that the eigenfunctions with which we have had to deal have this property. We will see that all the eigenfunctions with which we deal have this property. Before we prove this, let us give a few definitions.

Any set of real functions $\{\phi_n(x)\}$, $n = 1, 2, \ldots$, which has the property

$$\int_a^b \phi_n(x)\phi_m(x)\, dx = c_n\, \delta_{nm}, \tag{5.2}$$

where $c_n \neq 0$ is some real number, is said to form an *orthogonal* set on the interval $[a, b]$. If, in addition, $c_n = 1$ for all n, then the set is said to be orthogonal and *normalized* or more simply *orthonormal*. Any orthogonal set can be made orthonormal by replacing $\phi_n(x)$ by $\phi_n(x)/(c_n)^{1/2}$. Thus there is no loss of generality if we assume that every orthogonal set is orthonormal.

We can extend the concept of orthogonality to complex functions by replacing (5.2) by

$$\int_a^b \phi_n^*(x)\phi_m(x)\, dx = c_n\, \delta_{nm}, \tag{5.3}$$

where $c_n \neq 0$ is some real number and $\phi_n^*(x)$ is the complex conjugate of $\phi_n(x)$. We can obtain another generalization of (5.2) which will be useful in this chapter. A set of real functions $\{\phi_n(x)\}$, $n = 1, 2, \ldots$, which has the property

$$\int_a^b r(x)\phi_n(x)\phi_m(x)\, dx = c_n\, \delta_{nm}, \tag{5.4}$$

where $c_n \neq 0$ is some real number, is said to form an orthogonal set with respect to the *density function* $r(x)$ ($\neq 0$) on the interval $[a, b]$. Any set of functions orthogonal with respect to a density function $r(x)$ can be reduced to an orthogonal system in the original sense by multiplying each member of the set by $(r(x))^{1/2}$ if, as we shall assume, $r(x) \neq 0$ for all x in $[a, b]$.

We can obtain a *formal* expansion of a function $f(x)$ with respect to an orthogonal set of functions in a manner analogous to that used in obtaining Fourier series. Let us write

$$f(x) \sim \sum_{n=1}^{\infty} a_n \phi_n(x). \tag{5.5}$$

An equal sign has not been placed between the two parts of (5.5) because we do not yet know the conditions required for the series to converge to $f(x)$. If we multiply by $\phi_m(x)$ and integrate, we obtain

$$a_n = \left(\int_a^b f(x) \phi_n(x) \, dx \right) \Big/ \left(\int_a^b \phi_n^2(x) \, dx \right). \tag{5.6}$$

Thus, the orthogonality of the $\{\phi_n(x)\}$ permits us to calculate the a_n. However, the orthogonality is not sufficient to guarantee that the series converges to $f(x)$ or even converges at all. In particular, it may turn out that

$$\int_a^b f(x) \phi_n(x) \, dx = 0 \tag{5.7}$$

for all n, even though $f(x)$ is not everywhere zero in $[a, b]$. For this case $a_n = 0$ for all n even though $f(x)$ is nontrivial. For example, the set $\{\sin nx\}$ is orthogonal on $[0, 2\pi]$. However, for any even function (5.7) will be valid for all n.

A set of orthogonal functions $\{\phi_n\}$ must possess the further property of *completeness* before we can be certain that an expansion is meaningful. An orthogonal set of functions is *complete* if there exists no nontrivial function which is orthogonal to every member of the set.

We can now state the following:

Theorem If the formal expansion,

$$\sum_{n=1}^{\infty} a_n \phi_n(x),$$

of a function $f(x)$ in terms of a complete orthonormal set $\{\phi_n\}$ converges and can be integrated term by term, then the series converges to $f(x)$.

4. Boundary-Value Problems

Proof To prove this theorem we consider

$$g(x) = f(x) - \sum_{n=1}^{\infty} a_n \phi_n(x). \tag{5.8}$$

If we multiply by $\phi_m(x)$ and integrate, we obtain

$$\int_a^b \phi_m(x) g(x)\, dx = \int_a^b \phi_m(x) f(x) - \sum_{n=1}^{\infty} a_n \phi_n(x)\, dx$$

$$= \int_a^b \phi_m(x) f(x)\, dx - \sum_{n=1}^{\infty} a_n\, \delta_{nm} = a_m - a_m. \tag{5.9}$$

Thus, $g(x)$ is orthogonal to all the $\phi_n(x)$. Since $\{\phi_n(x)\}$ is complete, $g(x)$ must be identically zero. Thus

$$f(x) = \sum_{n=1}^{\infty} a_n \phi_n(x), \tag{5.10}$$

which completes the proof of the theorem.

A concept which is closely related to that of completeness is that of *closure*. If for every $f(x)$,

$$\lim_{m \to \infty} \int_a^b \left[f(x) - \sum_{n=1}^{m} a_n \phi_n(x) \right] dx = 0. \tag{5.11}$$

where

$$\sum_{n=1}^{\infty} a_n \phi_n(x)$$

is the expansion of the function $f(x)$ in terms of the members of an orthonormal set $\{\phi_n(x)\}$, then the set $\{\phi_n(x)\}$ is *closed*.

An important property of closed orthonormal sets is contained in:

Parseval's Theorem If

$$\sum_{n=1}^{\infty} a_n \phi_n(x)$$

is the expansion of the function $f(x)$ in terms of the members of a closed orthonormal set, then

$$\sum_{n=1}^{\infty} a_n^2 = \int_a^b [f(x)]^2\, dx. \tag{5.12}$$

Proof To prove this theorem we observe that because $\{\phi_n(x)\}$ is a closed set,

$$\lim_{m \to \infty} \int_a^b \left[f(x) - \sum_{n=1}^m a_n \phi_n(x) \right]^2 dx = 0. \tag{5.13}$$

Thus

$$\lim_{m \to \infty} \int_a^b \left[\{f(x)\}^2 - 2f(x) \sum_{n=1}^m a_n \phi_n(x) + \left\{ \sum_{n=1}^m a_n \phi_m(x) \right\}^2 \right] dx = 0 \tag{5.14}$$

or

$$\lim_{m \to \infty} \left\{ \int_a^b [f(x)]^2 \, dx - 2 \sum_{n=1}^m a_n^2 + \sum_{n=1}^m a_n^2 \right\} = 0. \tag{5.15}$$

Thus

$$\int_a^b [f(x)]^2 \, dx = \sum_{n=1}^\infty a_n^2, \tag{5.16}$$

which proves the theorem.

Finally, we have

Theorem A closed orthonormal set $\{\phi_n(x)\}$ is complete.

Proof To prove this theorem let us assume that $f(x)$ is orthogonal to each of the $\phi_n(x)$. Hence, $a_n = 0$, and thus from Parseval's theorem,

$$\int_a^b [f(x)]^2 \, dx = 0, \tag{5.17}$$

which means that $f(x)$ is identically zero and, therefore, the set is complete and the theorem is proved.

The converse of this theorem is also true. However, the proof is difficult and is not presented here.

B. Abstract Vector Space

As has no doubt been noted, we have used the terminology of linear vector spaces. This is quite intentional because a function is a very general type of vector. The finite-dimensional vectors with which we are familiar are merely functions that are defined only on a finite number of discrete points on some coordinate axis. Thus, a function $f(x)$ which is defined on the nondenumeral points of the x axis in the interval $[a, b]$

4. Boundary-Value Problems

is a vector in a very general infinite-dimensional vector space which we may call a *function space*.

Hence, the familiar definition of the *scalar product* or *inner product* of two finite-dimensional vectors **A** and **B**,

$$\mathbf{A} \cdot \mathbf{B} = (A, B) = \sum_{i=1}^{n} A_i B_i, \tag{5.18}$$

where, for simplicity, the components A_i and B_i have been assumed to be real, is seen to be a particular case of the general definition of the inner product over the interval $[a, b]$ of two real functions $f(x)$ and $g(x)$:

$$(f, g) = \int_a^b f(x) g(x) \, dx. \tag{5.19}$$

Just as in the more familiar finite-dimensional vector spaces, two vectors are orthogonal if their inner product is zero. Therefore, the trigonometric functions used in the Fourier series are just a set of orthogonal basis vectors spanning the vector space. Thus we can express any vector in the space in terms of its components along orthonormal axes. Hence

$$f = \sum_{n=1}^{\infty} (f, \phi_n) \phi_n, \tag{5.20}$$

which is just Eq. (5.5).

If we allow the functions in our space to be complex, then everything applies if we generalize (5.19) slightly to

$$(f, g) = \int_a^b f^*(x) g(x) \, dx, \tag{5.21}$$

where $f^*(x)$ is the complex conjugate of $f(x)$. A space of this type is often called a *Hilbert space*.

C. Sturm–Liouville Problem

Given some linear differential operator L, then the operator M is the *adjoint* to L if

$$zLy - yMz = (d/dx) F(y, y', \ldots, y^{(n-1)}, z, z', \ldots, z^{(n-1)}), \tag{5.22}$$

where F is some function and y and z are differentiable functions of x.

For example, it is easy to show that the adjoint to

$$L = p(x) \, d^n/dx^n \qquad (5.23)$$

is

$$M = (-1)^n \sum_{k=0}^{n} \binom{n}{k} \frac{d^k p}{dx^k} \frac{d^{n-k}}{dx^{n-k}}, \qquad (5.24)$$

where $\binom{n}{k}$ are the binomial coefficients. If $M = L$, then L is *self-adjoint*. To make the relation with linear vector spaces more transparent we could call L *symmetric* if we are dealing with real functions or *Hermitian* if we are dealing with complex functions. This usage is less common but not unknown.

For the particular case of a second-order operator, a self-adjoint operator has the form

$$L = \frac{d}{dx}\left[p(x)\frac{d}{dx}\right] + q(x). \qquad (5.25)$$

We can now establish the following:

Theorem Any second-order linear differential operator can be written in self-adjoint form.

To establish this result, we consider the general second-order linear operator

$$a(x)\frac{d^2}{dx^2} + b(x)\frac{d}{dx} + c(x).$$

Let us multiply by $r(x) \neq 0$. Thus our operator becomes

$$r(x)a(x)\frac{d^2}{dx^2} + r(x)b(x) + r(x)c(x).$$

This operator will be self-adjoint if

$$(d/dx)[r(x)a(x)] = r(x)b(x). \qquad (5.26)$$

Thus

$$\frac{1}{r}\frac{dr}{dx} = \frac{b}{a} - \frac{1}{a}\frac{da}{dx}. \qquad (5.27)$$

Integrating, we obtain

$$\ln r = \int (b/a) \, dx + \ln a + \ln c \qquad (5.28)$$

or
$$r(x) = \frac{c}{a(x)} \exp\left\{\int \frac{b(x)}{a(x)} dx\right\}, \qquad (5.29)$$

where c is some constant of integration. Thus, as long as the right-hand side of (5.29) exists, i.e., no singularities, then we can find $r(x)$ and can put the operator in self-adjoint form.

We can now state an important:

Theorem The eigenfunctions of a self-adjoint operator are orthogonal.

Proof We will prove this only for the most important case of the second-order linear operator. The eigenvalue equation is

$$Ly = \frac{d}{dx}\left[p(x)\frac{dy}{dx}\right] + q(x)y = \lambda y. \qquad (5.30)$$

Let us suppose that the homogeneous boundary conditions are

$$\alpha_1 y(a) - \beta_1 y'(a) = 0 \qquad (5.31)$$

and

$$\alpha_2 y(b) - \beta_2 y'(b) = 0, \qquad (5.32)$$

where α_1 and β_1 are not both zero and similarly for α_2 and β_2. Equations (5.31) and (5.32) encompass Dirichlet, Neumann, and intermediate boundary conditions. We now prove that any solutions of (5.30) which satisfy (5.31) and (5.32) are orthogonal if they belong to different eigenvalues. We assume

$$Ly_1 = \lambda_1 y_1 \qquad (5.33)$$

and

$$Ly_2 = \lambda_2 y_2. \qquad (5.34)$$

Multiply (5.33) by y_2 and (5.34) by y_1. Thus

$$y_2 L y_1 - y_1 L y_2 = (d/dx)[p(y_2 y_1' - y_1 y_2')] = (\lambda_1 - \lambda_2) y_1 y_2. \qquad (5.35)$$

Integrating,

$$p(y_2 y_1' - y_1 y_2')\Big|_a^b = (\lambda_1 - \lambda_2) \int_a^b y_1 y_2 \, dx. \qquad (5.36)$$

Now

$$\alpha_1 y_1 + \beta_1 y_1 = 0 \qquad (5.37)$$

and
$$\alpha_1 y_2 + \beta_1 y_2 = 0 \tag{5.38}$$
at $x = a$. Thus
$$y_1 y_2' - y_2 y_1' = 0 \tag{5.39}$$
at $x = a$. Similarly, (5.39) can be shown to be valid at $x = b$. Hence
$$(\lambda_1 - \lambda_2) \int_a^b y_1 y_2 \, dx = 0. \tag{5.40}$$

Therefore, if $\lambda_1 \neq \lambda_2$, y_1 and y_2 are orthogonal. We note that if $p(a)$ of $p(b)$ is zero, the corresponding boundary condition is irrelevant.

Thus, we have proved the theorem for the case of nondegenerate eigenvalues. In the case of degeneracy, we observe that if y_1 and y_2 are eigenfunctions corresponding to λ,
$$Ly_1 = \lambda y_1 \tag{5.41}$$
and
$$Ly_2 = \lambda y_2, \tag{5.42}$$
then
$$L(\alpha y_1 + \beta y_2) = \lambda(\alpha y_1 + \beta y_2). \tag{5.43}$$

Therefore, any linear combination of y_1 and y_2 is also an eigenfunction of L corresponding to λ. We now show that it is possible to take linear combinations of y_1 and y_2 which are orthogonal. Let the new eigenfunctions be ϕ_1 and ϕ_2. Further, let
$$\phi_1 = y_1 \tag{5.44}$$
and
$$\phi_2 = y_2 - by_1. \tag{5.45}$$
Thus
$$\int_a^b \phi_1 \phi_2 \, dx = \int_a^b y_1 y_2 \, dx - b \int_a^b y_1 y_1 \, dx. \tag{5.46}$$
Hence, if
$$b = \left(\int_a^b y_1 y_2 \, dx \right) \bigg/ \left(\int_a^b y_1 y_1 \, dx \right), \tag{5.47}$$
the new eigenfunctions will be orthogonal. This process of constructing ϕ_1 and ϕ_2 is just the Schmidt orthogonalization process introduced in Chapter I, Section II.C.2.

Thus the eigenfunctions of a self-adjoint operator are orthogonal or, in the case of degeneracy, they can be made orthogonal.

4. Boundary-Value Problems

Suppose that L is not self-adjoint. We have seen that L can be made self-adjoint by multiplying by some function $r(x)$. Thus, instead of (5.30), we have

$$\frac{d}{dx}\left[p(x)\frac{dy}{dx}\right] + q(x)y = \lambda r(x)y. \tag{5.48}$$

Thus, if this is the case

$$\int_a^b r(x)y_1 y_2\, dx = 0 \tag{5.49}$$

and the eigenfunctions are orthogonal with respect to the density function $r(x)$.

Equation (5.48) is called the *Liouville equation*, and the problem of determining the dependence of the general behavior of $y(x)$ on the parameter λ and the dependence of the eigenvalues of λ on the homogeneous boundary conditions imposed on $y(x)$ is called the *Sturm–Liouville problem*.

It is of interest to prove a further:

Theorem The eigenvalues of a self-adjoint operator are real.

Proof To prove this theorem, suppose that there is a complex eigenvalue λ with eigenfunction y. Then λ^* will be an eigenvalue with eigenfunction y^*. Thus

$$Ly = \lambda y \tag{5.50}$$

and

$$Ly^* = \lambda^* y^*. \tag{5.51}$$

Hence

$$y^* L y - y L y^* = (\lambda - \lambda^*) y^* y \tag{5.52}$$

and, using the boundary conditions (5.31) and (5.32),

$$(\lambda - \lambda^*) \int_a^b y^* y\, dx = 0. \tag{5.53}$$

But

$$\int_a^b y^* y\, dx \neq 0. \tag{5.54}$$

Moreover, the integral above is real. Thus

$$\lambda = \lambda^*, \tag{5.55}$$

and hence λ is real.

It is possible to prove additional theorems. We could, if we wished, determine the conditions under which the eigenvalues are positive. However, this takes us a bit far afield.

However, we will state one additional:

Theorem The eigenfunctions of a self-adjoint operator form a complete set.

We will not prove this theorem.

This completes our study of the fundamental characteristics of eigenvalues and eigenfunctions. Thus we see that it was the general properties of eigenfunctions and not the coincidental fact that Fourier series were involved which enabled us to satisfy the initial conditions of the problems we have considered.

D. Another Heat Problem

A slender rod of length l whose curved surface is perfectly insulated has its left end insulated and its right end radiating freely into air of temperature T. If the initial temperature of the rod is T_0 find the temperature of the rod at any point at any subsequent temperature.

Thus we must solve

$$k\, \partial^2 u/\partial x^2 = \partial u/\partial t \tag{5.56}$$

subject to the boundary conditions

$$\partial u(0, t)/\partial x = 0 \tag{5.57}$$

and

$$-\partial u(l, t)/\partial x = \lambda[u(l, t) - T], \tag{5.58}$$

and the initial condition

$$u(x, 0) = T_0. \tag{5.59}$$

Equation (5.58) is just Newton's law of cooling.

First we set aside the initial condition and then we choose $\psi = T$. Next we consider the homogeneous system. We try

$$\varphi = e^{\pm i\alpha x} e^{-\beta t}. \tag{5.60}$$

Thus

$$k\alpha^2 = \beta. \tag{5.61}$$

4. Boundary-Value Problems

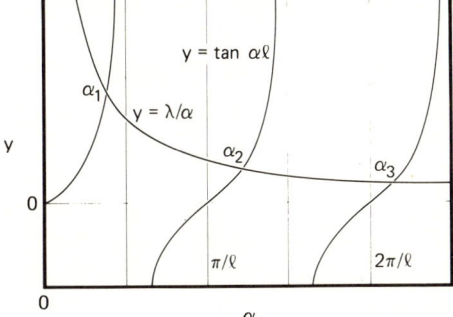

FIG. 1. Graphical solution of the equation $\tan \alpha l = \lambda/\alpha$.

Equation (5.57) requires that only $\cos \alpha x$ remain. Equation (5.58) requires that

$$\alpha \sin \alpha l = \lambda \cos \alpha l \tag{5.62}$$

or

$$\tan \alpha l = \lambda/\alpha. \tag{5.63}$$

Equation (5.63) has only a discrete set of solutions, as may be seen in Fig. 1. Let the solutions be α_n. Thus

$$u(x, t) = T + \sum_{n=0}^{\infty} A_n \exp(-k\alpha_n^2 t) \cos \alpha_n x. \tag{5.64}$$

The initial condition requires that

$$T_0 - T = \sum_{n=0}^{\infty} A_n \cos \alpha_n x. \tag{5.65}$$

If we multiply by $\cos \alpha_m x$, integrate, and make use of the orthogonality of $\cos \alpha_n x$, we obtain

$$A_n = (T_0 - T)\left[\left(\int_0^l \cos \alpha_n x \, dx\right) \middle/ \left(\int_0^l \cos^2 \alpha_n x \, dx\right)\right]. \tag{5.66}$$

Thus

$$A_n = [4(T_0 - T) \sin \alpha_n l]/(2\alpha_n l + \sin 2\alpha_n l). \tag{5.67}$$

VI. Boundary-Value Problems in Cylindrical Coordinates

Thus far we have dealt with boundary-value problems in which Cartesian coordinates were used. In this and the next section boundary-value problems in cylindrical and spherical coordinates are examined.

A. Laplace's Equation in Cylindrical Coordinates

In cylindrical coordinates

$$\nabla^2 u = \frac{\partial^2 u}{\partial r^2} + \frac{1}{r}\frac{\partial u}{\partial r} + \frac{1}{r^2}\frac{\partial^2 u}{\partial \phi^2} + \frac{\partial^2 u}{\partial z^2} = 0. \tag{6.1}$$

Since ϕ and z are cyclic, we attempt a solution of the form

$$u = f(r)e^{\pm \alpha z}e^{\pm in\phi}. \tag{6.2}$$

Thus we must solve the ordinary differential equation

$$r^2 \frac{d^2 f}{dr^2} + r\frac{df}{dr} + (\alpha^2 r^2 - n^2)f = 0. \tag{6.3}$$

We first examine the solutions of (6.3) and then we consider a few boundary-value problems in cylindrical coordinates.

B. Bessel Functions

The equation

$$x^2 \frac{d^2 y}{dx^2} + x\frac{dy}{dx} + (\lambda^2 x^2 - \nu^2)y = 0 \tag{6.4}$$

is called *Bessel's equation of order ν with parameter λ*. If we make the change of variable $t = \lambda x$, we obtain *Bessel's equation of order ν*:

$$t^2 \frac{d^2 y}{dt^2} + t\frac{dy}{dt} + (t^2 - \nu^2)y = 0. \tag{6.5}$$

Let us attempt a solution of (6.5) of the form

$$y(t) = \sum_{n=0}^{\infty} a_n t^{n+c}. \tag{6.6}$$

It will be assumed that the reader is familiar with this method (*the method of Frobenius*) of solving ordinary differential equations. In any case, if (6.6) is substituted into (6.5), it is found that $c = \pm \nu$. If the choice $c = \nu$ and

$$a_0 = [2^\nu \Gamma(\nu + 1)]^{-1} \tag{6.7}$$

is made, then

$$J_\nu(t) = \sum_{n=0}^{\infty} \frac{(-1)^n}{n!\Gamma(\nu + n + 1)} \left(\frac{t}{2}\right)^{\nu + 2n} \tag{6.8}$$

4. Boundary-Value Problems

is a solution of (6.5). This function is called the *Bessel function of the first kind of order v*. The series for $J_v(t)$ converges for all values of t if $v \geq 0$.

The function $\Gamma(x)$ is the *gamma function*:

$$\Gamma(x) = \int_0^\infty e^{-y} y^{x-1}\, dy, \tag{6.9}$$

which is discussed in detail in Chapter 3, Section VIII.A. In particular, we note that

$$\Gamma(x+1) = x\Gamma(x) \tag{6.10}$$

so that $\Gamma(x)$ is a generalization of the factorial $n!$ to nonintegral values. If n is an integer,

$$\Gamma(n+1) = n!. \tag{6.11}$$

We note that $J_{-v}(t)$ is also a solution of (6.5). If v is not an integer, $J_v(t)$ and $J_{-v}(t)$ are independent and can be taken as the solutions of (6.5). Since $J_{-v}(t)$ contains negative powers of t, it will diverge at $t = 0$. However, when $v = n$ is an integer,

$$J_{-n}(t) = (-1)^n J_n(t). \tag{6.12}$$

Hence, we cannot always take $J_{-v}(t)$ as the second solution of (6.5). It is conventional to take

$$Y_v(t) = [\cos v\pi J_v(t) - J_{-v}(t)]/\sin v\pi, \tag{6.13}$$

where v is not an integer, or

$$Y_n(t) = \lim_{v \to n} \{[\cos v\pi J_v(t) - J_{-v}(t)]/\sin v\pi\}, \tag{6.14}$$

where n is an integer, as the second solution to (6.5). Thus the general solution to (6.5) is

$$y(t) = c_1 J_v(t) + c_2 Y_v(t). \tag{6.15}$$

The function $Y_v(t)$ is called the *Bessel function of the second kind of order v*.

Some values of $J_0(t)$, $J_1(t)$, and $J_2(t)$, and of $Y_0(t)$, $Y_1(t)$, and $Y_2(t)$ are shown in Figs. 2 and 3, respectively. These graphs illustrate the following important points. First, these functions oscillate (although not with a constant period) so that there are infinitely many solutions to equations of the type

$$J_v(t) = 0. \tag{6.16}$$

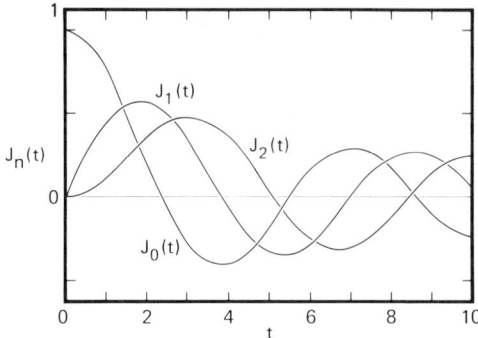

FIG. 2. Values of $J_0(t)$, $J_1(t)$, and $J_2(t)$.

Second, $J_\nu(0) = 0$ except for $\nu = 0$ and $J_0(0) = 1$. Third, $Y_\nu(0)$ diverges for all ν.

In view of the fact that the Bessel functions oscillate, it is not surprising that they have certain qualitative similarities to trigonometric functions. Two results of interest are

$$J_\nu(t) \sim (2/\pi t)^{1/2} \cos[t - (\pi/4) - (\nu\pi/2)] \qquad (6.17)$$

and

$$Y_\nu(t) \sim (2/\pi t)^{1/2} \sin[t - (\pi/4) - (\nu\pi/2)] \qquad (6.18)$$

for large t.

It is sometimes convenient to use another form for the solution of (6.5). If we define the *Hankel functions*,

$$H_\nu^{(1)}(t) = J_\nu(t) + iY_\nu(t) \qquad (6.19)$$

$$H_\nu^{(2)}(t) = J_\nu(t) - iY_\nu(t), \qquad (6.20)$$

then the general solution to (6.5) is

$$y(t) = c_1 H_\nu^{(1)}(t) + c_2 H_\nu^{(2)}(t). \qquad (6.21)$$

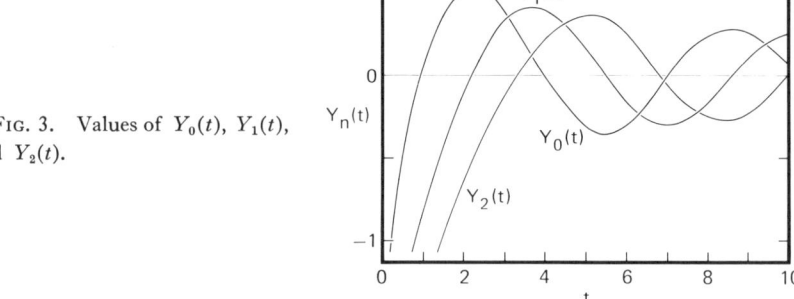

FIG. 3. Values of $Y_0(t)$, $Y_1(t)$, and $Y_2(t)$.

4. Boundary-Value Problems

In case these alternative forms for the solution of (6.5) seem a bit confusing, let us recall that the solution of the equation

$$y'' + y = 0 \tag{6.22}$$

can be written in terms of trigonometric functions or in terms of

$$e^{ix} = \cos x + i \sin x \tag{6.23}$$

$$e^{-ix} = \cos x - i \sin x. \tag{6.24}$$

Equations (6.19) and (6.20) are nothing more than analogies of (6.23) and (6.24).

C. Modified Bessel Functions

The equation

$$t^2 \frac{dy^2}{dt^2} + t \frac{dy}{dt} - (t^2 - v^2)y = 0 \tag{6.25}$$

is called the *modified Bessel equation of order* v. This is no more than the Bessel equation of order v with parameter $i = \sqrt{-1}$. Thus, we could write

$$y(x) = c_1 J_v(it) + c_2 Y_v(it). \tag{6.26}$$

However, this is an awkward form in which to write the solution. We note that

$$J_v(it) = i^v \sum_{n=0}^{\infty} [n!\Gamma(v + n + 1)]^{-1}(t/2)^{v+2n}. \tag{6.27}$$

Thus

$$I_v(t) = i^{-v} J_v(it) = \sum_{n=0}^{\infty} [n!\Gamma(v + n + 1)]^{-1}(t/2)^{v+2n} \tag{6.28}$$

is a solution of (6.25). This function is called the *modified Bessel function of the first kind of order* v. If v is not an integer, we could take $I_{-v}(t)$ as the second solution. However, it is conventional to use

$$K_v(t) = (\pi/2)\{[I_{-v}(t) - I_v(t)]/\sin v\pi\}, \tag{6.29}$$

where v is not an integer, or

$$K_n(t) = \lim_{v \to n}(\pi/2)\{[I_{-v}(t) - I_v(t)]/\sin v\pi\}, \tag{6.30}$$

where n is an integer, as the second solution to (6.25). Thus the general solution to (6.25) is

$$y(t) = c_1 I_\nu(t) + c_2 K_\nu(t). \tag{6.31}$$

The function $K_\nu(t)$ is called the *modified Bessel function of the second kind of order* ν.

Some values of $I_0(t)$, $I_1(t)$, and $I_2(t)$ and of $K_0(t)$, $K_1(t)$, and $K_2(t)$ are shown in Figs. 4 and 5, respectively. These graphs illustrate the following important points. First, $I_\nu(t)$ diverges as $t \to \infty$. Second, $I_\nu(0) = 0$ except for $\nu = 0$ and $I_0(0) = 1$. Third, $K_\nu(0)$ diverges for all ν.

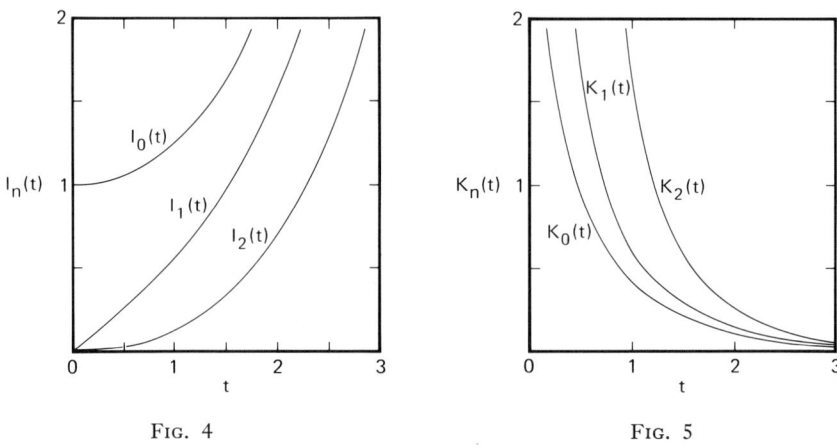

FIG. 4.

FIG. 5.

FIG. 4. Values of $I_0(t)$, $I_1(t)$, and $I_2(t)$.
FIG. 5. Values of $K_0(t)$, $K_1(t)$, and $K_2(t)$.

Just as the Bessel functions have certain qualitative similarities to trigonometric functions, the modified Bessel functions have certain qualitative similarities to exponential functions. For large t, the analogs of (6.17) and (6.18) are

$$I_\nu(t) \sim (2\pi t)^{-1/2} e^t \tag{6.32}$$

and

$$K_\nu(t) \sim (\pi/2t)^{1/2} e^{-t}. \tag{6.33}$$

D. SOME RELATIONS FOR BESSEL FUNCTIONS

Two useful relations are

$$(d/dt)[t^\nu J_\nu(t)] = t^\nu J_{\nu-1}(t) \tag{6.34}$$

4. Boundary-Value Problems

and
$$(d/dt)[t^{-\nu}J_\nu(t)] = -t^{-\nu}J_{\nu+1}(t). \tag{6.35}$$

These relations can easily be proved from the series expansion for $J_\nu(t)$. Performing the differentiations in (6.34) and (6.35) yields

$$dJ_\nu(t)/dt = J_{\nu-1}(t) - (\nu/t)J_\nu(t) \tag{6.36}$$

and

$$dJ_\nu(t)/dt = (\nu/t)J_\nu(t) - J_{\nu+1}(t). \tag{6.37}$$

Adding (6.36) and (6.37) yields

$$dJ_\nu(t)/dt = \tfrac{1}{2}[J_{\nu-1}(t) - J_{\nu+1}(t)]. \tag{6.38}$$

Subtracting (6.37) from (6.36) gives the recurrence relation

$$J_{\nu+1}(t) = (2\nu/t)J_\nu(t) - J_{\nu-1}(t). \tag{6.39}$$

If we integrate (6.34) and (6.35), we obtain

$$\int t^\nu J_{\nu-1}(t)\, dt = t^\nu J_\nu(t) + c \tag{6.40}$$

and

$$\int t^{-\nu} J_{\nu+1}(t)\, dt = -t^{-\nu} J_\nu(t) + c. \tag{6.41}$$

In particular,

$$\int t J_0(t)\, dt = t J_1(t) + c \tag{6.42}$$

and

$$\int J_1(t)\, dt = -J_0(t) + c. \tag{6.43}$$

In general, an integral of the form

$$\int t^m J_n(t)\, dt,$$

where m and n are integers, can be integrated completely if $m + n$ is odd by the use of (6.40) and (6.41) and integration by parts. However, if $m + n$ is even, the value of the integral will ultimately depend on the integral

$$\int J_0(t)\, dt,$$

for which there is no simple result. However, tabulations of this integral do exist.

Each of the formulas given above for $J_\nu(t)$ are applicable to $Y_\nu(t)$ and the Hankel functions. Similar results hold for $I_\nu(t)$ and $K_\nu(t)$. For example,

$$(d/dt)[t^\nu I_\nu(t)] = t^\nu I_{\nu-1}(t) \tag{6.44}$$

$$(d/dt)[t^{-\nu} I_\nu(t)] = t^{-\nu} I_{\nu+1}(t) \tag{6.45}$$

$$(d/dt)[t^\nu K_\nu(t)] = -t^\nu K_{\nu-1}(t) \tag{6.46}$$

and

$$(d/dt)[t^{-\nu} K_\nu(t)] = -t^{-\nu} K_{\nu+1}(t). \tag{6.47}$$

E. ORTHOGONALITY OF THE BESSEL FUNCTIONS

If we write Bessel's equation of order ν in the form

$$xy_\nu'' + y_\nu' + \left(\lambda^2 x - \frac{\nu^2}{x}\right)y_\nu = \frac{d}{dx}(xy_\nu') + \left(-\frac{\nu^2}{x} + \lambda^2 x\right)y_\nu = 0, \tag{6.48}$$

we see on comparison with (5.48) that

$$p(x) = x \tag{6.49}$$

$$q(x) = -\nu^2/x \tag{6.50}$$

and

$$r(x) = x. \tag{6.51}$$

Thus, if the boundary conditions are of the form

$$A_i y_\nu(\lambda x_i) - B_i \frac{dy_\nu(\lambda x_i)}{dx} = 0, \tag{6.52}$$

then, as we have seen in Section V.C,

$$\int_{x_1}^{x_2} xy_\nu(\lambda_i x) y_\nu(\lambda_j x)\, dx = 0 \tag{6.53}$$

if $\lambda_i \neq \lambda_j$. Thus *Bessel functions are orthogonal with density x.*

To solve boundary-value problems it is not enough to know that the Bessel functions are orthogonal. We must also know the value of

$$I = \int_{x_1}^{x_2} x\{y_\nu(\lambda x)\}^2. \tag{6.54}$$

4. Boundary-Value Problems

We will now prove the following:

Theorem The contribution to this integral is

$$\pm \tfrac{1}{2} x_i{}^2 \{y_{\nu+1}(\lambda x_i)\}^2,$$

if $B_i = 0$, and

$$\pm (1/2\lambda^2)\{y_\nu(\lambda x_i)\}^2 [\lambda^2 x_i{}^2 - \nu^2 + (x_i A_i/B_i)^2],$$

if $B_i \neq 0$.

The sign in the two expressions above is positive for x_2 and negative for x_1.

Proof We can obtain this result by observing that

$$t^2 y_\nu'' + t y_\nu' + (t^2 - \nu^2) y_\nu = 0. \tag{6.55}$$

Thus, multiplying by y_ν' and integrating gives

$$\int t^2 y_\nu' y_\nu'' \, dt + \int t(y_\nu')^2 \, dt + \int t^2 y_\nu y_\nu' \, dt - \nu^2 \int y_\nu y_\nu' \, dt = 0. \tag{6.56}$$

But integrating by parts yields

$$\int t^2 y_\nu' y_\nu'' \, dt = \tfrac{1}{2} t^2 (y_\nu')^2 - \int t(y_\nu')^2 \, dt \tag{6.57}$$

and

$$\int t^2 y_\nu y_\nu' \, dt = \tfrac{1}{2} t^2 (y_\nu)^2 - \int t(y_\nu)^2 \, dt. \tag{6.58}$$

Hence,

$$\int t\{y_\nu(t)\}^2 \, dt = \tfrac{1}{2}(t^2 - \nu^2)\{y_\nu(t)\}^2 + \tfrac{1}{2} t^2 \{y_\nu'(t)\}^2 \tag{6.59}$$

or

$$\int x\{y_\nu(\lambda x)\} \, dx = (1/2\lambda^2)[(\lambda^2 x^2 - \nu^2)\{y_\nu(\lambda x)\}^2 + x^2 \{dy_\nu(\lambda x)/dx\}^2]. \tag{6.60}$$

Thus, if $B_i = 0$, $y_\nu(\lambda x_i) = 0$ and the contribution to I from x_i is

$$(1/2\lambda^2) x_i{}^2 \{dy_\nu(\lambda x)/dx\}^2_{x=x_i}.$$

We may simplify this by recalling from the preceding section that $J_\nu(\lambda x)$, $Y_\nu(\lambda x)$, and the Hankel functions satisfy

$$x \, dy_\nu/dx = y_\nu(\lambda x) - \lambda x y_{\nu+1}(\lambda x). \tag{6.61}$$

Thus, if $B_i = 0$,

$$x_i \frac{dy_\nu(\lambda x_i)}{dx} = -\lambda x_i y_{\nu+1}(\lambda x_i), \tag{6.62}$$

which establishes the first of our results.

If $B_i \neq 0$, then we may substitute (6.52) in the form

$$dy_\nu(\lambda x_i)/dx = (A_i/B_i)y_\nu(\lambda x_i) \tag{6.63}$$

into (6.60) to obtain the second of our results.

As we have already seen in Section V.C, if $x_1 = 0$, then there is no contribution to I from the lower limit because $p(0) = 0$.

F. Three Examples

We saw in Section VI.A that we were usually led to Bessel functions when we attempted to solve boundary-value problems in cylindrical coordinates. However, we should not presume that Bessel functions arise only in such problems. For example, half-integer Bessel functions arise in problems with spherical symmetry. *Spherical Bessel functions* which contain these half-integer functions are discussed by Silverstone in Chapter 3, Section VIII.

Three problems which illustrate the solution of boundary-value problems in cylindrical coordinates are now considered.

1. Vibration of a circular membrane. We consider a membrane stretched over a fixed circular frame of diameter a in the $z = 0$ plane. Thus we have

$$c^2 \left\{ \frac{\partial^2 z}{\partial r^2} + \frac{1}{r} \frac{\partial z}{\partial r} + \frac{1}{r^2} \frac{\partial^2 z}{\partial \phi^2} \right\} = \frac{\partial^2 z}{\partial t^2} \tag{6.64}$$

with the boundary condition

$$z(a, \phi, t) = 0. \tag{6.65}$$

There are two additional boundary conditions which are implicit in the problem and are often not stated. However, it is perhaps helpful to state them explicitly. The first is that $z(r, \phi, t)$ be finite at $r = 0$ and the second is that $z(r, \phi, t)$ be a periodic function of ϕ.

Let us solve this problem subject to the initial conditions

$$z(r, \phi, 0) = f(r, \phi) \tag{6.66}$$

4. Boundary-Value Problems

and
$$\partial z(r, \phi, 0)/\partial t = 0. \tag{6.67}$$

We first set aside the initial conditions and use $\psi = 0$. Let us try

$$\varphi = e^{\pm i\alpha t} e^{\pm i\beta \phi} R(r). \tag{6.68}$$

Since φ must be periodic in ϕ, we have $\beta = n$, where $n = 0, 1, \ldots$. Substituting (6.68) into (6.64) gives

$$r^2 R'' + rR' + [(\alpha^2 r^2/c^2) - n^2]R = 0 \tag{6.69}$$

so that

$$R = C_n J_n(\alpha r/c) + D_n Y_n(\alpha r/c). \tag{6.70}$$

However, finiteness at the origin requires that $D_n = 0$. Equation (6.65) requires that

$$J_n(\alpha a/c) = 0. \tag{6.71}$$

Thus α is also restricted to a discrete set of values $\alpha_{n,k}$, $k = 1, 2, \ldots$. Hence

$$z(r, \phi, t) = \sum_{n=0}^{\infty} \sum_{k=1}^{\infty} J_n(\alpha_{nk} r/c)\{A_{nk} \cos n\phi + B_{nk} \sin n\phi\}$$
$$\times \{C_{nk} \sin \alpha_{nk} t + D_{nk} \cos \alpha_{nk} t\}. \tag{6.72}$$

We now apply the initial conditions. Equation (6.67) requires that $C_{nk} = 0$. We may absorb D_{nk} into A_{nk} and B_{nk}. Thus

$$z(r, \phi, t) = \sum_{n=0}^{\infty} \sum_{k=1}^{\infty} J_n(\alpha_{nk} r/c)\{A_{nk} \cos n\phi + B_{nk} \sin n\phi\} \cos \alpha_{nk} t. \tag{6.73}$$

Equation (6.66) requires that

$$f(r, \phi) = \sum_{n=0}^{\infty} \sum_{k=1}^{\infty} J_n(\alpha_{nk} r/c)\{A_{nk} \cos n\phi + B_{nk} \sin n\phi\}. \tag{6.74}$$

We multiply by $rJ_m(\alpha_{nk} r/c) \cos m\phi$ and $rJ_m(\alpha_{nk} r/c) \sin m\phi$ and integrate. The results are

$$\int_0^a r J_0(\alpha_{0k} r/c) \, dr \int_0^{2\pi} f(r, \phi) \, d\phi = A_{0k} 2\pi \int_0^a r J_0^2(\alpha_{0k} r/c) \, dr, \tag{6.75}$$

$$\int_0^a r J_n(\alpha_{nk} r/c) \, dr \int_0^{2\pi} f(r, \phi) \cos n\phi \, d\phi = A_{nk} \pi \int_0^a r J_n^2(\alpha_{nk} r/c) \, dr, \tag{6.76}$$

and

$$\int_0^a rJ_n(\alpha_{nk}r/c)\,dr \int_0^{2\pi} f(r,\phi)\sin n\phi\,d\phi = B_{nk}\pi \int_0^a rJ_n^2(\alpha_{nk}r/c)\,dr. \quad (6.77)$$

From the theorem which we proved in the preceding section, we have

$$\int_0^a rJ_n^2(\alpha_{nk}r/c)\,dr = (\pi/2)a^2 J_{n+1}^2(\alpha_{nk}a/c). \quad (6.78)$$

This completes the solution of the problem. However, it is worth noting that if our initial condition had been

$$z(r,\phi,0) = f(r) \quad (6.79)$$

instead of (6.66), $B_{nk} = 0$ and $A_{nk} = 0$ except for $n = 0$. Thus

$$z(r,t) = \sum_{k=1}^{\infty} A_k J_0(\alpha_k r/c)\cos\alpha_k t. \quad (6.80)$$

This is, of course, fairly obvious because if (6.79) replaces (6.66), then there is no angle dependence in the problem and β, and thus n, would never appear had we made use of the lack of angle dependence from the beginning. However, this illustrates a general principle that if there is no angle dependence, only zero-order Bessel functions enter the problem.

It is also worth pointing out that had we been dealing with a ring-shaped membrane bounded by the circles of diameter $a < b$, then we could not have excluded the Y_n from the problem.

Finally, we note that had we been dealing with a pie-shaped membrane, we would not necessarily have been able to restrict ourselves to integral-order Bessel functions.

2. Consider a right-circular cylinder of radius b and height h. Let us find the steady-state temperature distribution if the upper and lower faces of the cylinder are maintained at $0°$ and the curved surface is maintained at the temperature $u(b,z) = f(z)$.

Thus we must solve

$$\frac{\partial^2 u}{\partial r^2} + \frac{1}{r}\frac{\partial u}{\partial r} + \frac{\partial^2 u}{\partial z^2} = 0 \quad (6.81)$$

subject to

$$u(r,0) = u(r,h) = 0 \quad (6.82)$$

and
$$u(b, z) = f(z). \tag{6.83}$$

Again we have the implicit condition that $u(0, z)$ be finite.

First we set aside (6.83) and take $\psi = 0$. Next we attempt
$$\varphi = e^{\pm i\alpha z} R(r). \tag{6.84}$$

We expect the z dependence to be trigonometric rather than exponential because of (6.82). Substitution of (6.84) into (6.81) gives
$$r^2 R'' + rR' - \alpha^2 r^2 R = 0. \tag{6.85}$$
Thus
$$R = I_0(\alpha r). \tag{6.86}$$

The boundary conditions (6.82) require that only $\sin \alpha z$ remain and that
$$\alpha = n\pi/h, \quad n = 1, 2, \ldots. \tag{6.87}$$
Thus
$$u(r, z) = \sum_{n=1}^{\infty} A_n I_0(n\pi r/h) \sin(n\pi z/h). \tag{6.88}$$

We must now fit (6.83). This condition requires that
$$f(z) = \sum_{n=1}^{\infty} A_n I_0(n\pi b/h) \sin(n\pi z/h). \tag{6.89}$$

Thus, if we multiply by $\sin(m\pi/h)z$ and integrate we obtain
$$A_n = \frac{2}{h} \left(\int_0^h f(z) \sin(n\pi z/h) \, dz \right) \bigg/ \{I_0(n\pi b/h)\}. \tag{6.90}$$

This completes the solution.

If we has not anticipated some of the development of the solution and used $e^{\pm \alpha z}$ instead of $e^{\pm i\alpha z}$ in (6.84), we would have been led to the conclusion that α was imaginary and still would have arrived at (6.88).

3. Lest we think that all problems involving cylindrical coordinates lead to Bessel functions, consider a semicircular plate of radius a. Let us find the steady-state temperature when the curved boundary is maintained at T and the diameter is maintained at $0°$.

Thus we must solve
$$\frac{\partial^2 u}{\partial r^2} + \frac{1}{r} \frac{\partial u}{\partial r} + \frac{1}{r^2} \frac{\partial^2 u}{\partial \phi^2} = 0 \tag{6.91}$$

subject to
$$u(a, \phi) = T \tag{6.92}$$
and
$$u(r, 0) = u(r, \pi) = 0. \tag{6.93}$$

Again we have the important condition that $u(0, \phi)$ be finite.

First we set aside (6.92) and let $\psi = 0$. Next we attempt
$$u = e^{\pm i\alpha\phi} R(r). \tag{6.94}$$
Thus
$$r^2 R'' + rR' - \alpha^2 R = 0. \tag{6.95}$$

Equation (6.95) is Euler's equation and is satisfied by r^β. Thus, on substituting r^β into (6.95), we obtain
$$\beta(\beta - 1) + \beta - \alpha^2 = 0 \tag{6.96}$$

so that $\beta = \pm \alpha$. Equation (6.93) requires that only $\sin \alpha\phi$ remain and that $\alpha = n$, for $n = 1, 2, \ldots$. Finiteness at the origin requires that r^{-n} be rejected. Thus
$$u(r, \phi) = \sum_{n=1}^{\infty} A_n r^n \sin n\phi. \tag{6.97}$$

If we choose the A_n to satisfy
$$T = \sum_{n=1}^{\infty} A_n a^n \sin n\phi \tag{6.98}$$

in the usual way, we obtain
$$u(r, \phi) = \frac{4T}{\pi} \sum_{n=1}^{\infty} \frac{1 - (-1)^n}{n} \left(\frac{r}{a}\right)^n \sin n\phi. \tag{6.99}$$

VII. Boundary-Value Problems in Spherical Coordinates

A. Laplace's Equation in Spherical Coordinates

In spherical coordinates,
$$\nabla^2 u = \frac{1}{r^2} \frac{\partial}{\partial r}\left(r^2 \frac{\partial u}{\partial r}\right) + \frac{1}{r^2 \sin \theta} \frac{\partial}{\partial \theta}\left(\sin \theta \frac{\partial u}{\partial \theta}\right)$$
$$+ \frac{1}{r^2 \sin^2 \theta} \frac{\partial^2 u}{\partial \phi^2} = 0. \tag{7.1}$$

4. Boundary-Value Problems

Since ϕ is cyclic, we attempt a solution of the form

$$u = R(r)\Theta(\theta)e^{\pm im\phi}. \qquad (7.2)$$

Thus we must solve

$$\frac{1}{R}\frac{d}{dr}\left(r^2\frac{dR}{dr}\right) = \lambda \qquad (7.3)$$

and

$$\frac{1}{\Theta \sin\theta}\frac{d}{d\theta}\left(\sin\theta\frac{d\Theta}{d\theta}\right) - \frac{m^2}{\sin^2\theta} = -\lambda. \qquad (7.4)$$

It is convenient to write $\lambda = n(n+1)$. Indeed, we see in the next section that for Θ to be finite at $\theta = \pi$, n must be an integer. In any case (7.3) becomes

$$r^2 R'' + 2rR' - n(n+1)R = 0. \qquad (7.5)$$

This is Euler's equation and is satisfied by r^β. Substituting this result into (7.5) gives

$$\beta = \pm n. \qquad (7.6)$$

Thus

$$R = c_1 r^n + (c_2/r^n). \qquad (7.7)$$

With $\lambda = n(n+1)$, (7.4) becomes

$$\sin^2\theta\,\Theta'' + \sin\theta\cos\theta\,\Theta' + \{n(n+1)\sin^2\theta - m^2\}\Theta = 0. \qquad (7.8)$$

It is convenient to use $x = \cos\theta$ as the independent variable. With this change of variables, (7.8) becomes

$$(1-x^2)\frac{d^2\Theta}{dx^2} - 2x\frac{d\Theta}{dx} + \left\{n(n+1) - \frac{m^2}{1-x^2}\right\}\Theta = 0. \qquad (7.9)$$

Equation (7.9) is called the *associated Legendre equation*.

If the original problem is independent of ϕ, then $m = 0$ and (7.9) becomes

$$(1-x^2)\frac{d^2\Theta}{dx^2} - 2x\frac{d\Theta}{dx} + n(n+1)\Theta = 0. \qquad (7.10)$$

Equation (7.10) is called the *Legendre equation*.

We will now solve (7.9) and (7.10). The solutions are the associated Legendre functions and the Legendre polynomials which are discussed by Silverstone in Chapter 3, Section VIII.

B. LEGENDRE POLYNOMIALS

Consider the Legendre equation

$$(1 - x^2)P'' - 2xP' + \lambda P = 0. \tag{7.11}$$

This equation has singularities at $x = \pm 1$. Let us seek a series solution in powers of $(x - 1)$. That is, we will expand about the singular point $x = 1$. If we change variables to $t = x - 1$, (7.11) becomes

$$(t^2 - 2t)P'' + 2(t + 1)P' - \lambda P = 0. \tag{7.12}$$

Let us attempt a series solution of (7.12). Thus

$$P = \sum_{k=0}^{\infty} a_k t^{c+k}. \tag{7.13}$$

If we substitute (7.13) into (7.12), we find that $c = 0$ and

$$a_{k+1} = \frac{\lambda - k(k + 1)}{2(k + 1)^2} a_k. \tag{7.14}$$

Hence, $a_{k+1}/a_k \to -\tfrac{1}{2}$ as $k \to \infty$ so that the series converges for $|x - 1| < 2$. However, for $x = -1$ the series will not converge unless it is terminated. Thus, the only possible values of λ which will give a solution that converges for $-1 \leq x \leq 1$ are

$$\lambda = n(n + 1) \quad \text{where} \quad n = 0, 1, \ldots. \tag{7.15}$$

The corresponding solutions, $P_n(x)$, are called *Legendre polynomials*. The constant a_0 is chosen to be unity. Hence

$$P_n(1) = 1. \tag{7.16}$$

From (7.14) and (7.15) we have

$$P_0(x) = 1, \tag{7.17}$$

$$P_1(x) = 1 + (x - 1) = x, \tag{7.18}$$

$$P_2(x) = 1 + 3(x - 1) + \tfrac{3}{2}(x - 1)^2 = \tfrac{1}{2}(3x^2 - 1), \tag{7.19}$$

and so on. The Legendre polynomials $P_n(x)$ are polynomials of degree n. The other solution of (7.11), with $\lambda = n(n + 1)$, is denoted $Q_n(x)$. This function is not of interest here as it diverges when $x = 1$. However, there are applications where $Q_n(x)$ is of interest.

By straightforward differentiation it is easy to show that

$$P_n(x) = \frac{1}{2^n n!} \frac{d^n(x^2-1)^n}{dx^n}. \tag{7.20}$$

Equation (7.20) is called *Rodrigues's formula*. From (7.20) we note that

$$P_n(-x) = (-1)^n P_n(x). \tag{7.21}$$

By a straightforward application of (7.20), it is easy to obtain the recurrence relation

$$P'_{n+1}(x) - P'_{n-1}(x) = (2n+1)P_n(x). \tag{7.22}$$

Three other recurrence relations which can be obtained easily are

$$P'_{n+1}(x) - xP_n'(x) = (n+1)P_n(x) \tag{7.23}$$

$$xP_n'(x) - P'_{n-1}(x) = nP_n(x) \tag{7.24}$$

and

$$(x^2-1)P_n'(x) = nxP_n(x) - nP_{n-1}(x). \tag{7.25}$$

One further property of the Legendre polynomials which is of interest is

$$(1 - 2xy + y^2)^{-1/2} = \sum_{n=0}^{\infty} P_n(x) y^n. \tag{7.26}$$

The proof of (7.26) is quite straightforward and is not given here. If we differentiate (7.26) and equate like powers of y, we obtain the recurrence relation

$$(n+1)P_{n+1}(x) = (2n+1)xP_n(x) - nP_{n-1}(x). \tag{7.27}$$

Quite obviously (7.22)–(7.25) and (7.27) are not all independent.

C. Associated Legendre Functions

We define the *associated Legendre function* $P_n{}^m(x)$ by

$$P_n{}^m(x) = (1-x^2)^{m/2} \frac{d^m P_n(x)}{dx^m} = \frac{(1-x^2)^{m/2}}{2^n n!} \frac{d^{n+m}(x^2-1)}{dx^{n+m}} \tag{7.28}$$

where $0 \le m \le n$. The latter part of (7.28) gives a meaningful result for $P_n{}^m(x)$ even if m is negative. Indeed,

$$P_n^{-m}(x) = (-1)^m \frac{(n-m)!}{(n+m)!} P_n{}^m(x). \tag{7.29}$$

It is a straightforward matter to show that $P_n{}^m(x)$ satisfies the associated Legendre equation. Thus the $P_n{}^m(x)$ arise in problems that do not have axial symmetry. It should be noted that many authors use an alternative definition of the $P_n{}^m(x)$ which is obtained by multiplying the right-hand side of (7.28) by $(-1)^m$.

The other solution of (7.9) is denoted by $Q_n{}^m(x)$. This function diverges when $x = 1$ and is not considered here. However, applications which require the $Q_n{}^m(x)$ are not unknown.

A few associated Legendre functions are

$$P_1^1(x) = (1 - x^2)^{1/2} \qquad (7.30)$$

$$P_2^1(x) = 3(1 - x^2)^{1/2} \qquad (7.31)$$

and

$$P_2^2(x) = 3(1 - x^2). \qquad (7.32)$$

Recurrence relations similar to those which we displayed in the preceding section can be obtained quite straightforwardly. A few such relations are

$$(n - m + 1)P_{n+1}^m(x) = (2n + 1)xP_n{}^m(x) - (n + m)P_{n-1}^m(x) \qquad (7.33)$$

$$(1 - x^2)^{1/2} P_n^{m+1}(x) = 2mx P_n{}^m(x) - [n(n+1) - m(m-1)] \\ \times (1 - x^2)^{1/2} P_n^{m-1}(x) \qquad (7.34)$$

$$P_{n+1}^{m+1}(x) - P_{n-1}^{m+1}(x) = (2n + 1)(1 - x^2)^{1/2} P_n{}^m(x) \qquad (7.35)$$

and

$$(2n + 1)(1 - x^2)^{1/2} P_n^{m+1}(x) = (n + m)(n + m + 1)P_{n+1}^m(x) \\ - (n - m)(n - m + 1)P_{n+1}^m(x). \qquad (7.36)$$

Also

$$(1 - x^2)\frac{dP_n{}^m(x)}{dx} = (1 - x^2)P_n^{m+1}(x) - mxP_n{}^m(x). \qquad (7.37)$$

Substitution of the recurrence relations (7.33)–(7.36) into (7.37) gives

$$(1 - x^2)\frac{dP_n{}^m(x)}{dx}$$
$$= mxP_n{}^m(x) - [n(n+1) - m(m-1)](1 - x^2)^{1/2} P_n^{m-1}(x)$$
$$= (n + m)P_{n-1}^m(x) - nxP_n^m(x)$$
$$= (n + 1)xP_n{}^m(x) - (n - m + 1)P_{n+1}^m(x). \qquad (7.38)$$

D. Orthogonality of the Legendre Functions

If we write the associated Legendre equation in self-adjoint form, we have

$$\frac{d}{dx}\left[(1-x^2)\frac{dP_n{}^m}{dx}\right] + \left[n(n+1) - \frac{m^2}{1-x^2}\right]P_n{}^m = 0. \quad (7.39)$$

Similarly, we have

$$\frac{d}{dx}\left[(1-x^2)\frac{dP_{n'}^{m'}}{dx}\right] + \left[n'(n'+1) - \frac{m'^2}{1-x^2}\right]P_{n'}^{m'} = 0. \quad (7.40)$$

If $m = m'$, then multiplying (7.39) by $P_{n'}^m$ and (7.40) by $P_n{}^m$ and integrating yield

$$\left\{(1-x^2)\left[P_{n'}^m \frac{dP_n{}^m}{dx} - P_n{}^m \frac{dP_{n'}^m}{dx}\right]\right\}_{-1}^1$$

$$= [n'(n'-1) - n(n-1)] \int_{-1}^1 P_n{}^m(x) P_{n'}^m(x)\, dx. \quad (7.41)$$

The left-hand side of (7.41) vanishes because $1 - x^2$ is zero at $x = \pm 1$. Thus

$$\int_{-1}^1 P_n{}^m(x) P_{n'}^m(x)\, dx = 0, \quad (7.42)$$

unless $n = n'$. Similarly, if $n = n'$,

$$\int_{-1}^1 P_n{}^m(x) P_n{}^{m'}(x)[dx/(1-x^2)] = 0, \quad (7.43)$$

unless $m = m'$.

Using Rodriques's formula and repeated integration by parts, it is easy to show that

$$\int_{-1}^1 [P_n(x)]^2\, dx = 2/(2n+1). \quad (7.44)$$

Also dividing (7.37) by $(1-x^2)^{1/2}$, squaring and integrating the resulting identity, integrating by parts, and using (7.9), it is straightforward to show that

$$\int_{-1}^1 [P_n^{m+1}(x)]^2\, dx = (n-m)(n+m+1) \int_{-1}^1 [P_n{}^m(x)]^2\, dx. \quad (7.45)$$

Thus
$$\int_{-1}^{1} [P_n{}^m(x)]^2 \, dx = \frac{(n+m)!}{(n-m)!} \frac{2}{2n+1}. \tag{7.46}$$

It is often convenient to combine Θ with $(\Phi(\phi) = \exp\{im\phi\}$ and the appropriate normalization factors to obtain the *spherical harmonics*

$$Y_{nm}(\theta, \phi) = \left[\frac{2n+1}{4\pi} \frac{(n-m)!}{(n+m)!} \right]^{1/2} P_n{}^m(\cos\theta) e^{im\phi}. \tag{7.47}$$

We note that
$$Y_{n,-m}(\theta, \phi) = (-1)^m Y^*_{nm}(\theta, \phi), \tag{7.48}$$

where Y^*_{nm} is the complex conjugate of Y_{nm}. In terms of the spherical harmonics, the orthonormality condition becomes

$$\int Y^*_{n'm'}(\theta, \phi) Y_{nm}(\theta, \phi) \, d\Omega = \delta_{nn'} \delta_{mm'}, \tag{7.49}$$

where
$$d\Omega = \sin\theta \, d\theta \, d\phi. \tag{7.50}$$

Many authors define the spherical harmonics by multiplying the right-hand side of (7.47) by $(-1)^m$. For example, this is the definition used by Silverstone in Chapter 3.

E. The Addition Theorem for Spherical Harmonics

A result of considerable interest, called the addition theorem for spherical harmonics, can be obtained if we consider two vectors $\mathbf{r}_1$ and $\mathbf{r}_2$ with spherical coordinates (r_1, θ_1, ϕ_1) and (r_2, θ_2, ϕ_2) as shown in Fig. 6.

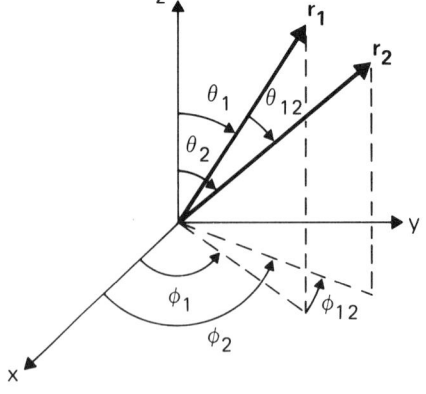

Fig. 6. Geometry relevant to addition theorem.

4. Boundary-Value Problems

The angle θ_{12} satisfies the trigonometric identity

$$\cos\theta_{12} = \cos\theta_1\cos\theta_2 + \sin\theta_1\sin\theta_2\cos(\phi_1 - \phi_2). \tag{7.51}$$

The addition theorem asserts that

$$P_n(\cos\theta_{12}) = 4\pi/(2n+1)\sum_{m=-n}^{n} Y_{nm}^*(\theta_2, \phi_2)Y_{nm}(\theta_1, \phi_1). \tag{7.52}$$

In terms of the associated Legendre functions the addition theorem is

$$P_n(\cos\theta_{12}) = P_n(\cos\theta_1)P_n(\cos\theta_2)$$
$$+ 2\sum_{m=1}^{n}\frac{(n-m)!}{(n+m)!}P_n^m(\cos\theta_1)P_n^m(\cos\theta_2)\cos m(\phi_1 - \phi_2). \tag{7.53}$$

To prove the addition theorem, we first consider the expansion of $f(\theta, \phi)$. We have

$$f(\theta_1, \phi_1) = Y_{nm}(\theta_1, \phi_1) \tag{7.54}$$

relative to (x, y, z) and

$$f(\theta_1, \phi_1) = \sum_{m=-n}^{n}\alpha_{nm}Y_{nm}(\theta_{12}, \phi_{12}) \tag{7.55}$$

relative to $\mathbf{r}_2$. Since $1 - x^2 = \sin^2\theta = 0$ when $\theta = 0$, $Y_{nm}(0, \phi) = 0$ when $\theta = 0$ unless $m = 0$. Thus, since $P_n(1) = 1$, we have at $\theta_{12} = 0$

$$f(\theta_1, \phi_1)|_{\theta_{12}=0} = \alpha_{n,0}[(2n+1)/4\pi]^{1/2}. \tag{7.56}$$

In addition, from (7.55)

$$\alpha_{nm} = \int f(\theta_1, \phi_1)Y_{nm}^*(\theta_{12}, \phi_{12})\,d\Omega. \tag{7.57}$$

Using (7.54),

$$\alpha_{nm} = \int Y_{nm}(\theta_1, \phi_1)Y_{nm}^*(\theta_{12}, \theta_{12})\,d\Omega. \tag{7.58}$$

Now let us expand $P_n(\cos\theta_{12})$ as follows:

$$P_n(\cos\theta_{12}) = \sum_{m=-n}^{n} a_{nm}Y_{nm}(\theta_1, \phi_1), \tag{7.59}$$

where

$$a_{nm} = \int P_n(\cos\theta_{12})Y_{nm}^*(\theta_1, \phi_1)\,d\Omega$$
$$= [4\pi/(2n+1)]^{1/2}\int Y_{n0}(\theta_{12}, \phi_{12})Y_{nm}^*(\theta_1, \phi_1)\,d\Omega. \tag{7.60}$$

Comparison of (7.58) and (7.60) gives

$$a_{nm}^* = [4\pi/(2n+1)]^{1/2}\alpha_{n0} = [4\pi/(2n+1)]f(\theta_1, \phi_1)|_{\theta_{12}=0}. \quad (7.61)$$

Now $\theta_1 \to \theta_2$ and $\phi_1 \to \phi_2$ when $\theta_{12} \to 0$. Thus

$$a_{nm} = [4\pi/(2n+1)]Y_{nm}^*(\theta_2, \phi_2). \quad (7.62)$$

Substitution of (7.62) into (7.59) gives (7.52). An alternative derivation of the addition theorem is given in Chapter 3.

The addition theorem can be put to use in obtaining the electrostatic potential at $\mathbf{r}_1$ due to a charge q at $\mathbf{r}_2$ in its most general form. We have

$$\phi = q/r_{12}, \quad (7.63)$$

where

$$r_{12} = (r_1^2 + r_2^2 - 2r_1 r_2 \cos\theta_{12})^{1/2}. \quad (7.64)$$

Let $x = \cos\theta_{12}$ and $y = r_</r_>$ where $r_<$ is the smaller of r_1 and r_2 and and $r_>$ is the larger of r_1 and r_2. Thus

$$r_{12} = r_>(1 - 2xy + y^2)^{1/2}. \quad (7.65)$$

Using (7.26) we obtain

$$\phi = q \sum_{n=0}^{\infty} (r_<^n/r_>^{n+1}) P_n(\cos\theta_{12}). \quad (7.66)$$

Substitution of (7.52) into (7.66) yields

$$\phi = 4\pi q \sum_{n=0}^{\infty} \sum_{m=-n}^{n} (2n+1)^{-1}(r_<^n/r_>^{n+1}) Y_{nm}^*(\theta_2, \phi_2) Y_{nm}(\theta_1, \phi_1). \quad (7.67)$$

F. THREE EXAMPLES

1. Let us calculate the electric potential inside and outside a sphere of radius a formed by two hemispheres at potential V and 0. Thus we must solve

$$\nabla^2 \phi = 0 \quad (7.68)$$

subject to the boundary conditions

$$\phi(a, \theta) = \begin{cases} V, & 0 \le \theta \le \pi/2 \\ 0, & \pi/2 \le \theta \le \pi, \end{cases} \quad (7.69)$$

and with the further condition that ϕ be finite at $r=0$ and $r=\infty$.

4. Boundary-Value Problems

We set aside (7.69) and take $\psi = 0$. Separating the variables and rejecting divergent terms, we obtain

$$\phi_1(r, \theta) = \sum_{n=0}^{\infty} A_n r^n P_n(\cos \theta) \tag{7.70}$$

for $r \leq a$, and

$$\phi_2(r, \theta) = \sum_{n=0}^{\infty} (B_n/r^{n+1}) P_n(\cos \theta) \tag{7.71}$$

for $r \geq a$.

Applying (7.69) we obtain

$$\phi_1(a, \theta) = \sum_{n=0}^{\infty} A_n a^n P_n(\cos \theta). \tag{7.72}$$

Multiplying (7.72) by $P_m(\cos \theta) \sin \theta$ and integrating

$$V \int_0^{\pi/2} P_n(\cos \theta) \sin \theta \, d\theta = A_n a^n [2/(2n+1)]. \tag{7.73}$$

Thus

$$A_n = \{(2n+1)V/2a^n\} \int_0^1 P_n(x) \, dx, \tag{7.74}$$

and therefore

$$\phi_1(r, \phi) = \tfrac{1}{2} V [1 + \tfrac{3}{2}(r/a) \cos \theta + \cdots]. \tag{7.75}$$

Similarly,

$$\phi_2(a, \theta) = \sum_{n=0}^{\infty} (B_n/a^{n+1}) P_n(\cos \theta). \tag{7.76}$$

Proceeding as before

$$B_n = A_n a^{2n+1} = [(2n+1)/2] V a^{n+1} \int_0^1 P_n(x) \, dx, \tag{7.77}$$

and therefore

$$\phi_2(r, \theta) = \tfrac{1}{2} V(a/r) [1 + \tfrac{3}{2}(a/r) \cos \theta + \cdots]. \tag{7.78}$$

2. Let us calculate the electric potential of a dielectric sphere of radius a in the presence of a point charge a distance $b > a$ from the center of the sphere. Thus we must solve

$$\nabla^2 \phi_1 = \nabla^2 \phi_2 = 0, \tag{7.79}$$

where ϕ_1 is the potential inside the sphere and ϕ_2 the potential outside the sphere. The boundary conditions are

$$\phi_1 = \phi_2 \tag{7.80}$$

and

$$\varepsilon\, \partial\phi_1/\partial r = \partial\phi_2/\partial r \tag{7.81}$$

at $r = a$. In addition, we have the conditions that ϕ_1 be finite at $r = 0$ and $\phi_2 \to q/r$ as $r \to \infty$.

As usual we set aside (7.80) and (7.81) and take $\psi = 0$. Separating variables and rejecting divergent terms, we obtain, if we choose the coordinate system so that the point charge lies on the z axis,

$$\phi_1 = \sum_{n=0}^{\infty} A_n r^n P_n(\cos\theta) \tag{7.82}$$

and

$$\phi_2 = q/R + \sum_{n=0}^{\infty} (B_n/r^{n+1}) P_n(\cos\theta). \tag{7.83}$$

Substituting (7.66) into (7.83) we obtain, for $r < b$,

$$\phi_2 = (q/b) \sum_{n=0}^{\infty} (r/b)^n P_n(\cos\theta) + \sum_{n=0}^{\infty} (B_n/r^{n+1}) P_n(\cos\theta). \tag{7.84}$$

Of course, (7.84) must be modified for $r > b$ but this is of no consequence as we will use (7.84) only to satisfy the boundary conditions.

If we apply (7.80) and (7.81), we obtain

$$A_n = \frac{2n+1}{n(\varepsilon+1)+1} \frac{q}{b^{n+1}} \tag{7.85}$$

and

$$B_n = q\, \frac{a^{2n+1}}{b^{n+1}} \frac{n(1-\varepsilon)}{n(\varepsilon+1)+1}. \tag{7.86}$$

It is of interest to consider the special case of a conducting sphere, i.e., $\varepsilon \to \infty$. For this case, if $r < b$,

$$\phi_1 = 0 \tag{7.87}$$

and

$$\phi_2 = (q/b) \sum_{n=0}^{\infty} (r/b)^n P_n(\cos\theta) - (aqr/b) \sum_{n=0}^{\infty} (a^2/br)^n P_n(\cos\theta). \tag{7.88}$$

4. Boundary-Value Problems

Thus the potential outside the sphere is the same as that due to the charge q at a distance b from the center of the sphere and an "image" charge on the same radius vector at a distance a^2/b from the center of the sphere.

3. Lest we conclude that all boundary-value problems involving spherical coordinates lead to Legendre functions let us calculate the temperature distribution in a sphere whose surface is maintained at T and whose initial temperature is given by $u(r, 0) = f(r)$.

Thus we must solve

$$k\left[\frac{\partial^2 u}{\partial r^2} + \frac{2}{r}\frac{\partial u}{\partial r}\right] = \frac{\partial u}{\partial t} \tag{7.89}$$

subject to the boundary conditions

$$u(a, t) = T \tag{7.90}$$

and $u(0, t)$ finite and subject to the initial condition

$$u(r, 0) = f(r). \tag{7.91}$$

Proceeding as usual, we set aside (7.91) and choose $\psi = T$. We attempt to separate the variables by trying

$$\varphi = e^{-\alpha t} R(r). \tag{7.92}$$

Thus

$$\frac{d^2 R}{dr^2} + \frac{2}{r} R = -\frac{\alpha}{k} R \tag{7.93}$$

or

$$\frac{d^2(rR)}{dr^2} = -\frac{\alpha}{k}(rR). \tag{7.94}$$

Hence

$$rR = A \sin(\alpha/k)^{1/2} r + B \cos(\alpha/k)^{1/2} r. \tag{7.95}$$

We must put $B = 0$ because otherwise there will be a divergence at $r = 0$. The homogeneous boundary condition at $r = a$ requires that

$$\sin(\alpha/k)^{1/2} a = 0 \tag{7.96}$$

and so $\alpha = n^2 \pi^2 k / a^2$, $n = 1, 2, 3, \ldots$.

Thus

$$u(r, t) = T + \sum_{n=1}^{\infty} A_n \exp(-n^2 \pi^2 k t/a^2) \sin(n\pi r/a). \tag{7.97}$$

The A_n can be determined in the usual way. The result is

$$A_n = (2T/n\pi)[(-1)^n - 1] + \int_0^a f(r) \sin(n\pi r/a)\, dr. \qquad (7.98)$$

VIII. Green's Functions

In this section we are able to give only a brief introduction to the techniques involving Green's functions. More extensive treatments can be found in Morse and Feshback (1953), Friedman (1956), Panofsky and Phillips (1955), and Jackson (1962).

A. Introductory Example

Consider an elastic string at rest under an external force. Thus

$$T\, d^2y/dx^2 = F(x) \qquad (8.1)$$

where $F(x)$ is the external force per unit length. The boundary conditions are

$$y(0) = y(l) = 0. \qquad (8.2)$$

Let us first consider the special problem where the external force is concentrated at $x = x'$. Thus

$$F(x) = F_0\, \delta(x - x') \qquad (8.3)$$

where $\delta(x)$ is the *delta function* described in Chapter 2. We must solve

$$d^2G(x, x')/dx^2 = \delta(x - x') \qquad (8.4)$$

subject to the boundary conditions

$$G(0, x') = G(l, x') = 0. \qquad (8.5)$$

We expect the solution to look like that shown in Fig. 7.

Fig. 7. Green's function for an elastic string.

4. Boundary-Value Problems

The function $G(x, x')$ is called the *Green's function* for the problem. The Green's functions we discuss here are to be distinguished from the statistical Green's functions treated by Mavroyannis in Chapter 8.

Except at $x = x'$,

$$d^2G/dx^2 = 0. \tag{8.6}$$

Integrating, we obtain

$$G(x, x') = ax + b \tag{8.7}$$

for $x < x'$, and

$$G(x, x') = cx + d \tag{8.8}$$

for $x > x'$. The boundary conditions require that $b = 0$ and that $cl + d = 0$. Also $G(x, x')$ must be continuous at x'. Thus

$$ax' = c(x' - l). \tag{8.9}$$

We need one additional condition to determine a and c. This is obtained by using the fact that the integral of $\delta(x)$ is a step function. Hence

$$\left.\frac{dG}{dx}\right|_{x=x_+'} - \left.\frac{dG}{dx}\right|_{x=x_-'} = 1. \tag{8.10}$$

Therefore,

$$c = 1 + a. \tag{8.11}$$

Solving (8.9) and (8.11) gives

$$a = (x' - l)/l \tag{8.12}$$

and

$$c = x'/l. \tag{8.13}$$

Thus

$$G(x, x') = -x(l - x')/l \tag{8.14}$$

for $0 \leq x \leq x'$, and

$$G(x, x') = -x'(l - x)/l \tag{8.15}$$

for $x' \leq x \leq l$. It should be noted that the Green's function is *symmetric* in x and x'; that is,

$$G(x, x') = G(x', x). \tag{8.16}$$

The solution to this special problem is

$$y(x) = (F_0/T)G(x, x'). \tag{8.17}$$

Now suppose that the string is subjected to two concentrated forces at x_1 and x_2. We can superimpose the two solutions. Thus

$$y(x) = (F_1/T)G(x, x_1) + (F_2/T)G(x, x_2). \tag{8.18}$$

Generalizing to an arbitrary number of concentrated forces, we have

$$y(x) = \sum_{k=1}^{n} (F_k/T)G(x, x_k), \tag{8.19}$$

where $G(x, x_k)$ is given by (8.14) and (8.15). Finally, generalizing to a continuous external force, we expect that

$$y(x) = \int_0^l G(x, x')(F(x')/T) \, dx', \tag{8.20}$$

where $G(x, x')$ is given by (8.14) and (8.15). Equation (8.20) can be verified by substitution into (8.1).

The Green's function is the response of the string to a unit force concentrated at x'. It is occasionally called the *response function*. We may draw an analogy with a system of linear algebraic equations. The solution of the nonhomogeneous system $Ay = f$ is $y = A^{-1}f$. Thus a Green's function may be regarded as a type of inverse of a differential operator.

We may also construct the Green's function by means of a Fourier expansion by attempting a solution of the form

$$G(x, x') = \sum_{n=1}^{\infty} A_n(x') \sin(n\pi x/l). \tag{8.21}$$

Substitution of (8.21) into (8.4) gives

$$d^2G/dx^2 = -\sum_{n=1}^{\infty} (n^2\pi^2/l^2)A_n(x') \sin(n\pi x/l) = \delta(x - x'). \tag{8.22}$$

We may expand $\delta(x - x')$ in a Fourier series. Thus

$$\delta(x - x') = \sum_{n=1}^{\infty} B_n(x') \sin(n\pi x/l), \tag{8.23}$$

where

$$B_n(x') = (2/l) \int_0^l \delta(x - x') \sin(n\pi x/l)\, dx = (2/l) \sin(n\pi x'/l). \tag{8.24}$$

Hence, combining (8.22), (8.23), and (8.24) gives

$$G(x, x') = -(2l/\pi^2) \sum_{n=1}^{\infty} (1/n^2) \sin(n\pi x'/l) \sin(n\pi x/l). \tag{8.25}$$

Substitution of (8.25) into (8.20) gives

$$y(x) = (l^2/\pi^2) \sum_{n=1}^{\infty} (1/n^2) \alpha_n \sin(n\pi x/l), \tag{8.26}$$

where

$$\alpha_n = (2/l) \int_0^l (F(x')/T) \sin(n\pi x'/l)\, dx'. \tag{8.27}$$

That is, the α_n are the Fourier coefficients of $F(x)/T$. This result could have been obtained directly from (8.1) by expanding $y(x)$ and $F(x)/T$ in Fourier series.

Generally speaking, the Green's function technique is most useful when the Green's function is obtained in closed form.

B. Green's Function for the Sturm–Liouville Operator

Consider the nonhomogeneous differential equation

$$\frac{d}{dx}\left[p(x) \frac{dy}{dx}\right] + q(x)y - \lambda r(x)y = f(x) \tag{8.28}$$

together with the boundary conditions

$$\alpha_1 y(a) - \beta_1 y'(a) = 0 \tag{8.29}$$

and

$$\alpha_2 y(a) - \beta_2 y'(b) = 0, \tag{8.30}$$

where α_1 and β_1 are not both zero and α_2 and β_2 are not both zero.

Let us find the Green's function that satisfies

$$\frac{d}{dx}\left[p(x) \frac{dG}{dx}\right] + q(x)y - \lambda r(x)y = \delta(x - x') \tag{8.31}$$

subject to the boundary conditions. In the preceding section we determined $G(x, x')$ when $p(x) = 1$ and $q(x) = r(x) = 0$.

We follow the same procedure as was used in the preceding section. That is, we first require that $G(x, x')$ satisfy the homogeneous equation

$$\frac{d}{dx}\left[p(x)\frac{dG}{dx}\right] + q(x)G - \lambda r(x)G = 0, \qquad (8.32)$$

except at $x = x'$. Second, we require that $G(x, x')$ satisfy the boundary conditions

$$\alpha_1 G(a, x') - \beta_1 G(a, x') = 0 \qquad (8.33)$$

and

$$\alpha_2 G(b, x') - \beta_2 G(b, x') = 0. \qquad (8.34)$$

If $y_1(x)$ is a nontrivial solution of (8.32) satisfying (8.33) and if $y_2(x)$ is a nontrivial solution of (8.32) satisfying (8.34), then

$$G(x, x') = c_1 y_1(x) \qquad (8.35)$$

for $a \leq x \leq x'$, and

$$G(x, x') = c_2 y_2(x) \qquad (8.36)$$

for $x' \leq x \leq b$. Third, we require that $G(x, x')$ be continuous at $x = x'$. Thus

$$c_1 y_1(x') = c_2 y_2(x'). \qquad (8.37)$$

Finally, we can integrate (8.31) to obtain the condition

$$\left.\frac{dG}{dx}\right|_{x=x_+'} - \left.\frac{dG}{dx}\right|_{x=x_-'} = \frac{1}{p(x')}. \qquad (8.38)$$

Thus

$$c_1 y_1'(x') - c_2 y_2'(x') = -1/p(x'). \qquad (8.39)$$

For a nontrivial solution of (8.37) and (8.39) for c_1 and c_2 we require that the *Wronskian* of y_1 and y_2 be nonzero at x'. Thus

$$W(x') = y_1(x')y_2'(x') - y_2(x')y_1'(x') \neq 0. \qquad (8.40)$$

That is, y_1 and y_2 must be independent. If this is not the case, the Green's function will not exist.

Thus, if we write (8.28) in the form

$$(L - \lambda r)y = f, \qquad (8.41)$$

4. Boundary-Value Problems

then the Green's function exists for all values of λ except those which are eigenvalues of the homogeneous system. This is in analogy with linear algebraic equations. The system $Ay = f$ has the solution $y = A^{-1}f$ only if A is nonsingular. However, $A = L - \lambda I$ is singular when λ is an eigenvalue of L and, for such values of λ, A^{-1} does not exist.

If (8.40) is satisfied, we may solve (8.37) and (8.39) for c_1 and c_2. The results are

$$c_1 = y_2(x')/[p(x')W(x')] \tag{8.42}$$

and

$$c_2 = y_1(x')/[p(x')W(x')]. \tag{8.43}$$

Thus

$$G(x, x') = [y_1(x)y_2(x')]/[p(x')W(x')] \tag{8.44}$$

for $a \leq x \leq x'$, and

$$G(x, x') = [y_2(x)y_1(x')]/[p(x')W(x')] \tag{8.45}$$

for $x' \leq x \leq b$. It is to be noted that $G(x, x')$ is symmetric in x and x'. This is a consequence of (8.33) and (8.34). For other types of boundary conditions $G(x, x')$ may not be symmetric. However, for the particular case of periodic boundary conditions, the Green's function is symmetric.

We may also construct $G(x, x')$ by means of an expansion in the eigenfunctions $\varphi_n(x)$ of the Sturm–Liouville operator. Thus, if we assume without loss of generality that the $\varphi_n(x)$ are an orthonormal set of eigenfunctions, we may expand the Green's function and get

$$G(x, x') = \sum_{n=1}^{\infty} A_n(x')\varphi_n(x) \tag{8.46}$$

where

$$\frac{d}{dx}\left[p(x)\frac{d\varphi_n}{dx}\right] + q(x)\varphi_n(x) = \lambda_n\varphi_n(x). \tag{8.47}$$

Similarly, we may expand $\delta(x - x')$ and get

$$\delta(x - x') = \sum_{n=1}^{\infty} B_n(x')\varphi_n(x) \tag{8.48}$$

where

$$B_n(x') = \int_a^b \delta(x - x')\varphi_n(x)\, dx = \varphi_n(x'). \tag{8.49}$$

If we substitute (8.46), (8.48), and (8.49) into (8.31), we obtain

$$A_n = \varphi_n(x')/(\lambda_n - \lambda). \tag{8.50}$$

Thus we obtain the *bilinear formula*

$$G(x, x') = \sum_{n=1}^{\infty} \frac{\varphi_n(x')\varphi_n(x)}{\lambda_n - \lambda}. \tag{8.51}$$

From (8.51) we see very clearly that if λ is one of the eigenvalues, i.e., $\lambda = \lambda_n$ for some n, $G(x, x')$ does not exist. In addition, we can see easily that $G(x, x')$ is symmetric.

We expect that the solution of (8.28) is

$$y(x) = \int_a^b G(x, x') f(x') \, dx'. \tag{8.52}$$

This may be verified by substituting (8.52) into (8.28). In doing this we must keep in mind that there are separate forms for $G(x, x')$ for $x < x'$ and $x > x'$. Thus, we write

$$y(x) = \int_a^x G(x, x') f(x') \, dx' + \int_x^b G(x, x') f(x') \, dx'. \tag{8.53}$$

Differentiating, we obtain

$$\frac{dy}{dx} = \int_a^x \frac{dG}{dx} f(x') \, dx' + \int_x^b \frac{dG}{dx} f(x') \, dx' \tag{8.54}$$

and

$$\frac{d^2y}{dx^2} = \int_a^x \frac{d^2G}{dx^2} f(x') \, dx' + \int_x^b \frac{d^2G}{dx^2} f(x') \, dx'$$
$$+ \frac{dG}{dx}\bigg|_{x'=x_-} - \frac{dG}{dx}\bigg|_{x'=x_+}. \tag{8.55}$$

The last two terms of (8.55) do not cancel, as did the corresponding terms in (8.54), because $\partial G/\partial x$ is not continuous. Using (8.39) we obtain

$$\frac{d^2y}{dx^2} = \int_a^x \frac{d^2G}{dx^2} f(x') \, dx' + \int_x^b \frac{d^2G}{dx^2} f(x') \, dx' + \frac{1}{p(x)}. \tag{8.56}$$

Thus

$$\frac{d}{dx}\left[p(x)\frac{dy}{dx}\right] + q(x)y - \lambda r(x)y$$
$$= \int_a^b \left\{\frac{d}{dx}\left[p(x)\frac{dG(x, x')}{dx}\right] + [q(x) - \lambda r(x)]G(x, x')\right\}$$
$$\times f(x') \, dx' + f(x). \tag{8.57}$$

4. Boundary-Value Problems

Because of (8.31) the integrand is zero except at $x' = x$. Thus the integral vanishes and (8.28) is satisfied. It only remains to show that (8.52) satisfies the boundary conditions. We have

$$\alpha_1 y(a) - \beta_1 y'(a) = \int_a^b \left\{ \alpha_1 G(a, x') - \beta_1 \frac{dG(a, x')}{dx} \right\} f(x') \, dx'. \tag{8.58}$$

The integrand in (8.58) is zero because $G(x, x')$ satisfies the boundary condition at $x = a$. Thus $y(x)$ satisfies the boundary condition at $x = a$. Similarly, $y(x)$ satisfies the boundary condition at $x = b$.

As an example of the use of the results just obtained, consider a stretched string subjected to forced vibrations by an external force which varies harmonically with time. Thus we have

$$\frac{\partial^2 y}{\partial x^2} = \frac{1}{c^2} \frac{\partial^2 y}{\partial t^2} + F(x) e^{-i\omega t}. \tag{8.59}$$

The boundary conditions are

$$y(0, t) = y(l, t) = 0. \tag{8.60}$$

We expect $y(x, t)$ to have the same time dependence. Thus

$$y(x, t) = f(x) e^{-i\omega t}, \tag{8.61}$$

where

$$d^2 f/dx^2 + k^2 f = F \tag{8.62}$$

and $k^2 = \omega^2/c^2$.

The Green's function satisfies

$$d^2 G/dx^2 + k^2 G = \delta(x - x') \tag{8.63}$$

and the boundary conditions

$$G(0, x') = G(l, x') = 0. \tag{8.64}$$

The solution of the homogeneous differential equation which satisfies the first boundary condition is $\sin kx$ and the solution which satisfies the second boundary condition is $\sin k(l - x)$. Thus the Wronskian is

$$W = -k \sin kl \tag{8.65}$$

and the Green's function is

$$G(x, x') = -[\sin kx \sin k(l - x')]/(k \sin kl) \tag{8.66}$$

for $0 \leq x \leq x'$, and

$$G(x, x') = -[\sin kx' \sin k(l - x)]/(k \sin kl) \tag{8.67}$$

for $x' \leq x \leq l$. We see that the Green's function diverges when $\sin kl = 0$, $k = n\pi/l$ for $n = 1, 2, \ldots$. In other words when ω is one of the characteristic frequencies of the string we have *resonance*.

The normalized eigenfunctions of the differential operator are $(2/l)^{1/2} \sin(n\pi x/l)$ and the eigenvalues are $n\pi/l$. Thus, if we use the bilinear formula, we obtain

$$G(x, x') = \frac{2}{l} \sum_{n=1}^{\infty} \frac{\sin(n\pi x/l) \sin(n\pi x'/l)}{k^2 - n^2\pi^2/l^2}. \tag{8.68}$$

The solution is

$$y(x, t) = e^{-i\omega t} \int_0^l G(x, x') F(x')\, dx'. \tag{8.69}$$

If we use (8.68), we obtain

$$y(x, t) = e^{-i\omega t} \sum_{n=1}^{\infty} [A_n/(k^2 - n^2\pi^2/l^2)] \sin(n\pi x/l), \tag{8.70}$$

where

$$A_n = (2/l) \int_0^l F(x') \sin(n\pi x'/l)\, dx' \tag{8.71}$$

are the Fourier coefficients of $F(x)$. Of course, we could have obtained (8.70) by expanding $f(x)$ and $F(x)$ in a Fourier series and solving (8.62) directly. However, the Green's function technique has the advantage of yielding an alternative form for the solution.

C. Solution of Potential Problems by Green's Functions

One of the most common uses of Green's functions is in the solution of electrostatic potential problems. For such problems, the Green's function $G(\mathbf{r}, \mathbf{r}')$ for a particular geometrical arrangement is the solution of the potential problem at $\mathbf{r}$ for this given geometrical arrangement of grounded conducting surfaces when the only charge present is a unit charge at the point $\mathbf{r}'$. It is to be noted that the grounded conducting surfaces may be at infinity.

We shall consider two types of problems which can be solved by Green's functions. In the first type of problem the charge distribution is given within a region bounded by a grounded conductor and in the second type of problem the potential distribution over a certain conducting surface, which encloses a charge-free region, is given. In the first type of problem the differential equation is nonhomogeneous. The second type of problem is slightly different from those problems which we have considered thus far in this section. It is the boundary condition that is nonhomogeneous. Of course we have dealt with such problems by the method of separation of variables.

The solutions of both types of problem can be given together by means of Green's theorem:

$$\int_V (\phi \nabla^2 \psi - \psi \nabla^2 \phi) \, dV = \int_S (\phi \nabla \psi - \psi \nabla \phi) \cdot d\mathbf{S}, \qquad (8.72)$$

where V is the volume enclosed by the surface S. This theorem is derived by Jacob at the end of Section III in Chapter 1. Let ϕ be the desired solution of the potential problem and $\psi = G$ be the Green's function. That is, ψ is the solution to the problem of a unit charge located within the surface which is grounded. Thus

$$G(\mathbf{r}, \mathbf{r}') = 1/R(\mathbf{r}, \mathbf{r}') + F(\mathbf{r}, \mathbf{r}'), \qquad (8.73)$$

where F is the potential due to the induced charge on S. We note that

$$\nabla^2 F = 0 \qquad (8.74)$$

and

$$\nabla^2 G = -4\pi \, \delta(R), \qquad (8.75)$$

where $R = |\mathbf{r} - \mathbf{r}'|$. Therefore,

$$\int_V (G \nabla^2 \phi - \phi \nabla^2 G) \, dV' = \int_V \{G \nabla^2 \phi + 4\pi \phi \, \delta(R)\} \, dV'$$

$$= \int_V (G \nabla^2 \phi) \, dV' + 4\pi \phi(\mathbf{r})$$

$$= \int_S (G \nabla \phi - \phi \nabla G) \cdot d\mathbf{S}'. \qquad (8.76)$$

By definition $G = 0$ on the surface S. Thus

$$\phi(\mathbf{r}) = -(1/4\pi) \int_V G(\mathbf{r}, \mathbf{r}') \nabla^2 \phi(\mathbf{r}') \, dV'$$

$$- (1/4\pi) \int_S \phi(\mathbf{r}')(\partial G(\mathbf{r}, \mathbf{r}')/\partial n') \, dS'. \qquad (8.77)$$

Now let us consider the two special cases mentioned earlier. In the first case the surface is grounded. Hence, $\phi = 0$ on S. Further, $\nabla^2\phi = -4\pi\varrho$. Thus

$$\phi(\mathbf{r}) = \int_V \varrho(\mathbf{r}')G(\mathbf{r}, \mathbf{r}')\,dV'. \tag{8.78}$$

In the second case there are no sources within V and $\nabla^2\phi = 0$. Hence

$$\phi(\mathbf{r}) = -(1/4\pi)\int \phi(\mathbf{r}')(\partial G(\mathbf{r}, \mathbf{r}')/\partial n')\,dS'. \tag{8.79}$$

Physically $(4\pi)^{-1}\,\partial G/\partial n$ is the surface charge density induced on the grounded conducting surface by a unit charge at $\mathbf{r}$.

D. Green's Function for a Sphere

We see from Eq. (7.88) that the potential for a unit charge in the presence of a grounded conducting sphere is

$$G(r, r') = \frac{1}{|\mathbf{r} - \mathbf{r}'|} - \frac{a/r'}{|\mathbf{r} - (a/r')^2\mathbf{r}'|}$$
$$= (r^2 + r'^2 - 2rr'\cos\gamma)^{-1/2} - (r^2r'^2/a^2 + a^2 - 2rr'\cos\gamma)^{-1/2}, \tag{8.80}$$

where γ is the angle between $\mathbf{r}$ and $\mathbf{r}'$. Thus

$$\cos\gamma = \cos\theta\cos\theta' + \sin\theta\sin\theta'\cos(\varphi - \varphi'). \tag{8.81}$$

We note again that $G(\mathbf{r}, \mathbf{r}') = G(\mathbf{r}', \mathbf{r})$.

We also need $\partial G/\partial n'$ at $r' = a$. If we remember that n' is the *outward* normal from the region of interest, we obtain

$$\left.\frac{\partial G}{\partial n'}\right|_{r'=a} = \frac{r^2 - a^2}{a(r^2 + a^2 - 2ar\cos\gamma)^{3/2}}. \tag{8.82}$$

Thus, for example, the potential outside a sphere with potential $V(\theta, \varphi)$ on its surface is

$$\phi(\mathbf{r}) = \frac{1}{4\pi}\int V(\theta', \varphi')\frac{a(r^2 - a^2)}{(r^2 + a^2 - 2ar\cos\gamma)^{3/2}}\,d\Omega', \tag{8.83}$$

where

$$d\Omega' = \sin\theta'\,d\theta'\,d\varphi' \tag{8.84}$$

is the element of solid angle. For the interior problem we change the sign of $\partial G/\partial n'$.

Hence, if we return to the problem of the sphere of radius a formed by two hemispheres at potential V and 0, we have for the potential outside the sphere

$$\phi(r, \theta) = \frac{V}{4\pi} a(r^2 - a^2) \int_0^{2\pi} \int_0^{\pi/2} \frac{d\Omega'}{(r^2 + a^2 - 2ar \cos \gamma)^{3/2}}. \qquad (8.85)$$

We cannot perform the integration in closed form. Expanding,

$$\phi(r, \theta) = \frac{Va(r^2 - a^2)}{4\pi(r^2 + a^2)^{3/2}} \int_0^{2\pi} \int_0^{\pi/2} (1 + 3\alpha \cos \gamma + \cdots) \, d\Omega', \qquad (8.86)$$

where $\alpha = ar/(r^2 + a^2)$. Thus

$$\phi(r, \theta) = \frac{Va(r^2 - a^2)}{2(r^2 + a^2)^{3/2}} \left[1 + 3\alpha \cos \theta \int_0^{\pi/2} \cos \theta' \sin \theta' \, d\theta' + \cdots \right]$$

$$= \frac{Va(r^2 - a^2)}{2(r^2 + a^2)^{3/2}} \left[1 + \frac{3}{2} \frac{ar}{(r^2 + a^2)} \cos \theta + \cdots \right]. \qquad (8.87)$$

On the other hand, the potential inside the sphere is obtained by replacing $r^2 - a^2$ in (8.85) by $(a^2 - r^2)$. Thus, expanding, we obtain

$$\phi(r, \theta) = \frac{Va(a^2 - r^2)}{2(r^2 + a^2)^{3/2}} \left[1 + \frac{3}{2} \frac{ar}{(a^2 + r^2)} \cos \theta + \cdots \right]. \qquad (8.88)$$

Equations (8.87) and (8.88) are alternative expressions to (7.78) and (7.75), respectively. We can see that they are equivalent by expanding (8.87) in powers of a/r and (8.88) in powers of r/a.

IX. Laplace Transform Methods

It is sometimes convenient to solve boundary-value problems using the *Laplace transform*

$$\mathscr{L}[f] = \int_0^\infty e^{-st} f(t) \, dt. \qquad (9.1)$$

The fundamental property of the Laplace transform is

$$\mathscr{L}[df/dt] = s\mathscr{L}[f] - f(0). \qquad (9.2)$$

If we continue, we also obtain

$$\mathscr{L}[d^2f/dt^2] = s^2\mathscr{L}[f] - sf(0) - df(0)/dt. \tag{9.3}$$

The technique of solving boundary-value problems by means of the Laplace transform is illustrated in the following two heat problems. For a more extensive treatment, the reader is referred to Churchill (1944).

A. A Heat Problem in a Semiinfinite Rod

Consider a semiinfinite rod whose ends are at $x = 0$ and $x = \infty$ and whose curved surface is perfectly insulated. Thus

$$k\, \partial^2 u/\partial x^2 = \partial u/\partial t. \tag{9.4}$$

Let us find the temperature distribution in the rod subject to the boundary condition

$$-k\, \partial u(0, t)/\partial x = \phi \tag{9.5}$$

and the initial condition

$$u(x, 0) = 0. \tag{9.6}$$

Let us take the Laplace transform of (9.4) and (9.5). Presuming that the order of the integrations and differentiations can be changed, we obtain

$$k\, d^2 \mathscr{L}/dx^2 = s\mathscr{L} - u(x, 0) \tag{9.7}$$

and

$$-k[d\mathscr{L}/dx]_{x=0} = \phi/s. \tag{9.8}$$

Thus, if we use the initial condition (9.6), we must solve the ordinary differential equation

$$(k\, d^2/dx^2)[\mathscr{L}] - s\mathscr{L} = 0. \tag{9.9}$$

Hence

$$\mathscr{L}[u] = A \exp\{(s/k)^{1/2}x\} + B \exp\{-(s/k)^{1/2}x\}. \tag{9.10}$$

If we are to have a finite solution, $A = 0$. Equation (9.8) requires that

$$B = \phi/s(ks)^{1/2}. \tag{9.11}$$

Thus

$$\mathscr{L}[u] = (\phi/s(ks)^{1/2}) \exp\{-(s/k)^{1/2}x\}. \tag{9.12}$$

4. Boundary-Value Problems

To complete the solution, Eq. (9.12) must be inverted. We recall that

$$\mathscr{L}[\phi] = \phi/s \tag{9.13}$$

and

$$\mathscr{L}[(\pi t)^{-1/2} \exp\{-x^2/4kt\}] = (s)^{-1/2} \exp\{-(s/k)^{1/2}x\}. \tag{9.14}$$

Thus

$$\mathscr{L}[u] = (k)^{-1/2} \mathscr{L}[\phi] \mathscr{L}[(\pi t)^{-1/2} \exp\{-x^2/4kt\}]. \tag{9.15}$$

Using the *convolution theorem*

$$\mathscr{L}\left[\int_0^t f(t-u)g(u)\,du\right] = \mathscr{L}[f]\mathscr{L}[g], \tag{9.16}$$

we obtain

$$u(x, t) = [\phi/(k\pi)]^{1/2} \int_0^t \exp\{-x^2/4ku\} u^{-1/2}\,du. \tag{9.17}$$

Changing the variable of integration,

$$u(x, t) = (x\phi/k\sqrt{\pi}) \int_{x/2(kt)^{1/2}}^{\infty} \exp(-\eta^2)\eta^{-2}\,d\eta. \tag{9.18}$$

Integrating by parts,

$$u(x, t) = (\phi/k)[2(kt/\pi)^{1/2} \exp\{-x^2/4kt\} - x\,\mathrm{erfc}\{x/2(kt)^{1/2}\}], \tag{9.19}$$

where

$$\mathrm{erfc}(y) = (2/\sqrt{\pi}) \int_y^{\infty} \exp(-\eta^2)\,d\eta = 1 - \mathrm{erf}(y) \tag{9.20}$$

is called the *complementary error function*.

B. A Heat Problem in a Finite Rod

Consider a finite rod whose ends are at $x = 0$ and $x = l$, whose curved surface is perfectly insulated, whose left-hand end is perfectly insulated, and whose right-hand end is maintained at 0°. Hence,

$$k\,\partial^2 u/\partial x^2 = \partial u/\partial t \tag{9.21}$$

and

$$\partial u(0, t)/\partial x = u(l, t) = 0. \tag{9.22}$$

Let us solve this problem subject to the initial condition

$$u(x, 0) = T_0. \tag{9.23}$$

If we take the Laplace transform, we obtain

$$k\, d^2\mathscr{L}/dx^2 - s\mathscr{L} = T_0 \tag{9.24}$$

and

$$[d\mathscr{L}/dx]_{x=0} = [\mathscr{L}]_{x=l} = 0. \tag{9.25}$$

Solving this system, we obtain

$$\mathscr{L}[u] = \frac{T_0}{s}\left[1 - \frac{\cosh\{(s/k)^{1/2}x\}}{\cosh\{(s/k)^{1/2}l\}}\right]. \tag{9.26}$$

The problem is now to invert (9.26). We will outline two different methods. First of all, we can expand the denominator in (9.26) and obtain

$$\mathscr{L}[u] = (T_0/s)\left[1 - \sum_{n=0}^{\infty}(-1)^n\{\exp(-\lambda(s/k)^{1/2}) + \exp(-\mu(s/k)^{1/2})\}\right], \tag{9.27}$$

where

$$\lambda = (2n+1)l - x \tag{9.28}$$

and

$$\mu = (2n+1)l + x. \tag{9.29}$$

Hence,

$$u(x, t) = T_0\left[1 - \sum_{m=0}^{\infty}(-1)^n\left\{\mathrm{erfc}\left[\frac{(2n+1)l - x}{2(kt)^{1/2}}\right] + \mathrm{erfc}\left[\frac{(2n+1)l + x}{2(kt)^{1/2}}\right]\right\}\right], \tag{9.30}$$

where we have used the fact that

$$\mathscr{L}[\mathrm{erfc}(x/2(kt)^{1/2})] = s^{-1}\exp\{-(s/k)^{1/2}x\}. \tag{9.31}$$

Second, we could use the inversion integral

$$f(x) = (2\pi i)^{-1}\int_{\gamma-i\infty}^{\gamma+i\infty} e^{st}\mathscr{L}[f]\, ds, \tag{9.32}$$

to invert (9.26). In using this method, we are interested in the places where $\cosh[(s/k)^{1/2}l] = 0$. Thus

$$(s/k)^{1/2}l = i(n + \tfrac{1}{2})\pi \tag{9.33}$$

or

$$s = -k(n + \tfrac{1}{2})^2\pi^2/l^2. \tag{9.34}$$

We can evaluate the inversion integral by the method of residues (Chapter 3). We obtain

$$u(x, t) = \frac{4T_0}{\pi} \sum_{n=0}^{\infty} \frac{(-1)^n}{2n + 1} \exp\left\{-k \frac{(2n + 1)^2\pi^2}{4l^2} t\right\} \cos \frac{(2n + 1)\pi x}{2l}. \tag{9.35}$$

We could also have obtained (9.35) by the method of separation of variables.

The two solutions (9.30) and (9.35) are equivalent. However, they are complementary. Equation (9.30) is useful for large t and (9.35) is useful for small t.

X. Conformal Mapping

Consider a function

$$w = f(z) = u(x, y) + iv(x, y) \tag{10.1}$$

of a complex variable $z = x + iy$. This function may be regarded as a *mapping* or *transformation* from the xy plane into the uv plane. It is shown by Silverstone in Chapter 3, Section III.G that if $f(z)$ is analytic, u and v are harmonic functions. That is, they satisfy Laplace's equation. Thus

$$\partial^2 u/\partial x^2 + \partial^2 u/\partial y^2 = 0 \quad \text{and} \quad \partial^2 v/\partial x^2 + \partial^2 v/\partial y^2 = 0. \tag{10.2}$$

A transformation is *conformal* at a point z if it preserves both the magnitude and sense of angles between every pair of curves through z. If a transformation is conformal at z, then the image of every small figure near z *conforms* to the original figure in the sense that it has approximately the same shape (although it may be magnified). However, large figures may transform into figures that bear no resemblance to the original figure.

In this section we briefly illustrate how conformal mappings may be used to solve some problems involving the two-dimensional Laplace's equation. The method is elegant and can be used to solve problems which cannot be analytically solved otherwise. However, it is limited to two-dimensional problems.

Our discussion is brief. We state a few theorems and give one simple example of the solution of boundary-value problems by means of conformal mapping. More extensive discussions can be found in Morse and Feshbach (1953) and Churchill (1963).

Theorem At each point z of a domain where $f(z)$ is analytic and $f'(z) = 0$, the mapping $w = f(z)$ is conformal.

Proof To establish this theorem, we observe that

$$f'(z) = \lim_{\Delta z \to 0} \frac{\Delta w}{\Delta z} = \lim \frac{|\Delta w| e^{i\phi}}{|\Delta z| e^{i\theta}}. \tag{10.3}$$

Hence,

$$|f'(z)| = \lim_{\Delta z \to 0} \frac{|\Delta w|}{|\Delta z|} \tag{10.4}$$

and

$$\lim_{\Delta z \to 0} (\phi - \theta) = \arg f'(z). \tag{10.5}$$

Thus, provided that $f'(z) \neq 0$, any curve passing through z is rotated by the angle $\arg f'(z)$. As a result, the magnitude and sense of angles between curves are preserved, the mapping is conformal, and the theorem is proved.

In addition, we see from (10.4) that the length of small lines passing through z is magnified by $|f'(z)|$ and that areas are magnified by $|f'(z)|^2$. We can also obtain the latter result by observing that, because of the Cauchy–Riemann equations, the Jacobian of the transformation, which determines the magnification of areas, is

$$\partial(u, v)/\partial(x, y) = (\partial u/\partial x)^2 + (\partial v/\partial x)^2 = |f'(z)|^2. \tag{10.6}$$

Points at which $f'(z) = 0$ are called critical points. Since a mapping will have a single-valued inverse when the Jacobian is nonzero, we see from (10.6) that a mapping has single-value inverse when $f'(z) \neq 0$.

As a corollary to the theorem above we have the result:

Corollary A conformal transformation maps orthogonal curves into orthogonal curves.

4. Boundary-Value Problems

In particular, the transformation (10.1) maps the curves $u(x, y) = c_1$ and $v(x, y) = c_2$ into the lines $u = c$ and $v = c_2$ in the uv plane. Since these lines are orthogonal, $u(x, y) = c_1$ is orthogonal to $v(x, y) = c_2$. Since $u(x, y)$ and $v(x, y)$ are harmonic functions as well as orthogonal functions, they can be used to represent equipotentials and lines of force in potential problems or other pairs of orthogonal functions that arise in the solution of Laplace's equation.

We can use the Cauchy–Riemann equations to construct one of these conjugate functions from the other. For example, if $u(x, y)$ is known, it is easy to verify that

$$v(x, y) = \int [\partial u/\partial x - \partial u/\partial y] + c. \tag{10.7}$$

Since $u(x, y)$ satisfies Laplace's equation, the line integral in (10.7) is independent of path.

Theorem Solutions of Laplace's equation remain solutions of Laplace's equation when subjected to a conformal mapping.

Proof To prove this theorem we assume that ϕ satisfies Laplace's equation. Thus

$$\partial^2\phi/\partial x^2 + \partial^2\phi/\partial y^2 = 0. \tag{10.8}$$

If we apply the conformal transformation (10.1), then it is straightforward but tedious to show that

$$\begin{aligned}\frac{\partial^2\phi}{\partial x^2} + \frac{\partial^2\phi}{\partial y^2} &= \left\{\frac{\partial^2\phi}{\partial u^2} + \frac{\partial^2\phi}{\partial y^2}\right\}\left\{\left(\frac{\partial u}{\partial x}\right)^2 + \left(\frac{\partial v}{\partial x}\right)^2\right\} \\ &= \left\{\frac{\partial^2\phi}{\partial u^2} + \frac{\partial^2\phi}{\partial y^2}\right\} |f'(z)|^2.\end{aligned} \tag{10.9}$$

Since the transformation is conformal $f'(z) \neq 0$ and

$$\partial^2\phi/\partial u^2 + \partial^2\phi/\partial v^2 = 0, \tag{10.10}$$

and the theorem is proved.

In order to apply conformal mapping to boundary-value problems, we must consider the transformation of boundary conditions. It is easy to show that the boundary conditions

$$\phi = c \tag{10.11}$$

or

$$\partial\phi/\partial n = 0 \qquad (10.12)$$

along some curve in the xy plane transform into the same conditions along the corresponding curve in the uv plane.

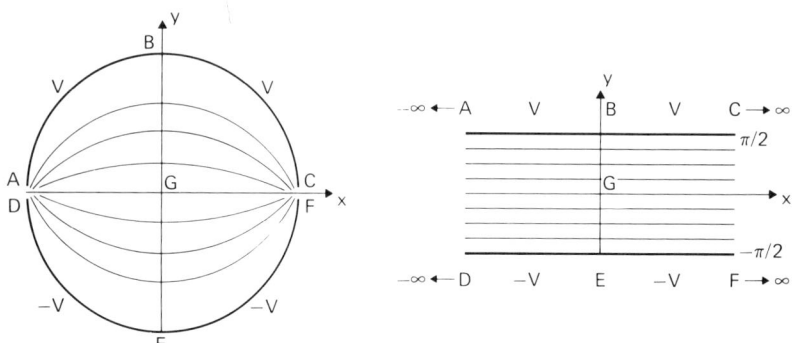

FIG. 8. Conformal mapping to solve problem of charged conduction semicircles.

The method of conformal mapping is illustrated by the following example. Let us suppose that two halves of a circle of unit radius are given the electric potentials V and $-V$ as is illustrated in Fig. 8. The potential may be easily calculated if we observe that the conformal transformation

$$w = \ln[(1+z)/(1-z)] \qquad (10.13)$$

maps the unit circle in the xy plane into the region between the lines $v = \pm\pi/2$ in the uv plane which is shown in Fig. 8. Thus the potential is

$$\phi = (2V/\pi)v. \qquad (10.14)$$

If we solve (10.13) for v, we obtain

$$\phi(x, y) = (2V/\pi)\tan^{-1}[2y/(1-x^2-y^2)]. \qquad (10.15)$$

The lines of constant potential are circles with center on the y axis.

Acknowledgments

This chapter was written while the author was a visiting professor in the Department of Physics, National University of La Plata, Argentina. The author is grateful to the Organization of American States for the financial support which made this visit possible and to Drs. A. Rodriguez and R. Caligaris for their hospitality.

References

Abramowitz, M., and Stegun, I. A. (1965). "Tables of Mathematical Functions." Dover, New York.
Arfken, G. (1966). "Mathematical Methods for Physicists." Academic Press, New York.
Bateman, H. (1944). "Partial Differential Equations of Mathematical Physics." Dover, New York.
Carslaw, H. S., and Jaeger, J. C. (1959). "Conduction of Heat in Solids," 2nd ed. Oxford Univ. Press, London and New York.
Churchill, R. V. (1944). "Modern Operational Mathematics in Engineering." McGraw-Hill, New York.
Churchill, R. V. (1960). "Complex Variables and Applications," 2nd ed. McGraw-Hill, New York.
Churchill, R. V. (1963). "Fourier Series and Boundary-Value Problems," 2nd ed. McGraw-Hill, New York.
Courant, R., and Hilbert, D. (1953). "Methods of Mathematical Physics," Vol. 1. Wiley (Interscience), New York.
Courant, R., and Hilbert, D. (1962). "Methods of Mathematical Physics," Vol. 2. Wiley (Interscience), New York.
Davis, H. F. (1963). "Fourier Series and Orthogonal Functions." Allyn and Bacon, Boston, Massachusetts.
Duff, G. F. D., and Naylor, D. (1966). "Differential Equations of Applied Mathematics." Wiley, New York.
Friedman, B. (1956). "Principles and Techniques of Applied Mathematics." Wiley, New York.
Hobson, E. W. (1931). "The Theory of Spherical and Ellipsoidal Harmonics." Cambridge Univ. Press, London and New York.
Jackson, J. D. (1962). "Classical Electrodynamics." Wiley, New York.
Jahnke, E., and Emde, F. (1945). "Tables of Higher Functions." Dover, New York.
Jeffreys, H., and Jeffreys, B. (1956). "Methods of Mathematical Physics," 3rd ed. Cambridge Univ. Press, London and New York.
Lebedev, N. N. (1965). "Special Functions and Their Applications." Prentice-Hall, Englewood Cliffs, New Jersey.
MacRobert, T. M. (1967). "Spherical Harmonics," 3rd ed. Pergamon, Oxford.
Margenau, H., and Murphy, G. M. (1956). "The Mathematics of Physics and Chemistry," 2nd ed. Van Nostrand Reinhold, Princeton, New Jersey.
Morse, P. M., and Feshbach, H. (1953). "Methods of Theoretical Physics," Vols. 1 and 2. McGraw-Hill, New York.
Panofsky, W. K. H., and Phillips, M. (1955). "Classical Electricity and Magnetism." Addison-Wesley, Reading, Massachusetts.
Sagan, H. (1961). "Boundary and Eigenvalue Problems in Mathematical Physics." Wiley, New York.
Smythe, W. R. (1950). "Static and Dynamic Electricity." McGraw-Hill, New York.
Sneddon, I. N. (1961). "Spherical Functions of Mathematical Physics and Chemistry." Oliver and Boyd, Edinburgh.
Sommerfeld, A. (1949). "Partial Differential Equations in Physics." Academic Press, New York.

TRANTER, C. J. (1968). "Bessel Functions With Some Physical Applications." English Univ. Press, London.

WATSON, G. N. (1944). "A Treatise on the Theory of Bessel Functions," 2nd ed. Cambridge Univ. Press, London and New York.

WEBSTER, A. G. (1955). "Partial Differential Equations of Mathematical Physics," 2nd ed. Dover, New York.

WHITTAKER, E. T., and WATSON, G. N. (1927). "A Course of Modern Analysis," 4th ed. Cambridge Univ. Press, London and New York.

WYLIE, C. R. (1966). "Advanced Engineering Mathematics," 3rd ed. McGraw-Hill, New York.

Chapter 5

Numerical Analysis

R. G. Stanton and W. D. Hoskins

    I. Introduction 337
   II. Approximation by Polynomial Interpolation 338
  III. Approximation by Spline Interpolation 346
  IV. Approximation by Least Squares 352
   V. Numerical Differentiation 355
  VI. Approximate Integration or Quadrature 359
 VII. Differential Equations 362
VIII. Equations in a Single Unknown 364
  IX. Systems of Linear Equations 366
   X. Special Methods for Solving Sparse Sets of Equations 369
       Appendix 371
       References 372

I. Introduction

The subject of numerical analysis is vast, and we can only hope, in a single chapter, to concentrate on some specific areas of general interest and importance. Our pattern of presentation is as follows.

First, we consider the problem of approximating functions whose values are known only at a few points; we indicate the three commonest approaches to such a problem, namely, by interpolating polynomials, splines, and least squares. Then we continue with a discussion of the use of such approximations in numerical differentiation, integration, and the solution of differential equations.

The topics just outlined frequently lead to a requirement for solving equations. This is our second main area of concentration; we discuss the solution of an equation in a single unknown, and continue on to consider systems of linear equations and special methods appropriate when the linear systems have a particular form.

At the end of the chapter, we indicate a few general and particular references for further amplification and consultation.

II. Approximation by Polynomial Interpolation

Frequently, we encounter the situation that a physical quantity is described by an unknown function $f(x)$. Observationally, we know $f(x)$ for a sequence of values $x_0, x_1, x_2, \ldots, x_n$ (often these values are equally spaced), and we wish to interpolate from the known values $f(x_0), \ldots, f(x_n)$, to obtain an approximation to the unknown value $f(x)$, where $x_0 < x < x_n$. This same situation arises if the values $f(x_i)$ are known from a tabulation, and we need to obtain $f(x)$, where x does not appear as a table argument.

We assume that $f(x)$ is a continuous function, and define a set of operators on f. These are Δ (the forward difference operator), E (the shift operator), and hD (the differentiation operator). Two subsidiary operators are δ (the central difference operator) and μ (the averaging operator). These operators simplify discussion of many formulae in common use.

In this section, we assume that $f(x)$ is given for equally-spaced values of the argument; thus

$$x_j = x_0 + jh, \quad j = 0, 1, \ldots, n,$$

where h is a fixed positive quantity, called the *interval of differencing*. Thus

$$x_n - x_{n-1} = \cdots = x_2 - x_1 = x_1 - x_0 = h.$$

For simplicity, it is convenient to abbreviate $f(x_j)$ as f_j in subsequent use, and we shall assume that j ranges from 1 to n unless otherwise stated.

The *forward difference operator* Δ is defined by the relation

$$\Delta f_j = f_{j+1} - f_j.$$

The quantity Δf_j is called the *first difference* of $f(x)$ at x_j, and is just the difference between the next functional value and the current functional value. Similarly, we define the *second difference* of $f(x)$ at x_j to be

$$\Delta^2 f_j = \Delta(\Delta f_j) = \Delta(f_{j+1} - f_j) = \Delta f_{j+1} - \Delta f_j$$
$$= (f_{j+2} - f_{j+1}) - (f_{j+1} - f_j) = f_{j+2} - 2f_{j+1} + f_j.$$

An analogous computation establishes that

$$\Delta^3 f_j = f_{j+3} - 3f_{j+2} + 3f_{j+1} - f_j,$$

and it is easy to prove, by mathematical induction, that the pth difference, $\Delta^p f_j$, which is defined by

$$\Delta^p f_j = \Delta^{p-1} f_{j+1} - \Delta^{p-1} f_j,$$

can be expressed in the form

$$\Delta^p f_j = \sum_{k=0}^{p} (-1)^{k+p} \binom{p}{k} f_{j+k},$$

where $\binom{p}{k} = p(p-1) \cdots (p-k+1)/k!$.

The *shift operator* or *displacement operator* E is defined by the relation

$$E f_j = f_{j+1}.$$

Thus, E can be regarded as moving the value f_j along to the next value f_{j+1}. Clearly, we have

$$E^2 f_j = E(E f_j) = E f_{j+1} = f_{j+2};$$

in general, a similar calculation shows that

$$E^p f_j = f_{j+p}.$$

We may also define the inverse operator E^{-1} by the relation

$$E^{-1} f_{j+1} = f_j.$$

The *averaging operator* μ and the *central difference operator* δ are defined by

$$\mu f_j = \tfrac{1}{2}(f_{j+\frac{1}{2}} + f_{j-\frac{1}{2}}), \qquad \delta f_j = f_{j+\frac{1}{2}} - f_{j-\frac{1}{2}}.$$

It is easy to verify that

$$\mu^2 f_j = \tfrac{1}{2}[\tfrac{1}{2}(f_{j+1} + f_j) + \tfrac{1}{2}(f_j + f_{j-1})] = \tfrac{1}{4}[f_{j+1} + 2f_j + f_{j-1}],$$
$$\delta^2 f_j = \delta(f_{j+\frac{1}{2}} - f_{j-\frac{1}{2}}) = f_{j+1} - 2f_j + f_{j-1}.$$

Our final operator hD is an extremely important one, the *differentiation operator*. We define

$$hDf_j = h[df(x)/dx]_{x=x_j} = hf_j',$$

where the prime denotes differentiation with respect to x; the factor h is included for later computational convenience.

Now any reasonable function $f(x)$ can be expanded in a Taylor Series to give

$$f_{j+1} = f_j + \frac{h}{1!} f_j' + \frac{h^2}{2!} f_j'' + \frac{h^3}{3!} f_j''' + \cdots,$$

where we use $f_j^{(k)}$ to denote

$$[d^k f(x)/dx^k]_{x=x_j}.$$

Thus

$$f_{j+1} = f_j + \frac{hD}{1!} f_j + \frac{(hD)^2}{2!} f_j + \frac{(hD)^3}{3!} f_j + \cdots$$

$$= \left[1 + \frac{hD}{1!} + \frac{(hD)^2}{2!} + \frac{(hD)^3}{3!} + \cdots\right] f_j = e^{hD} f_j.$$

This is the way in which the operator hD most frequently occurs; equivalently, we may write

$$f_{j-1} = e^{-hD} f_j.$$

We can now sum up our remarks on operators by introducing operator equivalence. We note that $\Delta f_j = f_{j+1} - f_j = (E - 1) f_j$, where we use 1 to indicate the identity operation; thus, the result of applying the operator Δ is equivalent to the result of using the operator $E - 1$. So we call Δ and $E - 1$ *equivalent operators*, and write

$$\Delta \equiv E - 1.$$

Similarly, Taylor's Theorem takes a simple operator form; the relation

$$E f_j = f_{j+1} = e^{hD} f_j$$

gives us the operator equivalence

$$E \equiv e^{hD}.$$

Other similar relations between operators, all obtained by simple manipulation from the definitions, are given in Table I.

Now that we have introduced the difference operator Δ, let us exemplify its use in interpolation. For definiteness, consider a function $f(x)$ tabulated at x_j, where $x_0 = 0$, $h = 2$, $f(x) = x^3 - 3x^2 + 5x + 7$. Then

TABLE I

Operator Equivalents

	E	Δ	δ, μ	hD
E	—	$1 + \Delta$	$1 + \mu\delta + \delta^2$	e^{hD}
Δ	$E - 1$	—	$\mu\delta + \delta^2/2$	$e^{hD} - 1$
δ	$E^{1/2} - E^{-1/2}$	$\Delta(1 + \Delta)^{-1/2}$	—	$2\sinh(hD/2)$
hD	$\log E$	$\log(1 + \Delta)$	$2\sinh^{-1}\delta/2$	—
μ	$\frac{1}{2}(E^{1/2} + E^{-1/2})$	—	$(1 + \delta^2/4)^{1/2}$	$\cosh(hD/2)$

the differences of $f(x)$ can be conveniently arranged in the format of Table II.

From this difference table, we see that the third differences of the given function are constant. This appears to indicate a relation between the degree of the polynomial (here, 3) and the order of the first constant difference. Indeed, we can establish this by writing

$$\Delta f_j = (E - 1)f_j = (e^{hD} - 1)f_j.$$

Then

$$\Delta^3 f_j = (e^{hD} - 1)^3 f_j = [hD + (h^2D^2/2) + (h^3D^3/6)]^3 f_j.$$

TABLE II

Difference Table for $x^3 - 3x^2 - 5x + 7$

x_j	$f(x_j)$	$\Delta f(x_j)$	$\Delta^2 f(x_j)$	$\Delta^3 f(x_j)$	$\Delta^4 f(x_j)$
0	7				
		6			
2	13		24		
		30		48	
4	43		72		0
		102		48	
6	145		120		0
		222		48	
8	367		168		
		390			
10	757				

Note that we can omit any terms which involve D^4 or higher powers since f is a cubic polynomial, and therefore the fourth derivative vanishes (as do all higher derivatives). By cubing, we simplify the last relation to

$$\Delta^3 f_j = h^3 D^3 f_j;$$

again we omit terms involving D^k for $k \geq 4$, since $D^4 f = 0$. Thus

$$\Delta^3 f_j = h^3 f_j''' = h^3(6) = 48,$$

and we have established the constancy of the third differences in Table II.

An exactly similar argument shows that, if $f(x)$ is a polynomial of degree m, say

$$f(x) = \sum_{i=0}^{m} a_i x^i,$$

then $\Delta^m f_j$ is a constant, and is given by the formula

$$\Delta^m f_j = (e^{hD} - 1)^m f_j = h^m f_j^{(m)} = h^m m! a_m.$$

We are now in a position to use Table II for interpolation (that is, the calculation of additional non-tabular values from the tabulated values of a given function). The simplest of many available interpolation formulae is due to Newton, and is easily obtained operationally as

$$f_{j+p} = E^p f_j = (1 + \Delta)^p f_j = \sum_{i=0}^{\infty} \binom{p}{i} \Delta^i f_j.$$

This equation is known as *Newton's Advancing Difference Formula*. If p is a positive integer, then $\Delta^i f_j = 0$ for $i > p$, and the series terminates; otherwise, we require that $\Delta^i f_j$ tend to zero rapidly as i increases.

If we apply Newton's Advancing Difference Formula to obtain $f(5.5)$ in Table II, we obtain

$$f(5.5) = f_{2.75} = E^{2.75} f_0 = E^{0.75} f_2.$$

Thus

$$f(5.5) = (1 + \Delta)^{3/4} f_2 = \left(1 + \frac{3\Delta}{4} - \frac{3\Delta^2}{32} + \frac{5\Delta^3}{128}\right) f_2$$

$$= 43 + \frac{3(102)}{4} - \frac{3(120)}{32} + \frac{5(48)}{128}$$

$$= 43 + 76.5 - 11.25 + 1.875 = 110.125.$$

5. Numerical Analysis

In general, not all higher differences of $f(x)$ will vanish identically, since usually $f(x)$ is not a polynomial. The most that can be hoped is that, for h sufficiently small, higher order differences are small enough that they can be neglected. The problem with higher order differences is illustrated in Table III.

TABLE III

$f(x) = 10^5 \tan x$

x_j°	f_j	Δf_j	$\Delta^2 f_j$	$\Delta^3 f_j$	$\Delta^4 f_j$	$\Delta^5 f_j$
35	70021					
		5334				
37	75355		289			
		5623		39		
39	80978		328		5	
		5951		44		4
41	86929		372		9	
		6323		53		4
43	93252		425		13	
		6748		66		4
45	100000		491		17	
		7239		83		4
47	107339		574		21	
		7813		104		
49	115052		678			
		8491				
51	123543					

The values tabulated in Table III are those of x and $f(x)$ for $x = 35°$, $37°, \ldots, 45°$; these values are differenced as far as possible, and produce the upper portion of the table. Interpolation can then be carried out by using the formula

$$f_{j+p} = \sum_{i=0}^{\infty} \binom{p}{i} \Delta^i f_j$$

with $j = 0$, $35 \leq p \leq 45$, $h = 2$. The assumption made in such an interpolation is that sixth differences are zero; this is tantamount to assuming that $f(x)$, as tabulated in the range $(35°, 45°)$ is represented by a quintic polynomial. Since fifth differences are small, such an assumption is reasonable, and we can expect excellent interpolatory results; what we are really doing is replacing the relation

$$f_{j+p} = \sum_{i=0}^{\infty} \binom{p}{i} \Delta^i f_j$$

by the truncated relation

$$f_{j+p} = \sum_{i=0}^{k} \binom{p}{i} \Delta^i f_j,$$

where $k = 5$ in this example.

On the other hand, if we wish to extrapolate beyond the limits of the original table, we assume that the fifth difference is constant, and work diagonally downward in Table III to give the lower portion of the table and the three values $f(47)$, $f(49)$, $f(51)$. These three new values are thus obtained, in the extended range (35°, 51°), by using the same quintic polynomial which represented the function well, within the original range; since we require the extra assumption that the same polynomial works well in the extended range, we expect extrapolation to produce less accurate values than interpolation. This may be checked by comparing the approximations to $f(44°)$ and $f(51°)$ with the values tabulated in a larger table.

Actually, if h is not small enough to make higher differences small, we can even run into trouble with interpolation (for example, if $f(x) = e^x$, $h = 1$, $x_j = j$ for $j = 0, 1, 2, \ldots, 5$, interpolation is not successful).

In summary, we can basically draw three conclusions.

1. If the function $f(x)$ is a polynomial, then, provided that enough differences are used, both interpolation and extrapolation are exact.

2. If the differences decrease rapidly towards zero, then interpolation is accurate; extrapolation is possible, but gets worse as we progress farther and farther past the table limits.

3. If the differences do not decrease, or decrease only slowly, then neither interpolation nor extrapolation is likely to give meaningful results. The solution of the difficulty is to decrease the value of h and try again.

Up till this stage, we have assumed a constant interval h between successive x values. We now drop this requirement, and consider the problem of representing tabulated values (x_j, f_j), where $j = 0, 1, \ldots, n$, by that unique polynomial of degree n which passes through the $n + 1$ tabular points. This polynomial is called the *Lagrange Interpolation Polynomial* of degree n.

From the definition, we see that the Lagrange Interpolation Polynomial can be written in the form

$$y(x) = \sum_{j=0}^{n} k_j(x) f_j,$$

where the polynomials $k_j(x)$ must satisfy the conditions that $k_j(x_j) = 1$, $k_j(x_i) = 0$ for $i \neq j$; these conditions guarantee that all points (x_j, f_j) lie on the graph of the polynomial. Thus $k_j(x)$ is a polynomial which possesses zeros at $x_0, x_1, \ldots, x_{j-1}, x_{j+1}, \ldots, x_n$, and so must have the form

$$k_j(x) = a_j \prod_{\substack{j=0 \\ i \neq j}}^{n} (x - x_i).$$

By putting $x = x_j$, we solve for the constant a_j, and find that

$$(a_j)^{-1} = \prod_{\substack{i=0 \\ i \neq j}}^{n} (x_j - x_i).$$

If we define $p_n(x) = \prod_{i=0}^{n} (x - x_i)$, we see that

$$k_j(x) = a_j p_n(x)/(x - x_j).$$

Furthermore,

$$p_n'(x_j) = [dp_n(x)/dx]_{x=x_j} = (a_j)^{-1}.$$

Thus we can write the Lagrange interpolation polynomial as

$$y(x) = \sum_{j=0}^{n} \frac{p_n(x)}{(x - x_j) p_n'(x_j)} f_j.$$

This representation of the interpolation polynomial does not require generation and storage of large difference tables; hence it is good for high-speed work. However, it does suffer from the disadvantage that there is no indication as to whether we have used a polynomial of excessive degree (for instance, a polynomial of degree 10 when $f(x)$ is really quadratic); also, it is difficult to improve the interpolated value without recalculation of a higher-order polynomial, whereas difference methods of interpolation merely require the addition of another term.

Note, in particular, that for $n = 1$, the Lagrange Interpolation Polynomial reduces to

$$y(x) = \sum_{i=0}^{1} \frac{p_1(x)}{(x - x_j) p_1'(x_j)} f_j = \frac{x - x_1}{x_0 - x_1} f_0 + \frac{x - x_0}{x_1 - x_0} f_1.$$

This is the ordinary linear interpolator between (x_0, f_0) and (x_1, f_1).

III. Approximation by Spline Interpolation

It may be that use of the Lagrange Interpolation Polynomial is undesirable because of the fact that a polynomial of degree n which exactly fits $n + 1$ tabular points can have a number of extrema and inflection points, and so may fluctuate too much between tabular points. This situation can be avoided by using a spline approximation.

Originally, a spline was a heavy plastic ruler constrained to fit a set of data points (x_i, f_i). Such a mechanism avoided abrupt inflections and extrema in the approximating function. Mathematically, *cubic splines* are most used, and we shall couch our discussion in terms of these, although the generalization to quintic and higher-order splines is obvious.

A cubic spline $s(x)$ can be defined in a variety of different but equivalent ways (cf. Powell, 1966; Ahlberg, et al., 1967; Schoenberg, 1969). However, all forms of definition describe the cubic spline in such a way that it passes exactly through some given set of points (x_0, y_0), (x_1, y_1), $\ldots$, (x_n, y_n), has continuous first and second derivatives, and is at most a cubic polynomial between any adjacent pair of points (x_{j-1}, y_{j-1}) and (x_j, y_j). The salient feature of the spline $s(x)$ is that different cubic polynomials are used between different pairs of points; this ensures greater flexibility than if a single polynomial were employed.

We can construct the spline $s(x)$ in the following manner. Let the second derivative of $s(x)$ be denoted by $s''(x)$, and let $s''(x_j)$ be abbreviated to s_j''. Since the function $s(x)$ is at most a cubic in the interval $[x_{j-1}, x_j]$, it follows that $s''(x)$ is linear in that interval. Thus $s''(x)$ is given by the ordinary linear relation

$$s''(x) = s''_{j-1}(x_j - x)/h_j + s''_j(x - x_{j-1})/h_j,$$

where $x_{j-1} \leq x \leq x_j$ and $h_j = x_j - x_{j-1}$. This equation can be integrated twice with respect to x to obtain the further relation

$$s(x) = s''_{j-1}(x_j - x)^3/6h_j + s''_j(x - x_{j-1})^3/6h_j + Ax + B,$$

where A and B are two, as yet undetermined, constants of integration. It is necessary that the final function $s(x)$ interpolate to the given (x, y) pairs; hence, the constants A and B can be determined by requiring that $s(x_j) = y_j$ and $s(x_{j-1}) = y_{j-1}$. Explicitly, this requirement produces the two simultaneous equations

$$Ax_j + B = y_j - h_j^2 s''_j/6 \quad \text{and} \quad Ax_{j-1} + B = y_{j-1} - h_j^2 s''_{j-1}/6$$

with solutions

$$A = (y_j - y_{j-1})/h_j - (h_j/6)(s_j'' - s_{j-1}'')$$
$$B = (x_j y_{j-1} - x_{j-1} y_j)/h_j - (h_j/6)(x_j s_{j-1}'' - x_{j-1} s_j'').$$

Substitution of these expressions for A and B gives the final form for the cubic spline in the interval $[x_{j-1}, x_j]$ as

$$s(x) = s_{j-1}'' \frac{(x_j - x)^3}{6h_j} + s_j'' \frac{(x - x_{j-1})^3}{6h_j} + \frac{(x_j - x)}{h_j}\left(y_{j-1} - s_{j-1}'' \frac{h_j^2}{6}\right)$$
$$+ \frac{(x - x_{j-1})}{h_j}\left(y_j - s_j'' \frac{h_j^2}{6}\right).$$

This function is defined in the interval $[x_{j-1}, x_j]$, and has the property that it interpolates to y_{j-1} and y_j at x_j and x_{j-1}; also, its second derivative interpolates to the quantities s_j'' and s_{j-1}'' at x_j and x_{j-1}. Now if we consider the compound function $s(x)$, defined in $[x_0, x_n]$, obtained by using each interval $[x_{j-1}, x_j]$ in turn for $j = 1, 2, 3, \ldots, n$, then two observations can be made. Each of the sections of $s(x)$ interpolates to the appropriate function values, that is, the function is continuous in $[x_0, x_n]$; also, since $s''(x)$ has been defined in such a way that the sections interpolate to the quantities s_j'' ($j = 0, 1, 2, \ldots, n$), the second derivative is continuous as well. However, we still have to arrange that the first derivative of $s(x)$ be continuous. Differentiation with respect to x in the interval $[x_{j-1}, x_j]$ gives the result

$$s'(x) = -s_{j-1}'' \frac{(x_j - x)^2}{2h_j} + s_j'' \frac{(x - x_{j-1})^2}{2h_j} - \frac{1}{h_j}\left(y_{j-1} - s_{j-1}'' \frac{h_j^2}{6}\right)$$
$$+ \frac{1}{h_j}\left(y_j - s_j'' \frac{h_j^2}{6}\right)$$

and we obtain the left-hand limit for $s'(x_j)$, denoted by $s'(x_j -)$, as

$$s'(x_j -) = (y_j - y_{j-1})/h_j + (h_j/6)s_{j-1}'' + (h_j/3)s_j'',$$

where j ranges from 1 to n. The right-hand limit for $s'(x_j)$, denoted by $s'(x_j +)$, can be obtained by replacing the subscript j by $j+1$ and evaluating the expression at x_j to yield

$$s'(x_j +) = (y_{j+1} - y_j)/h_{j+1} - (h_{j+1}/3)s_j'' - (h_{j+1}/6)s_{j+1}'',$$

where j ranges from 0 to $n-1$. It is clear that the requirement of continuity on $s'(x)$ at each of the points x_j ($j=1, 2, 3, \ldots, n-1$) implies that both left and right limits of the derivative be equal. Thus, $s'(x_i -) = s'(x_i +)$, that is,

$$\frac{h_i}{6} s''_{i-1} + \frac{(h_i + h_{i+1})}{3} s''_i + \frac{h_{i+1}}{6} s''_{i+1} = \frac{(y_{i+1} - y_i)}{h_{i+1}} - \frac{(y_i - y_{i-1})}{h_i},$$

for $i = 1$ to $n - 1$. This last equation is known as the continuity equation, and is an equation relating the unknown second derivatives of the spline with the given functional values. A count of unknowns indicates that there are $n + 1$ unknowns, $s''_0, s''_1, \ldots, s''_n$; however, there are only $n - 1$ equations. Two additional equations are thus necessary for unique determination of the cubic spline; for example, these might be *initial conditions* that the function $s(x)$ have given first and second derivatives $y'(x_0)$ and $y''(x_0)$. An alternative requirement could be that the spline must have a given first derivative both at x_0 and x_n; this would specify $y'(x_0)$ and $y'(x_n)$, and the two additional equations would be described as *boundary conditions*.

TABLE IV

i	x_i	$y(x)$
0	0	0
1	1	0
2	2	16
3	3	162

As a definite illustration of our discussion, let us consider the example given by the tabulated values in Table IV and the two boundary conditions $y'(x_0) = 0$ and $y'(x_n) = 297$. Here $n = 3$, and the boundary conditions give us the equations

$$y'(x_0 +) = (y_1 - y_0)/h_1 - (h_1/3)s''_0 - (h_1/6)s''_1 = 0$$

and

$$y'(x_3 -) = (y_3 - y_2)/h_3 + (h_3/6)s''_2 + (h_3/3)s''_3 = 297.$$

Since $h_j = x_j - x_{j-1}$, h_j is equal to 1 for all j for the data of Table IV. Thus we obtain the simplified equations

$$\tfrac{1}{3} s_0'' + \tfrac{1}{6} s_1'' = 0 \quad \text{and} \quad \tfrac{1}{6} s_2'' + \tfrac{1}{3} s_3'' = 151.$$

Also, in this example, the continuity equation becomes

$$\tfrac{1}{6} s_{i-1}'' + \tfrac{2}{3} s_i'' + \tfrac{1}{6} s_{i+1}'' = y_{i+1} - 2 y_i + y_{i-1}, \quad i = 1, 2.$$

We thus possess four simultaneous linear equations in the unknowns s_0'', s_1'', s_2'', s_3'', namely,

$$\tfrac{1}{3} s_0'' + \tfrac{1}{6} s_1'' \qquad\qquad\qquad = 0,$$
$$\tfrac{1}{6} s_0'' + \tfrac{2}{3} s_1'' + \tfrac{1}{6} s_2'' \qquad\quad = 16,$$
$$\tfrac{1}{6} s_1'' + \tfrac{2}{3} s_2'' + \tfrac{1}{6} s_3'' = 130,$$
$$\tfrac{1}{6} s_2'' + \tfrac{1}{3} s_3'' = 151.$$

It is easy to verify that the solution is

$$s_0'' = -0.4, \quad s_1'' = 0.8, \quad s_2'' = 93.2, \quad s_3'' = 406.4.$$

More general boundary conditions for spline approximations can be discussed in an anologous fashion. If boundary conditions are given as

$$a y_0' + b y_0'' = c \quad \text{and} \quad d y_n' + e y_n'' = f,$$

where a, b, c, d, e, f, are specified numerical quantities, then the boundary conditions give us

$$a\left(\frac{y_1 - y_0}{h_1} - \frac{h_1}{3} s_0'' - \frac{h_1}{6} s_1'' \right) + b y_0'' = c$$

and

$$d\left(\frac{y_n - y_{n-1}}{h_n} + \frac{h_n}{6} s_{n-1}'' + \frac{h_n}{3} s_n'' \right) + e y_n'' = f$$

as the two extra equations (the other $n - 1$ equations are still obtained from the continuity equation). A Fortran algorithm for the determination of the quantities $s_0'', s_1'', \ldots, s_n''$, when given the quantities (x_0, y_0), $(x_1, y_1), \ldots, (x_n, y_n)$, and general conditions of the form $a y_0' + b y_0'' = c$, $d y_n' + e y_n'' = f$, is given in the Appendix.

Frequently, we wish to use a spline which is periodic with period $x_n - x_0$, that is, $s(x) = s(x + x_n - x_0)$. In such a case, the continuity equation can be written in the form

$$\frac{h_i}{6} s''_{i-1} + \frac{(h_i + h_{i+1})}{3} s''_i + \frac{h_{i+1}}{6} s''_{i+1} = \frac{(y_{i+1} - y_i)}{h_{i+1}} - \frac{(y_i - y_{i-1})}{h_i}$$

for $i = 1, 2, 3, \ldots, n$. Since the periodicity condition requires that $y_0 = y_n$ and $s''_0 = s''_n$, these equations represent n equations in the n unknowns $s''_1, s''_2, s''_3, \ldots, s''_n$. An algorithm for solving these equations and thus determining the periodic spline has been published (Hoskins and King, 1972).

An interesting practical application of periodic splines arises if we consider the problem of fitting a curve, smooth in the sense that its first and second derivatives are continuous, through an ordered periodic sequence of points in the plane. We illustrate this by considering the set of ordered points given in Table V and illustrated in Fig. 1. The problem

TABLE V

j	x_j	y_j
0	1	1
1	2	0.5
2	2	1.5
3	1	1
4	0.5	0.5
5	0.5	1.5
6	1	1

of fitting a smooth curve through the data of Table V in the given order is not practicable using the simple cubic splines previously described. The difficulty arises because the curve to be fitted possesses vertical tangents. A solution is found by expressing x and y in terms of some parameter, say, the cumulative chordal distance squared. We then obtain, as an intermediate step, the data in Table VI. Then, we carry out a final step of fitting a periodic spline to x against t and another periodic spline

5. Numerical Analysis

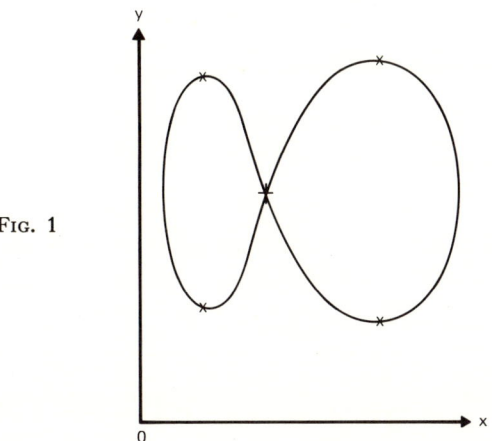

Fig. 1

to y against t. The curve of y against x is then obtained in the form $(t_{j-1} \leq t \leq t_j)$:

$$z(t) = z''_{j-1} \frac{(t_j - t)^3}{6h_j} + z''_j \frac{(t - t_{j-1})^3}{6h_j} + \frac{(t_j - t)}{h_j} \left(x_{j-1} - z''_{j-1} \frac{h_j^2}{6} \right)$$
$$+ \frac{(t - t_{j-1})}{h_j} \left(x_j - z''_j \frac{h_j^2}{6} \right),$$

$$s(t) = s''_{j-1} \frac{(t_j - t)^3}{6h_j} + s''_j \frac{(t - t_{j-1})^3}{6h_j} + \frac{(t_j - t)}{h_j} \left(y_{j-1} - s''_{j-1} \frac{h_j^2}{6} \right)$$
$$+ \frac{(t - t_{j-1})}{h_j} \left(y_j - s''_{j-1} \frac{h_j^2}{6} \right).$$

TABLE VI

j	x_j	y_j	$t_j = \sum_{k=0}^{j} [(x_{k+1} - x_k)^2 + (y_{k+1} - y_k)^2]$
0	1	1	0
1	2	0.5	1.25
2	2	1.5	2.25
3	1	1	3.50
4	0.5	0.5	4.00
5	0.5	1.5	5.00
6	1	1	5.50

In these equations, $h_j = t_j - t_{j-1}$, and $z(t)$ and $s(t)$ represent the x spline and the y spline, respectively; derivatives are taken with respect to t.

The resulting fit to the data of Table VI is shown in Fig. 2; we call the approximating curve a *periodic parametric spline*.

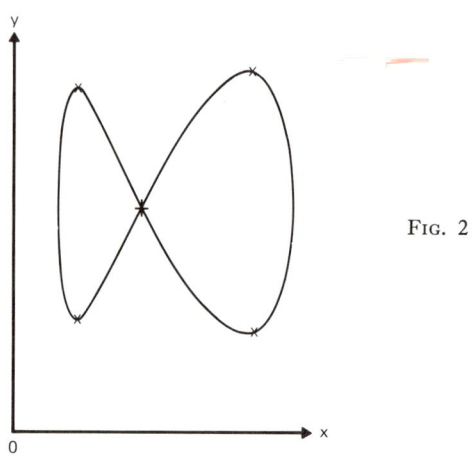

Fig. 2

IV. Approximation by Least Squares

It frequently occurs that we encounter a situation where n data points $(x_1, y_1), \ldots, (x_n, y_n)$ are known; usually the x_i are assigned values, and the y_i are obtained by measurement, and so are subject to random fluctuation. Further, one often knows, from theoretical considerations, that these observations ideally lie upon a certain curve $y = f(x; \alpha, \beta, \gamma, \ldots)$, where the $\alpha, \beta, \gamma, \ldots$, are free parameters. Then, one is basically confronted with an optimization procedure. To handle this case, let us make a model of the physical situation.

We make observations at assigned values $x_1, x_2, \ldots, x_n$, of the independent variable; suppose that the observation at x_i is y_i. We then have n data points (x_i, y_i) which we can plot in the plane, and which theoretically should lie on the curve $y = f(x; \alpha, \beta, \ldots)$. Thus, for any i, we have two y values, namely,

$$y_i, \quad \text{the observed value}$$
$$f_i = f(x_i; \alpha, \beta, \ldots), \quad \text{the theoretical value.}$$

If we form the sum of squared deviations between these two values, we obtain a quantity

$$S = \sum_{i=1}^{n} (y_i - f_i)^2$$

which would be zero if all the data points were to lie exactly on the theoretical curve. Because of random fluctuation (experimental deviations), it is unlikely that the data points lie exactly on a curve of the desired form. Hence, we can not expect S to be zero; and so we rather try to make S as small as possible (*principle of least squares*). Since S is solely a function of the parameters, α, β, ..., we select these parameters so that S is a minimum. This can of course be done by setting

$$\partial S/\partial \alpha = \partial S/\partial \beta = \cdots = 0,$$

and these equations, being equal in number to the number of parameters, just suffice to determine estimates of the parameters α, β, γ, ...

Two examples should make the procedure clear. In the first example, we suppose that we are given that n data points $(x_1, y_1), \ldots, (x_n, y_n)$ lie approximately on the theoretical curve $y = \alpha + \beta x$. We wish to determine the two parameters α and β so that the line $y = \alpha + \beta x$ best describes the n data points (in the sense of least squares).

Here $f_i = \alpha + \beta x_i$, and the sum of squared deviations between observed and theoretical y values is

$$S = \sum_{i=1}^{n} (y_i - \alpha - \beta x_i)^2.$$

Then

$$\partial S/\partial \alpha = \sum_{i=1}^{n} - 2(y_i - \alpha - \beta x_i) = 0,$$

$$\partial S/\partial \beta = \sum_{i=1}^{n} - 2x_i(y_i - \alpha - \beta x_i) = 0.$$

These two equations simplify (all summations from $i = 1$ to $i = n$) to

$$\sum y_i - n\alpha - \beta \sum x_i = 0, \qquad \sum x_i y_i - \alpha \sum x_i - \beta \sum x_i^2 = 0.$$

If we write these equations in the normal form

$$n\alpha + \beta \sum x_i = \sum y_i, \qquad \alpha \sum x_i + \beta \sum x_i^2 = \sum x_i y_i,$$

we see that we have two linear equations for the two unknowns α and β

(the coefficients n, $\sum x_i$, $\sum x_i^2$, $\sum y_i$, $\sum x_i y_i$, all being readily computable from the given data).

As a second example, let us suppose that the n data points of the preceding example should theoretically lie on the curve $y = \alpha + \beta x + \gamma x^2$; let us develop the equations for α, β, γ.

The procedure of the preceding example immediately produces the quantity

$$S = \sum (y_i - \alpha - \beta x_i - \gamma x_i^2)^2,$$

and the equations $\partial S/\partial \alpha = \partial S/\partial \beta = \partial S/\partial \gamma = 0$ simplify to

$$\sum y_i = n\alpha + \beta \sum x_i + \gamma \sum x_i^2,$$
$$\sum x_i y_i = \alpha \sum x_i + \beta \sum x_i^2 + \gamma \sum x_i^3,$$
$$\sum x_i^2 y_i = \alpha \sum x_i^2 + \beta \sum x_i^3 + \gamma \sum x_i^4.$$

In this case, we have three linear equations for the three unknowns α, β, γ.

As a numerical illustration, suppose that observations are made for $x = 1, 2, \ldots, 10$, and that the corresponding y values are 2.54, 6.46, 12.02, 19.19, 29.76, 41.91, 54.67, 70.88, 89.34, 108.62. Then one readily calculates

$$n = 10, \quad \sum x_i = 55, \quad \sum x_i^2 = 385, \quad \sum x_i^3 = 3025, \quad \sum x_i^4 = 25333,$$
$$\sum y_i = 435.39, \quad \sum x_i y_i = 3368.53, \quad \sum x_i^2 y_i = 28010.05.$$

The resulting system of linear equations is then found to be

$$11\alpha + 55\beta + 385\gamma = 435.39,$$
$$55\alpha + 385\beta + 3025\gamma = 3368.53,$$
$$385\alpha + 3025\beta + 25333\gamma = 28010.05.$$

By the methods of a later section, we immediately solve these equations to get

$$\alpha = 0.37622, \quad \beta = 0.860517, \quad \gamma = 0.9972027,$$

and our best-fitting curve of the form $y = \alpha + \beta x + \gamma x^2$ is simply

$$y = 0.37622 + 0.860517x + 0.9972027x^2.$$

From this relation, we easily compute the values $f_1, f_2, \ldots, f_{10}$ for this curve (for example, $f_{10} = 0.37622 + 8.60517 + 99.72027 = 108.70166$), and can then calculate the minimal S value as $\sum (y_i - f_i)^2 = 1.4239$.

In some cases, work can be avoided by using a transformation before applying the principle of least squares. For example, suppose that the data are known, on theoretical grounds, to follow an exponential law $y = \alpha \exp \beta x$. Then we may take logarithms, and write $Y = \gamma + \beta x$, where $Y = \log y$, $\gamma = \log \alpha$. The method of our first example is then applicable to the linear relation $Y = \gamma + \beta x$, where our new "data points" are (x_i, Y_i), Y_i being $\log y_i$. After finding the best values for γ and β, we obtain an estimate of α from the relation $\gamma = \log \alpha$.

Another common relationship which can be simplified by taking logarithms is $y = \alpha x^\beta$.

V. Numerical Differentiation

An interpolation formula provides a continuous polynomial approximation to a function which is given only for discrete values of the independent variable. If this formula is differentiated, then we obtain an approximation to the derivatives of the function. For example, Newton's Advancing Difference Formula is given by

$$f_{j+x} = \sum \binom{x}{i} \Delta^i f_j$$
$$= f_j + \frac{x}{1!} \Delta f_j + \frac{x(x-1)}{2!} \Delta^2 f_j + \frac{x(x-1)(x-2)}{3!} \Delta^3 f_j + \cdots,$$

and, if this result is differentiated with respect to x, we obtain an expression for the first derivative of $f(x)$ in the form

$$f'_{j+x} = \Delta f_j + \frac{2x-1}{2} \Delta^2 f_j + \frac{3x^2 - 6x + 2}{6} \Delta^3 f_j + \cdots .$$

This procedure can be applied to any general interpolation formula. Usually, the simplest and most convergent expressions occur for the odd derivatives at the "half-way points" $x_i + h/2$, and for the even derivatives at the tabular points x_i.

Simple expressions for differences in terms of derivatives can be derived by using the Taylor series expansion. We merely write

$$f_{j+1} = f_j + hf'_j + (h^2/2!)f''_j + (h^3/3!)f'''_j + \cdots$$

and

$$f_{j-1} = f_j - hf'_j + (h^2/2!)f''_j - (h^3/3!)f'''_j + \cdots .$$

Subtraction of these two equations yields

$$\tfrac{1}{2}(f_{j+1} - f_{j-1}) = hf_j' + (h^3/3!)f_j''' + \cdots;$$

in terms of the central-difference operators μ and δ, this equation becomes

$$\mu\delta f_j = hf_j' + (h^3/3!)f_j''' + \cdots;$$

or simply

$$\mu\delta f_j \approx hf_j',$$

if we approximate to terms of order h^3.

Similarly, addition of the expressions for f_{j+1} and f_{j-1} produces first

$$\tfrac{1}{2}(f_{j+1} - 2f_j + f_{i-1}) = (h^2/2!)f_j'' + (h^4/4!)f_j'''' + \cdots,$$

and then, using the definition of the operator δ,

$$\tfrac{1}{2}\delta^2 f_j = (h^2/2!)f_j'' + (h^4/4!)f_j'''' + \cdots.$$

Finally, to terms of order h^4, we can write the approximation

$$\delta^2 f_j \approx h^2 f_j''.$$

The results $hf_j' \approx \mu\,\delta f_j$ and $h^2 f_j'' \approx \delta^2 f_j$ are the most commonly used approximations for the first and second derivatives of a function. However, more systematic and general methods of obtaining expansions of the form

$$h^k f_j^{(k)} = \sum_{i=(k-1)/2}^{\infty} a_i \mu\, \delta^{2i+1} f_j \qquad (k \text{ odd})$$

and

$$h^k f_j^{(k)} = \sum_{i=k/2}^{\infty} a_i\, \delta^{2i} f_j \qquad (k \text{ even}),$$

for $k = 1, 2, 3, \ldots$, are possible using an operator approach. We now illustrate this operator approach.

Consider the definition of the operator hD, namely,

$$hf_j' = hDf_j.$$

Now, from Table I, we may write

$$hD \equiv 2\sinh^{-1} \delta/2 \qquad \text{and} \qquad \mu \equiv (1 + \delta^2/4)^{1/2}.$$

Then we obtain the result

$$hf_j' = \mu\mu^{-1}hDf_j = \mu\mu^{-1}2(\sinh^{-1}\delta/2)f_j = \mu(1+\delta^2/4)^{-1/2}2(\sinh^{-1}\delta/2)f_j.$$

Now, using the Binomial Theorem,

$$(1 + \delta^2/4)^{1/2} = 1 - \delta^2/8 + 3\delta^4/128 + \cdots,$$

and, from Abramowitz and Stegun (1968),

$$2\sinh^{-1}\delta/2 = \delta - \delta^3/24 + 3\delta^5/640 + \cdots.$$

By substitution of these series and collection of terms, we obtain the result

$$hf_j' = \{\mu\delta - \mu\delta^3/6 + \mu\delta^5/30 + \cdots\}f_j.$$

It is slightly easier to obtain the relation for $h^2 f_j''$. In this case, a similar approach yields

$$h^2 f'' = (hD)^2 f_j = \{2\sinh^{-1}\delta/2\}^2 f_j$$
$$= \{\delta - \delta^3/24 + 3\delta^5/640 + \cdots\}^2 f_j = \{\delta^2 - \delta^4/12 + \delta^6/90 + \cdots\}f_j.$$

Formulae for higher-order odd and even derivatives follow the patterns established by these formulae for hf_j' and $h^2 f_j''$, and are given extensively in *Interpolation and Allied Tables* [Her Majesty's Stationery Office, 1956].

If we are given the tabulated values $f_0, f_1, \ldots, f_n$, we may require the derivatives $hf_0', \ldots, hf_n', \ldots$, and $h^2 f_0'', \ldots, h^2 f_n''$, in terms of the tabulated values of f_j. In this case, the formulae we have developed fail at the extremities of the table, since they introduce the values of $f(x)$ at non-tabulated values. To handle this difficulty, we obtain an expansion of hf_0' in terms of powers of Δ. Thus, using Table I,

$$hf_0' = hDf_0 = \log(1 + \Delta)f_0 = (\Delta - \Delta^2/2 + \Delta^3/3 - \Delta^4/4 + \cdots)f_0.$$

This expression gives hf_0' in terms of tabulated values of $f(x)$, and the second derivative can be obtained by an analogous procedure as

$$h^2 f_0'' = (hD)^2 f_0 = \{\log(1 + \Delta)\}^2 f_0$$
$$= \left\{\Delta - \frac{\Delta^2}{2} + \frac{\Delta^3}{3} - \frac{\Delta^4}{4} + \cdots\right\}^2 f_0$$
$$= \{\Delta^2 - \Delta^3 + \tfrac{11}{12}\Delta^4 - \tfrac{5}{6}\Delta^5 + \cdots\}f_0.$$

Formulae at the other boundary of the table for hf_n' and $h^2 f_n''$ are obtained by using these same two expressions with the tabular values of $f_0, f_1, \ldots, f_n$ as $f_n, f_{n-1}, \ldots, f_0$, respectively.

We conclude the section by a brief illustration of the procedure. In Table VII, we exhibit the difference table for a function $f(x)$ which is

TABLE VII

$f(x) = \sinh x$

x	$f(x)$	$\Delta f(x)$	$\Delta^2 f(x)$	$\Delta^3 f(x)$
0.50	0.52110			
		5705		
0.55	0.57815		145	
		5850		15
0.60	0.63665		160	
		6010		13
0.65	0.69675		173	
		6183		18
0.70	0.75858		191	
		6374		14
0.75	0.82232		205	
		6579		17
0.80	0.88811		222	
		6801		17
0.85	0.95612		239	
		7040		17
0.90	1.02652		256	
		7296		20
0.95	1.09948		276	
		7572		
1.00	1.17520			

not cubic (actually, it is a table of $\sinh x = (e^x - e^{-x})/2$). We calculate f' by use of the approximation

$$hf_j' = (\mu\delta - \mu\delta^3/6)f_j.$$

For example,

$$0.05f_2' = \mu\delta f_2 - \mu\delta^3 f_2/6$$
$$= \tfrac{1}{2}\{0.05850 + 0.06010\} - \tfrac{1}{12}\{0.00013 + 0.00015\};$$

hence $f_2' = 1.18554$. Similarly, $h^2 f_j'' = \delta^2 f_j$; thus

$$0.0025 f_1'' = \delta^2 f_1 = \Delta^2 f_0 = 0.00145.$$

Hence $f_1'' = 0.58000$. For comparison, we note that the exact values of the derivatives are $\cosh 0.60 = 1.18547$ and $\sinh 0.55 = 0.57815$ (note that higher derivatives become less accurate).

VI. Approximate Integration or Quadrature

The most common problem encountered in approximate integration is that, given a table of values $\{(x_j, f_j); j = 0, 1, 2, \ldots, n\}$, of finding an estimate for

$$I = \int_{x_0}^{x_n} f(x)\, dx.$$

Pictorially, the situation is illustrated for $n = 5$ in Fig. 3, and it is necessary to find an expression for an approximation to the integral as a function of the values (x_j, y_j).

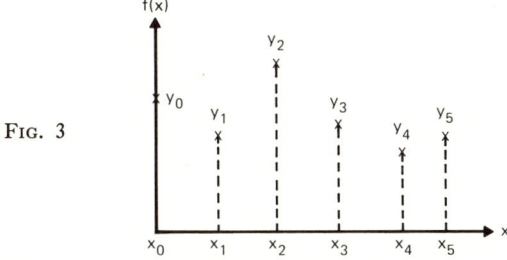

Fig. 3

If we suppose that $x_j = x_0 + jh$, then use of rectangles gives the approximation

$$\int_{x_0}^{x_5} f(x)\, dx \approx \sum_{j=0}^{4} h f_j,$$

where the function $f(x)$ is represented by the shaded area in Fig. 4. This is not a particularly good approximation to a smooth function, and a better one would be to join adjacent ordinates by straight lines as in Fig. 5. Each of the sections in Fig. 5 is a trapezoid, and therefore the

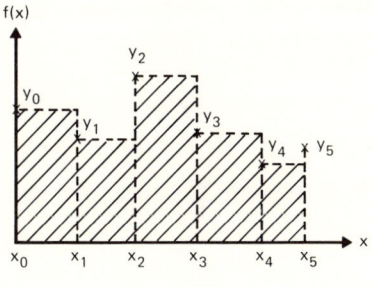

Fig. 4

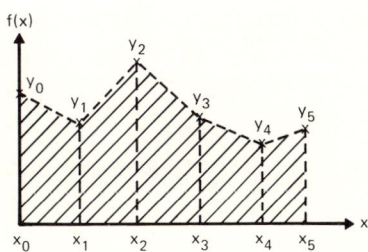

Fig. 5

total shaded area provides the approximation

$$\int_{x_0}^{x_5} f(x)\, dx \approx \sum_{j=0}^{4} \tfrac{1}{2} h(f_j + f_{j+1}).$$

The two approximations just obtained are called, respectively, the Rectangle Rule and the Trapezoidal Rule. The accuracy with which these two methods work is of paramount interest as soon as applications are considered, and then the question of generating formulae of higher accuracy arises. Systematic methods for producing such formulae employ the operators of Section II.

Let us consider

$$I = \int_{x_{j-1}}^{x_{j+1}} f(x)\, dx,$$

and recall that $hDf_j = hf_j'$. We can regard the inverse operator D^{-1} as defined by $D^{-1}f_j' = f_j$, in which case

$$\int_{x_{j-1}}^{x_{j+1}} f(x)\, dx = D^{-1}(f_{j+1} - f_{j-1}).$$

This equation is an *exact* relationship, and all integration formulae over two intervals can be identified as different methods of approximating to the inverse operator D^{-1}.

Using the definitions for μ and δ, this equation can be written more conveniently as

$$\int_{x_{j-1}}^{x_{j+1}} f(x)\, dx = 2h(hD)^{-1}\mu\delta f_j.$$

However,

$$hD \equiv 2 \sinh^{-1} \delta/2 = 2\{\delta/2 - \delta^3/48 + 3\delta^5/1280 + \cdots\},$$

and

$$\mu \equiv (1 + \delta^2/4)^{1/2}.$$

Consequently,

$$\int_{x_{j-1}}^{x_{j+1}} f(x)\, dx = 2h\{\delta - \delta^3/24 + 3\delta^5/640 + \cdots\}^{-1} \delta(1 + \delta^2/4)^{1/2} f_j.$$

Use of the Binomial Theorem and collection of terms produces the general result

$$\int_{x_{j-1}}^{x_{j+1}} f(x)\, dx = 2h\{1 + \delta^2/6 - \delta^4/180 + \delta^6/1512 + \cdots\} f_j.$$

If this equation is truncated to neglect terms of order δ^4 and higher, then we have

$$\int_{x_{j-1}}^{x_{j+1}} f(x)\,dx \approx 2h\{1 + \delta^2/6\}f_j = (h/3)(f_{j+1} + 4f_j + f_{j-1}).$$

This latter equation is known as Simpson's Rule; since its leading error term is of order $\delta^4 f_j$, it will produce an exact result if $f(x)$ is at most a cubic polynomial.

If Simpson's Rule is applied to evaluating the integral

$$I = \int_{x_0}^{x_n} f(x)\,dx,$$

then n must be even, and we find that, with $n = 2m$,

$$I \approx (h/3) \sum (f_{i+2} + 4f_{i+1} + f_i) \qquad i = 0, 2, 4, \ldots, 2m - 2.$$

Simpson's Rule is of such great importance that we conclude this section with an alternative development for it. Suppose that we consider Fig. 6, and pass a vertical parabola through the three points (x_{j-1}, f_{j-1}), (x_j, f_j), and (x_{j+1}, f_{j+1}). Then, for any reasonably smooth curve, this

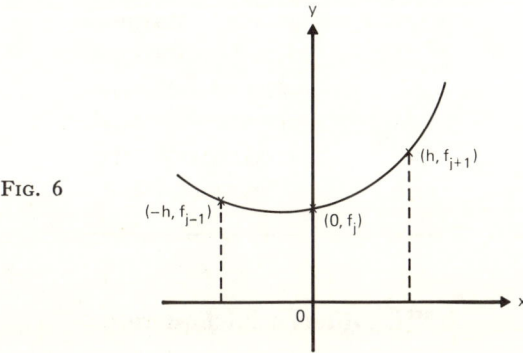

Fig. 6

parabola provides an excellent approximation. If we place a new y axis through the mid-ordinate, then the parabola can be written as $y = \alpha x^2 + \beta x + \gamma$, and we have

$$\int_{x_{j-1}}^{x_{j+1}} f(x)\,dx \approx \int_{-h}^{h} (\alpha x^2 + \beta x + \gamma)\,dx = (\alpha/3)(2h^3) + 2\gamma h$$
$$= (h/3)(2\alpha h^2 + 6\gamma).$$

However,

$$f_{j+1} + 4f_j + f_{j-1} = (\alpha h^2 + \beta h + \gamma) + 4\gamma + (\alpha h^2 - \beta h + \gamma)$$
$$= 2\alpha h^2 + 6\gamma.$$

Substitution thus produces Simpson's Rule.

In practice, one normally applies Simpson's Rule with a specified h, and gets an excellent idea of its accuracy by repeating the calculation with h successively replaced by $h/2$, $h/4$, $h/8$, etc. For instance, if $y = (1+x^2)^{-1}$, we can apply Simpson's Rule with $h = 0.5, 0.25, 0.125, \ldots$, to obtain the successive approximations to $\int_0^1 (1 + x^2)^{-1} dx$ as shown in Table VIII, where n is the number of intervals.

TABLE VIII

n	Simpson approximation
2	0.7833333333333333
4	0.785392156862745
8	0.7853981256146767
16	0.7853981628062051
32	0.7853981633882086
64	0.7853981633973034
128	0.7853981633974444
256	0.7853981633974434
512	0.7853981633974412
1024	0.7853981633974401

VII. Differential Equations

The simplest form of a differential equation consists of a relation

$$dy/dx = f(x, y),$$

together with an initial condition that, when $x = a$, $y = b$. The solution will then be a curve in the xy plane which passes through the point (a, b).

In general, it is either difficult or impossible to describe the solution curve by a simple equation, and we really need, for any practical purpose, only a method whereby we can tabulate the values of y for a set of values

$x_1, x_2, \ldots, x_n$, where $x_1 = a$, and where (normally) $x_{i+1} - x_i = h$, a constant. We now explain two methods for achieving such a table of values.

If we choose h to be quite small, then the chord joining the points (x_1, y_1) and (x_2, y_2) is very close to the tangent at $P_1(x_1, y_1)$. Consequently, we may write

$$\frac{y_2 - y_1}{x_2 - x_1} \approx \frac{dy}{dx}\bigg|_{x_1} = f(x_1, y_1) = f_1.$$

But $x_2 - x_1 = h$, and so we obtain the *Euler approximation formula*

$$y_2 \approx y_1 + hf_1.$$

Clearly, this relation merely connects adjacent y values, and so we may write the generalization

$$y_{i+1} = y_i + hf_i.$$

As an example of the use of the Euler formula, let us tabulate the solution of

$$dy/dx = x^2 - y$$

in the interval $[1, 2]$ with $h = 0.1$, given the initial condition that the curve passes through $(1, 0)$.

Here we have $x_1 = 1.0$, $x_2 = 1.1$, $x_3 = 1.2, \ldots, x_{11} = 2.0$. Our computation proceeds through a looping process, for which we give the first few stages; the computation is merely a set of applications of the basic formula.

$x_1 = 1,$ $y_1 = 0,$ $f_1 = 1^2 - 0 = 1,$ $y_2 = 0.1.$
$x_2 = 1.1,$ $y_2 = 0.1,$ $f_2 = 1.21 - 0.10 = 1.11,$ $y_3 = 0.211.$
$x_3 = 1.2,$ $y_3 = 0.211,$ $f_3 = 1.229,$ $y_4 = 0.3339.$
$x_4 = 1.3,$ $y_4 = 0.3339,$ $f_4 = 1.3561,$ $y_5 = 0.46951.$

Obviously, this procedure is well suited to machine computation.

The Euler method, though it possesses the great advantage of extreme simplicity, is rather crude, and we normally use a very accurate refinement of it known as the Runge–Kutta process. The analogue of the Euler formula is the set of five relations

$$k_1 = hf(x_1, y_1), \qquad k_2 = hf(x_1 + h/2, y_1 + k_1/2),$$
$$k_3 = hf(x_1 + h/2, y_1 + k_2/2), \qquad k_4 = hf(x_1 + h, y_1 + k_3),$$
$$y_2 = y_1 + (k_1 + 2k_2 + 2k_3 + k_4)/6.$$

This set of relations allows us to compute the next functional value y_{i+1} in terms of y_i, h, and various values of f (we pay for increased accuracy by having to compute several values of f). For a development of the Runge–Kutta formula, we refer the reader to Stanton (1961, pp. 151–155).

We now apply the Runge–Kutta process to the previous example. The results are shown in Table IX.

TABLE IX

RUNGE–KUTTA SOLUTION

x	y	x	y
1.0	0.00000	1.6	0.81119
1.1	0.10516	1.7	0.99342
1.2	0.22127	1.8	1.19067
1.3	0.34918	1.9	1.40343
1.4	0.48968	2.0	1.63212
1.5	0.64347		

There are many other developments in the numerical solution of differential equations. One may have a second-order differential equation with two given conditions; or one may have a partial differential equation. In any case, however, the idea of a numerical solution is to present the answer by a table of values. For details, we must refer to a specialized treatise, such as Milne (1953).

VIII. Equations in a Single Unknown

In any numerical problem, one sooner or later may confront the problem of solving an equation of the form $f(x) = 0$. Of the various approaches possible, *Newton's Method* is the most generally applicable.

By computer tabulations of $f(x)$, it is easy to obtain a value x_i which is near the desired root a. Consider Fig. 7, in which we have drawn the graph of $y = f(x)$; $f(a) = 0$, and x_i is a value near a. We draw the tangent at $P(x_i, f(x_i))$, and let it meet the x axis at Q, a point with abscissa x_{i+1}. It is clear that x_{i+1} is a closer approximation to the value of a.

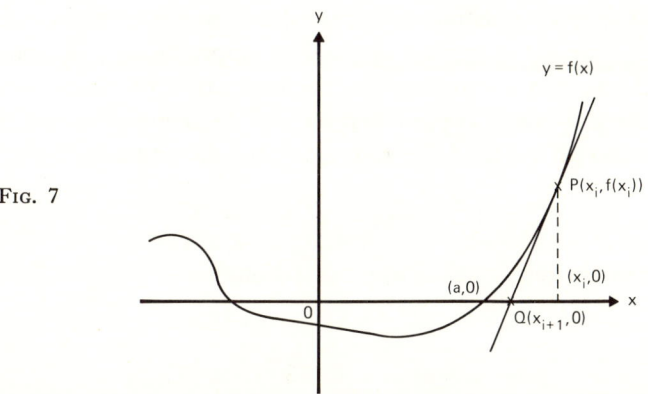

Fig. 7

From Fig. 7, we see that the slope of QP is given by $f'(x_i)$ and by $f(x_i)/(x_i - x_{i+1})$. Equate these two expressions, and we immediately find that
$$x_{i+1} = x_i - f(x_i)/f'(x_i).$$
This result, usually known as Newton's Formula, provides a rapid iterative scheme for obtaining a root of any equation in a single unknown. For example, suppose that we wish a root of the equation
$$x^6 - 5x^5 + 20x^2 + 50x + 12 = 0.$$
If we select $x_1 = 4$, then a few iterations give us an excellent approximation via the sequence

$\qquad$ 4.
$\qquad$ −6.695652173913043
$\qquad$ −5.491718712273415
$\qquad$ −4.494209813010194
$\qquad$ −3.669133302829003
$\qquad$ −2.988824807884936
$\qquad$ −2.431947780794708
$\qquad$ −1.984437344047138
$\qquad$ −1.641288564662818
$\qquad$ −1.40700596294052
$\qquad$ −1.286341611852411
$\qquad$ −1.25430397805492
$\qquad$ −1.252223472197575
$\qquad$ −1.252215059430044
$\qquad$ −1.252215059292918.

This sequence illustrates the fact that, once the iteration gets under way, the number of correct decimals is approximately doubled at each stage. Thus, convergence of the sequence is very rapid.

Newton's method applies just as well to non-polynomial functions. For instance, if we need the root near 2 of the equation

$$e^{3x} - 8x^4 - 20x^3 - 10x^2 - 10x - 32 = 0,$$

we obtain the sequence of approximations

$$x_1 = 2.$$
$$x_2 = 1.9647308838759$$
$$x_3 = 1.9615747092050$$
$$x_4 = 1.9615511849671$$
$$x_5 = 1.9615511836699$$
$$x_6 = 1.9615511836699.$$

IX. Systems of Linear Equations

In the physical world, a mathematical model of a process frequently leads to a representation where the quantities describing the discrete model are obtainable only as the solution of a set of simultaneous linear equations. Examples of this situation occur in Section III, where the parameters for a least-squares fit are given by the solution of a set of linear equations, and in Section IV where the $n + 1$ parameters of an interpolating cubic spline are determined by the solution of $n + 1$ simultaneous equations.

A set of equations in three unknowns can be used to illustrate the methods of solution, namely,

$$a_{11}x_1 + a_{12}x_2 + a_{13}x_3 = b_1,$$
$$a_{21}x_1 + a_{22}x_2 + a_{23}x_3 = b_2,$$
$$a_{31}x_1 + a_{32}x_2 + a_{33}x_3 = b_3,$$

or, in matrix notation, $AX = B$, where A is the matrix with elements a_{ij} ($i, j = 1, 2, 3$), X is the vector with components x_1, x_2, x_3, and B is a vector of constants b_1, b_2, b_3. It is required to determine the quantities x_1, x_2, x_3.

5. Numerical Analysis

Existence of the solution is assured *mathematically* provided that the determinant of A is non-zero. However, computationally, since the working length of our arithmetic is restricted in all but trivial cases, propagation of rounding errors is of serious consequence, and various strategies have been developed to eliminate or minimize their effect. Initially, we consider *Gaussian Elimination* applied to the matrix equation $AX = B$.

The purpose of Gaussian Elimination is to reduce a given set of linear equations to an upper triangular form

$$a_{11}x_1 + a_{12}x_2 + a_{13}x_3 = b_1,$$
$$a_{22}^{(1)}x_2 + a_{23}^{(1)}x_3 = b_2^{(1)},$$
$$a_{33}^{(2)}x_3 = b_3^{(2)}.$$

This can be achieved by adding a multiple of the first row to the second and third, and then a multiple of the new second row to the third. Solution of this "triangular" set of equations is then easily carried out in a reverse manner; the last equation provides a solution for x_3, the penultimate equation then gives x_2, and the first equation then gives x_1 in terms of the (now) known values x_2 and x_3.

In detail, this process is as follows. In the equation $AX = B$, the multiple $-a_{21}/a_{11}$ of the first equation is added to the second and the multiple $-a_{31}/a_{11}$ of the first equation is added to the third. This procedure produces the modified set

$$a_{11}x_1 + a_{12}x_2 + a_{13}x_3 = b_1,$$
$$\left(a_{22} - \frac{a_{21}}{a_{11}}a_{12}\right)x_2 + \left(a_{23} - \frac{a_{21}}{a_{11}}a_{13}\right)x_3 = b_2 - \frac{a_{21}}{a_{11}}b_1,$$
$$\left(a_{32} - \frac{a_{31}}{a_{11}}a_{12}\right)x_2 + \left(a_{33} - \frac{a_{31}}{a_{11}}a_{13}\right)x_3 = b_3 - \frac{a_{31}}{a_{11}}b_1.$$

If we abbreviate the new coefficients appropriately to $a_{ij}^{(1)}$ and $b_i^{(1)}$, then this last set of equations can be written as

$$a_{11}x_1 + a_{12}x_2 + a_{13}x_3 = b_1,$$
$$a_{22}^{(1)}x_2 + a_{23}^{(1)}x_3 = b_2^{(1)},$$
$$a_{32}^{(1)}x_2 + a_{33}^{(1)}x_3 = b_3^{(1)}.$$

Finally, if the multiple $-a_{32}^{(1)}/a_{22}^{(1)}$ of the second equation is added to the third equation, we obtain the "triangular" set of equations

$$a_{11}x_1 + a_{12}x_2 + a_{13}x_3 = b_1,$$
$$a_{22}^{(1)}x_2 + a_{23}^{(1)}x_3 = b_2^{(1)},$$
$$a_{33}^{(2)}x_3 = b_3^{(2)}.$$

Back-substitution proceeds as before. For a larger number of equations, the process required is just a natural generalization.

The method just described will fail if, at any time, a zero pivot is encountered, that is, if, for a set of n simultaneous equations in n unknown values, any value $a_{kk}^{(k-1)}$ ($1 \leq k \leq n$) is zero. Simple manual rearrangement of the order of the equations will remove this problem, and consideration of good arrangements leads to the ideas of partial pivoting and complete pivoting.

Let us start from the equation $AX = B$. In Gaussian Elimination, we initially add a multiple of the first row of A to all subsequent rows to produce zeros in the first column. With partial pivoting, we look first for the element in the first column which has largest modulus. Suppose this element is a_{41}; then the fourth equation is interchanged with the first equation, and the first step of Gaussian Elimination is performed, using multiples of the new first equation. This gives a set of equations of the form

$$\begin{pmatrix} x & x & x & x & x & x \\ 0 & x & x & x & x & x \\ 0 & x & x & x & x & x \\ 0 & x & x & x & x & x \\ 0 & x & x & x & x & x \\ 0 & x & x & x & x & x \end{pmatrix} X = B_1.$$

In the second column of the new set of equations, a search is now conducted for the element of largest modulus (elements in the first row being excluded this time), and the procedure is repeated. Ultimately, one again reaches a "triangular" set of equations, and back-substitution produces the vector X of unknowns.

Full pivoting is an analogous process; however, at the kth stage we search for the element of largest modulus, not in the kth column, but in the entire submatrix formed by the last $n - k + 1$ columns and $n - k + 1$ rows. We can briefly summarize the steps in full pivoting which are required at stage k.

Let the array at the start of stage k be $A = (a_{pq}^{(k-1)})$.

1. Then, a search is conducted in the sub-array

$$a_{pq}^{(k-1)} \quad (p = k, k+1, \ldots, n; \quad q = k, k+1, \ldots, n)$$

for the element with the largest modulus—suppose it is $a_{mr}^{(k-1)}$.

2. The complete mth row is interchanged with the complete kth row, and the complete rth column with the complete kth column.

3. Elimination is performed in the kth column using the *new* kth row.

4. The value of k is increased to $k + 1$, and steps 1, 2, and 3 are repeated. Back-substitution proceeds as before.

For further discussion and for working algorithms, attention is directed to Wilkinson (1965) and Reinsch and Wilkinson (1971).

X. Special Methods for Solving Sparse Sets of Equations

Usually, the solution of second-order ordinary or partial differential equations leads to a set of linear equations, say $AX = B$, where the matrix A of coefficients has a large number of zero entries. The matrix A is then said to be sparse, and we try to achieve simplicity by selecting numerical techniques so that, in solving $AX = B$, we retain a large number of zero entries at each stage of the work. The advantages of such an arrangement are twofold: (a) if entries remain zero, they take no part in the arithmetic and can be ignored; (b) if entries are zero and remain so, then they have no effect on other entries and do not even have to be stored in the machine.

For definiteness, let us consider an example in which we have n equations given schematically by

$$AX = \begin{bmatrix} x & x & 0 & 0 & 0 & \cdots & 0 & 0 & 0 & 0 \\ x & x & x & 0 & 0 & \cdots & 0 & 0 & 0 & 0 \\ 0 & x & x & x & 0 & \cdots & 0 & 0 & 0 & 0 \\ 0 & 0 & x & x & x & \cdots & 0 & 0 & 0 & 0 \\ & \cdots & & & \cdots & & & \cdots & & \\ & & & & & & & & & \\ 0 & 0 & 0 & 0 & 0 & \cdots & 0 & x & x & x \\ 0 & 0 & 0 & 0 & 0 & \cdots & 0 & 0 & x & x \end{bmatrix} X = B,$$

where B is a constant vector, X is the vector of unknowns, and each x in the matrix A denotes a non-zero entry (a matrix A of this form is often called a *band matrix*).

Straightforward application of Gaussian Elimination produces the equivalent system

$$\begin{bmatrix} x & x & 0 & 0 & \cdots & 0 & 0 & 0 \\ 0 & x & x & 0 & \cdots & 0 & 0 & 0 \\ 0 & 0 & x & x & \cdots & 0 & 0 & 0 \\ & \cdots & & & \cdots & & \cdots & \\ & \cdots & & & \cdots & & \cdots & \\ 0 & 0 & 0 & 0 & \cdots & 0 & x & x \\ 0 & 0 & 0 & 0 & \cdots & 0 & 0 & x \end{bmatrix} X = B^{(n-1)},$$

and it can be seen that the number of elements in A which actually are used is $3n - 2$; however, if the complete matrix A is stored, n^2 entries are recorded. If $n = 1000$ (not an unusual value in a physical problem of any significance), then the percentage use of the computer storage is $2998/10^4 \approx 0.3\%$; this makes for extremely inefficient and costly computation. Thus, we obviously should devise a method which only stores the non-zero entries of A and performs some modified version of Gaussian Elimination.

For example, let the elements a_{ii} of A be stored in the vector $C = (c_i)$, and the elements $a_{i+1,i}$ and $a_{i,i+1}$ be stored in vectors $D = (d_i)$ and $E = (e_i)$. We still denote our column of constants by the vector $B = (b_i)$. With this type of storage, Gaussian Elimination can be performed by using the following algorithm (c_1, e_1, and b_1 remain unchanged; the notation $i \leftarrow 2$ is used to assign the value 2 to i):

(a) $i \leftarrow 2$
(b) $c_i \leftarrow c_i - e_{i-1}d_{i-1}/c_{i-1}$
(c) $b_i \leftarrow b_i - b_{i-1}d_{i-1}/c_{i-1}$
(d) $i \leftarrow i + 1$
(e) repeat steps (a) to (d) until $i + 1 > n$.

The equations are now effectively in triangular form, and back-substitution can be performed by using the following statements:

(f) $i \leftarrow n$
(g) $x_n \leftarrow b_n/c_n$
(h) $x_{i-1} \leftarrow (b_{i-1} - x_i e_i)/c_{i-1}$
(i) $i \leftarrow i - 1$
(j) stop if $i = 1$, otherwise repeat steps (h) to (j).

More general procedures for dealing with band matrices can be found in Reinsch and Wilkinson (1971).

Another good example of a real improvement in use of storage is furnished by solution of the equation $AX = B$ when A is a symmetric matrix, that is, a matrix $A = (a_{ij})$ for which $a_{ij} = a_{ji}$. It can easily be verified that for Gaussian Elimination the resulting set of equations is always symmetric. Thus, it is unnecessary to store the original elements a_{ij} where $i > j$; for, if such an element is required in the elimination process, the corresponding symmetrical element obtained by interchanging the suffixes is accessed. The storage saving in this case is

$$\sum_{i=1}^{n-1} i = n(n-1)/2,$$

or approximately 50% of the storage area.

Additional material on sparse sets of equations is contained in Reid (1971).

Appendix

```
      SUBROUTINE TINV(A,B,C,D,N)
      DIMENSION A(N),B(N),C(N),D(N)
COMMENT THIS SUBROUTINE SOLVES A LINEAR SET OF
COMMENT SIMULTANEOUS EQUATIONS, WHERE THE
COMMENT COEFFICIENT MATRIX IS TRIDIAGONAL AND OF
COMMENT DEGREE N.  D IS THE CONSTANT VECTOR
COMMENT AND A,B,C ARE VECTORS CONTAINING SUB, DIAGONAL, AND
COMMENT SUPER DIAGONAL ENTRIES RESPECTIVELY OF THE COEFFICIENT MATRIX.
      DO 1 K=2,N
      I=N+2-K
      J=I-1
      Z=C(J)/B(I)
      B(J)=B(J)-Z*A(J)
1     D(J)=D(J)-Z*D(I)
      B(1)=D(1)/B(1)
      DO 2 I=2,N
2     B(I)=(D(I)-A(I-1)*B(I-1))/B(I)
      RETURN
      END


      SUBROUTINE SPLINE(H,X,Y,Y2,N,P)
      DIMENSION P(6),X(N),Y(N)    ,Y2(N),H(N),A( ),C( ),D( )
COMMENT THIS ROUTINE SETS UP THE COEFFICIENT MATRIX AND
COMMENT CONSTANT VECTOR FOR THE SET OF LINEAR
COMMENT SIMULTANEOUS EQUATIONS WHICH DEFINE THE
COMMENT SECOND DERIVATIVES, Y2(I) OF THE CUBIC SPLINE
COMMENT INTERPOLATING TO THE N POINTS X(I),Y(I).
COMMENT THE VECTOR P(1),P(2),...,P(6)
COMMENT MUST CONTAIN THE COEFFICIENTS FOR THE
COMMENT GENERAL BOUNDARY CONDITIONS,
COMMENT P(1)*Y'_1+P(2)*Y''_1=P(3)
COMMENT P(4)*Y'_N+P(5)*Y''_N=P(6).
```

```
COMMENT THE VECTORS A,C,D MUST BE DIMENSIONED
COMMENT SO AS TO HAVE LENGTH EQUAL TO THE LARGEST VALUE OF N
COMMENT USED.  ON EXIT, X,Y,P, RETAIN THEIR
COMMENT PREVIOUS VALUES AND H(J) CONTAINS
COMMENT THE QUANTITIES X(J+1)-X(J) NEEDED
COMMENT FOR INTERPOLATION WITH THE N
COMMENT CALCULATED VALUES OF THE CUBIC
COMMENT SPLINES SECOND DERIVATIVES Y2(I).
      M=N-1
      DO 6 J=1,M
6     H(J)=X(J+1)-X(J)
      M=M-1
      DO 7 J=1,M
      Y2(J+1)=(H(J+1)+H(J))/3.
      A(J)=H(J)/6.
      C(J)=A(J)
7     D(J+1)=(Y(J+2)-Y(J+1))/H(J+1)-(Y(J+1)-Y(J))/H(J)
      M=M+1
      Y2(1)=P(2)-H(1)*P(1)/3.
      C(1)=-P(1)*C(1)
      D(1)=P(3)-P(1)*(Y(2)-Y(1))/H(1)
      D(N)=P(6)+P(4)*(Y(M)-Y(N))/H(M)
      Y2(N)=P(5)+H(M)*P(4)/3.
      C(M)=H(M)/6.
      A(M)=P(4)*C(M)
      CALL TINV(A,Y2,C,D,N)
      RETURN
      END
```

Special Reference

Abramowitz, M., and Stegun, I. A. (1968). "Handbook of Mathematical Functions." Dover, New York.

Ahlberg, J. H., Nilson, E. N., and Walsh, J. L. (1967). "The Theory of Splines and Their Applications." Academic Press, New York.

Her Majesty's Stationery Office. (1956). "Interpolation and Allied Tables." London.

Hoskins, W. D., and King, P. R. (1972). *Comput. J.* **15**, No. 3, 282–283.

Milne, W. E. (1953). "Numerical Solution of Differential Equations." Wiley, New York.

Powell, M. J. D. (1966). On Best L_2 Spline Approximation, TP264. A.E.R.E., Harwell, England.

Reid J. K. (ed.) (1971). "Large Sparse Sets of Linear Equations." Academic Press, New York.

Reinsch, C., and Wilkinson, J. H. (1971). "Linear Algebra." Springer Verlag, Berlin and New York.

Schoenberg, I. J. (ed.) (1969). "Approximations with Special Emphasis on Spline Functions." Academic Press, New York.

Stanton, R. G. (1961). "Numerical Methods for Science and Engineering." Prentice-Hall, Englewood Cliffs, New Jersey.

Wilkinson, J. H. (1965). "The Algebraic Eigenvalue Problem." Oxford Univ. Press, London and New York.

Chapter 6

Group Theory

A. T. Amos

I. Introduction 374
II. Definitions 375
III. Symmetry Operators 376
    A. Spatial Symmetry 376
    B. Permutation Symmetry 381
    C. Classes 383
IV. Group Representation Theory 385
    A. Effect of the Symmetry Operators on Wave Functions 385
    B. Abelian Groups 388
    C. Non-Abelian Groups 391
    D. Character Tables 393
    E. Basis Functions and Projection Operators 395
    F. Direct Product of Two Representations 399
    G. Direct Product of Two Groups 400
V. Some Applications in Molecular Quantum Mechanics 401
    A. Matrix Elements 401
    B. Transition Probabilities 404
    C. Computation of Molecular Wave Functions 405
    D. Normal Modes of Vibration 406
VI. The Permutation Group and Spin 411
    A. Irreducible Representations of the Permutation Group . . . 411
    B. Spin Functions 414
    C. The Pauli Principle 415
VII. Continuous Groups 417
VIII. Group Theory and the Solid State 421
    A. Crystal Symmetry and Space Groups 421
    B. Crystal-Field Splitting 423
    C. Bloch Wave Functions 425
    References 426

I. Introduction

Group theory, on the face of it a very abstract branch of pure mathematics, has become an important tool for physicists and chemists because of the role it plays in quantum mechanics. The basic problem of quantum mechanics is to solve the Schroedinger equation

$$H\psi = E\psi \qquad (1.1)$$

for the system of interest (e.g., atom, molecule, or crystal) in order to find the wave functions and energy levels. If the system contains any symmetry, this will be included implicitly in the Hamiltonian H and, mathematically, this can be expressed by finding operators associated with the symmetry and showing that they commute with H. From a fundamental theorem of quantum mechanics it then follows that the eigenfunctions of H will also be eigenfunctions of the symmetry operators, except in cases where degeneracy complicates the issue. Physically this simply means that the wave functions have to reflect, in some way, the underlying symmetry of the system.

It turns out that the symmetry operators of a system form the elements of a mathematical group. Starting from this fact, it is possible to appeal to the results of a particular branch of group theory, called group representation theory, which show how to find the type of functions associated with the underlying symmetries of the group elements. Hence, via the use of group theory, it is possible to determine the symmetries of the wave functions without having to find them explicitly by solving (1.1).

By symmetry it is natural to think of the spatial symmetry due to the position of nuclei in molecules (e.g., NH_3) or the regular array of crystals. This is certainly important, and group theory allows the wave functions of such systems to be classified and certain of their properties to be deduced. However, equally important is the permutation symmetry associated with the indistinguishability of electrons. The group theoretical properties which follow from this, when combined with the Pauli principle, give rise to the spin multiplicity of states and physically allowed and unallowed wave functions.

It is clearly impossible in a relatively short chapter to deal with all the applications of group theory to physics and chemistry. In what follows the main consideration is to discuss the more important topics in some detail with sufficient examples to show how the fundamental theory can be applied. The mathematically more difficult topics and those of use only in very special situations are either ignored or mentioned just briefly so as to enable the interested reader to follow them up in the books cited in the References at the end of the chapter.

II. Definitions

A group consists of a number of elements $A, B, C, \ldots$, together with a binary operation which allows any two elements to be combined together, uniquely, to form a third in such a way that certain axioms are satisfied. Before stating these axioms the terms "elements" and "binary operation" should be examined more specifically. From a mathematical point of view the terms are best left as abstract as possible so that the theory can encompass many different situations but, in this chapter, an element will usually be a symmetry operator, although on occasions it will mean a matrix. The binary operation will invariably be multiplication and is referred to as such.

The axioms which must be satisfied by the elements in order that they form a group are as follows:

(i) Multiplication is closed; i.e., the product of any two elements in the group is also in the group.

(ii) Multiplication is associative; i.e., $A(BC) = (AB)C = ABC$.

(iii) There exists an identity element I in the group which satisfies

$$IA = AI = A, \quad \text{all} \quad A.$$

(iv) Every element A of the group has an inverse element A^{-1}, also belonging to the group, and satisfying $AA^{-1} = A^{-1}A = I$.

Using these axioms only it is possible to obtain a surprisingly large number of other results and, in this respect, group theory provides a beautiful example of deductive reasoning.

The main property that distinguishes one particular group as opposed to another is the manner in which any two elements combine to form a third. This can be found from the group multiplication table. Two examples are given in Table I. One is a group with four elements: I, C_2, σ_v, σ_v', and the other has six: I, P_{12}, P_{23}, P_{13}, P_{123}, P_{132}. (For the moment the meaning of these symbols does not matter; they are explained later). Using the multiplication tables it is relatively easy to verify, for both groups, that axioms (i), (iii), and (iv) are satisfied. As is usually the case, it takes longer to verify that (ii) is satisfied also.

TABLE I

Group Multiplication Tables for the Point Group C_{2v} and the Symmetric Group of Degree 3, $S(3)$[a]

C_{2v}	I	C_2	σ_v	σ_v'	$S(3)$	I	P_{12}	P_{23}	P_{13}	P_{123}	P_{132}
I	I	C_2	σ_v	σ_v'	I	I	P_{12}	P_{23}	P_{13}	P_{123}	P_{132}
C_2	C_2	I	σ_v'	σ_v	P_{12}	P_{12}	I	P_{123}	P_{132}	P_{23}	P_{13}
σ_v	σ_v	σ_v'	I	C_2	P_{23}	P_{23}	P_{132}	I	P_{123}	P_{13}	P_{12}
σ_v'	σ_v'	σ_v	C_2	I	P_{13}	P_{13}	P_{123}	P_{132}	I	P_{12}	P_{23}
					P_{123}	P_{123}	P_{13}	P_{12}	P_{23}	P_{132}	I
					P_{132}	P_{132}	P_{23}	P_{13}	P_{12}	I	P_{123}

[a] To find the product AB look along the row beginning A until it intersects the column headed B. The element at that position will equal the product AB.

It is important not to confuse the associative property of multiplication [which a group must possess by axiom (ii)] with the commutative property. The latter means that, for any two elements, $AB = BA$. Some groups possess this property but most do not. In Table I it is easy to see that the four elements I, C_2, σ_v, σ_v' do commute. A group whose elements commute is called *Abelian* or commutative. The multiplication table shows that the second group is *non-Abelian*. A particularly simple Abelian group can be formed if there is an element A such that $A^n = I$ where n is an integer. The elements I, A, A^2, ..., A^{n-1} form an Abelian group which is called a *cyclic* group.

The characteristic properties of any particular group are derived from its multiplication table. From this it follows that, if the elements of two separate groups can be put into a one-to-one (1 : 1) correspondence in such a way that their multiplication tables take identical forms, then their group theoretical properties will be the same. The groups are then said to be *isomorphic*.

Finally, the *order* of a group is the number of distinct elements in the group.

III. Symmetry Operators

A. Spatial Symmetry

If H is the Hamiltonian for an atom, molecule, or crystal, it will contain implicitly the symmetry of the system. This fact can be expressed mathematically by finding operators associated with this symmetry, and these operators will commute with H. To begin with we consider operators corresponding to the spatial symmetry of molecules.

In the Born–Oppenheimer approximation a molecule can be considered as a rigid framework of fixed nuclei about which the electrons move. Therefore, the Hamiltonian and hence the wave functions will contain the positions of the nuclei as parameters. As an example consider the water molecule shown in Fig. 1. The two hydrogen atoms can be labeled H(1)

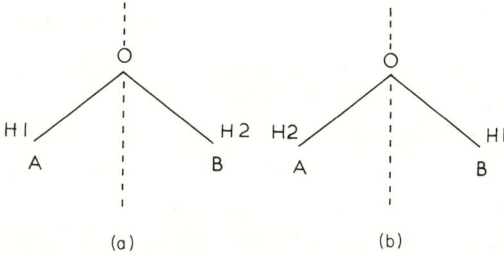

FIG. 1. The water molecule; (a) initial configuration; (b) after rotation through π.

and H(2); suppose initially that they are at positions A and B. Now let the molecular framework be rotated by 180° about an axis which passes through the oxygen molecule, lies in the molecular plane, and bisects the HOH angle. The result is that H(2) takes the place of H(1) at position A while H(1) moves to position B. Since H(1) and H(2) are both protons and are only distinguished by the labels 1 and 2, the new configuration is physically indistinguishable from the old. Consequently the Hamiltonian for both configurations will be the same because, if it is written down for the initial configuration (Fig. 1a), terms arising from the proton H(1) will involve the position A but there will be equivalent terms arising from proton H(2) which involve position B. Exactly the same terms will occur when writing down the Hamiltonian for the final configuration (Fig. 1b), the only difference being that those which came from proton H(1) now come from H(2), and vice versa. Thus H is invariant to the rotation, so that if we introduce the *symmetry operator* C_2 to represent the operation of rotation, we can write $HC_2 = C_2H$; i.e., the Hamiltonian and the symmetry operator commute.

Clearly this type of result is not special to the rotation operator C_2 but there are other operators which, for certain molecules, leave the Hamiltonian invariant. In general, symmetry operations, which correspond to a spatial change of the nuclear framework for which the initial and final conformations are equivalent and physically indistinguishable, will leave H invariant and the symmetry operators will commute with H.

Before considering these operations in detail, it should be noted that any uniform translation of a molecule will leave all wave functions invariant since it is equivalent only to a change of origin of the coordinate system which can have no physical significance. To eliminate this from the problem it is usual to consider the center of gravity of the molecule as fixed and consider only symmetry operations which leave it unchanged. There are then five distinct types of symmetry operators that arise:

(i) C_p: p an integer, rotation through an angle $2\pi/p$ about a fixed symmetry axis.

(ii) i: inversion through the center; i.e., if a nucleus has coordinates (x, y, z) relative to the center, then inversion changes it to position $(-x, -y, -z)$.

(iii) σ: reflection in a plane. If this plane contains the principal axis (see below), the reflection is written as σ_v. If the plane is perpendicular to this, it is written as σ_h. A special case of a σ_v reflection occurs when the plane bisects two symmetry axes in the plane perpendicular to the principal axis and in this case the symbol used is σ_d.

(iv) S_p: rotation through an angle $2\pi/p$ about a symmetry axis followed by reflection in a plane perpendicular to it. This is clearly equivalent to the two separate operations C_p and σ_h, the order in which they are performed being, in this special case, immaterial. Thus

$$S_p = C_p \sigma_h = \sigma_h C_p.$$

(v) I: the identity operator, which leaves the system unchanged.

As was anticipated in (iii), a system can have more than one symmetry rotation axis, thus giving rise to several operators of the type C_p. The p-fold axis corresponding to the largest value of p is called the principal axis. There can also be several axes with the same p, as often happens when $p = 2$. In this case primes are used to distinguish the symmetry operators, e.g., C_2, C_2', C_2'', and similarly with the other operators, σ_v, σ_v', S_p, S_p', and so on.

To illustrate some of these operations first consider the water molecule (Fig. 1). In addition to the rotation symmetry (C_2), there are two other symmetry operators: (a) reflection in the molecular plane, σ_v', and (b) reflection in the plane perpendicular to the molecular plane and intersecting it along the twofold rotation axis, σ_v. Since both reflection planes contain the symmetry axis the σ has the subscript v and the prime on the first σ operator is simply to distinguish it from the second. Adding the identity operator to these three gives the four symmetry operators I, C_2, σ_v, σ_v' for the water molecule, and each operator leaves the molecular Hamiltonian invariant and commutes with it. If A and B represent any two of these, then since $AH = HA$ and $BH = HB$, $ABH = A(BH) = A(HB) = AHB = (AH)B = (HA)B = HAB$ so that their product commutes with H. Thus, having obtained a set of symmetry operators, more can be generated by considering products of pairs of them. In the case of the four operators of the water molecule, however, no new operators can be formed in this way but only one of the four already obtained. The multiplication table for the operators $I, C_2, \sigma_v, \sigma_v'$ is, in fact, that given in Table I, and this shows that the operators form the elements of a group.

This result is quite general and holds for the symmetry operators associated with the spatial symmetry of any molecule. To prove this it is only necessary to show that all the axioms of a group are satisfied. It is easy to show that spatial symmetry operators obey the associative law of multiplication and the symmetry operators for a molecule clearly must include the identity. They must be complete, for the product of any two symmetry operators commutes with the Hamiltonian and therefore must itself be a symmetry operator. Finally each symmetry operator must have an inverse since, having performed any particular symmetry operation, there must be another operation which will restore the molecule to its original position, and this operation satisfies all the criteria for, and therefore must itself be, a symmetry operation.

As an illustration of the operators S_p and i and as an example of a molecule with a rather complicated symmetry, benzene (Fig. 2) can be considered. There are 24 different symmetry operations in this molecule:

(i) I, the identity.

(ii) C_6, rotation by 60° about an axis perpendicular to the molecular plane and passing through the center. There are five operators of this type corresponding to the powers of C_6: $C_6, C_6^2, C_6^3, C_6^4, C_6^5$, and there are only five distinct ones since $C_6^6 \equiv I$. Since C_6^2 corresponds to two rotations of 60°, i.e., one of 120°, it can be written as C_3. Similarly, $C_6^4 = C_3^2$ and $C_6^3 = C_2$.

(iii) The sixfold axis of symmetry just considered is the principal axis. Perpendicular to it and lying in the molecular plane are six different

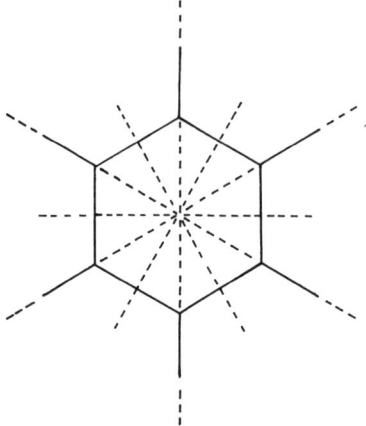

FIG. 2. Benzene molecule showing the six twofold axes.

twofold axes (C_2). Three of these pass through carbon atoms and three through midpoints of opposite bonds, as shown by dashed lines in Fig. 2. This gives six C_2 operators.

(iv) Planes perpendicular to the molecular plane and including any one of the six twofold axes will be reflection planes, giving six operators of the type σ_v.

(v) The molecular plane itself is a reflection plane σ_h.

(vi) i, inversion through the center.

(vii) Rotations of 60°, 120°, 240°, and 360° about the sixfold axis followed by reflection in the molecular plane give four operators, i.e., $\sigma_h C_6 = S_6$, $\sigma_h C_6^5 = S_6'$, $\sigma_h C_3 = S_3$, and $\sigma_h C_3^2 = S_3'$. Notice that there is no $S_2 = \sigma_h C_2$ since this is equivalent to i.

In a similar fashion, the symmetry operations of any given molecule can be found and the operators will form a group. Any two distinct molecules which have the same set of symmetry operators will be associated with the same group, which is simply an indication that their basic symmetries are the same, and hence properties depending on this will be the same even though the molecules themselves may be built up of very different atoms. The groups corresponding to the symmetry operators of molecules are called *point groups*. The word point is used because one point—the center—of the molecule is kept fixed. There is a system of classification of the various point groups according to the type of symmetry operators, the number of rotation axes, and so on, associated with them. A brief résumé of this is given in Table II.

TABLE II
Classification of Point Groups

Symbol	Operators and axes
C_n	One axis of symmetry of nth order. $I, C_n, C_n^2, \ldots, C_n^{n-1}$
C_{nh}	n-fold axis plus reflection plane perpendicular to it. $I, C_n, \ldots, C_n^{n-1}, \sigma_h, C_n\sigma_h, C_n^2\sigma_h, \ldots, C_n^{n-1}\sigma_h$
C_{nv}	n-fold axis plus n reflection planes passing through it. $I, C_n, \ldots, C_n^{n-1}$, plus n σ_v operators
D_n	n-fold axis plus n twofold axes perpendicular to it. $I, C_n, \ldots, C_n^{n-1}$, plus n C_2 operators
D_{nh}	D_n with reflection plane perpendicular to n-fold axis. D_n operators plus σ_h and products
D_{nd}	D_n with reflection planes bisecting angles between twofold axes (σ_d). D_n operators plus n σ_d operators and products
S_n	C_n plus rotation–reflection operators S_n
T, T_d, T_h	T: the group of the tetrahedron. T_d obtained from it by adding planes of reflection and T_h by adding inversion
O, O_h	O: the octahedron group. O_h obtained from it by adding inversion

Before concluding this section a further remark must be made concerning molecules which are axially symmetric (e.g., diatomics). Since a rotation of any arbitrary angle θ, $0 \leq \theta \leq 2\pi$, about this axis will be a symmetry operation, the point group of the molecule will contain infinitely many elements. A group of this type is called continuous. For the moment, molecules with this type of symmetry are excluded and a discussion of continuous groups is deferred until Section VII.

B. Permutation Symmetry

The spatial symmetry operators are not the only ones which have to be considered. Another less obvious type of symmetry arises in atoms and molecules due to the indistinguishability of the electrons. Suppose $H(1, 2, \ldots, n)$ is the Hamiltonian for a system for n electrons which are labeled from 1 to n. Now this labeling is purely a mathematical contrivance and has no physical significance. Therefore, a Hamiltonian of exactly the same form would be produced if these labels were rearranged, say, by relabeling electron 1 as 2 and electron 2 as 1. To represent the effect of this relabeling an operator P_{12}, called a permutation operator, can be introduced which simply interchanges the electron labels 1 and 2 when it acts on a function. Since $P_{12}^2 = I$, the identity, it follows that the operators I and P_{12} form a group which is called the symmetric or permutation group of degree 2.

As the number of electrons is increased the number of permutation operators also increases but much more rapidly. If there are three electrons, labeled 1, 2, and 3, the labeling on any two can be interchanged by the operators P_{12}, P_{13}, P_{23}, but in addition all three can be interchanged at once. This can be done cyclically, so that 1 becomes 2, 2 becomes 3, 3 becomes 1, as effected by the operator P_{123}, or anticyclically using the operator P_{132}. If the identity I is included, this gives six operators, $I, P_{12}, P_{13}, P_{23}, P_{123}, P_{132}$, whose multiplication table is in Table I. These permutation operators therefore form a group, the permutation group of degree 3 with order 6.

In general the symmetric (or permutation) group of degree n has as its elements the $n!$ permutation operators which interchange the labels of an n-electron system. To see that the order of the group is indeed $n!$, it is convenient to represent a permutation operator as

$$P \equiv \begin{pmatrix} n_1 & n_2 & n_3 & \cdots & n_n \\ 1 & 2 & 3 & \cdots & n \end{pmatrix}$$

where the n integers $n_1, \ldots, n_n$ are all distinct and range from 1 to n. The effect of the permutation represented by P is to change 1 to n_1, 2 to n_2, and so on. There are clearly $n!$ possible ways to choose the values $n_1, \ldots, n_n$, all of which are distinct and, therefore, represent different permutation operators. Note that one of these (with $n_1 = 1, n_2 = 2, \ldots$) is the identity.

The product of any two permutation operators is itself a permutation operator. This can best be seen by an example. Suppose $n = 4$ and that the two operators to be multiplied are

$$P(1) = \begin{pmatrix} 2 & 1 & 4 & 3 \\ 1 & 2 & 3 & 4 \end{pmatrix} \quad \text{and} \quad P(2) = \begin{pmatrix} 3 & 1 & 2 & 4 \\ 1 & 2 & 3 & 4 \end{pmatrix}.$$

Since the meaning of the symbol $P(1)$ is unaltered if the ordering of the columns is rearranged, $P(1)$ is also

$$P(1) = \begin{pmatrix} 4 & 2 & 1 & 3 \\ 3 & 1 & 2 & 4 \end{pmatrix}$$

so that the product $P(1)P(2)$ is

$$\begin{pmatrix} 4 & 2 & 1 & 3 \\ 1 & 2 & 3 & 4 \end{pmatrix},$$

which is also a permutation operator. Every permutation operator has an inverse; for example,

$$[P(1)P(2)]^{-1} = \begin{pmatrix} 1 & 2 & 3 & 4 \\ 4 & 2 & 1 & 3 \end{pmatrix} = \begin{pmatrix} 3 & 2 & 4 & 1 \\ 1 & 2 & 3 & 4 \end{pmatrix}.$$

These results can be generalized to the case of n arbitrary and it can also be shown that the multiplication of the operators is associative. This shows that the assumption that the permutation operators form a group is correct.

Although the bracket notation can be useful, it is more common to use the $P_{abc\cdots z}$ notation introduced earlier. This is to be interpreted as a permutation in which a is replaced by b, b by c, c by d, and so on, and finally z by a. The two notations are quite equivalent but notice that not all permutation operations can be represented by a single P operator. For example, $P(1)$ is equivalent to the product $P_{12}P_{34}$ while $P(2)$ is P_{132}.

C. Classes

If the 24 symmetry operations associated with benzene are considered, then it must be clear that they can be divided into sets of operations that are physically very similar. For example, the three C_2 operators corresponding to rotation about the three axes passing through two carbon atoms are exactly the same type of operation. Similarly, the six σ_v operations can be divided into two sets of three. These sets of physically similar operations are called *classes*.

Mathematically a class is defined in the following way. Let A be a group element. Then the set of elements GAG^{-1}, where G is put in turn equal to every element in the group, is called the class of A. In forming GAG^{-1}, n elements, where n is the order of the group, are obtained but this does not mean that the class of A consists of n elements since not all these are distinct. For example, suppose the group is Abelian. Then $GA = AG$ for all A and G so that $GAG^{-1} = AGG^{-1} = A$. Hence each of the n elements will be the same and equal to A. Thus an Abelian group contains n classes, each consisting of a single element of the group.

It is only in the case of non-Abelian groups that the concept of class becomes interesting. Even here there will be one class with only one element, namely, the class obtained from the identity. To illustrate this

TABLE III

COMPUTATION OF THE CLASSES OF $S(3)^a$

G	GIG^{-1}	$GP_{12}G^{-1}$	$GP_{13}G^{-1}$	$GP_{23}G^{-1}$	$GP_{123}G^{-1}$	$GP_{132}G^{-1}$
I	I	P_{12}	P_{13}	P_{23}	P_{123}	P_{132}
P_{12}	I	P_{12}	P_{23}	P_{13}	P_{132}	P_{123}
P_{13}	I	P_{23}	P_{13}	P_{12}	P_{132}	P_{123}
P_{23}	I	P_{13}	P_{12}	P_{23}	P_{132}	P_{123}
P_{123}	I	P_{23}	P_{12}	P_{13}	P_{123}	P_{132}
P_{132}	I	P_{13}	P_{23}	P_{12}	P_{123}	P_{132}

[a] There are three classes: (i) I, the class of identity; (ii) P_{12}, P_{13}, P_{23}, the class of P_{12}, of P_{13}, and of P_{23}; and (iii) P_{123}, P_{132}, the class of P_{123} and of P_{132}.

concept Table III shows how the classes of the permutation group $S(3)$ (whose multiplication table is given in Table I) are computed. The results illustrate the following:

Theorems (a) If B belongs to the class of A, then A belongs to the class of B. (b) Unless A and B belong to the same class, the elements in the class of A are entirely different from those in the class of B. (c) Every element belongs to its own class.

Thus the division of a group into its component classes is a division into closely associated operators. For the permutation group the division is into (i) the identity, which stands on its own; (ii) permutation operators involving a single interchange P_{12}, P_{13}, P_{23}; and (iii) permutation operators involving triple interchanges P_{123}, P_{132}. As will be seen later operators belonging to the same class have many properties in common. In listing the operators in a particular group this fact is used to simplify; i.e., only one of the operators in each class is given. Thus the operators in $S(3)$ would be given as I, $3P_{12}$, $2P_{123}$, the numbers 3 and 2 indicating the number of elements in each class.

This can be done for spatial symmetry as well as permutation symmetry; the classes will divide up into physically similar operators. For example, the division for benzene is

(i) I
(ii) C_2, $2C_3$, $2C_6$
(iii) $3C_2'$, $3C_2''$
(iv) $3\sigma_v$, $3\sigma_v'$
(v) σ_h
(vi) i
(vii) $2S_6$, $2S_3$.

The numbering (i)–(vii) corresponds to that used in the detailed discussion of benzene earlier in Section IIIA. In (iii) and (iv) the division is between axes passing through carbon atoms and those passing through bond midpoints. In (ii) $2C_3$ is the class made up of C_3 and C_3^2 and $2C_6$ of C_6, C_6^5, and similarly, in (vii).

IV. Group Representation Theory

A. Effect of the Symmetry Operators on Wave Functions

The symmetry operators discussed in the preceding section are operators on functions. The effect of the permutation operators is easy to write down. For example, if $\psi(1, 2)$ is a two-electron wave function and $\psi = \phi_A(1)\phi_B(2)$, then

$$P_{12}\psi(1, 2) = P_{12}\{\phi_A(1)\phi_B(2)\} = \phi_A(2)\phi_B(1) = \psi(2, 1).$$

The effect of the spatial symmetry operators depends on the form of the functions involved. A typical approximate molecular wave function is built up of atomic orbitals associated with the nuclei, and the effect of the symmetry operators is to change these orbitals from one nuclei to another. For example, a 1s hydrogen atomic orbital centered on the H(1) proton of the water molecule is left unchanged by the action of the operators I and σ_v, while C_2 and σ_v' transfer it to the H(2) proton.

Now consider a wave function ψ_i satisfying the Schroedinger equation (1.1). If R is a symmetry operator, then

$$R\{H\psi_i\} = RH\psi_i = HR\psi_i$$

since R leaves H invariant and commutes with it. Since $H\psi_i = E_i\psi_i$,

$$R\{H\psi_i\} = E_i R\psi_i,$$

so that

$$H\{R\psi_i\} = E_i\{R\psi_i\},$$

which shows that $R\psi_i$ is an eigenfunction of H with eigenvalue E_i. So the effect of the symmetry operators on molecular wave functions is to give molecular wave functions. There are two possibilities to be considered:

(a) The eigenvalue of E_i is nondegenerate. In this case $R\psi_i$ must be a scalar multiple of ψ_i, i.e.,

$$R\psi_i = r_i\psi_i,$$

so that the effect of the symmetry operator is just to multiply the wave function by a scalar, i.e., the wave function is an eigenfunction of the symmetry operator R with eigenvalue r_i.

(b) The eigenvalue E_i is n-fold degenerate. Suppose $\psi_i{}^1, \psi_i{}^2, \ldots, \psi_i{}^n$ are the degenerate eigenfunctions. Then, by the same argument as before, $R\psi_i{}^1, R\psi_i{}^2, \ldots, R\psi_i{}^n$ are also eigenfunctions so that each of them must be a linear combination of the $\psi_i{}^1, \ldots, \psi_i{}^n$. Thus the effect of the symmetry operators on a degenerate set of eigenfunctions is to transform them among themselves.

Using the eigenfunctions $\{\psi_i\}$ it is possible to set up a representation in a quantum mechanical sense. The idea is that any operator Q can be represented by a matrix $\mathbf{Q}$ with elements $\langle \psi_j | Q | \psi_i \rangle$. If this is done for the symmetry operators of the molecule, the matrices representing them take very simple forms. For the elements of the ith column of the matrix representing the operator R discussed above are $\langle \psi_j | R | \psi_i \rangle$, and if ψ_i is nondegenerate, it will be an eigenfunction of R so that $\langle \psi_j | R | \psi_i \rangle = r_i \langle \psi_j | \psi_i \rangle = r_i \, \delta_{ij}$ where the last step follows because of the orthonormality property of eigenfunctions. Hence the elements of the ith column will be zero except for the diagonal element which will equal the eigenvalue r_i. If i corresponds to a degenerate eigenvalue, the columns corresponding to the degenerate set $\psi_i{}^1, \ldots, \psi_i{}^n$ must be considered together, and these contain matrix elements of the form $\langle \psi_j | R | \psi_i{}^k \rangle$. But $R\psi_i{}^k$ gives only a linear combination of the degenerate set so the matrix elements vanish unless ψ_j is one of the $\psi_i{}^1, \ldots, \psi_i{}^n$. Therefore the elements of the columns corresponding to the degenerate eigenfunctions will be zero except for the diagonal block of size $n \times n$. Therefore the matrices representing the symmetry operators will have the

TABLE IV

AN EXAMPLE OF A 6×6 BLOCK-DIAGONAL MATRIX[a]

$$\begin{bmatrix} [1] & 0 & 0 & 0 & 0 & 0 \\ 0 & [2] & 0 & 0 & 0 & 0 \\ 0 & 0 & \begin{bmatrix} 1 & 2 \\ 2 & 3 \end{bmatrix} & 0 & 0 \\ 0 & 0 & & & 0 & 0 \\ 0 & 0 & 0 & 0 & \begin{bmatrix} 1 & 2 \\ 2 & 1 \end{bmatrix} \\ 0 & 0 & 0 & 0 & & \end{bmatrix}$$

[a] The blocks are of size 1×1, 1×1, 2×2, 2×2.

block-diagonal form (an example of which is given in Table IV). All the elements outside the diagonal blocks are zero; the blocks of size 1×1 correspond to nondegenerate eigenvalues and the blocks of size $n \times n$ correspond to eigenvalues which are n-fold degenerate.

Clearly each matrix representing one of the symmetry operators of the group will have the same block-diagonal form although the numerical values of the elements in the diagonal blocks will alter. The question arises whether these matrices can be simplified still further by making them of completely diagonal form rather than just block diagonal. For this to be done it is obviously necessary that the degenerate eigenfunctions $\psi_i^1, \ldots, \psi_i^n$ should be eigenfunctions of the symmetry operators. The $\{\psi_i^k\}$ are not unique since they can be transformed by a unitary transformation into a new set of eigenfunctions that are also degenerate eigenfunctions of H. For any operator which commutes with H, as the symmetry operator R does, this property can be used to choose the degenerate eigenfunctions to be eigenfunctions of R. Hence, it is possible to choose the eigenfunctions of H in such a way that the matrix representing R is completely diagonal. But this is only true for one of the symmetry operators, and the same basis set of eigenfunctions which completely diagonalize the matrix of R may not do the same for the matrices of the other symmetry operators. In fact any operator which commutes with R will, in this basis, have a diagonal matrix but one which does not commute with R will not. Therefore, the situation depends on whether or not the group to which the symmetry operators belong is Abelian. If the group is Abelian, the eigenfunctions of H can be chosen so that they are simultaneously eigenfunctions of all the symmetry operators of the group, and the matrices representing the operators will be completely diagonal. If the group is non-Abelian, this cannot be done and the eigenfunctions of H divide into degenerate sets which are such that the action of any one of the symmetry operators is to transform the functions of a given set among themselves. The matrices in this case will be of block-diagonal form and it is not possible to bring them all to completely diagonal form by the same transformation of the basis orbitals.

The situation as presented so far, while clear cut, is, unfortunately, the opposite way around to that required in practical applications. For, instead of having a complete set of eigenfunctions of H available with which to determine the functions which diagonalize or block-diagonalize the matrices representing the symmetry operators, it is necessary to use the symmetry operators and their properties to determine the type of functions that diagonalize the matrix of H, i.e., the wave functions.

In other words the group theoretical properties of the symmetry operators must be used to find the general forms of functions which bring the matrices representing the operators to diagonal or block-diagonal form since the eigenfunctions of H have this form. This problem is now considered in two stages, first for Abelian groups and then for non-Abelian groups.

B. Abelian Groups

Consider a system whose symmetry operators belong to an Abelian group and suppose ψ_1 is an arbitrary function of the electronic coordinates of the system but which depends parametrically on the nuclear coordinates, so that under the action of the spatial operators of the group it is transformed into another function. Let each of the symmetry operators in turn be applied to ψ_1 so that if there are n of them (including the identity), the n functions $\psi_1, \ldots, \psi_n$ are obtained. Provided ψ_1 is sufficiently general these functions will be linearly independent. If not, the process can still be carried through although some information is lost. Therefore, suppose the most general case occurs and the n functions are linearly independent.

Because of the closure property of a group it follows that the application of any of the symmetry operators of the group to any of the $\psi_1, \ldots, \psi_n$ will result in functions which are just linear combinations of the $\{\psi_i\}$ so that the functions themselves are closed with respect to the symmetry operators. Thus, if R is an operator:

$$R\psi_i = \sum_{j=1}^{n} R_{ji}\psi_j \tag{4.1}$$

where the R_{ji} are numbers. It is rather convenient to have a set of functions that are orthonormal and there is no loss of generality in assuming this to be the case for the $\{\psi_i\}$. From this it follows that $R_{ji} = \langle \psi_i \mid R \mid \psi_j \rangle$. If, therefore, the $n \times n$ matrix with elements R_{ij} is set up, the situation is very similar to matrix representations of operators in quantum mechanics. There is one important difference, however. In quantum mechanics the basis functions which are used to set up the representation have to form a complete set. In group theory they need not do so. However, the same nomenclature is used, the **R** matrix is called the representative of R and the functions $\{\psi_i\}$, the basis of the representation. Many results analogous to those in quantum mechanics can be obtained. The most important is that the matrix representing the product RS of two operators is just the product of the matrices representing each operator individually so that, if $T = RS$,

$$T_{ij} = \sum_k R_{ik} S_{kj}. \tag{4.2}$$

Because of this it follows that the set of n matrices, size $n \times n$, representing each of the group symmetry operators, will satisfy the same multiplication table as the operators and so the matrices themselves must form a group isomorphic with the original one. Thus, starting with a group of operators, this process leads to an equivalent group whose elements are matrices.

To give an example of this, consider the water molecule with symmetry operators I, C_2, σ_v, σ_v'. The four functions are $\psi_1 = I\psi_1$, $\psi_2 = C_2\psi_1$, $\psi_3 = \sigma_v\psi_1$, $\psi_4 = \sigma_v'\psi_1$. To obtain the effect of C_2, for example, on these functions, the group multiplication table (Table I) is used and gives

$$C_2\psi_1 = \psi_2 \quad \text{(by definition)}$$
$$C_2\psi_2 = C_2C_2\psi_1 = I\psi_1 = \psi_1$$
$$C_2\psi_3 = C_2\sigma_v\psi_1 = \sigma_v'\psi_1 = \psi_4$$
$$C_2\psi_4 = C_2\sigma_v'\psi_1 = \sigma_v\psi_1 = \psi_3.$$

Hence the matrix representing C_2 is

$$\begin{bmatrix} 0 & 1 & 0 & 0 \\ 1 & 0 & 0 & 0 \\ 0 & 0 & 0 & 1 \\ 0 & 0 & 1 & 0 \end{bmatrix}$$

and, in a similar way, those representing I, σ_v, and σ_v' are

$$\begin{bmatrix} 1 & 0 & 0 & 0 \\ 0 & 1 & 0 & 0 \\ 0 & 0 & 1 & 0 \\ 0 & 0 & 0 & 1 \end{bmatrix} \quad \begin{bmatrix} 0 & 0 & 1 & 0 \\ 0 & 0 & 0 & 1 \\ 1 & 0 & 0 & 0 \\ 0 & 1 & 0 & 0 \end{bmatrix} \quad \begin{bmatrix} 0 & 0 & 0 & 1 \\ 0 & 0 & 1 & 0 \\ 0 & 1 & 0 & 0 \\ 1 & 0 & 0 & 0 \end{bmatrix}.$$

Notice that in forming these matrices the explicit forms of the functions are not required, only the group multiplication table. It is not surprising, therefore, that the matrices themselves satisfy the group multiplication table, as is easily verified.

Returning to the general theory, consider a change of basis from the original functions $\psi_1, \ldots, \psi_n$ to a new set related to it by a linear transformation

$$\bar{\psi}_i = \sum_j S_{ji}\psi_j,$$

where the $n \times n$ matrix $\mathbf{S}$ must have an inverse $\mathbf{S}^{-1}$ for the transformation to be 1 : 1, in which case

$$\psi_i = \sum_j (\mathbf{S}^{-1})_{ji}\bar{\psi}_j.$$

Using these relations the matrices representing the operators in the new basis can be found. For, if $\mathbf{R}$ and $\bar{\mathbf{R}}$ are the matrices for R in the $\{\psi_i\}$ and $\{\bar{\psi}_i\}$ bases, respectively, i.e.,

$$R\bar{\psi}_i = \sum_l \bar{R}_{li}\bar{\psi}_l \quad \text{and} \quad R\psi_i = \sum_l R_{li}\psi_l,$$

then

$$R\bar{\psi}_i = R\sum_j S_{ji}\psi_j = \sum_j S_{ji}R\psi_j = \sum_{jk} S_{ji}R_{kj}\psi_k$$
$$= \sum_{jkl} S_{ji}R_{kj}(\mathbf{S}^{-1})_{lk}\bar{\psi}_l = \sum_l (\mathbf{S}^{-1}\mathbf{R}\mathbf{S})_{li}\bar{\psi}_l.$$

Hence $\bar{\mathbf{R}} = \mathbf{S}^{-1}\mathbf{R}\mathbf{S}$ and a similar relation holds for every matrix in the group. Because the group is Abelian the original matrices will be real and symmetric and will commute because the original operators do. Therefore, by a well-known theorem of matrix algebra, it is always possible to find a transformation matrix $\mathbf{S}$ which simultaneously brings them all to diagonal form. Because of this the new functions $\bar{\psi}_i$ will be eigenfunctions of each of the symmetry operators, the corresponding eigenvalues being given by the entries in the appropriate diagonal positions of the matrices.

In the example of the water molecule $\mathbf{S}$ is the matrix

$$\begin{bmatrix} 1 & 1 & 1 & 1 \\ 1 & 1 & -1 & -1 \\ 1 & -1 & 1 & -1 \\ 1 & -1 & -1 & 1 \end{bmatrix}$$

and in the new basis $\{\bar{\psi}_i\}$ the matrices corresponding to the operators I, C_2, σ_v, σ_v' are

$$\begin{bmatrix} 1 & 0 & 0 & 0 \\ 0 & 1 & 0 & 0 \\ 0 & 0 & 1 & 0 \\ 0 & 0 & 0 & 1 \end{bmatrix} \begin{bmatrix} 1 & 0 & 0 & 0 \\ 0 & 1 & 0 & 0 \\ 0 & 0 & -1 & 0 \\ 0 & 0 & 0 & -1 \end{bmatrix} \begin{bmatrix} 1 & 0 & 0 & 0 \\ 0 & -1 & 0 & 0 \\ 0 & 0 & 1 & 0 \\ 0 & 0 & 0 & -1 \end{bmatrix} \begin{bmatrix} 1 & 0 & 0 & 0 \\ 0 & -1 & 0 & 0 \\ 0 & 0 & -1 & 0 \\ 0 & 0 & 0 & 1 \end{bmatrix}.$$

The new functions are

$$\begin{aligned}
\bar{\psi}_1 &= \psi_1 + \psi_2 + \psi_3 + \psi_4 = (I + C_2 + \sigma_v + \sigma_v')\psi_1 \\
\bar{\psi}_2 &= \psi_1 + \psi_2 - \psi_3 - \psi_4 = (I + C_2 - \sigma_v - \sigma_v')\psi_1 \\
\bar{\psi}_3 &= \psi_1 - \psi_2 + \psi_3 - \psi_4 = (I - C_2 + \sigma_v - \sigma_v')\psi_1 \\
\bar{\psi}_4 &= \psi_1 - \psi_2 - \psi_3 + \psi_4 = (I - C_2 - \sigma_v + \sigma_v')\psi_1.
\end{aligned} \quad (4.3)$$

C. Non-Abelian Groups

If the same procedure is applied to a non-Abelian group, it will be found that it can be followed until the very last step. That last step involves simultaneously diagonalizing all the matrices and can be achieved only because the matrices commute. In the case of a non-Abelian group the matrices do not commute and so they cannot all be made diagonal by a transformation of the functions.

An example of this would be the permutation group $S(3)$. There will be six basis functions ψ_1, $\psi_2 = P_{12}\psi_1$, $\psi_3 = P_{23}\psi_1$, $\psi_4 = P_{13}\psi_1$, $\psi_5 = P_{123}\psi_1$, $\psi_6 = P_{132}\psi_1$, and the matrices for the six operators can be set up in this basis in exactly the same way as was done for the C_{2v} group. Although it is not possible to find a transformation of the $\{\psi_i\}$ into the $\{\bar{\psi}_i\}$ in such a way that all the $\bar{\mathbf{R}}$ matrices are diagonal it is possible to choose it so that one of the matrices is reduced to diagonal form. This can always be done no matter which matrix is chosen to have the diagonal form. For the sake of argument suppose it is that corresponding to P_{12}. The transformation which does this is given by

$$\begin{aligned}
\bar{\psi}_1 &= (I + P_{12} + P_{23} + P_{13} + P_{123} + P_{132})\psi_1 \\
\bar{\psi}_2 &= (I - P_{12} - P_{23} - P_{13} + P_{123} + P_{132})\psi_1 \\
\bar{\psi}_3 &= (I + P_{12} - \tfrac{1}{2}P_{23} - \tfrac{1}{2}P_{13} - \tfrac{1}{2}P_{123} - \tfrac{1}{2}P_{132})\psi_1 \\
\bar{\psi}_4 &= (P_{23} - P_{13} - P_{123} + P_{132})\psi_1 \\
\bar{\psi}_5 &= (P_{23} - P_{13} + P_{123} - P_{132})\psi_1 \\
\bar{\psi}_6 &= (I - P_{12} + \tfrac{1}{2}P_{23} + \tfrac{1}{2}P_{13} - \tfrac{1}{2}P_{123} - \tfrac{1}{2}P_{132})\psi_1.
\end{aligned} \quad (4.4)$$

Although these new basis functions produce diagonal matrices for the operators I and P_{12} the matrices for the remaining ones take up a block-diagonal form exactly as would be expected in view of the argument in Section IV.A. The pattern of the block diagonalization is that given in Table IV, i.e., two 1×1 blocks corresponding to the functions $\bar{\psi}_1$ and $\bar{\psi}_2$ and two 2×2 blocks corresponding to the pair $\bar{\psi}_3$, $\bar{\psi}_4$ and the pair $\bar{\psi}_5$, $\bar{\psi}_6$.

The matrix elements corresponding to $\bar{\psi}_1$ are all unity. Those for $\bar{\psi}_2$ are 1, −1, −1, −1, 1, 1, corresponding to the (2, 2) matrix elements in the matrices representing I, P_{12}, P_{23}, P_{13}, P_{123}, P_{132}, respectively. The 2×2 diagonal matrices corresponding to the pair of functions $\bar{\psi}_3$, $\bar{\psi}_4$ are, again in the order I, P_{12}, P_{23}, P_{13}, P_{123}, P_{132},

$$\begin{bmatrix} 1 & 0 \\ 0 & 1 \end{bmatrix} \quad \begin{bmatrix} 1 & 0 \\ 0 & -1 \end{bmatrix} \quad \begin{bmatrix} -\tfrac{1}{2} & \tfrac{1}{2}\sqrt{3} \\ \tfrac{1}{2}\sqrt{3} & \tfrac{1}{2} \end{bmatrix}$$

$$\begin{bmatrix} -\tfrac{1}{2} & -\tfrac{1}{2}\sqrt{3} \\ -\tfrac{1}{2}\sqrt{3} & \tfrac{1}{2} \end{bmatrix} \quad \begin{bmatrix} -\tfrac{1}{2} & \tfrac{1}{2}\sqrt{3} \\ -\tfrac{1}{2}\sqrt{3} & -\tfrac{1}{2} \end{bmatrix} \quad \begin{bmatrix} -\tfrac{1}{2} & -\tfrac{1}{2}\sqrt{3} \\ \tfrac{1}{2}\sqrt{3} & -\tfrac{1}{2} \end{bmatrix}.$$

The 2×2 diagonal matrices corresponding to the pair $\bar{\psi}_5$, $\bar{\psi}_6$ are exactly the same.

If the functions $\bar{\psi}_3$, $\bar{\psi}_4$ are used by themselves as a basis of a representation, the matrices will be those given above. Therefore the effect of any of the operators of the group on $\bar{\psi}_3$ and $\bar{\psi}_4$ is to give linear combinations of $\bar{\psi}_3$ and $\bar{\psi}_4$, so that $\bar{\psi}_3$ and $\bar{\psi}_4$ are transformed between themselves by the action of the group operations. Moreover it is clear that there is no possibility of transforming $\bar{\psi}_3$, $\bar{\psi}_4$ by a linear transformation into two functions which are each eigenfunctions of all the operators of the group, for this would involve diagonalizing all the 2×2 matrices simultaneously and this cannot be done. Therefore $\bar{\psi}_3$, $\bar{\psi}_4$ cannot be reduced into any simpler components and, for this reason, the representation of which they are the basis is said to be irreducible and $\bar{\psi}_3$, $\bar{\psi}_4$ are said to form the basis of an irreducible representation. In the same way $\bar{\psi}_5$, $\bar{\psi}_6$ form the basis of an irreducible representation and, since the matrices are the same, the irreducible representations are the same. The number of functions needed to form the basis of an irreducible representation is called its dimension; in the case just considered the dimension is two. Irreducible representations of dimension one also occur; for example, $\bar{\psi}_1$ forms the basis of one and $\bar{\psi}_2$ of another. The function which forms the basis of a one-dimensional irreducible representation has the special property that it is an eigenfunction of all the operators of the group. The functions which form the basis of an irreducible representation of dimension greater than one are transformed among themselves by the action of the group operations. These are just the types of functions the considerations in Section IV.A suggested ought to be obtained.

The full theory behind the block diagonalization of the matrices and the determination of the irreducible representations and their bases is not given here, partly due to lack of space but also because tables are available which give practically all the information of interest on the irreducible representations of the groups of physical and chemical importance so there is seldom any need to work through the details. In these tables, however, it is not usual to give the full matrices for the irreducible representations but only their traces which in group theory are called characters and are indicated by the symbol $\chi(R)$.

Before examining the properties of the character tables, as they are called, two more results are required. The first concerns the number of irreducible representations of a particular group. It can be shown that this equals the number of classes in the group. Following from this their dimensions can be considered, so suppose there are r of them with dimension $d_1, d_2, \ldots, d_r$. If there are n elements in the group, it is possible to show that

$$d_1^2 + d_2^2 + \cdots + d_r^2 = n. \tag{4.5}$$

In the permutation group $S(3)$, for instance, there are three classes and, therefore, three irreducible representations, as is already known. The dimensions of these are 1, 1, and 2 which do indeed satisfy $1^2 + 1^2 + 2^2 = n = 6$. In an Abelian group of order n there will be n basis functions and all the matrices will be diagonal. Hence there will be n irreducible representations of dimension one.

D. Character Tables

Three character tables are given in Table V. The first is for the Abelian spatial symmetry group C_{2v}. There are four elements and, therefore, four irreducible representations labeled A_1, B_2, A_2, B_1 (details of the notation

TABLE V

Character Tables for C_{2v}, C_{3v}, and $S(3)$

C_{2v}	I	C_2	σ_v	σ_v'	C_{3v}	I	$2C_3$	$3\sigma_v$	$S(3)$	I	$3P_{12}$	$2P_{123}$
A_1	1	1	1	1	A_1	1	1	1	$[3^1]$	1	1	1
A_2	1	1	−1	−1	A_2	1	1	−1	$[1^3]$	1	−1	1
B_1	1	−1	1	−1								
B_2	1	−1	−1	1	E	2	−1	0	$[1^1 2^1]$	2	0	−1

TABLE VI

Classification of Irreducible Representations[a]

	A	One-dimensional representation symmetric with respect to rotation of $2\pi/n$ about principal axis; i.e., $\chi(I) = 1$, $\chi(C_n) = 1$
	B	One-dimensional representation antisymmetric with respect to rotation of $2\pi/n$ about principal axis; i.e., $\chi(I) = 1$, $\chi(C_n) = -1$
	E	Two-dimensional irreducible representation, $\chi(I) = 2$
	T	Three-dimensional irreducible representation, $\chi(I) = 3$
Superscript	'	Symmetric with respect to reflection in σ_h, $\chi(\sigma_h) = 1$
Superscript	"	Antisymmetric with respect to reflection in σ_h, $\chi(\sigma_h) = -1$
Subscript	g	Symmetric with respect to inversion, $\chi(i) = 1$
Subscript	u	Antisymmetric with respect to inversion, $\chi(i) = -1$
Subscript	1	Symmetric with respect to reflection in σ_v, $\chi(\sigma_v) = 1$
Subscript	2	Antisymmetric with respect to reflection in σ_v, $\chi(\sigma_v) = -1$

[a] Spatial symmetry only. For permutation symmetry see Section VI.

are given in Table VI). Corresponding to each irreducible representation, the character $\chi(R)$ of each operator is given in the table. In the case of the Abelian group these characters are the same as the eigenvalues of the operators.

The two non-Abelian groups whose character tables are given in Table V, are C_{3v} and $S(3)$. These groups are isomorphic so their tables are the same, but the notations for the irreducible representations are different. There are three irreducible representations and the characters are just the traces of the matrices. This table illustrates two quite general results. The first is that the character of the identity element $\chi(I)$ is the same as the dimension of the irreducible representation. This is because the matrix representing the identity is just the unit matrix whatever the basis, and therefore its trace, which equals $\chi(I)$ is the number of functions in the basis. The second result is that the characters of two elements in the same class are equal. For if A and B are in the same class, there exists another element G such that $A = GBG^{-1}$. In an irreducible representation the operators become matrices with $\mathbf{A} = \mathbf{GBG^{-1}}$. Since the trace of the product of two matrices is independent of order, $\text{tr}(\mathbf{A}) = \text{tr}(\mathbf{BG^{-1}G}) = \text{tr } \mathbf{B}$. This result is made use of by listing the characters of only one member of each class and is a further indication that members of the same class are physically similar operators. A final point concerns the irreducible representation whose characters are all unity. Every group has such an irreducible representation since there are always functions which are eigenfunctions of all the group operators with eigenvalue unity, namely the functions left invariant under the action of operators (e.g., functions that are constants). Such an irreducible representation is called the identity or symmetric representation. Naturally since $\chi(I) = 1$, it is one dimensional.

E. Basis Functions and Projection Operators

In the first part of this section, a representation was set up using the eigenfunctions of H as basis. Because the symmetry operators of the system left H invariant it turned out that the eigenfunctions of H were either eigenfunctions of the symmetry operators also or else, as happened in the case of degeneracy, belonged to a set that was transformed within itself by the action of the operators. By using the results of group representation theory in Sections IV.B–D, this problem was approached from the opposite direction and it has been shown that sets of functions which are transformed among themselves by the action of the group operators and which cannot be subdivided into any simpler sets with this property form the basis of an irreducible representation of the group. Therefore, it can be concluded that to each energy level of the system, there corresponds an irreducible representation of the group of symmetry operations. The dimension of the irreducible representation equals the degeneracy of the level and the degenerate wavefunctions form a basis for the irreducible representation.[†] The behavior of the wave functions under the action of the symmetry operators (i.e., the symmetry properties of the wave functions) is determined from the matrices of the irreducible representation so that in many cases, examples of which are given below, the exact forms of the wave functions are not required, only their symmetry properties. In some cases, however, the particular form of the basis functions is required, and this is now discussed.

The problem is this. Given an arbitrary function ψ, how is it possible to obtain from it functions which form the basis of an irreducible representation? To do this a result is needed concerning the matrix elements

---

[†] In a few very rare cases it can happen that sets of functions belonging to different irreducible representations can have the same energy levels. This is termed "accidental degeneracy."

of the irreducible representations. Let $\mathbf{D}^i(R)$ be the matrix representing the operator R in the ith irreducible representation. Take the p, q element from $\mathbf{D}^i(R)$ and the r, s element from $\mathbf{D}^j(R)$, i.e., the matrix representing the same operator in the jth irreducible representation. Keeping (p, q) and (r, s) fixed, take the complex conjugate of one of these elements and multiply by the other, and do the same for every matrix, i.e., for all the operators R. The sum of these terms is

$$\sum_R D^i_{pq}(R)^* D^j_{rs}(R) = (n/d_i)\, \delta_{ij}\, \delta_{pr}\, \delta_{qs}. \tag{4.6}$$

The sum is zero unless the two irreducible representations are the same, and even when they are the same the sum is nonzero only when the same elements are taken. This extraordinarily powerful result requires a very lengthy proof, which is not given here. However, it is used to obtain the basis functions for a particular irreducible representation.

Suppose i is a particular irreducible representation and ψ_k^i is a function satisfying

$$\psi_k^i = (d_i/n) \sum_R D^i_{kk}(R)^* R\psi_k^i. \tag{4.7}$$

Then ψ_k^i is said to transform like, or to belong to, the kth row of the irreducible representation. Using ψ_k^i it is possible to form a basis for the irreducible representation. One of the basis functions will be ψ_k^i itself and the others are given by

$$\psi_l^i = (d_i/n) \sum_R D^i_{lk}(R)^* R\psi_k^i \tag{4.8}$$

for $l = 1, \ldots, d_i$, the choice $l = k$ simply reproducing ψ_k^i by (4.7). To prove this result it is only necessary to show that the functions are transformed among themselves by the action of the group operators, which is done by multiplying (4.8) by $D^i_{lj}(S)$ and summing to give

$$\sum_{l=1}^{d_i} D^i_{lj}(S)\psi_l^i = (d_i/n) \sum_{l=1}^{d_i} \sum_R D^i_{lj}(S) D^i_{lk}(R)^* R\psi_k^i$$

$$= (d_i/n) \sum_R \left\{ \sum_{l=1}^{d_i} D^i_{jl}(S^{-1})^* D^i_{lk}(R)^* \right\} R\psi_k^i,$$

where to obtain the term in braces the fact that $\mathbf{D}(S^{-1})\mathbf{D}(S) = \mathbf{D}(S^{-1}S) = \mathbf{D}(I) = \mathbf{I}$ is used (since the matrix representing the product of two operators is the product of the matrices representing the separate operators). In the same way the term can be further reduced to $D^i_{jk}(S^{-1}R)^*$. Putting $S^{-1}R = T$ and noting that a sum of $S^{-1}R$ over R is equivalent to a sum over all elements T gives

$$\sum_{l=1}^{d_i} D^i_{lj}(S)\psi_l{}^i = (d_i/n) S \sum_R D^i_{jk}(S^{-1}R)^* S^{-1} R \psi_k{}^i$$
$$= (d_i/n) S \sum_T D^i_{jk}(T)^* T \psi_k{}^i = S \psi_j{}^i,$$

which holds for any operator S of the group and which, therefore, proves the result.

Since there are d_i equations of the form (4.7) but with different values of the $D^i_{kk}(R)^*$ matrix elements corresponding to the d_i choices of the index k, it is implied that there are d_i distinct functions $\psi_k{}^i$ from each of which a different basis for the irreducible representation can be found. All told there will be $d_i{}^2$ functions and the total number of all the separate functions must equal the total number of linearly independent functions, namely, n, which can be obtained from a general arbitrary function by the action of the group operators. This is the content of Eq. (4.5).

Using similar arguments based on the relation (4.6), it is possible to show that the operator

$$(d_i/n) \sum_R D^i_{kk}(R)^* R, \qquad (4.9)$$

when operating on any of the $d_i{}^2$ functions of the ith irreducible representation, will give zero unless the function is $\psi_k{}^i$. Moreover it will give zero when operating on any function which belongs to a different irreducible representation. Since an arbitrary function ψ can be written as a sum of basis functions belonging to different irreducible representations, it follows that the action of (4.9) on ψ is to select the component $\psi_k{}^i$ so that

$$\psi_k{}^i = (d_i/n) \sum_R D^i_{kk}(R)^* R \psi. \qquad (4.10)$$

Having obtained $\psi_k{}^i$ in this way the remaining functions which form the basis belonging to the kth row of the ith irreducible representation can be obtained from (4.8).

These results are confirmed by the example of $S(3)$ treated in Section IV.C. The basis functions $\{\bar\psi\}$ are in fact obtained from ψ precisely by the method just described (except for a trivial numerical factor) and for the two-dimensional irreducible representation there are two sets of basis functions belonging to the first and second row, respectively.

A single completely arbitrary function will always contain d_i functions belonging to the ith irreducible representation, one belonging to each of the d_i rows as is implied by (4.10). If the function ψ is not completely arbitrary, as is often the case in practice, it may contain only some of these functions or even none. The same applies for a set of functions. Suppose the functions are used as the basis of a representation. Normally this will not be irreducible but, on the contrary, will be reducible, and the problem is to reduce it to the irreducible representations and to determine how many times the various irreducible representations are contained in it. Let $\chi(R)$ be the character representing the operator R in the reducible representation. Now let the matrices be brought to block-diagonal form corresponding to a division into the irreducible representations. Since this is effected by a similarity transformation the traces of the matrices are invariant and the characters $\chi(R)$ remain the same. However, if the ith irreducible representation is contained a_i times in the reducible one, its contribution to the total $\chi(R)$ will be just $a_i \chi_i(R)$, where $\chi_i(R)$ is the character of R in the irreducible representation. Hence it follows that

$$\chi(R) = \sum_i a_i \chi_i(R), \qquad (4.11)$$

where the sum is over all the irreducible representations. This equation holds for all R, giving a set of relations which allow the a_i to be determined.

To determine from the original set the actual functions which transform like the rows of the irreducible representation the methods developed earlier in this section may be used. Unfortunately these require a knowledge of the matrices $\mathbf{D}^i(R)$ and normally only the character tables are available. If, as is often the case, it is enough to know the sum of the functions which transform like the rows of the irreducible representation, this can be obtained using characters only. For

$$\sum_{k=1}^{d_i} \psi_k^i = (d_i/n) \sum_{k=1}^{d_i} \sum_R D_{kk}^i(R)^* R\psi = (d_i/n) \sum_R \chi_i^*(R) R\psi,$$

since $\sum_{k=1}^{d_i} D_{kk}^i(R)^*$ is just the character $\chi_i^*(R)$. Thus the operator

$$(d_i/n) \sum_R \chi_i^*(R) R \qquad (4.12)$$

selects from ψ a function which transforms like the ith irreducible representation.

F. DIRECT PRODUCT OF TWO REPRESENTATIONS

Suppose two sets of functions $\{\psi_k\}$ and $\{\phi_k\}$ separately form a basis for the representation of a particular group. In most practical applications the functions will form bases of irreducible representations and, if they do not, they can be subdivided until they do so. Therefore, assume $\{\psi_r\}$, $r = 1, \ldots, d_i$, to be a basis for the ith irreducible representation and $\{\phi_s\}$, $s = 1, \ldots, d_j$, a basis for the jth. The $d_i d_j$ functions obtained by multiplying the two sets together, i.e., $\{\eta_k\} \equiv \{\psi_r \phi_s\}$, will serve as a new basis for a representation of the group. The new representation is called the direct product of the original two.

In the new representations the matrices $\mathbf{R}^\eta$ representing the operators of the group must satisfy

$$R\eta_k = \sum_{l=1}^{d_i d_j} R^\eta_{lk} \eta_l,$$

while the transformations of the original functions satisfy

$$R\psi_r = \sum_{p=1}^{d_i} D^i_{pr}(R) \psi_p, \qquad R\phi_s = \sum_{q=1}^{d_j} D^j_{qs}(R) \phi_q$$

so that

$$R\{\psi_r \phi_s\} = \{R\psi_r\}\{R\phi_s\} = \sum_{p=1}^{d_i} \sum_{q=1}^{d_j} D^i_{pr}(R) D^j_{qs}(R) \psi_p \phi_q$$

and, therefore,

$$R^\eta_{lk} = D^i_{pr}(R) D^j_{qs}(R), \tag{4.13}$$

where l and k are single indices corresponding to the pairs (p, q) and (r, s). Thus the new representation is related to the old in a fairly simple manner. However, unless $\mathbf{R}^\eta$ is equal to the matrix $\mathbf{D}^k(R)$ of an irreducible representation for all the operators R, the new representation will not be irreducible. In fact for this to occur one of the i and j irreducible representations must have dimension unity so that one or the other of the $\mathbf{D}^i(R)$, $\mathbf{D}^j(R)$ matrices has only a single element.

If the new representation is not irreducible, it can be reduced into its irreducible components. To do this only the characters are needed. From (4.13) it follows that

$$\chi_\eta(R) = \sum_{l=1}^{d_i d_j} R^\eta_{ll} = \sum_{p,q} D^i_{pp}(R) D^j_{qq}(R)$$

$$= \sum_p D^i_{pp}(R) \sum_q D^j_{qq}(R) = \chi_i(R) \chi_j(R) \tag{4.14}$$

so that the new characters are the products of the old. This result can clearly be extended to the product of more than two representations. As an example, suppose $\psi_1, \ldots, \psi_4$ are functions, two of which, ψ_1 and ψ_2, transform like the A_1, A_2 irreducible representations of the $\mathbf{C}_{3v}$ group and two, ψ_3 and ψ_4, like the E irreducible representation, and let ϕ_1, ϕ_2 be two different functions belonging to the E irreducible representation. Dividing $\psi_1, \ldots, \psi_4$ into the three sets of ψ_1, ψ_2 and (ψ_3, ψ_4), the eight functions divide into three sets also. The set $\psi_1\phi_1$, $\psi_1\phi_2$ has characters

$$\chi(I) = 2, \quad \chi(C_3) = -1, \quad \chi(\sigma_v) = 0,$$

which is found by using (4.14) and the characters from Table VI. These two functions clearly belong to the E irreducible representation. The set $\psi_2\phi_1$, $\psi_2\phi_2$ has the same characters and hence belongs to the same irreducible representation. Finally the set of four functions obtained by multiplying the bases of the same irreducible representation together has characters

$$\chi(I) = 4, \quad \chi(C_3) = 1, \quad \chi(\sigma_v) = 0,$$

which does not correspond to a single irreducible representation but is a sum of them. Suppose the four functions can be transformed into a_1 functions belonging to A_1, a_2 to A_2, and $2a_3$ to E (i.e., E is contained a_3 times, giving $2a_3$ functions since E is doubly degenerate). Then from (4.14),

$$\chi_\eta(R) = a_1\chi_{A_1}(R) + a_2\chi_{A_2}(R) + a_3\chi_E(R).$$

This gives three equations (since only one operator per class need be considered, the others simply repeating the same equation) to determine a_1, a_2, a_3. By inspection $a_1 = 1$, $a_2 = 1$, $a_3 = 1$, so there are two functions belonging to E and one each to A_1 and A_2.

G. DIRECT PRODUCT OF TWO GROUPS

The direct product of two representations should not be confused with the direct product of two groups, which is now defined. Let **A** and **B** be two groups with elements $A_1, A_2, \ldots, A_n$ and $B_1, B_2, \ldots, B_m$, such that, except for identity, the two groups have no element in common. Suppose, in addition, every element from group **A** commutes with every element of group **B**. (**A** and **B** need not themselves be Abelian groups.)

By multiplying every element of **A** by every element of **B** a set of nm elements $A_1B_1, A_1B_2, A_1B_3, \ldots, A_2B_1, A_2B_2, \ldots,$ is formed. It is then trivial to show that these elements form a group called the direct product of **A** and **B**, which is written **A** × **B**.

It is easy to show that, if $\{\psi_k^i\}$ and $\{\phi_l^j\}$ are the basis functions for the ith irreducible representation of group **A** and the jth irreducible representation of group **B**, then the set $\{\psi_k^i \phi_l^j\}$ formed from their products is the basis for an irreducible representation of group **A** × **B**. Thus the set of irreducible representations of **A** × **B** can be obtained by multiplying the basis functions of the irreducible representations of **A** and **B**. The characters of the direct product group turn out to be the products of the characters of the separate groups **A** and **B**.

V. Some Applications in Molecular Quantum Mechanics

A. Matrix Elements

The basic idea that underlies the preceding section is that the eigenfunctions of a molecular Hamiltonian, i.e., the molecular wave functions, form the basis of a representation of the symmetry group of the molecule. Moreover, wave functions that are nondegenerate have to be eigenfunctions of each symmetry operator that commutes with H. Thus each nondegenerate wave function forms the basis of a one-dimensional irreducible representation of the symmetry group and, therefore, has the symmetry properties of that irreducible representation. On the other hand the wave functions corresponding to a d-fold degenerate set are transformed among themselves by the action of the group operators, and so form the basis of a d-dimensional irreducible representation. Thus every wave function can be classified by the irreducible representation to which it belongs.

Suppose ψ_r^i and ψ_s^j are two wave functions belonging to the i and j irreducible representations, respectively. In many quantum mechanical problems it is necessary to evaluate matrix elements of the type

$$\int \psi_r^{i*} f \psi_s^j \, d\tau \tag{5.1}$$

where f is some operator and the integral is over the configuration space. Suppose first that $f \equiv 1$ and consider the scalar product or orthogonality integral between the two wave functions. Since the integral equals a scalar and the integration is over all space, it must be invariant to the action of the symmetry operators of the group. Applying the operator R gives

$$\int \psi_r^{i*} \psi_s^j \, d\tau = \int \{R\psi_r^i\}^* \{R\psi_s^j\} \, d\tau$$

$$= \sum_{p=1}^{d_i} \sum_{q=1}^{d_j} \int \{D_{pr}^i(R)\psi_p^i\}^* \{D_{qs}^j(R)\psi_q^j\} \, d\tau$$

$$= \sum_{p=1}^{d_i} \sum_{q=1}^{d_j} D_{pr}^i(R)^* D_{qs}^j(R) \int \psi_p^{i*} \psi_q^j \, d\tau.$$

If this is done for all the operators of the group in turn and the n resulting equations summed, we obtain

$$n \int \psi_r^{i*} \psi_s^j \, d\tau = \sum_{p=1}^{d_i} \sum_{q=1}^{d_j} \left\{ \sum_R D_{pr}^i(R)^* D_{qs}^j(R) \right\} \int \psi_p^{i*} \psi_q^j \, d\tau$$

$$= \sum_{p=1}^{d_i} \sum_{q=1}^{d_j} \delta_{ij} \delta_{pq} \delta_{rs} (n/d_i) \int \psi_p^{i*} \psi_q^j \, d\tau$$

using Eq. (4.6). Hence,

$$\int \psi_r^{i*} \psi_s^j \, d\tau = \delta_{ij} \delta_{rs} d_i^{-1} \sum_{p=1}^{d_i} \int |\psi_p^i|^2 \, d\tau. \tag{5.2}$$

For wave functions this simply shows that they are orthogonal. However, the proof does not depend on the $\{\psi_r^i\}$ being wave functions and holds for any functions which belong to irreducible representations of the symmetry group. It therefore shows that functions belonging to different irreducible representations are orthogonal, as indeed are functions belonging to different rows ($r \neq s$) of the same irreducible representation.

In particular suppose ψ_r^i is constant and equal to unity. Since it is constant it will be invariant to the action of the group operators and so must belong to the symmetric or identity irreducible representation which can be labeled by $i = 1$. Hence,

$$\int \psi^{1*} \psi_s^j \, d\tau = \int \psi_s^j \, d\tau = 0 \quad \text{unless} \quad j = 1, \tag{5.3}$$

so that the integral $\int \psi_s^j \, d\tau$ of a function belonging to an irreducible representation is zero unless the irreducible representation is the symmetric one. Now suppose ψ is an arbitrary function and it is decomposed into the sum of functions transforming like rows of irreducible representations, i.e.,

$$\psi = a_1\psi^1 + \sum_{j \neq 1} \sum_{r=1}^{d_j} a_{jr}\psi_r^j,$$

where the coefficients a_1 and a_{jr} are unity or zero, depending on whether or not ψ contains components belonging to the appropriate irreducible representations. It follows that

$$\int \psi \, d\tau = a_1 \int \psi^1 \, d\tau + \sum_{j \neq 1} \sum_{r=1}^{d_j} a_{jr} \int \psi_r^j \, d\tau = a_1 \int \psi^1 \, d\tau$$

so that the integral is nonzero only if $a_1 \neq 0$; i.e., only if ψ contains a component transforming like the symmetric representation.

This result can now be used to discuss the matrix element (5.1). Suppose that the operator f transforms like the kth irreducible representation. Thus $\psi_r^{i*} f \psi_s^j$ is the direct product of three representations and can be decomposed into components transforming like different irreducible representations by the methods of Section IV.F. Only if the identity representation is among these will the integral differ from zero.

As an example, suppose the components of the electric polarizability of a molecule belonging to the $\mathbf{C}_{3v}$ point group are to be computed for a state with wave function ψ_{A_2} belonging to the A_2 irreducible representation. The fg component will be given by

$$2 \sum_X \frac{\langle \psi_{A_2} | f | \psi_X \rangle \langle \psi_X | g | \psi_{A_2} \rangle}{E_X - E_{A_2}} \tag{5.4}$$

where the sum is over all the molecular states X except the initial one and E_X and E_{A_2} are the energies. The operators f and g can be x, y, or z. For example, $f = x$, $g = y$ gives the xy component of the polarizability α_{xy}, $f = z$, $g = z$ the zz component α_{zz}, and so on. If z is chosen to be in the direction of the threefold axis of the $\mathbf{C}_{3v}$ group, it is easy to verify, by applying the group operators, that z transforms like A_1, the symmetric representation, while x and y belong to the rows of the two-dimensional E irreducible representation.

The matrix elements involved in the calculation of α are

$$\langle \psi_{A_2} | f | \psi_X \rangle = \int \psi_{A_2}^* f \psi_X \, d\tau$$

and these will be zero unless the symmetry of f and ψ_X is such that the product $\psi_{A_2}^* f \psi_X$ contains a component transforming like A_1. The wave functions ψ_X must transform like one of A_1, A_2, and E. Using the methods of Section IV.F it is easy to show that the product $\psi_{A_2}^* \psi_X$ transforms like A_2 if ψ_X belongs to A_1, like A_1 if ψ_X belongs to A_2, and like E if ψ_X belongs to E. It is possible to show that the product of two functions belonging to irreducible representations contains the identity representation only if these two irreducible representations are the same, and this result can be easily verified for $\mathbf{C}_{3v}$. Thus $\psi_{A_2}^* f \psi_X$ will contain a component transforming like A_1 only if the product $\psi_{A_2}^* \psi_X$ and the operator f belong to the same irreducible representation. If f is x or y and transforms like E, then so must $\psi_{A_2}^* \psi_X$, which means that ψ_X must transform like E. Similarly, if f is z, then ψ_X must belong to A_2. This simplifies expression (5.4) considerably since only wave functions of appropriate symmetry need be included in the sum, the others giving zero values for the matrix elements. It also shows that α_{xz} and α_{yz} are zero since the types of functions ψ_X for which $\langle \psi_{A_2} | x | \psi_X \rangle$ and $\langle \psi_{A_2} | y | \psi_X \rangle$ are nonzero are those for which $\langle \psi_X | z | \psi_{A_2} \rangle$ is zero.

B. Transition Probabilities

The probability of a molecule making a transition from a state ψ_I (usually the ground state) to another state ψ_F is governed by the integral

$$\mathbf{\mu} = \int \psi_I^* \mathbf{r} \psi_F \, d\tau$$

where $\mathbf{r}$ is the position vector (x, y, z). Since ψ_I and ψ_F are wave functions they belong to an irreducible representation of the molecular symmetry group, and similarly for the components of $\mathbf{r}$. Thus the vanishing or nonvanishing of the matrix element on symmetry grounds can be examined by the methods of the previous section.

In the example of the $\mathbf{C}_{3v}$ group, it follows from Section V.A that if ψ_I belongs to A_2, $\mathbf{\mu}$ has zero z component but nonzero x and y components if ψ_F belongs to E, while it has zero x and y components but a nonzero z component if ψ_F belongs to A_2. On the other hand, if ψ_F belongs to A_1, $\mathbf{\mu} = 0$ so that the transition $A_2 \to A_1$ is said to be symmetry forbidden. For transitions that are not symmetry forbidden, the direction of $\mathbf{\mu}$ determines the direction of polarization of the radiation required to effect the transition. Thus, in the $\mathbf{C}_{3v}$ example, a transition between two states belonging to A_2 requires the radiation to be polarized in the z direction. If the radiation is polarized in the x or y direction, there will be no transition.

C. Computation of Molecular Wave Functions

The most common method of obtaining molecular wave functions is by way of the variational principle. Suppose $\{\phi_r\}$, $r = 1, \ldots, s$, is a set of many-electron functions, and it is required to find a wave function in the form of a linear combination of these functions:

$$\Psi = \sum a_r \phi_r$$

where the a_r are constants. If the $\{a_r\}$ are written as a column vector **a**, then the variational method can be used to show that the best choice of **a** will satisfy

$$\mathbf{Ha} = E\mathbf{Sa} \qquad (5.5)$$

where **H** and **S** are $s \times s$ matrices with matrix elements $H_{tu} = \langle \phi_t | H | \phi_u \rangle$ (H being the molecular Hamiltonian) and $S_{tu} = \langle \phi_t | \phi_u \rangle$ (the overlap matrix). The constant E will be the variational energy. Since (5.5) is an eigenvalue type of equation there will be s different solutions, each corresponding to a different wave function.

To set up Eq. (5.5) and to solve it can be quite a difficult problem, which can, however, be simplified somewhat by using symmetry. Suppose the functions ϕ_r are chosen in such a way that they belong to the various irreducible representations of the point group of the molecule. If we indicate this by the label i, e.g., ϕ_r^i for a function belonging to the ith irreducible representation, then, from Section V.A, it follows that because H is left invariant by the action of the operators of the group the matrix elements $\langle \phi_t^i | H | \phi_u^j \rangle$ and $\langle \phi_t^i | \phi_u^j \rangle$ are zero unless i and j are the same. Thus the full equation (5.5) has block-diagonal form and so may be divided into a number of simpler equations corresponding to each irreducible representation and each row thereof. Hence the symmetry of the molecule can be used to simplify the calculation of its wave function. Nor, incidentally, is this method restricted to spatial symmetry; it can also be used in the case of permutation symmetry where it amounts to dividing the $\{\phi_r\}$ into functions with the same spin quantum numbers.

As a rule the many-electron functions $\{\phi_r\}$ are built up from one-electron functions $\{\omega_v\}$, say, written in the form of a Slater determinant

$$\phi = |\omega_1(1), \ldots, \omega_p(p)|$$

for a p-electron function. The ω_p are called molecular (spin) orbitals and are usually obtained from the Hartree–Fock equations. In their turn the molecular orbitals are written as a linear combination of atomic orbitals centered on the various atoms of the molecule. This leads to an equation very like (5.5) except that H is the Hartree–Fock Hamiltonian. However, this also is invariant to the action of the symmetry operators of the molecular point group so that the previous method can be used to simplify the calculation. As a result the molecular orbitals have to belong to the irreducible representations of the point group.

The question next arises as to the symmetry properties of the Slater determinant in this case. Although ϕ is a determinant it can be written as the sum of products, so the idea of the direct product of different representations can be used. Thus, if R is a group operator and $\chi(R)$ is the character of the representation built up from ϕ, then from Section IV.F.

$$\chi(R) = \chi_1(R)\chi_2(R). \ldots . \chi_p(R)$$

where $\chi_1(R), \ldots, \chi_p(R)$ are the characters of R with respect to $\omega_1, \omega_2, \ldots, \omega_p$. Since the molecular orbitals belong to irreducible representations these latter will be known. Thus the characters $\chi(R)$ for all the operators in the group can be found, and hence it is possible to deduce either the irreducible representation to which ϕ belongs or the set of functions, each belonging to a different irreducible representation, into which ϕ can be decomposed. In the latter case ϕ itself cannot be a proper wave function and only its irreducible components are used. Often this procedure is worked backwards. That is, we know which symmetry the approximate wave function ϕ should possess and so the required characters $\chi(R)$ are known. The problem then is to find to which irreducible representations the $\{\omega_v\}$ must belong in order that the product $\chi_1(R). \ldots . \chi_p(R)$ of their characters gives the proper result. Normally, of course, there is no unique answer and which particular one of the several possible choices is finally made has to be determined on other grounds.

D. NORMAL MODES OF VIBRATION

During a molecular vibration the nuclei are displaced from their equilibrium position; this can be specified by the displacement coordinates (x, y, z) of each nuclei. There will therefore be a set of $3N$ functions, i.e., three displacement coordinates for each of the N nuclei, and these can be used to set up a representation for the operators of the molecular point

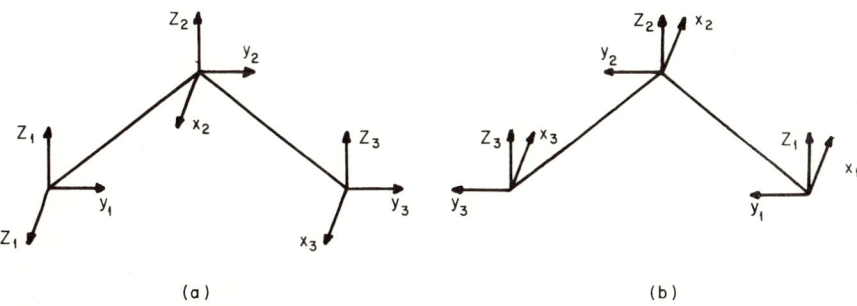

FIG. 3. Displacement coordinates for the nuclei of the water molecule (a) before and (b) after the action of the C_2 operator.

symmetry group. An example, that of the water molecule, is given in Fig. 3a and the effect of the rotation C_2 is shown in Fig. 3b. Since, for example, the effect of C_2 on x_1 is to change it into $-x_3$, we can write

$$C_2 x_1 = -x_3$$

while C_2 changes x_2 into $-x_2$ so that

$$C_2 x_2 = -x_2.$$

Using these and similar results, it follows from Eq. (4.1) that the matrix representing C_2 in the basis of $x_1, y_1, z_1, x_2, y_2, z_2, x_3, y_3, z_3$ will be

$$\begin{bmatrix} 0 & 0 & 0 & 0 & 0 & 0 & -1 & 0 & 0 \\ 0 & 0 & 0 & 0 & 0 & 0 & 0 & -1 & 0 \\ 0 & 0 & 0 & 0 & 0 & 0 & 0 & 0 & 1 \\ 0 & 0 & 0 & -1 & 0 & 0 & 0 & 0 & 0 \\ 0 & 0 & 0 & 0 & -1 & 0 & 0 & 0 & 0 \\ 0 & 0 & 0 & 0 & 0 & 1 & 0 & 0 & 0 \\ -1 & 0 & 0 & 0 & 0 & 0 & 0 & 0 & 0 \\ 0 & -1 & 0 & 0 & 0 & 0 & 0 & 0 & 0 \\ 0 & 0 & 1 & 0 & 0 & 0 & 0 & 0 & 0 \end{bmatrix}$$

The trace of this matrix is just the character of C_2 in the representation so that $\chi(C_2) = -1$. Since only the diagonal elements of the matrix contribute to the character, only those coordinates which are not changed into other coordinates need be considered. These will be the coordinates of nuclei lying on the axis of rotation since those lying off this axis must be transformed into other positions, which leads to off-diagonal elements only in the matrix. Moreover, of the three coordinates of any nucleus lying on the rotation axis two will change in sign while the remaining one, that in the direction of the axis, is left unchanged. So each nucleus of this type contributes three entries along the diagonal of the representation matrix, namely -1, -1, and $+1$, a net contribution to the total character of -1. Hence, if there are N_A nuclei lying on the axis, $\chi(C_2) = -N_A$, a result that confirms the detailed calculation on the water molecule.

In a similar fashion only the nuclei N_R in a plane of reflection contribute to $\chi(\sigma)$. Only one coordinate, that perpendicular to the plane, is changed in sign, the other two lying in the plane being left unaltered. Hence $\chi(\sigma) = N_R$. Since there are $3N$ basis functions in the representation, then $\chi(I) = 3N$.

For rotation operators C_n the result is modified slightly and becomes $\chi(C_n) = N_A(1 + 2\cos 2\pi/n)$. For operators S_n it becomes $\chi(S_n) = 0$, if there is no nucleus at the center, and $\chi(S_n) = -1 + 2\cos 2\pi/n$ if there is, since only in the latter case can a nucleus be left unchanged in position by the action of S_n. The case S_2 corresponds to inversion i.

These results enable the characters of the $3N$ representation to be found very easily. However, the $3N$ variables are more than are actually required for vibrational motion and include six corresponding to rotation and translation.[†] To obtain the characters corresponding to the $(3N-6)$-dimensional representation of the vibrational coordinates only it is not necessary to find the proper combinations of the $3N$ original coordinates which divide into the $3N-6$ vibrational coordinates, the three translation and three rotational ones. Instead, and this is much easier, the characters of the translational and rotational coordinates can be found using quite general arguments without the need for the explicit form of these coordinates. The translational coordinates can be considered as a triad fixed to the center of the molecule and, therefore, behave in exactly the same way as if there were a nucleus there.[§] Thus the characters due to this will be $\chi(C_n) = (1 + 2\cos 2\pi/n)$, $\chi(\sigma) = 1$, $\chi(S_n) = -1 + 2\cos 2\pi/n$, $\chi(I) = 3$. The rotational coordinates can also be considered as a vector at the molecular center. However, corresponding to each of the three components there is a sense of rotation determined by the right-hand rule. Taking this extra factor into account means that $\chi(C_n) = 1 + 2\cos 2\pi/n$, $\chi(\sigma) = -1$, $\chi(S_n) = 1 - 2\cos 2\pi/n$, $\chi(I) = 3$. The sum

---

[†] For linear molecules there are only five so that the number of vibrational coordinates is $3N-5$. This special case is not, however, considered here.

[§] If there is, of course, its coordinates will actually be the translation coordinates.

of these two effects, therefore, is zero for the S_n and σ operators and $\chi(I)$ = 6 and $\chi(C_n) = 2(1 + 2\cos 2\pi/n)$. Subtracting these from the characters of the $3N$-dimension representation shows that the characters of the $(3N - 6)$-dimensional one, i.e., that corresponding to vibrational coordinates only, will be $\chi(I) = 3N - 6$; $\chi(C_n) = (N_A - 2)(1 + 2\cos 2\pi/n)$; $\chi(\sigma) = N_R$; $\chi(S_n) = 0$, or $-1 + 2\cos 2\pi/n$.

Once these characters are known it is possible by the methods of Section IV.F to decompose the reducible $(3N - 6)$-dimensional representation into its irreducible components. Moreover, using the projection operators, the various combinations of the original coordinates which transform like the irreducible representations can be found. In the case of the water molecule, for example, since $N = 3$ the characters are $\chi(I) = 3$, $\chi(C_2) = 1$, $\chi(\sigma_v') = 3$ (since all three nuclei lie in the reflection plane), and $\chi(\sigma_v) = 1$ (since only the oxygen nucleus lies on this reflection plane). It is easily seen, from the character table, that this representation contains the irreducible representations A_1, twice over, and B_2, once. The projection operator for functions belonging to A_1 is $\frac{1}{4}(I + C_2 + \sigma_v + \sigma_v')$. Applying this in turn to each of the coordinates gives the three functions $(z_1 + z_2)$, z_2, and $(y_1 - y_3)$. The reason there are three and not two is that the three coordinates corresponding to translation form the basis of a representation that contains A_1 once. If we assume that the center of mass is at the oxygen nucleus, which is quite a reasonable approximation, the function corresponding to this will be z_2. Thus the vibrational coordinates belonging to the A_1 irreducible representation will be $(z_1 + z_3)$ and $(y_1 - y_3)$. These are usually combined together to correspond to the type of coordinates illustrated in Figs. 4a and 4b. In a similar fashion the coordinates for the B_2 irreducible representation are shown in Fig. 4c. For more complicated molecules this type of analysis becomes increasingly tedious. However, in many

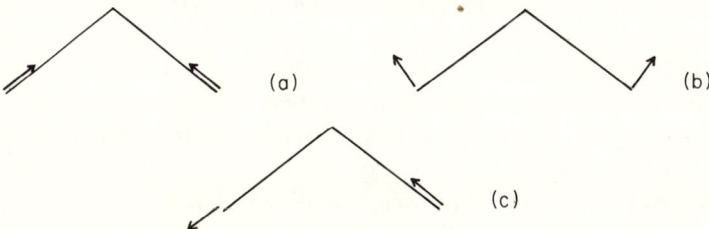

FIG. 4. Symmetries of the normal modes of vibration of the water molecule; (a) A_1 symmetry; (b) A_1 symmetry; (c) B_2 symmetry. (The oxygen nucleus is assumed to be the center of mass.)

cases it can be simplified by a judicious choice of coordinates initially. With experience it is often possible to write down the vibrational coordinates belonging to the irreducible representations by inspection without the need to work through the details.

Using these group theoretical ideas, therefore, the displacement coordinates may be transformed into a set of $3N-6$ coordinates associated with the molecular vibration and which belong to the various irreducible representations of the molecular point group. It turns out that these new coordinates are closely related to, and, as far as symmetry properties are concerned, are the same as the normal coordinates $\{Q_i\}$ of the molecular nuclei. These are the coordinates for which the vibrational Hamiltonian takes the separable form

$$-\tfrac{1}{2}\sum_i (\partial^2/\partial Q_i^2) + \tfrac{1}{2}\sum_i \lambda_i Q_i^2$$

where the λ_i are constants related to the force constants of the bonds in the molecule. The vibrational wave function is, therefore, the product of functions $\psi = \psi_1(Q_1)\psi_2(Q_2)\ldots$, where the ψ_i are harmonic oscillator wave functions. These correspond to the normal modes of vibration of the molecule, the symmetries of which are determined by the symmetry of the normal coordinates. Thus, in the water molecule, the three possible normal modes have symmetries A_1, A_1, and B_2 as indicated in Fig. 4. There are three normal coordinates so that the symmetry of the total vibrational wave function $\psi = \psi_1(Q_1)\psi_2(Q_2)\psi_3(Q_3)$ depends on the symmetry of the individual components. It can be shown that if $\psi_i(Q_i)$ is a harmonic oscillator wave function with quantum number n_i, then the character $\chi_i(R)$ of any group operator R with respect to ψ_i will be $[\chi_{Q_i}(R)]^{n_i}$ where $\chi_{Q_i}(R)$ is the character of R in the irreducible representation to which Q_i belongs. Thus the characters with respect to ψ are given by

$$\chi(R) = [\chi_{Q_1}(R)]^{n_1}[\chi_{Q_2}(R)]^{n_2}[\chi_{Q_3}(R)]^{n_3},$$

and from this the irreducible representation to which a given vibrational wave function ψ belongs can be determined in the usual way. Once this is known the various selection rules for transitions to other vibrational states can be determined in a way similar to that used for electronic transitions. For example, the lowest vibrational state has $n_1 = n_2 = n_3 = 0$ so that $\chi(R) = 1$, for all R, and the symmetry is A_1. If Q_1, Q_2, Q_3 correspond to the symmetries of Figs. 4a, 4b, and 4c, respectively, then $\chi_{Q_1} = \chi_{A_1} = 1$ for all R, and similarly for χ_{Q_2}. Thus the symmetry of ψ is determined by the characters $[\chi_{B_2}(R)]^{n_3}$. If n_3 is even, this means ψ has symmetry A_1, and if n_3 is odd, the symmetry is B_2. Therefore the transitions from the ground state correspond to $A_1 \to A_1$ (n_3 even), $A_1 \to B_2$ (n_3 odd). For the $\mathbf{C}_{2v}$ group the operators x, y, z belong to the irreducible representations B_1, B_2, and A_1, respectively. The y-polarized radiation induces the transition $A_1 \to B_2$ and the z-polarized the transition $A_1 \to A_1$. Radiation of vibrational frequencies polarized in the x direction induces no simple vibronic transition.

VI. The Permutation Group and Spin

A. Irreducible Representations of the Permutation Group

The permutation group $\mathbf{S}(2)$ is an Abelian group with the two elements I and P_{12}. There are just two irreducible representations, therefore, each of dimension unity, and if ψ is an arbitrary function, $\frac{1}{2}(I + P_{12})\psi$ and $\frac{1}{2}(I - P_{12})\psi$ are functions which belong to them.

The permutation group $\mathbf{S}(3)$ was discussed in Section IV.C and it turned out to have three irreducible representations, two of one dimension and one of dimension two. The corresponding basis functions for that group are given in Eq. (4.4). The increase in complexity in going from $\mathbf{S}(2)$ to $\mathbf{S}(3)$ is pronounced and as the degree of the permutation group is increased the number and type of the irreducible representations increase alarmingly. Surprisingly enough, however, it is possible to classify these for the general case.

A basic concept is that of a partition of a positive integer n. This is a set of integers whose sum is n so that if, in a given partition, 1 is included s_1 times, 2 s_2 times, and so on, then

$$n = 1s_1 + 2s_2 + 3s_2 + \cdots$$

and the partition would be written $[1^{s_1} 2^{s_2} 3^{s_3}, \ldots]$, any integer with $s = 0$ being omitted. For example, if $n = 3$, there are three possible partitions, $[1^3], [1^1 2^1], [3^1]$. If $n = 4$, there are five $[1^4], [1^2 2^1], [1^1 3^1], [4^1], [2^2]$. It can be proved that the number of classes, and therefore the number of irreducible representations of the permutation group of degree n, is equal to the number of partitions of n. The basic reason for this is that the classes are made up of operators of the same type and the different types of operators correspond to the different possible partitions. For example, one class consists of operators that permute pairs of indices, leaving the remainder unaltered, while another is made up of operators that permute only disjoint pairs of indices, and yet another of operators that permute three indices, leaving the remainder unchanged, and therefore if an unaltered index is represented by 1, the interchange of a pair by 2, and so on, the three types of operators just mentioned correspond to $[1^{n-2}2]$, $[1^{n-4}2^2]$, and $[1^{n-3}3^1]$. The remaining partitions correspond in a similar way to different types of operators and, hence, different classes.

The properties of the irreducible representations are closely related to the partitions, and the partitions can be used to label them (cf. Table V). Each partition, and hence each irreducible representation, can be represented by a Young diagram. If the partition is $[\lambda_1^{s_1}\lambda_2^{s_2} \cdots \lambda_i^{s_i}]$ with $\lambda_1 < \lambda_2 < \cdots < \lambda_i$, the Young diagram consists of rows of blocks with s_i rows of λ_i blocks at the top, followed by s_{i-1} rows of λ_{i-1} blocks, and ending with s_1 rows of λ_1 blocks. The Young diagrams for the five partitions of the $S(4)$ group are shown in Fig. 5. A prescription for constructing

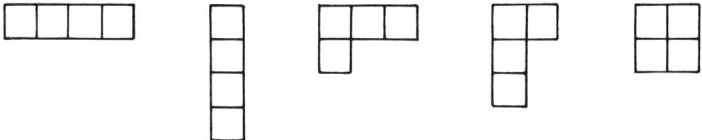

FIG. 5. Young diagrams for $S(4)$.

the irreducible representations and the projection operators for the basis functions can be given based on the use of these diagrams. For full details consult the books cited in the References since here only a few points are made. The prescription is based on filling the diagrams with the integers from 1 to n [in the case of $S(n)$], one integer in each block. This must be done such that (a) in any row the numbers increase from left to right and (b) in any column the numbers increase from top to bottom. Each different way this can be done is called a standard Young tableau and the number of these depends on the particular form for the diagram, and hence initially on the particular partition that gives rise to it. It turns out that the number of possible ways equals the dimension of the irreducible representation. There will always be two one-dimensional irreducible representations corresponding to the partitions $[1^n]$ and $[n^1]$ since the Young diagram for the first consists of a single column and for the second a single row, and both can only be filled in one way. The irreducible representation $[n^1]$ has all its characters equal to unity, so that the basis function is completely symmetric and the projection operator is just $\sum P$, the sum of all the operators. The irreducible representation [1^n] has characters $\chi(P) = +1$ if P is an even permutation, and $\chi(P) = -1$ is P is an odd permutation. Thus the basis of the representation is an antisymmetric function $\sum_p (-1)^p P\psi$ where p is the parity of the operator P.

There are several ways of finding functions which transform like the basis functions of the other irreducible representations. One way is first to find the matrices representing the operators in the various irreducible representations, and once these are known the operators which generate the basis functions can be obtained. There are systematic ways in which this can be done based on the properties of the Young tableau. For details see the References at the end of the chapter. An alternative method which generates the basis functions directly is based on algebraic ideas. However, in this section, we do not require the basis functions explicitly but only one particular property. This concerns functions $\phi(1, 2, \ldots, n)$ of n variables which are symmetrical with respect to the first m (say) and last $n - m$ variables, i.e., functions which contain the first m variables symmetrically and the last $n - m$ symmetrically. An example would be if ϕ were the product of m functions $v(1) \cdots v(m)$ and $n - m$ functions $\omega(m + 1) \cdots \omega(m)$ so that

$$\phi(1, 2, \ldots, n) = v(1) \cdots v(m)\omega(m + 1) \cdots \omega(n).$$

Clearly ϕ is left unaltered except by operators which interchange the first n coordinates and the last $n - m$. In other words only permutation operators of the form P_{ij} ($i \leq m, j > m$), i.e., permutations of two variables and products of these, change ϕ. It can be shown that functions of this type can belong only to irreducible representations whose Young diagrams have two rows. If the functions v and ω are identical, then ϕ is left unaltered under the action of all the permutation operators and so belongs to the symmetric representation [n^1]; i.e., its Young diagram has just one row.

Before leaving the general theory of the permutation group we need one more result. Suppose we consider two irreducible representations obtained from two Young diagrams which are related in that the diagram for the second is obtained from that of the first by interchanging rows and columns. Such representations are said to be conjugate. For example, the irreducible representations obtained from the partitions [$1^2 2^1$] and [$1^1 3^1$] in $\mathbf{S}(4)$ are conjugate. For $\mathbf{S}(n)$ the two one-dimensional irreducible representations $[1^n]$ and $[n^1]$ are conjugate. Conjugate irreducible representations are closely related. Suppose $\mathbf{D}(P)$ and $\mathbf{D}^c(P)$ are the matrices representing the permutation operator P in two conjugate irreducible representations. Then, if p is the parity of the permutation,

$$\mathbf{D}^c(P) = (-1)^p \tilde{\mathbf{D}}(P), \tag{6.1}$$

where the tilde means the transposed matrix, so that knowing the matrices for one of the conjugate pair the matrices for the other can be obtained easily.

B. Spin Functions

In addition to its position coordinates an electron has spin coordinates associated with the observables S_x, S_y, S_z. These observables satisfy the same commutation relations as the observables representing the components of angular momentum so that it follows that a spin representation can be set up using the eigenkets of S_z. It turns out that there are only two such eigenkets, which are usually written as α and β and which correspond to the eigenvalues of $\frac{1}{2}\hbar$ and $-\frac{1}{2}\hbar$ of S_z. The observables S_x and S_y do not have α and β as eigenkets but the operator for the square of the total spin $S^2 = S_x^2 + S_y^2 + S_z^2$ does.

For an n-electron system the operators $S_x(1, 2, \ldots, n), S_y(1, 2, \ldots, n)$, and $S_z(1, 2, \ldots, n)$ are obtained simply by adding the operators for each individual electron, e.g.,

$$S_z(1, 2, \ldots, n) = \sum_{i=1}^{n} S_z(i).$$

S^2 is obtained as in the one-electron case. Since each electron can be associated with either α or β there are 2^n linearly independent functions

$$\eta = \theta_1(1)\theta_2(2) \cdots \theta_n(n), \tag{6.2}$$

where each θ can be either α or β. Operating with S_z it follows that each of the η is an eigenfunction of S_z with eigenvalue $m\hbar$ where m equals $\frac{1}{2}$(number of α functions in η − number of β functions in η), and so ranges in value from $\frac{1}{2}n$ to $-\frac{1}{2}n$ in integral steps. For a given m the number of linearly independent spin functions will be $^nC_{\frac{1}{2}n-m}$.

Although eigenfunctions of S_z, the η functions are not eigenfunctions of S^2. However, by taking appropriate combinations, the functions may be transformed into a new set whose components are eigenfunctions of S^2 and S_z simultaneously. In this way we obtain a set of functions $\theta_{s,m,k}$ that belong to the eigenvalues $s(s+1)\hbar^2$ of S^2 and $m\hbar$ of S_z. The index k is used to label any degenerate eigenfunctions.

The operators S^2 and S_z operate on the spin coordinates of every electron equivalently. It therefore follows that the permutation operators of the group $\mathbf{S}(n)$ leave S^2 and S_z unchanged, i.e., they commute with S^2 and S_z. Hence, when the functions $\theta_{s,m,k}$ are divided into sets each corresponding to fixed values of s and m, each of these sets must form the basis of an irreducible representation of the permutation group. In view of the remarks at the end of the preceding subsection it follows that these irreducible representations must correspond to Young diagrams with at most two rows.

C. The Pauli Principle

Suppose we have solved the Schroedinger equation for an n-electron system. For a given energy level suppose there are f spatial wave functions $\psi_1, \ldots, \psi_f$. The permutation operators of the group $\mathbf{S}(n)$ commute with the Hamiltonian, and therefore the functions $\psi_1, \ldots, \psi_f$ form the basis of an irreducible representation of $\mathbf{S}(n)$. The effect of an operator P^r on one of the functions will be given by

$$P^r \psi_i = \sum_{j=1}^{f} D_{ji}(P) \psi_j \tag{6.3}$$

where the superscript r on P^r is to show that the operator acts only on spatial coordinates.

The functions $\psi_1, \ldots, \psi_f$ cannot be wave functions because they do not include any spin coordinates. To form a proper wave function we must multiply by spin functions. Suppose $\phi_1, \ldots, \phi_f$ are arbitrary spin functions. Then the most general combination we can take is

$$\Psi = \sum_{i=1}^{f} \psi_i \phi_i. \tag{6.4}$$

However, this combination must be such that Ψ satisfies the Pauli principle, which states that electronic wave functions must be antisymmetric; i.e.,

$$P\Psi = (-1)^p \Psi, \tag{6.5}$$

where P is a permutation operator on space and spin coordinates and p

is its parity. Introducing P^σ to indicate a permutation operator which acts on spin coordinates only, since $P = P^r P^\sigma$ the Pauli principle requires that

$$P\Psi = \sum_{i=1}^{f} (P^r \psi_i)(P^\sigma \phi_i) = (-1)^p \sum_{i=1}^{f} \psi_i \phi_i.$$

From (6.3)

$$\sum_{i=1}^{f} (P^r \psi_i)(P^\sigma \phi_i) = \sum_{i=1}^{f} \sum_{j=1}^{f} (P^\sigma \phi_i) D_{ji}(P) \psi_j,$$

and since the ψ_i are linearly independent,

$$(-1)^p \phi_k = \sum_{i=1}^{f} D_{ki}(P)(P^\sigma \phi_i).$$

Inverting these linear equations gives

$$P^\sigma \phi_k = \sum_{i=1}^{f} (-1)^p D_{ki}(P) \phi_i, \tag{6.6}$$

and this holds for every operator P of the group. This equation shows that the ϕ_i also form a basis for an irreducible representation of the permutation group. However, it is not the same as that for which the spatial functions form a basis, because the matrices are not the same. They are related, though, and Eq. (6.6) shows that, in an obvious notation, $\mathbf{D}^\phi(P) = (-1)^p \tilde{\mathbf{D}}^\psi(P)$. In view of Eq. (6.1), it follows that the two irreducible representations are conjugate.

Therefore, the Pauli principle forces us to write

$$\Psi = \sum_{i=1}^{f} \psi_i \theta_{s,m;i}, \tag{6.7}$$

which shows that the wave functions are always eigenfunctions of S^2. As chosen here they are also eigenfunctions of S_z but this is not necessary, for each of the sets of functions $\theta_{s,-s,i}, \theta_{s,-s+1,i}, \ldots, \theta_{s,s,i}$ forms a basis for the same irreducible representation. Thus there are $2s + 1$ functions represented by (6.7), with m taking $2s + 1$ different values. These are degenerate and so may be combined together without violating the Pauli principle. These would give new functions which are not eigenfunctions of S_z.

As was pointed out in Section VI.B the $\theta_{s,m,i}$ belong to reducible representations which are derived from Young diagrams with at most two rows. Since the ψ_i belong to the conjugate representation this must be derived from a Young diagram with at most two columns. But this will not exhaust all the possibilities. There will be degenerate spatial eigenfunctions of the Hamiltonian which belong to irreducible representations associated with Young diagrams with more than two columns. However, it will not be possible to combine these with spin functions so as to form wave functions which satisfy the Pauli principle. These functions, therefore, correspond to nonphysical solutions and have no practical significance.

VII. Continuous Groups

So far the groups we have considered have had a finite number of elements. However, in the case of diatomic molecules, a rotation $C(\alpha)$ through an arbitrary angle α about the internuclear axis leaves the diatomic unchanged so that the number of elements in the symmetry group of importance for diatomics, the group $\mathbf{C}_{\infty v}$, is no longer finite. In a similar way, the three-dimensional rotation group, which has as its symmetry elements arbitrary rotations about arbitrary axes passing through a fixed point, and which is related to the angular momentum of an atom, is also not a finite group. Thus one is led to a continuous group whose elements depend on parameters which vary continuously in a given range. Luckily in the case of diatomics and atoms, it is possible to obtain most of the results that are needed by simple extensions of the ideas of finite groups without any need to explore the mathematical complexities involved in a proper treatment of continuous groups.

Let us begin by considering properties of the symmetry group $\mathbf{C}_\infty$ which contains the symmetry operators $C(\alpha)$ for rotation about the internuclear axis through an arbitrary angle α, $0 \leq \alpha \leq 2\pi$. This group can be considered as the finite group $\mathbf{C}_n$ with elements $I, C_n, C_n^2, \ldots, C_n^{n-1}$ in the limit as $n \to \infty$: $\mathbf{C}_n$ is Abelian so that the irreducible representations are one dimensional, and since $(C_n)^n = I$, it follows that the eigenvalues of C_n are the n roots of unity. Using this result the character table is easily obtained, and is given in Table VII. If we let $n \to \infty$ and α be the small angle $2\pi/n$, it follows that there are infinitely many one-dimensional irreducible representations $\Gamma(m)$, $m = 0, \pm 1, \pm 2, \ldots$, and that the character of $C(\alpha)$ in the mth irreducible representation is $\exp im\alpha$.[†] Thus, the character table of $\mathbf{C}_\infty$ is easily obtained from that of $\mathbf{C}_n$ and has the form shown in Table VII.

[†] This result actually does not depend on α being small; for, if $\beta = N\alpha$ is large, i.e., if N is a large integer, then

$$\chi_m[C(\beta)] = \chi_m[C(\alpha)^N] = \{\chi_m[C(\alpha)]\}^N = \exp iNm\alpha = \exp im\beta.$$

TABLE VII

CHARACTER TABLES FOR $C_n{}^a$ AND $C_\infty{}^b$

C_n	I	C_n	C_∞	I	$C(\alpha)$
$\Gamma(0)$	1	1	$\Gamma(0)$	1	1
$\Gamma(m)$	1	$\exp im(2\pi/n)$	$\Gamma(m)$	1	$\exp im\alpha$

[a] $m = 0, \pm 1, \pm 2, \ldots, \pm n/2$ (n even).
[b] $m = 0, \pm 1, \pm 2, \ldots$.

Using this result it is now possible to find the condition that an arbitrary function $f(r, \theta, \phi)$, in terms of spherical polars (r, θ, ϕ) where ϕ is the azimuthal angle relative to the axis of symmetry, transforms like the mth irreducible representation of C_∞. The condition is that

$$C(\alpha)f(r, \theta, \phi) = \exp(im\alpha)f(r, \theta, \phi)$$

which, since

$$C(\alpha)f(r, \theta, \phi) = f(r, \theta, \phi + \alpha),$$

is equivalent to

$$f(r, \theta, \phi + \alpha) = \exp(im\alpha)f(r, \theta, \phi)$$

which must be satisfied for all operators $C(\alpha)$, i.e., for all angles α. Hence, by considering $\alpha = -\phi$,

$$f(r, \theta, \phi) = \exp(im\phi)f(r, \theta, 0)$$

where $f(r, \theta, 0)$ is an arbitrary function of r and θ only. This familiar result is usually obtained by finding the eigenfunctions of L_z, the angular momentum about the axis.

The point group for a diatomic or linear molecule is $C_{\infty v}$, obtained by adding to the rotation operators of the group C_∞ operators for reflection in any plane containing the molecular axis. (If there is a center of symmetry, as in H_2, there are the additional operators i and σ_h giving the group $D_{\infty h}$.)

Using the functions $\exp im\phi$, $m = \pm 1, \pm 2, \ldots$, as a basis, we can set up a representation for the operators of the group $C_{\infty v}$. The rotation operators will have a diagonal representation in this basis, but the reflection operators will not. If the reflection plane makes an angle Θ with the plane $\phi = 0$, then

$$\sigma_v(\Theta) \exp im\phi = \exp im(2\Theta - \phi)$$

so that $\sigma_v(\Theta)$ transforms $\exp im\phi$ into $\exp -im\phi$ and vice versa. Thus the pair of functions $\exp im\phi$ and $\exp -im\phi$ are transformed into each other by the action of every one of the infinitely many reflection operators σ_v and so must form the basis of a two-dimensional representation which is easily seen to be irreducible. Clearly the 2×2 matrix for each σ_v in this representation has diagonal elements of zero so that $\chi(\sigma_v) = 0$; while the operator $C(\alpha)$ has diagonal elements of $\exp im\alpha$ and $\exp -im\alpha$ whose sum gives that $\chi[C(\alpha)] = 2 \cos m\alpha$. This result holds for all $m = 1, 2, \ldots$, giving an infinite number of two-dimensional irreducible representations. There are two more irreducible representations of this group, both one dimensional. The first is the symmetric one with all its characters unity. The final one has $\chi(I) = \chi[C(\alpha)] = 1$ and $\chi(\sigma_v) = -1$.

In the case of atoms the rotation operators which commute with the Hamiltonian form the three-dimensional rotation group. To obtain the character table and discuss the irreducible representations of this group from purely group theoretical arguments is quite difficult. It is much easier to adopt the opposite point of view and use the known properties of atomic wave functions to determine the properties of the rotation group.

We know from the quantum mechanical theory of atoms and the theory of angular momentum that one-electron wave functions take the form

$$\psi(r, \theta, \phi) = R(r) Y_{l,m}(\theta, \phi) \qquad (7.1)$$

where $R(r)$ is some function of the radial coordinate and $Y_{l,m}(\theta, \phi)$ is a spherical harmonic. Each state of angular momentum quantum number l is $(2l + 1)$-fold degenerate, corresponding to m taking the values $-l, -l+1, \ldots, l$. Since the radial function $R(r)$ is irrelevant as far as rotation properties are concerned, it follows that the set of functions $Y_{l,-l}, Y_{l,-l+1}, \ldots, Y_{l,l}$ for every nonnegative integer l belongs to an irreducible representation of the three-dimensional rotation group. Thus the indices $l \,(= 0, 1, 2, \ldots)$ label the irreducible representations of this group, the lth one being of dimension $2l + 1$. The next step is to find the character of the operator $C(\alpha)$ corresponding to rotation through an angle α about an arbitrary axis. This will be independent of the particular axis so we can make the most convenient choice, which is the z axis. However, the rotation about this axis is determined by the fact that

$$Y_{l,m}(\theta, \phi) = P_{l,m}(\theta) \exp im\phi$$

where the function $P_{l,m}(\theta)$ depends on θ only so that

$$C_z(\alpha) Y_{l,m}(\theta, \phi) = P_{l,m}(\theta) \exp[im(\phi + \alpha)] = \exp(im\alpha) Y_{l,m}(\theta, \phi).$$

Therefore, the spherical harmonics $Y_{l,m}(\theta, \phi)$ belong also to the mth irreducible representation of the C_∞ group, and hence for a fixed l the $(2l+1)$-dimensional matrix representing $C_z(\alpha)$ will be diagonal with elements $\exp im\alpha$, $m = -l, \ldots, l$. The character of $C_z(\alpha)$ will be the trace of this matrix, so that

$$\chi^l[C(\alpha)] = \sum_{m=-l}^{l} \exp im\alpha = \frac{\exp[i(l+1)\alpha] - \exp(-il\alpha)}{\exp i\alpha - 1}$$
$$= \frac{\sin(l + \tfrac{1}{2})\alpha}{\sin \tfrac{1}{2}\alpha} \tag{7.2}$$

on summing the geometric series. The subscript z on $C_z(\alpha)$ has been dropped because the end result is independent of the particular axis of rotation.

Equation (7.1) represents the wave function for a single electron with angular momentum quantum number l. The wave function for the atom will be obtained by combining together the wave functions for individual electrons, but this must be done in such a way that the combined system also belongs to an irreducible representation of the rotation group. For example, suppose we consider two quite general systems which can consist of one electron or several electrons each, and let one have angular momentum quantum number l_1 and the other l_2. Thus associated with the first will be $(2l_1 + 1)$ functions belonging to the l_1 irreducible representation of the rotation group, and with the second $(2l_2 + 1)$ functions belonging to the l_2 irreducible representation. For the composite system there will be $(2l_1 + 1)(2l_2 + 1)$ product functions which form the basis of a product representation. Except in the trivial case of l_1 or $l_2 = 0$, this will not be irreducible but it can be divided into its irreducible components by the methods of Section IVF.

The character of $C(\alpha)$ in the product representation will be the product of its characters in the individual systems so that

$$\chi[C(\alpha)] = \chi^{l_1}(C) \times \chi^{l_2}(C) = \frac{\sin(l_1 + \tfrac{1}{2})\alpha \sin(l_2 + \tfrac{1}{2})\alpha}{\sin^2 \tfrac{1}{2}\alpha}.$$

It is simple[†] to show that the expression on the right equals

$$\frac{\sin(l_1 + l_2 + \tfrac{1}{2})\alpha}{\sin \tfrac{1}{2}\alpha} + \frac{\sin(l_1 + l_2 - 1 + \tfrac{1}{2})\alpha}{\sin \tfrac{1}{2}\alpha} + \cdots$$
$$+ \frac{\sin(|l_1 - l_2| + \tfrac{1}{2})\alpha}{\sin \tfrac{1}{2}\alpha}, \qquad (7.3)$$

which shows that

$$\chi[C(\alpha)] = \chi^{l_1+l_2}[C(\alpha)] + \chi^{l_1+l_2-1}[C(\alpha)] + \cdots + \chi^{|l_1-l_2|}[C(\alpha)],$$

so the product functions can be combined into functions which belong, respectively, to the $(l_1 + l_2)$, $(l_1 + l_2 - 1)$, $\ldots$, $|l_1 - l_2|$ irreducible representations. Hence the possible values of the angular momentum quantum number for the composite system will be $|l_1 - l_2|, \ldots, l_1 + l_2$. The more difficult problem is to actually find the combinations of the product functions that belong to the various irreducible representations. The calculation is long and tedious but a relatively simple formula for the required coefficients of the product functions can be obtained. These are the famous Clebsch–Gordon coefficients.

The situation as described here is simpler than usually occurs in practice. This is because we have not taken spin into account. When spin is included the orbital angular momentum has to be replaced by total angular momentum (orbital plus spin) and if the spin has a half-integral value, then l will not be an integer. The methods we have used to obtain the irreducible representations of the rotation group are meaningful only if l is integral. For half-integral l values, each matrix representing the elements of the group can have two distinct forms. This lack of uniqueness, which basically arises from the fact that the spin of an electron can take two values, can be dealt with using a concept known as a double group. Details can be found in the books cited in the References.

VIII. Group Theory and the Solid State

A. Crystal Symmetry and Space Groups

The characteristic property of a solid in the crystalline state is the repeated periodic pattern formed by the constituent atoms or molecules. The effect of this can be studied using symmetry operators and group

---

[†] Since $2 \sin \tfrac{1}{2}\alpha \sin(n + \tfrac{1}{2})\alpha = \cos n\alpha - \cos(n + 1)\alpha$, on summing from $n = |l_1 - l_2|$ to $n = l_1 + l_2$ we obtain $2 \sin \tfrac{1}{2}\alpha \sum \sin(n + \tfrac{1}{2})\alpha = \cos |l_1 - l_2|\alpha - \cos(l_1 + l_2 + 1)\alpha$, which is $2 \sin(l_1 + \tfrac{1}{2})\alpha \sin(l_2 + \tfrac{1}{2})\alpha$.

theory but, although the basic ideas are the same as those we have been using in the case of molecules, there are two changes which have to be made in the types of symmetry operators that are relevant.

The basic repeating unit in the crystal is the unit cell and the requirement that these can be packed together in a regular array so as to fill completely the three-dimensional space occupied by the crystal very much restricts the possible shapes. All told there are 14 possible types of unit cell giving the Bravais lattices. This includes unit cells with the same basic shape but with the crystal atoms arranged differently. For example, the cubic cell can have one, two, or four lattice points (or atoms) giving the simple, body-centered and face-centered cubic. Corresponding to each of these unit cells will be a number of possible point symmetry groups, but, due to the restricted shapes of the cells, only certain symmetry operations can occur. For example, the only axes of rotation to arise are one-, two-, three-, four- and six-fold axes so that the operators C_5, C_7, etc., are never needed. These restrictions reduce the possible point groups to 32 in total, called the crystallographic point groups.

To complement these types of operations we must introduce those associated with the translational symmetry of the crystal lattice. If $\mathbf{b}_1$, $\mathbf{b}_2$, $\mathbf{b}_3$ are the basis vectors along the sides of a unit cell and $\mathbf{a}(0, 0, 0)$ is an arbitrary point in a particular unit cell, then

$$\mathbf{a}(n_1, n_2, n_3) = \mathbf{a}(0, 0, 0) + n_1\mathbf{b}_1 + n_2\mathbf{b}_2 + n_3\mathbf{b}_3,$$

where n_1, n_2, n_3 are integers, will be an equivalent point in another unit cell. Thus we can define the translation operator $T(n_1, n_2, n_3)$ by

$$T\mathbf{a}(0, 0, 0) = \mathbf{a}(n_1, n_2, n_3). \tag{8.1}$$

Clearly the operator T can be written as the product of powers of the translation operators U, V, W defined by

$$U\mathbf{a}(0, 0, 0) = \mathbf{a}(1, 0, 0), \quad V\mathbf{a}(0, 0, 0) = \mathbf{a}(0, 1, 0),$$
$$W\mathbf{a}(0, 0, 0) = \mathbf{a}(0, 0, 1), \tag{8.2}$$

and representing unit translations in the direction of the basis vectors. The operators U, V, and W commute and $T = U^{n_1}V^{n_2}W^{n_3}$. By considering the set of operators of this form where n_1, n_2, n_3 take all integral values, we can generate from $\mathbf{a}(0, 0, 0)$ the set of points equivalent to itself in all the unit cells in the crystal. In particular the whole set of lattice points can be obtained from just one by the use of the T operators, which therefore generate the complete lattice from just one unit cell.

The operations of the point group can be combined with those of translation to obtain two new types of symmetry operations. One of these is called a glide plane and is made up of a reflection across a plane followed by a translation parallel to the plane. In order to generate only the lattice by a repetitive application of this, the distance of translation must be a simple fraction (as a rule one-half) of a principal length in the unit cell. The second new operation is called a screw axis and consists of a rotation about a symmetry axis followed by a translation in the direction of the axis. Again the distance of the translation must be a simple fraction of a primitive translation.

When the translational symmetry operators are added to those of the point symmetry groups, the possible number of groups that can be obtained is 230; these are called the space groups. The effect of symmetry on X-ray diffraction patterns to some extent allows a particular crystal to be allocated to its proper space group, and from this many of its properties follow. As a final comment it should be noted that the notation used by crystallographers for the various symmetry operators and groups is not the same as that used in this article, which is the conventional one for a discussion of molecular symmetry. It is, however, fairly simple to make the transition from one to the other.

B. Crystal-Field Splitting

As an example of the use of point group symmetry in solid-state problems we consider the splitting of the energy levels of a free ion in a crystal field. The model adopted is that of an ionic crystal in which, as a first approximation, each ion is assumed to be unaffected by the remainder of the crystal. The effect of the remainder of the crystal is then allowed for by adding to the potential energy of the ion an extra term which represents the electrostatic potential energy due to the entire lattice (except for the ion under consideration) at the site of that ion. This is called the crystal field, and its exact magnitude is not so important. The significant fact is that since it is a property of the lattice it must reflect the symmetry of the lattice. In other words it must be invariant to the transformations of the point group of the lattice site.

Before the crystal field is considered, the Hamiltonian for the ion will have the symmetry properties of the rotation group. When the crystal-field potential is added the new Hamiltonian will no longer be invariant to arbitrary rotations because the crystal field has a much lower symmetry than this. Instead it will be invariant to the operations of the point group of the lattice site and so the wave functions must belong to the irreducible representations of that point group rather than to those of the rotation group as is the case for the free ion. Since the dimensions of irreducible representations of the point groups are usually lower than those of the rotation group this causes a splitting of the degenerate states of the free ion, the so-called crystal-field splitting.

TABLE VIII

CHARACTER TABLE FOR THE OCTAHEDRAL GROUP O

O	I	$8C_3$	$3C_2$	$6C_2'$	$6C_4$
A_1	1	1	1	1	1
A_2	1	1	1	−1	−1
E	2	−1	2	0	0
T_1	3	0	−1	−1	1
T_2	3	0	−1	1	−1

As a typical example consider the effect of a crystal field with cubic symmetry. The symmetry operators in this case are I, two types of C_2 rotations, and C_3 and C_4 rotations, all of which form the cubic or octahedral group O. This group has five irreducible representations, the characters of which are given in Table VIII. Suppose, before the crystal field is considered, the free ion has angular momentum quantum number l so that there are $(2l + 1)$ degenerate wave functions belonging to the l irreducible representation of the rotation group. When the crystal field is considered, these wave functions will form the basis of a representation, but not an irreducible one, of the group O. From Section VII we know the character of the rotation $C(\alpha)$ in the l irreducible representation of the rotation group. Hence it follows that the characters of the operators of the O group in a representation whose basis is made up of the $(2l + 1)$ degenerate wave functions of the free ion will be

$$\chi(C_2) = \chi^l[C(\alpha = \pi)] = (-1)^l$$
$$\chi(C_3) = \chi^l[C(\alpha = 2\pi/3)] = \{\sin(l + \tfrac{1}{2})2\pi/3\}/\sin(\pi/3)$$
$$\chi(C_4) = \chi^l[C(\alpha = \pi/2)] = \{\sin(l + \tfrac{1}{2})\pi/2\}/\sin(\pi/4)$$

and, since the number of functions is $2l + 1$,

$$\chi(I) = 2l + 1.$$

The first interesting case to arise is the ^{5}D state with $l = 2$. The characters here are $\chi(I) = 5$, $\chi(C_2) = 1$, $\chi(C_3) = -1$, and $\chi(C_4) = -1$. Thus the representation is not irreducible but it can be decomposed into its irreducible components by the methods of Section IV. Suppose that the five free ion wave functions can be combined to give a_1 functions belonging to the A_1 irreducible representation of **O**, a_2 to A_2, and so on. Then from Eq. (4.11) it follows that

$$\chi(R) = \sum a_i \chi^i(R),$$

where R is taken in turn to be the operators I, C_2, C_3, and C_4, $\chi^i(R)$ are the characters in Table VIII, and $\chi(R)$ are the values just obtained. By inspection the solution is seen to be $a_1 = a_2 = a_5 = 0$, $a_3 = a_4 = 1$, and so the five functions can be divided into a set of two functions forming a basis for E and a set of three forming a basis for T_1. Because these two sets belong to different irreducible representations they will be affected differently by the crystal field and thus have different energy levels. Hence the five fold degenerate energy ^{5}D of the free ion is split into two separate energy levels by the action of the crystal field.

This approach can be used for other energy levels and for other symmetries of the crystal field, the most important being hexagonal symmetry. Complications arise, however, if spin–orbit interaction is large and if there are an odd number of electrons, and the proper treatment requires the use of double groups. The results obtained in this way are, of course, of great importance in the theory of the magnetic properties of ionic crystals.

C. Bloch Wave Functions

In this section we show how the translation operators U, V, W introduced in Section VIII.A can be used to obtain the Bloch wave functions. Considering the operators U, V, W and their powers only, suppose that periodic boundary conditions are to be imposed so that after, say, n unit cells the wave functions repeat themselves, i.e., $U^n = I$, $V^n = I$, $W^n = I$. It follows that the group with elements I, U, U^2, ..., U^{n-1}, V, V^2, ..., V^{n-1}, W, W^2, ..., W^{n-1} will be Abelian. The irreducible representations will be one dimensional and the characters will be the n roots of unity. Each irreducible representation can be labeled by the three integers m_1, m_2, m_3, each lying between zero and $n - 1$, so that

$$\chi(U) = \exp im_1\pi/n, \qquad \chi(V) = \exp im_2\pi/n, \qquad \chi(W) = \exp im_3\pi/n.$$

It is usually convenient to introduce a vector **k** in the reciprocal lattice which is defined by $\mathbf{k} \cdot \mathbf{b}_1 = m_1\pi/n$, $\mathbf{k} \cdot \mathbf{b}_2 = m_2\pi/n$, $\mathbf{k} \cdot \mathbf{b}_3 = m_3\pi/n$ so that, for example, $\chi(U) = \exp i\mathbf{k} \cdot \mathbf{b}_1$.

Now let us obtain a wave function that transforms like the irreducible representation labeled **k** of the group formed by the translation operators. If this is $\psi_\mathbf{k}(\mathbf{r})$, it must satisfy

$$U\psi_\mathbf{k}(\mathbf{r}) = \exp i\mathbf{k} \cdot \mathbf{b}_1 \psi_\mathbf{k}(\mathbf{r}).$$

But $U\psi_\mathbf{k}(\mathbf{r}) = \psi_\mathbf{k}(\mathbf{r} + \mathbf{b}_1)$. If we write $\psi_\mathbf{k}(\mathbf{r}) = \exp i\mathbf{k} \cdot \mathbf{r} \phi_\mathbf{k}(\mathbf{r})$, it follows that $\phi_\mathbf{k}(\mathbf{r} + \mathbf{b}_1) = \phi_\mathbf{k}(\mathbf{r})$, and similarly for the effect of the V and W operators, so that

$$\phi_\mathbf{k}(\mathbf{r} + \mathbf{b}_2) = \phi_\mathbf{k}(\mathbf{r}) \quad \text{and} \quad \phi_\mathbf{k}(\mathbf{r} + \mathbf{b}_3) = \phi_\mathbf{k}(\mathbf{r}).$$

The $\psi_\mathbf{k}(\mathbf{r})$ are just the Bloch wave functions which exhibit the translational symmetry of the lattice. The wave vector **k** can take any value within the appropriate Brillouin zone; values of **k** lying outside the zone simply repeat the wave functions obtained from those values of **k** within the zone.

GENERAL REFERENCES

There are numerous books on the subject of group theory and its applications. The following list is a very brief selection.

A. *General Textbooks*

[1] HALL, G. G. (1967). "Applied Group Theory." Longmans, London.
[2] HAMMERMESH, M. (1962). "Group Theory and its Applications to Physical Problems." Addison-Wesley, Reading, Massachusetts.
[3] HEINE, V. (1964). "Group Theory in Quantum Mechanics: An Introduction to its Present Usage." Pergamon, Oxford.
[4] McWeeny, R. (1964). "Symmetry: An Introduction to Group Theory and its Applications." Pergamon, Oxford.
[5] SCHONLAND, D. (1965). "Molecular Symmetry." Van Nostrand Reinhold, Princeton, New Jersey.
[6] WIGNER, E. P. (1959). "Group Theory and Its Application to the Quantum Mechanics of Atomic Spectra." Academic Press, New York.

B. *Point Groups and Molecular Symmetry*

As well as [4], [5], [6] the following deal with this topic.

COTTON, F. A. (1963). "Chemical Applications of Group Theory." Wiley, New York.
HALL, L. H. (1969). "Group Theory and Symmetry in Chemistry." McGraw-Hill, New York.

6. Group Theory

HERZBERG, G. (1945). "Molecular Spectra and Molecular Structure." Van Nostrand Reinhold, Princeton, New Jersey.
JAFFE, H. H., and ORCHIN, M. (1962). "Theory and Applications of Ultra Violet Spectroscopy." Wiley, New York.
JAFFE, H. H., and ORCHIN, M. (1965). "Symmetry in Chemistry." Wiley, New York.
URCH, D. S. (1970). "Orbitals and Symmetry." Penguin, London.
WILSON, E. B., DECIUS, J. C., and CROSS, P. C. (1955). "Molecular Vibrations." McGraw Hill, New York.

C. *Permutation Symmetry and Spin*

This is treated extensively in [2] and [6] as well as in

KOTANI, M., AMEMIYA, A., ISHIGURO, E., and KIMURA, T. (1963). "Tables of Molecular Integrals." Maruzen, Tokyo.
WEYL, H. (1931). "Theory of Groups and Quantum Mechanics." Princeton Univ. Press, Princeton, New Jersey.

D. *Rotation Groups and Angular Momentum*

Three of the standard books are

EDMONDS, A. E. (1957). "Angular Momentum in Quantum Mechanics." Princeton Univ. Press, Princeton, New Jersey.
FANO, U., and RACAH, G. (1959). "Irreducible Tensorial Sets." Academic Press, New York.
ROSE, M. E. (1957). "Elementary Theory of Angular Momentum." Wiley, New York.

E. *Crystal Symmetry*

Many of the books cited under B also deal with this topic. More specialized textbooks are

BUERGER, M. J. (1956). "Elementary Crystallography: An Introduction to the Fundamental Geometrical Features of Crystals." Wiley, New York.
BUNN, C. W. (1946). "Chemical Crystallography." Oxford Univ. Press (Clarendon), London and New York.
GRIFFITH, J. S. (1961). "The Theory of Transition Metal Ions." Cambridge Univ. Press, London and New York.
WHEATLEY, P. J. (1959). "The Determination of Molecular Structure." Oxford Univ. Press (Clarendon), London and New York.

Chapter 7

Density Matrices

F. DAVID PEAT

I.	Introduction	430
II.	The Full Density Matrix	431
III.	The Reduced Density Matrix	435
	A. Definitions	435
	B. Properties of the Reduced Density Matrix	436
	C. The Transition Matrix	441
IV.	The N-Representability Problem	442
	A. N-Representability of the 1-Matrix	444
	B. N-Representability of the 2-Matrix	445
V.	The Single-Particle Reduced Density Matrix	447
	A. Natural Orbitals and Their Eigenvalues	447
	B. The Dirac and Bloch Density Matrices	450
	C. The Energy Functional Method	455
	D. The X-Ray Form Factor	457
	E. Trial 1-Matrices	458
VI.	The Second-Order Reduced Density Matrix	458
	A. Introduction	458
	B. The 2-Matrix in a Single-Particle Basis	459
	C. The 2-Matrix in a Correlated Basis	461
	D. The Grimley–Peat Approximation	467
	E. Hamiltonian-Dependent Conditions	472
VII.	General Geminal Wave Functions	474
VIII.	Condensation Phenomena	480
	A. Introduction	480
	B. Superconductivity	482
	C. Superfluidity	483
	References	485

I. Introduction

The subject of this chapter is the density matrix, and in particular its importance to problems in physical chemistry. While space prevents a detailed discussion of the applications of density matrices, it is the author's intention to give a feeling for this particular approach and to outline those areas in which it has been employed.

The philosophy basic to the density matrix is that of averaging, that the precise motions of all parts of a large system may not be important in determining its overall behavior. Therefore the averaging procedure is carried out over many of the degrees of freedom, and a resulting quantity, the density matrix, assumes an important role. The density matrix or statistical matrix is discussed in many books on statistical mechanics and is touched on briefly in this chapter. Within quantum theory a similar overwhelming quantity of information presents itself in the form of the wave function, and much of this information may be averaged to yield the reduced density matrix, which is the main subject of this chapter.

In the first section, the importance of the statistical density matrix is indicated and its relationship to the reduced density matrix illustrated. In Section III the properties and definitions concerning the reduced density matrix are developed and it is noted that these general properties are insufficient to ensure that a trial reduced density matrix is derivable from a full N-particle density matrix. This latter characterization of the class of reduced density matrices is a major problem if approximation schemes are to be used, as discussed in Section IV. Since the emphasis of this book is on techniques rather than formal theory the contents of this section are treated only briefly, but sufficient references are given to enable the reader to supplement the treatment.

Two main forms of the reduced density matrix predominate, the first-order reduced density matrix, which expresses the effects due to details of the electron density, and the second-order reduced density matrix, which is concerned with the details of the correlated behavior of pairs of electrons. The first-order density matrix is rather simpler to obtain in practice and is of use in the discussion of those properties of atoms, molecules, and solids which result from a particular density distribution. Applications of the first-order density matrix to such systems are discussed in Section V.

Section VI is devoted to the calculation and application of the second-order density matrix. It is found that applications of the second-order density matrix have not been as wide as with the first-order form. The eigenfunctions of the second-order density matrix are pair functions or "geminals" and provide a natural basis for the discussion of electronic correlation, for which reason Section VII is devoted to a discussion of many-body wave functions built up of such functions.

General properties of the density matrix reflect general characteristics of the many-body system. One such property is the size of the eigenvalues of the reduced density matrix which is related to the formation of a condensed phase. Such phenomena are discussed in Section VIII.

It will appear from this chapter that, although the density matrix has been of use in certain limited areas of physical chemistry, it has not yet been fully exploited. The difficulties of the N-representability problem are severe. Insufficient theoretical work has been done on calculating accurate density matrices, with the powerful modern tools of many-body physics, under the guide of physical argument related to the particular system under consideration. It is true that a large amount of numerical work on computers involves the reduced density matrix. Unfortunately much of this work does not illuminate the fundamental problems involved or provide an understanding of their solution.

II. The Full Density Matrix

The density matrix, or statistical matrix, provides a convenient starting point in the study of quantum mechanics and in the consideration of the problems inherent in the interpretation of this theory. Classical physics, as a result of its masterful quantification of the physical world, gave rise to a world view which remained unchallenged until the start of this century. Systems could be analyzed with an arbitrary precision and ideally considered as isolated from the measuring apparatus and the remainder of the universe. In turn a system could be considered as composed of simpler elements in interaction and its properties computed according to deterministic laws.

With the advent of experiments directed toward the constituents of matter, it became apparent that interactions of the measuring apparatus could not be ideally removed and that the quantum mechanical system and its apparatus must be considered as a whole. Since an elementary system, having definite properties independent of external conditions, had lost its independent validity, a new description of matter was necessary to give emphasis to invariance, conservation laws, and symmetries. An elementary entity is retained, in the form of a wave function, which may be calculated according to the Schroedinger equation but it is worth recalling the words of Bohr:

> In the treatment of atomic problems, actual calculations are most conveniently carried out with the help of a Schrödinger state function, from which the statistical laws governing observables under specified conditions can be deduced by definite mathematical operations, it must be recognised, however, that we are here dealing with purely symbolic procedure, the unambiguous physical interpretation of which in the last resort requires a reference to a complete experimental arrangement.

The interconnection between apparatus and system may be more clearly indicated by use of the density matrix. For example, a particular experimental situation defines a quantum mechanical system; specifically, with respect to the apparatus, the system is defined by an Hermitian, normalizable, positive-definite operator $\hat{D}$. That is, the density matrix is formed by averaging over all variables other than those of the particular system under investigation. Given a Hamiltonian $\hat{H}$ for the system, the temporal development of the density matrix obeys the well-known equation

$$d\hat{D}/dt = [\hat{D}, \hat{H}], \qquad (2.1)$$

in the Schroedinger representation.

In some suitable orthonormal basis the matrix elements d_{nm} of the density matrix may be computed. (Note that there is some ambiguity of nomenclature since the term "density matrix" is used for both the operator and its matrix.) With the aid of these matrix elements the expectation value of the Hamiltonian, or any other operator, may be computed:

$$E = \langle \hat{H} \rangle_{\hat{D}} = \operatorname{tr} \hat{H}\hat{D} = \sum_{nm} H_{nm} d_{nm} \qquad (2.2)$$

where H_{nm} are the matrix elements of the Hamiltonian in the same basis.

In addition to calculating the density matrix from (2.1), given some particular Hamiltonian, one must be able to determine the density matrix and hence characterize a system using the results of series of measurements. For example, if measurements are made on a series of observables, corresponding to the operators $\hat{A}_i$, and giving results a_i, then the density matrix may be expanded in terms of these operators:

$$\hat{D} = \sum_i a_i \hat{A}_i. \qquad (2.3)$$

Similarly, the matrix elements d_{nm} in a particular basis may be given in terms of the expectation values of the operators $\hat{\mathbf{A}}_i$ in the same basis:

$$d_{nm} = \sum_i a_i \langle n \hat{\mathbf{A}}_i m \rangle. \tag{2.4}$$

In a system possessing N quantum states the density matrix will contain N^2 elements; of course, as a result of the normalization condition and Hermiticity $d_{nm} = d_{mn}$ the number of independent elements is less than N^2.

In general one finds that the density matrix (2.3) is not idempotent

$$\operatorname{tr} \hat{\mathbf{D}}^2 \leq 1; \tag{2.5}$$

that is, the inequality

$$d_{nm} d_{mn} \geq |d_{nm}|^2 \tag{2.6}$$

holds. In the particular cases of idempotence it is clear that the matrix elements must be of the form $d_{nm} = d_n d_m{}^*$ and the density matrix is a projection operator

$$\hat{\mathbf{D}} = |i\rangle\langle i|. \tag{2.7}$$

That is, in the case of an idempotent density matrix, the system is described by a single state function, and in such a situation the density matrix is called a "pure state density matrix." Such density matrices minimize the uncertainty in the simultaneous measurement of two non-commuting observables (Stoler and Newman, 1972). In general a system is not, however, described by a single state function and the density matrix may be said to describe a mixed state

$$\hat{\mathbf{D}} = \sum_i a_i |i\rangle\langle i| \tag{2.8}$$

having matrix elements, with respect to the basis $\{|n\rangle, |m\rangle, \ldots\}$,

$$d_{nm} = \sum_i \langle n|i\rangle a_i \langle i|m\rangle. \tag{2.9}$$

It is frequently stated that the density matrix involves two distinct averaging procedures: an ensemble average over all degrees of freedom other than that of the system, which is similar to that found in statistical mechanics, and a quantum mechanical averaging. Considerations of the nature of those procedures given in the first paragraphs of this section, on the inseparability of systems, indicate that the averaging is not performed in such a mutually exclusive fashion.

The time dependence of the density matrix illuminates its structure and relationship to the Green's function. From Eq. (2.1)

$$d\hat{D}(xx't)/dt = [\hat{H}\ \hat{D}] \tag{2.10}$$

with the boundary condition

$$\hat{D}(xx'0) = \hat{D}(xx'), \tag{2.11}$$

where x stands for the space and spin coordinates of all particles, one obtains the solution

$$\hat{D}(xx't) = \exp(i\hat{H}t)\hat{D}(xx') = \sum_{k=0}^{\infty} \frac{(-it)^k}{k!} \hat{H}^k \hat{D}(xx')$$

$$= \sum_{k=0}^{\infty} \frac{(-it)^k}{k!} \hat{D}^k(xx'). \tag{2.12}$$

$\hat{D}^k(xx')$ is the kth moment of the density matrix, that is

$$\hat{D}^k(xx') = \hat{H}^k D(xx') = \sum_i \omega_i E_i^k \hat{D}_i(xx') \tag{2.13}$$

where $\hat{D}_i(xx')$ are the individual pure states and ω_i their corresponding weights in the density matrix $\hat{D}(xx')$. The following function may be defined:

$$F(\omega) = \int_{-\infty}^{\infty} dt\ e^{-i\omega t}\ \mathrm{tr}\ D(xx't). \tag{2.14}$$

Therefore

$$F(\omega) = 2\pi \sum_{n=1} \omega_n\, \delta(\omega - E_n). \tag{2.15}$$

That is, the density matrix passes through its pure states during its time development; the energies of these states E_n and their weights ω_n are revealed in the transformed function $F(\omega)$.

It is clear that the discussions of the wave function given in any book on quantum mechanics may be applied to the density matrix, but in addition the density matrix is used in the discussion of more general systems. In addition to its importance in statistical physics, for example in the description of optically coherent states, the density matrix is a useful tool in the discussion of the foundations of quantum mechanics and the measurement problem. It is unfortunate that limitations of space do not permit a development of these topics.

III. The Reduced Density Matrix

A. Definitions

Consider a system of N identical bosons or fermions characterized by the density matrix $\hat{D}(1, 2, 3, 4, \ldots, N: 1', 2', 3', 4', \ldots, N')$ and possessing the operators $\hat{A}_i(1, 2, 3, \ldots, N)$. The expectation value of these operators may be calculated according to the equation

$$\langle A_i \rangle = \text{tr }\hat{A}_i \hat{D}. \tag{3.1}$$

In general one is intested in operators that can be expressed as the sum of, at most, one- and two-particle terms. For example, the Hamiltonian of an atom may be written

$$\hat{H}(1, 2, \ldots, N) = \sum_{i=1}^{N} \hat{H}(i) + \sum_{i<j}^{N} \hat{V}(ij) \tag{3.2}$$

and, on defining the two-body operator $\hat{h}(1, 2)$, referred to as the two-body reduced Hamiltonian,

$$\hat{h}(1, 2) = \hat{H}(1) + \hat{H}(2) + (N-1)\hat{V}(1, 2), \tag{3.3}$$

the total Hamiltonian may be written as

$$\hat{H} = (N-1)^{-1} \sum_{i,j} \hat{h}(i, j). \tag{3.4}$$

Therefore, if $\hat{Q}$ is the projection operator formed from all even permutations $\hat{P}$ of the N particles,

$$\hat{Q} = (2/N!) \sum_{\text{even}} \hat{P}, \tag{3.5}$$

then

$$\hat{H} = (N/2)\hat{Q}\hat{h}. \tag{3.6}$$

The expectation value of the Hamiltonian may now be written

$$E = \text{tr }\hat{H}\hat{D} = (N/2)\text{ tr }\hat{h}\hat{D} = (N/2)\text{ tr }\hat{h}\hat{D}^{(2)} \tag{3.7}$$

where we have used the property $\hat{Q}\hat{D} = \hat{D}$ and $\hat{D}^{(2)}$ is the two-particle reduced density matrix, or 2-matrix, defined by

$$\hat{D}^{(2)}(1, 2 : 1', 2') = \operatorname*{tr}_{3,\ldots,N} \hat{D}(1, 2, \ldots, N : 1', 2', \ldots, N'). \tag{3.8}$$

Therefore, provided that at most only two-body forces operate within the system, the two-particle or second-order reduced density matrix defines the expectation values of the system. It will be noted that the definition of the full density matrix for a system, by averaging over the redundant degrees of freedom, is analogous to the definition of the reduced density matrix, by the averaging of the redundant degrees of freedom represented by the variables 3, ..., N. In general the pth-order reduced density matrix is defined as

$$\hat{D}^{(p)}(1, 2, \ldots, p : 1', 2', \ldots, p')$$
$$= \operatorname*{tr}_{p+1,\ldots,N} \hat{D}^{(N)}(1, 2, \ldots, N : 1', 2', \ldots, N'). \qquad (3.9)$$

It will be noted from the definition above that all density matrices are normalized to unity; alternative normalizations occur in the literature and may be more convenient in certain applications.

In the case of time-independent problems the 2-matrix is defined by four parameters, or 12 variables, while the wave function is defined by $3N$ variables and the full density matrix by $6N$ variables. Bearing in mind that the wave function is generally calculated by computer using some numerical scheme, it is apparent that as the size of a system increases it becomes not only increasingly difficult to perform the calculations but indeed to characterize the function itself. The alternative possibility, that the "shorthand" notation of the reduced density matrix enables a system to be characterized in an economic fashion, gave rise to much interest in the application of the density matrix to chemical systems. However, when the problem of N-representability is discussed it will be realized that any advantage gained in using the reduced density matrix is obtained at the expense of satisfying a complicated set of restrictions.

In addition it must be recalled that physical chemistry does not simply consist in the calculation of the expectation values of operators for a system in a particular state, rather one is frequently interested in reactions and transitions between various states and in their matrix elements. Such calculations may be performed using the Green's function which is discussed by Mavroyannis in Chapter 8, the initial value of which is the density matrix.

B. Properties of the Reduced Density Matrix

From the definition of Eq. (3.9), many of the properties of the reduced density matrix follow. The pth order reduced density matrix $\hat{D}^p(x, x')$

(where x stands for the variables $1, 2, \ldots, p$ and y for the variables $p+1, \ldots, N$) has the following properties:

1. Hermiticity:
$$\hat{D}^p(x : x') = \hat{D}^{*(p)}(x' : x). \qquad (3.10)$$

2. Positive semidefinite:
$$\langle \Psi \hat{D}^p \Psi \rangle \geq 0. \qquad (3.11)$$

3. Unit trace:
$$\text{tr } \hat{D}^p = \int_x D^p(x : x') = 1. \qquad (3.12)$$

4. It possesses the permutational symmetry of the full density matrix
$$\hat{P}_{ij} \hat{D}^p(x : x') = \theta \hat{D}^p(x : x'), \qquad 1 \leq i < j \leq p, \qquad (3.13)$$

where $\theta = +1$ for bosons and -1 for fermions.

5. Eigenfunctions of the reduced density matrix and associated real eigenvalues may be defined:
$$\hat{D}^{(p)}(x : x') g_i^*(x) = \int_{x'} D^{(p)}(x : x') g^*(x') = \lambda_i^{(p)} g_i^*(x) \qquad (3.14)$$

$$g_i(x') \hat{D}^{(p)}(x : x') = \int_x g_i(x) D^{(p)}(x : x') = \lambda_i^{(p)} g_i(x'). \qquad (3.15)$$

The density matrix may be expanded in terms of these eigenfunctions
$$D^{(p)}(x : x') = \sum \lambda_i g_i^*(x) g_i(x') \qquad (3.16)$$

which illustrates its projection properties:
$$\hat{D}^{(p)}(x : x') f(x) = \sum_{i=1} \langle f g_i \rangle \lambda_i g_i(x). \qquad (3.17)$$

The eigenvalues of the p-matrix are referred to as the "natural p states"; in the case $p = 2$ they are referred to as "natural geminals" and for $p = 1$ as "natural orbitals." The natural p states may be ordered according to their eigenvalues
$$\lambda_1^{(p)} \geq \lambda_2^{(p)} \geq \lambda_3^{(p)} \geq \cdots,$$

which are in turn subject to the inequalities

$$0 \leq \lambda_i^{(1)} \leq 1/N$$
$$0 \leq \lambda_i^{(2)} \leq 1/N, \quad N \text{ odd} \quad (3.18)$$
$$0 \leq \lambda_i^{(2)} \leq 1/(N-1), \quad N \text{ even}$$

for fermions, and

$$0 \leq \lambda_i^{(p)} \leq 1 \quad (3.19)$$

for bosons, all of which follow from the permutational symmetry of the N-particle system. In the special case of an N fermion state function expanded in terms of a sum of Slater determinants $\{X_j\}$,

$$\Psi(1, 2, \ldots, N) = \sum_{j=1}^{K} C_j X_j(1, 2, \ldots, N), \quad (3.20)$$

the bounds on the eigenvalues of the p-matrix derived from Ψ are

$$\lambda_i^{(p)} \leq K[p(N-p)!/N!]. \quad (3.21)$$

Thus the eigenvalues of the 2-matrix are bounded by $[2/N(N-1)]$ in the Hartree–Fock approximation.

The density matrix may be expanded in a complete orthonormal set of functions of p particles (having the correct permutational symmetry) $\{f_n(x)\}$

$$\mathbf{D}^{(p)}(x:x') = \sum B_{nm}^{(p)} f_n^*(x) f_m(x') \quad (3.22)$$

where

$$B_{nm}^{(p)} = \langle f_n \hat{D}^{(p)} f_m \rangle = \sum_i \lambda_i^{(p)} \langle f_n g_i \rangle \langle g_i f_m \rangle = \sum_i \lambda_i^{(p)} C_{ni}^* C_{mi}^* \quad (3.23)$$

and

$$f_n(x) = \sum C_{ni} g_i(x). \quad (3.24)$$

The weights $B_{nm}^{(p)}$ are real and the diagonal elements satisfy the inequality

$$0 \leq B_{nn}^{(p)} \leq \lambda_1^{(p)} \quad (3.25)$$

while the off-diagonal elements satisfy

$$B_{nm}^2 \leq B_{nn} \cdot B_{mm}, \quad (3.26)$$

and may be of either sign.

6. An interesting and important result which is given here without proof is the Carson–Keller theorem. This states that the reduced density matrices $\hat{D}^{(p)}$ and $\hat{D}^{(N-p)}$ have the same nonzero eigenvalues. An obvious corollary is that the eigenvalues of the matrix $\hat{D}^{(N-1)}$ are bound by $1/N$, and those of $\hat{D}^{(N-2)}$ are bound by $1/N$ for N even and $1/(N-1)$ for N odd. An application of this theorem is the expansion of an N-particle wave function in the eigenfunctions of the p- and $(N-p)$-matrices:

$$\Psi(xy) = \sum_i (\lambda_i)^{1/2} g_i(x) h_i(y). \tag{3.27}$$

This function automatically posesses the correct permutational symmetry in all variables.

7. In the case of those operators that can be written in the form

$$\hat{A}^{(N)}(xy) = \hat{A}^{(p)}(x) + \hat{A}^{(N-p)}(y), \tag{3.28}$$

$\hat{A}^{(p)}$ commutes with $\hat{D}^{(p)}$ whenever $\hat{A}^{(N)}$ commutes with $\hat{D}^{(N)}$. In such cases the natural p states will be eigenfunctions of the operator $\hat{A}^{(p)}$. In general operators may not be factored in this way. For example, a Hamiltonian with interelectronic interaction terms will not obey Eq. (3.28) and therefore the eigenfunctions of the reduced Hamiltonian do not diagonalize the reduced density matrix. Similarly, the natural geminals may not be classified as singlet or triplet functions. (The specific exceptions to this rule are when the total spin is zero or $N/2$. In such a case the natural geminals may be classified as singlets and triplets.)

Operators that can be expanded in terms of one-particle operators, for example the z component of spin, can always commute with reduced density matrices.

It is often convenient to expand the density matrix in terms of eigenfunctions $\{\Phi_n\}$ of the reduced Hamiltonian or some other reduced operator of physical interest:

$$D(x:x') = \sum B_{nm} \Phi_n^*(x) \Phi_m(x'). \tag{3.29}$$

In such cases one can obtain useful rules concerning the eigenvalues since in general several important properties of the system are supposed to be known. For example, suppose that the total spin of the system is S. Then the reduced density matrix must satisfy condition

$$S(S+1) = (N/2) \operatorname{tr} \hat{s}^2 \hat{D}^{(2)} \tag{3.30}$$

where $\hat{s}^2$ is the reduced two-particle spin operator

$$\hat{s}^2 = \hat{S}_1{}^2 + \hat{S}_2{}^2 + 2(N-1)\hat{S}_1 \cdot \hat{S}_2. \qquad (3.31)$$

$\hat{s}^2$ may be expressed in terms of the total spin operator for two electrons

$$\hat{\sigma}^2 = \hat{S}_1{}^2 + \hat{S}_2{}^2 + 2\hat{S}_1 \cdot \hat{S}_2, \qquad (3.32)$$

that is,

$$\hat{s}^2 = (N-1)\hat{\sigma}^2 - (N-2)(\hat{S}_1{}^2 + \hat{S}_2{}^2). \qquad (3.33)$$

The eigenvalues of $\hat{s}^2$ are $-\tfrac{3}{2}N + 3$ for singlets and $\tfrac{1}{2}N + 1$ for triplets; substitution of these values in the sum (3.30) gives the conditions

$$\sum_{\text{singlets}} B_{nn} = [N(N+2) - 4S(S+1)]/4N(N-1) \qquad (3.34)$$

$$\sum_{\text{triplets}} B_{nn} = [3N(N-2) + 4S(S+1)]/4N(N-1) \qquad (3.35)$$

on the geminal expansion coefficients. Similar sum rules may be obtained as a result of knowledge of the total orbital angular momentum.

9. In the case of an expansion of the 2-matrix in eigengeminals of the reduced Hamiltonian a particularly simple expression is given for the energy expectation value

$$E = \langle \hat{H}^{(N)} \rangle_{\hat{D}^{(N)}} = (N/2)\,\text{tr}\,\hat{h}\hat{D}^{(2)} = (N/2) \sum B_{nn}\varepsilon_n \qquad (3.36)$$

where ε_n are the eigenvalues of the reduced Hamiltonian. Such a sum proves attractive since it is a sum over correlated energies. That is, the solution of the many-body problem is reduced to an appropriate linear combination of the solutions to a two-body problem. One is normally concerned in the many-body problem with a single-particle basis; any correlation is treated by including more terms in the basis. Eq. (3.36) indicates, however, that one may begin with a correlated pair basis and deal with a system in terms of excitations of correlated pairs of particles.

At first sight one would therefore direct one's computations to the two-body problem

$$\hat{h}\Phi_n = \varepsilon_n \Phi_n \qquad (3.37)$$

and after inserting the appropriate eigenvalues in the sum (3.36) minimize the energy by varying the coefficients B_{nn} subject to the restrictions outlined in this section. The problem, however, is far more difficult since one becomes involved in problems of N-representability, that is, the interrelationship between the B_{nn} and the Φ_n, the nature of which is discussed in Section IV.

C. The Transition Matrix

Physical chemistry consists of much more than the study of small molecules or atoms in their ground state, and the pure state density matrix, while specifying the expectation values of a particular state, does not give information concerning transitions. It is useful, therefore, to define a transition matrix $\hat{D}_{AB}^{(p)}(x:x')$ between two states $\Psi_A(xy)$ and $\Psi_B(xy)$:

$$\hat{D}_{AB}^{(p)}(x:x') = \int \Psi_A^*(xy)\Psi_B(x'y)\,dy. \tag{3.38}$$

With the aid of the reduced transition density matrix the transition matrix elements of operators may be obtained, e.g.,

$$\langle \hat{O} \rangle_{AB} = (N/2)\,\mathrm{tr}\,\hat{O}^{(2)}\hat{D}_{AB}^{(2)}(x:x'). \tag{3.39}$$

From the orthogonality between states,

$$\mathrm{spur}\,\hat{D}_{AB}(x:x') = 0; \tag{3.40}$$

that is, the condition

$$\sum_n B_{nn}^{AB} = 0 \tag{3.41}$$

holds on the diagonal elements.

By Schurr's lemma the following inequalities hold for the elements:

$$|B_{nm}^{AB}|^2 \leq B_{nm}^{AA} \cdot B_{nm}^{BB}. \tag{3.42}$$

In addition to describing transitions between states, the transition density matrix would be of use in a perturbation scheme. The investigation of a transition density matrix between a correlated and uncorrelated wave function has been suggested (Peat, 1972) as a device for introducing correlation into a calculation without the considerable increase in numerical difficulty in performing the integrals involved.

IV. The N-Representability Problem

In Section III reduced density matrices were defined and certain of their properties discussed. Assuming that a set of trial density matrices were available for the treatment of a certain many-body system, how would one choose that reduced density matrix which best represented the system? Faced with such a problem in the case of a wave function, after applying conditions related to the spin and spatial symmetries of the system, one would be guided to select that function which gave rise to the lowest expectation value for the energy of the system. If such a procedure should apply in the case of the choice of reduced density matrix, then the variational method would indeed be attractive. Consider Eq. (3.36) in which the energy expectation value of a system is expressed in terms of the eigenvalues of the reduced Hamiltonian and the diagonal elements B_{nn} of the 2-matrix, in the representation of eigengeminals of the reduced Hamiltonian. This expression may be minimized upon varying the B_{nn} subject to the restriction of Section III when it is found that the expectation value falls drastically below the lowest energy of the system. That is, the reduced density matrices of Section III are too wide a class; indeed, in the case of pure state density matrices, they are derivable from wave functions of the form

$$\Psi(1, 2, \ldots, N) = \sum_n \Phi_n(1, \ldots, p) f_n(p+1, \ldots, N) \qquad (4.1)$$

where the functions $\{\Phi_n\}$ possess the correct permutational symmetry but the operations $-\hat{\mathbf{P}}_{pi}$ $(i > p)$ and $-\hat{\mathbf{P}}_{ij}$ $(p < i < j)$ of the permutation group do not leave the form of the wave function invariant.

Restricting our attention to a system containing a constant number N of particles and a particular permutational symmetry, the problem of characterizing the proper reduced density matrices for this system is known as the "N-representability" problem (Coleman, 1963).

One normally considers Bose (or Fermi) systems as described by symmetric (or antisymmetric) wave functions but one may equally well work in a spin-free scheme in which the wave function is partitioned into a linear combination of products of functions of the spin variables only and functions of the space variables. Integration over all spin variables and $(N-p)$ space variables results in the spin-free reduced p-matrices and in many problems it may be of advantage to work with such forms. The various spatial wave functions, corresponding to different spin values, are obtained as different representations of the permutation group. The N-representability problem therefore is concerned not only with symmetric or antisymmetric functions but with all representations of the permutation group; however, most of the research performed on this problem has been concerned with antisymmetric representations alone and our discussion is confined mainly to this case.

The set of all $\hat{\mathbf{D}}^N$ is identical to that of all positive Hermitian operators $\hat{\mathbf{S}}^{(N)}$ having unit trace defined on a Hilbert space of given permutational symmetry. The set $\hat{\mathbf{S}}^{(N)}$ is convex and therefore characterized by its extreme points. (A set is convex if $C = \alpha A + \beta B$ is a member of the set $\hat{\mathbf{S}}^N$ where $A \in \hat{\mathbf{S}}^{(N)}$, $B \in \hat{\mathbf{S}}^N$, and α, β are nonnegative real numbers such that $\alpha + \beta = 1$. If $C = \alpha A + \beta B$ is extreme, then A and B are multiples of C.) The extreme points of $\hat{\mathbf{S}}^N$ correspond to the pure state density matrices. The set $\mathbf{S}^{(p)}$ is a subset of $\mathbf{S}^{(N)}$ where $p < N$.

The definitions of the preceding sections indicate that the N-representable $\hat{\mathbf{D}}_N^{(p)}$ are a subset of $\mathbf{S}^{(p)}$ which will be called $\mathbf{S}_N^{(p)}$. The set $\mathbf{S}_N^{(p)}$ is convex and characterized by its extreme points which are those p-matrices derived from the extreme points in $\mathbf{S}^{(N)}$.

Hence the N-representability problem is that of characterizing the set $\mathbf{S}_N^{(p)}$ as a subset of $\mathbf{S}^{(p)}$ and the problem can be reduced to a discussion of the extreme points of $\mathbf{S}_N^{(p)}$; these extreme points are those p-matrices derived from N-particle pure state density matrices defined on a Hilbert space of the correct symmetry. That is, the problem of ensemble N-representability can be reduced to that of pure state N-representability.

The solution of this problem has in general been confined to functions defined on an antisymmetric Hilbert space, and in the case $p = 1$ the characterization of ensemble N-representable 1-matrices has been solved. For $p > 1$ the problem appears to be more difficult and it is apparent that for the time being one must be content with a set of necessary but not sufficient conditions for N-representability.

The increased difficulty of solving the N-representability problem for the 2-matrix, over the 1-matrix, can be seen from a consideration of the unitary invariants of N-representability. If the p-matrix is expanded in a complete orthonormal set of orbitals $\psi_i^{(1)}$ with coefficients $C_{i_1, i_2, \ldots, i_p, i_1', i_2', \ldots, i_p'}$, antisymmetric in the indices $i_1, \ldots, i_p$ and $i_1', \ldots, i_p'$, the N-representability conditions are invariant with respect to unitary transformations of the basis. That is, the N-representability conditions can be expressed in terms of these unitary invariants and the number N. In the case of the 1-matrix the eigenvalues $\lambda_i^{(1)}$ provide a complete set of unitary invariants, and the N-representability conditions for $\hat{\mathbf{D}}_N^{(1)}$ are expressible in the set $\{\lambda_i^{(1)}\}$ and the number N. In the case $p = 2$ the unitary invariants are not so simple and it would appear that they involve both the $\{\lambda_n^{(2)}\}$ and the basis set $\{\Phi_n(12)\}$ in addition to N.

It may be mentioned that in addition to consideration of the N-representability problem for systems of fixed N one may seek to characterize that subset of $\mathbf{S}^{(p)}$ which is contracted from an ensemble of sets $\mathbf{S}^{(N)}$ over different N.

Before terminating this general discussion of N-representability it should be noted that in addition to applying to the density matrix, these conditions apply also to the p-particle Green's function. That is, the general one- or two-particle Green's function may not be N-representable unless special precautions are taken. However, in the application of Green's function techniques to solid-state problems one does not normally note special attention being paid to such problems and it is not too difficult to see why. Since one is presented with a translationally invariant system for which a single-particle basis set, often of plane waves, is normally employed, and since with such a set the Pauli conditions are normally automatically applied by virtue of the formalism, ensemble N-representability is frequently sufficiently ensured. Again one is generally concerned with perturbation corrections to some zero-order, and therefore probably N-representable, Green's function. However, in employing the Green's function attention should always be given to the problem of N-representability and the possibility of falling into the trap of working with "unphysical" functions.

The N-representability conditions that have been derived for the 1-matrix and the 2-matrix, are discussed briefly below.

A. N-Representability of the 1-Matrix

As was indicated above, the N-representability of the 1-matrix may be expressed in terms of unitary invariants, the eigenvalues, and the number N.

Ensemble N-representability for fermions is ensured by demanding that the eigenvalues of $\mathbf{D}^{(1)}$ are each bounded by $1/N$, that is, that $\hat{\mathbf{I}} - N\hat{\mathbf{D}}^{(1)}$ be positive semidefinite, where $\hat{\mathbf{I}}$ is the identity operator (Coleman, 1963).

In the case of bosons it is sufficient to demand that the $\lambda_n^{(1)}$ be bounded from above by unity.

The conditions for N-representability by pure states are not so simple and the problem has not been completely solved. A useful sufficient condition for pure fermion N-representability is that in addition to the eigenvalues of $\hat{D}^{(1)}$ satisfying the above bounds, there exists an integer which divides both N and the degeneracies of all eigenvalues (Smith, 1966).

The necessary condition on N-representability, that the eigenvalues be less than $1/N$, has an obvious physical interpretation in terms of the Pauli restrictions—explicitly that in building up a state the occupation number of any spin orbital should not exceed unity. Since use is not made of any addition conditions in this chapter, they are not mentioned. The reviews by Coleman (1968) and Smith (1968) discuss most of the important results in this field.

B. N-Representability of the 2-Matrix

A solution of the N-representability problem in the case of 2-matrices has been obtained by Garrod and Percus (1964) (see also Kummer, 1967); however, it appears in the form of an infinite set of conditions.

If $\{\hat{A}\}$ is the set of positive operators on the space of symmetric or antisymmetric N-particle functions,

$$\hat{A} = \sum_i B(i) + \sum_{i<j}^N C(ij), \qquad (4.2)$$

then $D^{(2)}$ is ensemble N-representable if and only if

$$\text{tr } \hat{A}^{(2)} D^{(2)} > 0$$

for all $\hat{A}$ in $\{\hat{A}\}$, where $\hat{A}^{(2)}$ is the reduced two-particle operator corresponding to $\hat{A}$. The conditions have a simple interpretation in that there exists some Hamiltonian for which a nonensemble N-representable density matrix calculates its expectation value as below the ground-state energy. In its complete form the condition is therefore of little practical use. Nevertheless, a particular choice of the operator $\hat{A}$ will provide a necessary condition on the 2-matrix.

A powerful necessary condition which follows from the above is that the operators

$$\hat{Q}(1, 2 : 1', 2') = N(N-1)\hat{D}^{(1)}(1', 2 : 1, 2') + N\,\delta(1-1')\hat{D}^{(1)}(2 : 2')$$
$$- N^2\hat{D}^{*(1)}(1 : 2)\hat{D}^{(1)}(1' : 2') \qquad (4.3)$$

and

$$\begin{aligned}\hat{Q}(1,2:1',2') = {} & N(N-1)\hat{D}^{(2)}(1',2':1,2) - N\,\delta(2-2')\hat{D}^{(1)}(1:1') \\ & + N\,\delta(1-2')\hat{D}^{(1)}(1':2) + N\,\delta(2-1')\hat{D}^{(1)}(2':1) \\ & - N\,\delta(1-1')\hat{D}^{(1)}(2':2) + \delta(1-1')\,\delta(2-2') \\ & - \delta(1-2')\,\delta(2-1') \end{aligned} \qquad (4.4)$$

be nonnegative. It is believed that these conditions are strong in that trial 2-matrices which satisfy them may be "almost N-representable."

A large class of inequalities on the 2-matrix are related to the conditions of Garrod and Percus (Davidson, 1969), but these and several other conditions are not discussed here.

As an alternative to attempting a general solution or set of conditions which apply generally, attention may be confined to certain classes of wave functions or to extreme points.

For example, in the case of the Hartree–Fock wave functions the N-representability conditions are rather simple and one may generalize to simple wave functions containing only a limited number of orbitals or to sums of Slater determinants.

A wave function that has received much attention is the antisymmetrized geminal product (see Section VI) (Coleman 1965, 1972):

$$\Psi(1,2,\ldots,N) = \hat{A}g(12)g(34)\cdots g(N-1,N). \qquad (4.5)$$

In the case where the eigenvalues of the 1-matrix of $g(12)$ are all equal, the wave function is of extreme type and the eigenvalue of the 2-matrix to the wave function approaches its upper bound as the 1-rank of (4.5) increases. In Section VIII it is shown that such extreme wave functions may be associated with condensation phenomena. Since the 2-matrices corresponding to these wave functions describe some of the extreme points of $\hat{S}_N^{(2)}$ and the set is characterized by its extreme points, the study of the antisymmetrized geminal product (AGP) is of great interest. It should be noted, however, that *any* wave function of the form (4.5) is not necessarily extreme, nor does it necessarily characterize a condensed system; for example, a single Slater determinant can always be written in the form (4.5).

The 2-matrix corresponding to the AGP function has therefore been characterized and discussed (Coleman, 1965, 1972) and attempts have been made to generalize it. Again a detailed discussion of the situation is beyond the scope of this chapter.

The N-representability problem is of course connected with the general problem of the statistics of p-particle functions, their creation and annihilation operators, and the construction of N-particle wave functions out of p-particle group functions. A discussion of this problem in the case $p = 2$, that is, of general geminal wave functions, is given in Section VII.

V. The Single-Particle Reduced Density Matrix

A. Natural Orbitals and Their Eigenvalues

The single-particle reduced density matrix $\mathbf{D}(1:1')$, also written $\gamma(1:1')$ in the literature, is the reduced density matrix of most familiarity to the chemist. For example, the particle density of a particular system is given by the 1-matrix

$$p(\mathbf{r}) = \mathbf{D}^{(1)}(\mathbf{r}:\mathbf{r}). \tag{5.1}$$

The eigenvalues of the 1-matrix correspond to the occupation numbers of the various natural orbitals of the system (or the squares of the occupation numbers, depending on the convention used). That is, $\lambda_i^{(1)}$ is the probability that an electron is associated with the natural orbital $\phi_i^{(1)}$.

The calculation of ground-state properties of atoms or molecules is frequently performed using a configurational interaction scheme whereby the correlation effects are taken into account by increasing the size of the orbital basis set. It is of interest that the natural orbitals to a particular system provide the most rapidly convergent basis in the sense that

$$\| \Delta \|^2 = \left\| \Psi - \sum_n \lambda_n \phi_n(1) f_n(2, \ldots, N) \right\|^2 \tag{5.2}$$

is minimized when ϕ_n are the natural orbitals of the 1-matrix of Ψ. Therefore it is the eigenvalues of the 1-matrix which have an important physical relevance, rather than the expansion coefficients of an arbitrary basis which may be changed under a unitary transformation. It becomes possible to compare numerical wave functions by correlating the eigenvalues of their 1-matrices. Similarly, one may be able to systematize the various properties of atoms and molecules according to the occupation numbers of the various natural orbitals. Such a procedure involves the computer analysis of accurate wave functions, for atoms and molecules, and has become quite popular. Of course it must be remembered that the properties of the ground states of atoms or molecules represent only a small portion of the field of quantum chemistry and such analyses by themselves are of little value unless they illuminate a wider area.

Examples of such analyses indicate how well the eigenvalues of the natural orbitals of accurate wave functions conform to the intuitive picture of occupation numbers devised by earlier theorists. For example (Davidson, 1963), a 44-term trial function containing functions of the interelectronic separation r_{12}, for the helium atom, which was in error in its calculation of the ground-state expectation energy by only 0.1 cm^{-1}, was analyzed into its natural orbitals. It was found that the total weight associated with the S states lay very close to unity, 0.995847, while the weight for the $L = 1$ states was 0.003941 and for the $L = 2$ states, 0.000174.

An indication of the rapidity of convergence of the natural orbitals is is given in Table I. Truncation after the sixth term of the natural orbital series yields an energy superior to that obtained with the original wave function containing 20 configurations.

TABLE I

The Energy and Overlap with the Accurate Wave Function of a Natural Orbital Series of Differing Expansion Length in the Case of the Helium Atom Ground State

Expansion length	Overlap with accurate wave function	Energy (a.u.)
1	0.99599	−2.861653
2	0.99790	−2.882044
3	0.99979	−2.897484
4	0.99987	−2.899243
5	0.99994	−2.900903
6	0.99997	−2.901689
∞	1.00000	−2.903724

The simple picture of the beryllium atom in its ground state is of two pairs of electrons in singlet states, one pair in the K shell and the other in the L shell. The extent to which this is borne out by a natural orbital analysis is indicated in Table II (recall that the sum of all the eigen-

TABLE II

Occupation Numbers of the Natural Orbitals to an Accurate Trial Function for the Beryllium Atom Ground State (Barrett et al., 1965)

Eigenvalue number	Eigenvalue magnitude	Degeneracy	Identification
1,2	0.249577	2	1s
3,4	0.229643	2	2s
5–10	0.006758	6	2p
11,12	0.000180	2	3s
13–18	0.000070	6	3p

values is normalized to unity.) The first two eigenvalues may be identified with the 1s functions, and to a high approximation the $1s^2$ shell is totally occupied. This approximation is not quite so good as regards the $2s^2$ shell where admixtures of higher orbitals, the eigenvalues 5–18, intrude.

Investigation of the 1-matrix therefore provides a tool for the justification of the rules of atomic and molecular structure. Of course these analyses have been confined to relatively simple systems and it would be of greater interest to investigate, for example, the occupation numbers of the "d" electrons in transition metals, or the change in magnitude of the eigenvalues of the 1-matrix under various perturbations, for example the formation of atomic and molecular groups.

Computer investigations of natural orbitals have been concerned mainly with the calculation of the 1-matrix of some previously computed wave function and the subsequent diagonalization of this matrix. However, it has been pointed out (Kutzelnigg, 1963) that in the case of a two-electron system the natural orbitals may be obtained from a coupled set of integrodifferential equations. In addition it is possible to extend these equations to larger systems provided that the state function for such systems can be written in the form

$$\Psi(1, 2, \ldots, N) = N\hat{A}g_1(1, 2)g_2(3, 4) \cdots g_{N/2}(N-1, N) \quad (5.3)$$

where $\hat{A}$ is the antisymmetrization operator and the geminals $\{g_n(12)\}$ satisfy the condition of "strong orthogonality"

$$\int_1 g_n^*(1, 2)g_m(1, 2') = \delta_{nm}. \quad (5.4)$$

B. The Dirac and Bloch Density Matrices

1. Introduction

The Hartree–Fock method provides the most common starting point for the study of systems in physics and chemistry and is frequently used as the first term in some approximation scheme. It is appropriate, therefore, that consideration be given to the 1-matrix which corresponds to a wave function of the form

$$\Psi(1, 2, \ldots, N) = (N!)^{-1/2} \det \psi_i(r_j). \tag{5.5}$$

It will be recalled from Eq. (3.21) that the appropriate 1-matrix is written

$$\gamma(1 : 1') = (1/N) \sum_{i=1}^{N} \psi_i^*(1)\psi_i(1'). \tag{5.6}$$

This is referred to as the Dirac density matrix and, for convenience, the symbol γ is reserved for this particular form. In the case of wave functions of the form (5.5) the 2-matrix is determined by the Dirac matrix

$$2\mathbf{D}^{(2)}(1, 2 : 1', 2') = \gamma(1 : 1')\gamma(2 : 2') - \gamma(2 : 1')\gamma(1 : 2'). \tag{5.7}$$

The Dirac matrix will of course commute with Hamiltonians of the form

$$\hat{\mathbf{H}}(\mathbf{r}) = -\tfrac{1}{2}\nabla_r^2 + V(\mathbf{r}) \tag{5.8}$$

and this enables the 1-matrix corresponding to the Hamiltonian $\hat{\mathbf{H}}(\mathbf{r})$ to be calculated from the Bloch equation

$$\hat{\mathbf{H}}\mathbf{C} = -\partial \mathbf{C}/\partial \beta. \tag{5.9}$$

$\mathbf{C}(\mathbf{r} : \mathbf{r}', \beta)$ is the Bloch density matrix, a Laplace transform of the Dirac matrix

$$\mathbf{C}(\mathbf{r} : \mathbf{r}', \beta) = (\beta/N) \int_0^\infty \gamma(\mathbf{r} : \mathbf{r}', \varepsilon_\mathrm{F}) \exp(-\beta \varepsilon_\mathrm{F}) \, d\varepsilon_\mathrm{F} \tag{5.10}$$

where ε_F is the energy corresponding to the highest occupied orbital in the Dirac matrix, the Fermi energy; β is the familiar symbol for $k_\mathrm{B} T$, T being the absolute temperature and k_B the Boltzmann constant. $\mathbf{C}(\mathbf{r}, \mathbf{r}', \beta)$, then, is the solution to the Bloch equation subject to the boundary condition

$$\mathbf{C}(\mathbf{r} : \mathbf{r}', 0) = \delta(\mathbf{r} - \mathbf{r}'), \tag{5.11}$$

a result following from the completeness of the eigenvalues. The general solution is of the form

$$C(\mathbf{r}:\mathbf{r}',\beta) = (1/N) \sum_i \psi_i^*(1)\psi_i(r') \exp(-\varepsilon_i\beta). \tag{5.12}$$

This formalism enables one to calculate the changes in the 1-matrix and hence in the electron density as a result of some perturbation of a single-particle Hamiltonian.

In the case of an atom or small molecule the replacement of the Dirac matrix [Eq. (5.6)] by a single analytic function will in general be an extremely difficult problem and one would have to be content with the calculation of the effect of the perturbation $V(\mathbf{r})$ on each of the eigenfunctions separately. However, in the case of a long molecule or solid one may be able to expand in plane waves. For example, suppose the electron gas is quantized in a cube of volume Ω; then the Dirac matrix becomes

$$\gamma(\mathbf{r}:\mathbf{r}') = (2/\Omega) \sum_{k<k_{\mathrm{f}}} \exp[-i\mathbf{k}(\mathbf{r}'-\mathbf{r})] \tag{5.13}$$

$$= (2/8\pi^3) \int_{k<k_{\mathrm{f}}} \exp[i\mathbf{k}\cdot(\mathbf{r}'-\mathbf{r})]\,d\mathbf{k}. \tag{5.14}$$

Expanding the plane waves in spherical waves,

$$\gamma(\mathbf{r}:\mathbf{r}') = \pi^{-2} \int_0^{k_{\mathrm{f}}} k \frac{\sin(k\,|\,\mathbf{r}'-\mathbf{r}\,|)}{\mathbf{r}'-\mathbf{r}}\,dk, \tag{5.15}$$

which results in the expression

$$\gamma(\mathbf{r}:\mathbf{r}') = \frac{k_{\mathrm{f}}^3}{\pi^2} \frac{j_1(k_{\mathrm{f}}\,|\,\mathbf{r}-\mathbf{r}'\,|)}{k_{\mathrm{f}}\,|\,\mathbf{r}-\mathbf{r}'\,|}, \tag{5.16}$$

where $\mathbf{k}_{\mathrm{f}}$ is the wave vector corresponding to the highest filled orbital and $j_1(x)$ is the first-order Bessel function; that is,

$$j_1(x) = (\sin x - x\cos x)/x. \tag{5.17}$$

The particle density of N electrons in a volume $\Omega = N/\varrho_0$ is regained in the limit $\mathbf{r}\to\mathbf{r}'$

$$\gamma(\mathbf{r}:\mathbf{r}) = \varrho_0 = k_{\mathrm{f}}^3/3\pi^2. \tag{5.18}$$

The diagonal elements of the 2-matrix are

$$D^{(2)}(\mathbf{r}_1,\mathbf{r}_2:\mathbf{r}_1,\mathbf{r}_2) = \frac{k_{\mathrm{f}}^6}{18\pi^4} - \frac{k_{\mathrm{f}}^6}{4\pi^4} \left\{ \frac{j_1(k_{\mathrm{f}}\,|\,\mathbf{r}_1-\mathbf{r}_2\,|)}{k_{\mathrm{f}}\,|\,\mathbf{r}_1-\mathbf{r}_2\,|} \right\}^2 \tag{5.19}$$

and the pair correlation function, for this noninteracting electron gas, is defined with unit normalization as $|\mathbf{r}_1 - \mathbf{r}_2| \to \infty$; from the 2-matrix

$$g(r) = g(|\mathbf{r}_1 - \mathbf{r}_2|) = 1 - \tfrac{9}{2}\{j_1(k_f r)/k_f r\}^2, \quad (5.20)$$

while the free electron Bloch matrix is

$$\mathbf{C}(\mathbf{r} : \mathbf{r}', \beta) = (2\pi\beta)^{-3/2} \exp(-|\mathbf{r}' - \mathbf{r}|^2/2\beta). \quad (5.21)$$

With the knowledge of the free electron density matrices for a metal or long-chain molecule it now becomes possible to calculate the perturbations in electron density or pair correlation function as a result of some external one-particle potential $V(\mathbf{r})$. The Bloch equation may be written as an integral equation

$$\mathbf{C}(\mathbf{r} : \mathbf{r}', \beta) = \mathbf{C}_0(\mathbf{r} : \mathbf{r}', \beta) - \int d\mathbf{r}_1 \int_0^\beta d\beta_1 \, \mathbf{C}_0(\mathbf{r} : \mathbf{r}_1, \beta - \beta_1)$$
$$\times V(\mathbf{r}_1)\mathbf{C}(\mathbf{r} : \mathbf{r}', \beta_1), \quad (5.22)$$

where $\mathbf{C}_0(\mathbf{r} : \mathbf{r}', \beta)$ is the unperturbed Bloch density matrix, and can be solved by an iterative procedure:

$$\mathbf{C}(\mathbf{r} : \mathbf{r}', \beta) = \sum_{n=0}^{\infty} \mathbf{C}_n(\mathbf{r} : \mathbf{r}', \beta) \quad (5.23)$$

where

$$\mathbf{C}_n(\mathbf{r} : \mathbf{r}', \beta) = (2\pi\beta)^{-3/2} \int \prod_{l=1}^{n} \{-d\mathbf{r}_l V(\mathbf{r}_l)/2\pi\} \left(\sum_{l=1}^{n+1} |\mathbf{r}_l - \mathbf{r}_{l-1}|\right)$$
$$\times \exp\left\{(-1/2\beta)\left(\sum_{l=1}^{n+1} |\mathbf{r}_l - \mathbf{r}_{l-1}|^2\right)\right\} \prod_{l=1}^{n+1} |\mathbf{r}_l - \mathbf{r}_{l-1}|. \quad (5.24)$$

In the case where the perturbing potential varies slowly over a characteristic wavelength one may replace the variable $\mathbf{r}_l$ in $V(\mathbf{r}_l)$ by $\mathbf{r}$. The integrations may then be completed directly, giving, for the Dirac matrix,

$$\gamma(\mathbf{r} : \mathbf{r}, k_F) = \frac{k_F^3}{6\pi}\left(1 - \frac{3V(\mathbf{r})}{k_F^2} + \frac{3V(\mathbf{r})^2}{2k_f^4} + \frac{V(\mathbf{r})^2}{2k_F^6} + \cdots\right)$$
$$= \frac{1}{6\pi^2}(k_f^2 - 2V(\mathbf{r}))^{3/2}. \quad (5.25)$$

2. The Fermi–Thomas Approximation

If E is the highest occupied orbital in the Dirac density matrix, then Eq. (5.25) yields the following functional relationship between the electron density and the energy and perturbing potential, in the case of a slowly varying perturbation:

$$\varrho(\mathbf{r}, E) = (2^{3/2}/6\pi^2)[E - V(\mathbf{r})]^{3/2}. \tag{5.26}$$

The approximation devised by Thomas and Fermi attempts to determine the equilibrium distribution of electrons in a large atom. The interelectronic interactions are ignored and the electrons are assumed to move in the screened Coulombic field due to the nucleus of charge Ze. This screened potential may be written in the form

$$V(r) = Z(r)e^2/r \tag{5.27}$$

where $Z(1) \to Z$ as $r \to 0$ and $Z(1) \to 0$ as $r \to \infty$. Assuming that the potential changes sufficiently slowly it may be related self-consistently to the density via Eq. (5.26). However, it is necessary that this relationship be indeed self-consistent, that is, that such a potential results from the density distribution $\varrho(\mathbf{r}, E)$; this condition of self-consistency may be imposed using Poisson's equation

$$\nabla^2 V = -4\pi e^2 \varrho(r). \tag{5.28}$$

Close to the nucleus the self-consistent potential is of the form

$$V(r) \approx -(2e^2/r) + 1.80(eZ^{4/3}/a_0),$$

the second term being the potential due to the distribution of electrons. The electron density has its maximum value at a distance from the nucleus proportional to $a_0 Z^{-1/3}$, and the binding energy per electron is found to be

$$\varepsilon = -0.769 Z^{7/3} e^2/a_0. \tag{5.29}$$

In the case of a long molecule or solid this method may be extended to determine, self-consistently, the screened interelectronic interaction and, with a periodic perturbation, the longitudinal dielectric constant.

These techniques may be applied to calculate the response to an impurity or a point defect in a metal (March and Murray, 1961) or the effect of an atomic or molecular substitution in some long-chain molecule. In the event that the perturbation is not slowly varying, then the integral in Eq. (5.24) must be evaluated directly or in some approximation and the result expressed as a perturbation sum to the required order.

3. Effective Potential Method

In the case of a long-chain molecule, or electron gas in a lattice, it may be desired to improve upon the plane wave expansion by using, for example, Bloch waves. Such a procedure would greatly complicate the solution of the Bloch equation in its integral equation form. An alternative mode of solution therefore is to employ an effective potential matrix **U** and write (Hilton, et al., 1967) the perturbed Bloch matrix in the form

$$\mathbf{C} = \mathbf{C}_0 e^{-\beta \mathbf{U}}. \tag{5.30}$$

Substitution in the Bloch equation yields a nonlinear differential equation for the effective potential

$$\tfrac{1}{2}\beta \nabla^2 \mathbf{U} - \tfrac{1}{2}\beta^2 (\nabla \mathbf{U})^2 - (\mathbf{r} - \mathbf{r}_0) \cdot \nabla \mathbf{U} + V = \beta(\partial \mathbf{U}/\partial \beta) + \mathbf{U}. \tag{5.31}$$

Solutions to this equation may be obtained directly, if the nonlinear term $\tfrac{1}{2}\beta^2(\nabla \mathbf{U}^2)$ is dropped, in the form

$$\mathbf{U}'(\mathbf{r} : \mathbf{r}_0, \beta) = \int G(\mathbf{r}, \mathbf{r}_0, \mathbf{r}_1) V(\mathbf{r}_1) \, d\mathbf{r}_1, \tag{5.32}$$

where the Green's function satisfies the linearized form of (5.31) in the region $\mathbf{r} \neq \mathbf{r}_1$ and with $V = 0$.

The results obtained using this method are exact in first order and contain contributions from all higher orders. In the case of the Thomas–Fermi approximation the effective potential matrix is simply the potential V.

4. Thermodynamic Quantities

Assuming that the 1-matrix has been calculated for some system, the thermodynamic quantities may then be computed (Boardman and March, 1964). For example, the partition function is available from the Bloch density matrix

$$\sum_i \exp(-\beta E_i) = \int C(\mathbf{r} : \mathbf{r}, \beta) \, d\mathbf{r}. \tag{5.33}$$

In addition other quantities may be calculated with the aid of the function $P(E)$:

$$\int d\mathbf{r} \, C(\mathbf{r} : \mathbf{r}, \beta) = \beta^2 \int_0^\infty dE \, \exp(-\beta E) P(E). \tag{5.34}$$

The free energy is

$$F = NE_F + 2\int_0^\infty dE\, P(E)(\partial f/\partial E) \tag{5.35}$$

where $f(E)$ is the Fermi–Dirac distribution function. This function is used to evaluate the entropy

$$S = -\left(\frac{\partial F}{\partial T}\right)_{E_F} - \left(\frac{\partial F}{\partial E_F}\right)\cdot\frac{dE_F}{dT} \tag{5.36}$$

and in turn the specific heat

$$C_V = T\left(\frac{\partial S}{\partial T}\right)_V. \tag{5.37}$$

C. The Energy Functional Method

In the preceding section the electron density of a system was expressed as a functional of the energy and of some perturbing potential. An alternative statement of that approach is to write the total energy of the system as a functional of the electron density, as was done by Honenberg and Kohn (1964). (Of course if one were to know the *exact* 1-matrix of a system, it would be possible to calculate exactly the corresponding energy of the system in equilibrium since the potential energy is related to the kinetic energy through the Virial theorem.)

If the Hamiltonian for the unperturbed system is of the form

$$\hat{H} = \hat{T} + \hat{U}, \tag{5.38}$$

then a universal function F of the density is defined:

$$F[\varrho(\mathbf{r})] = \langle \Psi(\hat{T} + \hat{U})\Psi\rangle. \tag{5.39}$$

In the presence of some external potential $V(\mathbf{r})$ the energy is given as a functional of the density

$$E_V(\varrho) = \int V(\mathbf{r})\varrho(\mathbf{r})\,d\mathbf{r} + F(\varrho). \tag{5.40}$$

E_V will have its minimum in the case of the *exact* density $\varrho(r)$, provided that one varies over a class of N-representable densities $\varrho(r)$ subject to the restriction

$$\int \varrho(\mathbf{r})\,d\mathbf{r} = N. \tag{5.41}$$

The basis of the energy functional method is to obtain an approximation to the universal functional $F(\varrho)$ in some sufficiently convenient form and then calculate the ground-state energy in the presence of the perturbing potential by minimization over a class of N-representable densities.

The functional may be written as a sum of the classical Coulombic interaction and another functional $G(\varrho)$

$$F(\varrho) = \frac{1}{2} \int \frac{\varrho(\mathbf{r})\varrho(\mathbf{r}')}{|\mathbf{r}-\mathbf{r}'|} d\mathbf{r}\, d\mathbf{r}' + G(\varrho) \tag{5.42}$$

where, from Eq. (5.39), $G(\varrho)$ has the form

$$G(\varrho) = \frac{1}{2} \int \nabla_{\mathbf{r}} \nabla_{\mathbf{r}'} D^{(1)}(\mathbf{r}:\mathbf{r}')\, d\mathbf{r} + \frac{1}{2} \int \frac{C(\mathbf{r}:\mathbf{r}')}{|\mathbf{r}-\mathbf{r}'|} d\mathbf{r}\, d\mathbf{r}'. \tag{5.43}$$

The two-particle correlation function in the second term is defined as

$$C(\mathbf{r}:\mathbf{r}') = D^{(2)}(\mathbf{r},\mathbf{r}':\mathbf{r},\mathbf{r}') - D^{(1)}(\mathbf{r}:\mathbf{r}')D^{(1)}(\mathbf{r}:\mathbf{r}'). \tag{5.44}$$

Since one does not in general know the second-order density matrix it is necessary to introduce various approximations into the scheme at this point. For example, in the case of an almost constant electron density

$$\varrho(\mathbf{r}) = \varrho_0 + \varrho'(\mathbf{r}), \tag{5.45}$$

the function $G(\varrho)$ may be expanded as

$$G(\varrho) = G(\varrho_0) + \int K(\mathbf{r}-\mathbf{r}')\varrho'(\mathbf{r})\varrho'(\mathbf{r}')\, d\mathbf{r}\, d\mathbf{r}'$$
$$+ \int L(\mathbf{r}:\mathbf{r}':\mathbf{r}'')\varrho'(\mathbf{r})\varrho'(\mathbf{r}')\varrho'(\mathbf{r}'')\, d\mathbf{r}\, d\mathbf{r}'\, d\mathbf{r}''. \tag{5.46}$$

Neglect of all correlation terms results in the Thomas–Fermi approximation.

The kernel in the second term is related to the electronic polarizability, for on writing

$$K(\mathbf{r}-\mathbf{r}') = (1/\Omega) \sum_{\mathbf{q}} K(\mathbf{q}) \exp[i\mathbf{q} \cdot (\mathbf{r}-\mathbf{r}')], \tag{5.47}$$

one has

$$K(q) = (2\pi/q^2)[\varepsilon(q) - 1]^{-1} \tag{5.48}$$

where $\varepsilon(q)$ is the dielectric constant.

Using such an expansion, or an expansion in gradients of the electron density, one may, for example, calculate the energy in a molecule or solid in the presence of some background potential.

The energy of an inhomogeneous electronic distribution may be calculated in the presence of correlation by a correlation functional which is computed for the *homogeneous* gas by assuming that the changes in density introduced by the perturbing function do not radically alter its form (Kohn and Sham, 1965). The final density may be determined in a self-consistent manner since it may be used to construct a one-particle effective Hamiltonian, the eigenfunctions of which must reproduce this density.

While not yet applied to atomic or molecular systems the energy functional method has been applied to a system of considerable interest to the physical chemist, the surface of a metal (Lang and Kohn, 1970). Since theories of chemisorption and surface catalysis depend on an accurate knowledge of the electronic density in the neighborhood of a metal surface it is encouraging to know that numerical calculations have been made. The density does not fall away from the surface as a simple exponential decay but exhibits the characteristic Friedel oscillations of the perturbed electron gas.

The Hohenberg and Kohn technique may be applied with some advantage to the calculation of the energies of large atoms where the number of electrons becomes prohibitive to the calculation of Hartree–Fock energies. Such calculations have been performed with success for heavy nuclei (Breuckner *et al.*, 1968).

D. The X-Ray Form Factor

The 1-matrix may be employed in a description of the x-ray structure of atoms since the form factor for x-ray scattering is the Fourier transform of the first-order reduced density matrix

$$F(\boldsymbol{\mu}) = \int \exp(i\boldsymbol{\mu} \cdot \mathbf{r}) D^{(1)}(\mathbf{r} : \mathbf{r}) \, d\mathbf{r} \tag{5.49}$$

where $\boldsymbol{\mu} = (\mu, \alpha, \beta)$, μ being the momentum transfer in scattering and α, β the angles. The total scattering intensity is likewise given in terms of the 2-matrix

$$I(\mu)/I_{\text{Cl}} = N + N(N-1) \int D^{(2)}(\mathbf{r}_1, \mathbf{r}_2 : \mathbf{r}_1, \mathbf{r}_2) \exp(i\boldsymbol{\mu} \cdot \mathbf{r}_{12}) \, d\mathbf{r}_1 \, d\mathbf{r}_2 \tag{5.50}$$

where I_{Cl} is the classical scattering intensity. In the case of a spherically symmetric system the following asymptotic form is obtained for the form factor:

$$F(\mu) - 4Z\varrho_0(0)\mu^{-4} + O(\mu^{-6}), \qquad (5.51)$$

in the limit of large μ, where use has been made of the so-called cusp condition,

$$(\partial\varrho_0(r)/\partial_r)_{r=0} = -2Z\varrho_0(0). \qquad (5.52)$$

Calculations have been performed (Benesch and Smith, 1971) on the form factors using these formulas for a series of atoms, based on knowledge of the respective 1-matrices.

E. TRIAL 1-MATRICES

The first-order reduced density matrix occupies a unique position in the hierarchy of reduced density matrices (with the obvious exception of the matrix $\mathbf{D}^{(N-1)}$) since its N-representability conditions are known and are not overcomplicated to apply. It therefore becomes possible to write down a trial parameterized N-representable density matrix for an atom or molecule and determine the parameters variationally. If the density matrix is expanded in terms of some orbital bases, then the preference for its use over that of the wave function becomes merely a matter of storage and calculation of matrix elements, and to this extent the problem is not of particular interest to anyone other than a numerical analyst. In other cases it may be possible to express all or part of the density matrix in a closed form, such as the Dirac matrix of Eq. (5.16), and in such cases the 1-matrix is of obvious advantage.

To date, however, the use of trial density matrices has been confined to their determination by the perturbation schemes of Section V.B or the density functional methods of Section V.C.

VI. The Second-Order Reduced Density Matrix

A. INTRODUCTION

In Section V it was shown that the properties of a system of noninteracting electrons, in the presence of a background or other perturbing potential, could be treated using the single-particle reduced density matrix. In addition, effects of interelectronic interactions could be accounted for in various approximations. However, it would seem apposite to employ the two-particle reduced density matrix when considerations of particle correlations become relevant. Nevertheless it must be borne in mind that the many-body problems in physics and chemistry have met with many successes using single-particle techniques. That is, whether one deals with the one-particle or two-particle density matrix, considerable progress may be made while staying with a single-particle basis. While it is attractive to work with a basis in which part of the correlation is specifically included, for example by using correlated geminals, consideration of the N-representability problem indicates that many difficulties will have to be overcome in constructing proper trial 2-matrices.

If one is to deal with larger systems which may be quantized in a box or along a line, then it becomes possible to work in a plane wave basis. In such cases perturbation calculations are greatly simplified and sums of matrix elements in different orders may be evaluated as the result of the analytic properties of the basis. For example, the success of the Green's function in solid-state physics and chemistry is aided by the properties of the plane wave basis used. In the case of atoms and small molecules the bases that may be used do not have such desirable properties and one may be inclined to face the difficulties of working with a correlated two-particle basis.

B. The 2-Matrix in a Single-Particle Basis

It is not too difficult to construct trial 2-matrices from N-particle wave functions provided that the geminals are expanded in the same single-particle basis $[\psi_i]$ as the wave function. An obvious example is in a single Slater determinant approximation in which the geminals have the form

$$\Phi_{ij}(12) = [i, j : 1, 2] \qquad (6.1)$$

where the symbol $[i, j : 1, 2]$ stands for a normalized Slater determinant consisting of the particles 1 and 2 distributed over the orbitals ψ_i and ψ_j; the corresponding weight of these geminals in the density matrix is $2/(N(N-1))$. Density matrices corresponding to configurational interaction wave functions, that is, to wave functions composed of sums of Slater determinants, may be similarly constructed using single determinantal geminals and weights which are appropriately less that $2/(N(N-1))$.

For example, using the 2-matrix

$$D^{(2)}(1, 2 : 1', 2') = \sum a_{k_1 k_2 l_1 l_2} \Phi^*_{k_1 k_2}(12) \Phi_{l_1 l_2}(1', 2') \tag{6.2}$$

and imposing the Pauli condition in the form

$$0 \leq a_{k_1 k_2 l_1 l_2} \leq 2/N(N-1) \tag{6.3}$$

the ground-state energy of a gas of interacting electrons in a solid or large molecule possessing the Hamiltonian

$$\hat{H} = \frac{1}{2} \sum_{i=1}^{N} \nabla_i^2 + \sum_{i<j} \frac{1}{|\mathbf{r}_i - \mathbf{r}_j|} - \frac{3}{4\pi r_s^3} \sum_i^N \int \frac{d\mathbf{r}_0}{|\mathbf{r}_i - \mathbf{r}_0|}$$
$$+ \frac{1}{2} \left(\frac{3}{4\pi r_s^3}\right)^2 \iint \frac{d\mathbf{r}_0 \, d\mathbf{r}_i}{|\mathbf{r}_0 - \mathbf{r}_i|} \tag{6.4}$$

may be calculated. The first two terms in the Hamiltonian are the electronic kinetic and interelectronic energy, respectively. The third term represents the interaction between the electrons and the background lattice of ions that make up the solid or molecule and is a function of the mean interparticle spacing $\mathbf{r}_s$, and the final term is the energy of interaction of these positive ions. In the case of a large molecule or solid translational invariance may be assumed, that is,

$$D^{(1)}(\mathbf{r} : \mathbf{r}') = f(\mathbf{r} - \mathbf{r}') \quad \text{where} \quad f(0) = 1$$
$$D^{(2)}(\mathbf{r}, \mathbf{r} : \mathbf{r}', \mathbf{r}') = F(\mathbf{r} - \mathbf{r}'). \tag{6.5}$$

Performing a variation calculation over the geminals and restricting the orbital basis to the first N, one obtains the Hartree–Fock result. A generalization of this technique is to employ a more general geminal basis. Young and March (1960) suggested the use of "transformed orbitals" by writing the geminal in the form

$$\phi_{k_1 k_2}(\mathbf{r}_1, \mathbf{r}_2) = [J(\mathbf{r}_1 - \mathbf{r}_2)]^{1/2}$$
$$\times \begin{vmatrix} \psi_{k_1}[\frac{1}{2}(\mathbf{r}_1 + \mathbf{r}_2) + \frac{1}{2}T(\mathbf{r}_1 - \mathbf{r}_2)] & \psi_{k_2}[\frac{1}{2}(\mathbf{r}_1 + \mathbf{r}_2) + \frac{1}{2}T(\mathbf{r}_1 - \mathbf{r}_2)] \\ \psi_{k_1}[\frac{1}{2}(\mathbf{r}_1 + \mathbf{r}_2) - \frac{1}{2}T(\mathbf{r}_1 - \mathbf{r}_2)] & \psi_{k_2}[\frac{1}{2}(\mathbf{r}_1 + \mathbf{r}_2) - \frac{1}{2}T(\mathbf{r}_1 - \mathbf{r}_2)] \end{vmatrix}$$
(6.6)

where $\mathbf{J}$ is the Jacobian of the transformation $\mathbf{T}$. By limiting the range of k_i to

$$-\tfrac{1}{2}(N-1) \leq k_1 - k_2 \leq \tfrac{1}{2}(N-1), \quad -(N-1) \leq k_1 + k_2 \leq N-1 \tag{6.7}$$

the Pauli conditions are satisfied. In this form the conditions are not convenient to apply and the restriction

$$-\tfrac{1}{2}(N-1) \leq k_1, k_2 \leq \tfrac{1}{2}(N-1) \tag{6.8}$$

is easier to use in an actual calculation, but it is not clear that it results in complete satisfaction of the Pauli conditions.

The technique has been applied to a calculation of the correlation energy of a gas of electrons in the limits of high and low densities with accurate results.

In the case of an atom or small molecule, calculations may be performed using a 2-matrix with uncorrelated geminals which is one that corresponds to a configurational wave function. The relationships between the various geminals in the density matrix become rather complicated and simplifications may be used which result in a more manageable density matrix. One such simplifying suggestion (Weinhold and Wilson, 1967) is to use trial 2-matrices derivable from configurational interaction wave functions for which each orbital is doubly occupied; that is, both spin orbitals corresponding to a particular spatial orbital are occupied. The resulting density matrix has the appearance of the 2-matrix in the Hartree–Fock approximation which has been generalized to take account of "intrashell" correlation.

C. The 2-Matrix in a Correlated Basis

Just as the natural orbitals provide the most rapidly convergent orbital basis, so do the natural geminals provide the most rapidly convergent set of correlated pair functions. Therefore accurate wave functions for atoms or molecules may be integrated to form the 2-matrix, and this matrix diagonalized to give the natural geminals and their eigenvalues. As an example of this analysis the ground state of the beryllium atom is illustrated in Table III. These results (Smith and Fogel, 1965; Kutzelnigg, 1965) are derived from an accurate 37-term configurational interaction wave function. The $1s^2$ and $2s^2$ levels are almost totally occupied, which explains the success in describing the ground state of a beryllium atom as being a $1s^22s^2$ configuration. Such analyses, which to date have been performed only for relatively simple systems, support the conventional notions of atomic and molecular structure. However, as was mentioned in the discussion of natural orbital analysis of wave functions, systems of more interesting structure to the chemist remain to be treated. The physical significance of the natural geminals is not as clear as with the case of

TABLE III

THE OCCUPATION NUMBERS OF THE NATURAL GEMINALS TO THE BERYLLIUM ATOM GROUND STATE[a]

Eigenvalue number n	Value of λ_n	Degeneracy	Label	Predicted λ_n
1	1.0085	1	$1s^2\ ^1S$	1.00000
2	0.99893	1	$2s^2\ ^1S$	1.00000
3	0.91704	1	$1s2s\ ^1S$	0.91702
4–6	0.91697	3	$1s2s\ ^3S$	0.91702
7–9	0.02700	3	$1s2p\ ^1P$	0.02699
10–18	0.02700	9	$1s2p\ ^3P$	0.02699
14	0.00064	1	$1s3s\ ^1S$	0.00066
20–22	0.00064	3	$2s3s\ ^3S$	0.00060
23–25	0.00026	3	$2s3p\ ^1P$	0.00026
26–34	0.00026	4	$2s3p\ ^3P$	0.00026
35–39	0.000065	5	$1s3d\ ^1D$	0.000066
40–54	0.000065	15	$1s3d\ ^3D$	0.000066

[a] For clarity the normalization of the density matrix is to six.

natural orbitals, for the result III.B.7 indicates that this basis does not diagonalize many of the important two-particle operators relevant to the system.

In the discussion of the 1-matrix the form of the corresponding wave function was well known. In dealing with the 2-matrix it is necessary to refer to wave functions for the N-particle system which contain, or are composed of, geminals. Since such functions have not been intensively used in theoretical chemistry, a discussion is given of some of the main forms that geminal wave functions take. In addition the properties of the particular density matrices related to such functions are included.

1. Antisymmetrized Product of Geminals

It would be an advantage to explain the properties and reactions of chemical systems in terms of their constituent molecules and the interactions between them. For example, a molecule may be formed by the interaction of two molecules A and B, possessing the ground-state wave functions $\Phi_A(r_1, r_2, \ldots, r_A)$ and $\Phi_B(r_{A+1}, \ldots, r_{A+B})$, respectively. According to this philosophy a suitable trial wave function for the new molecule would be of the form

$$\Psi(r_1, \ldots, r_{A+B}) = \hat{\mathbf{A}}\Phi_A(r_1, \ldots, r_A)\Phi_B(r_{A+1}, \ldots, r_{A+B}) \quad (6.9)$$

where $\hat{\mathbf{A}}$ is an antisymmetrization operator permuting the electronic coordinates r_1 and r_{A+1} and therefore ensuring that the wave function is totally antisymmetric. To the extent that such a function represents a good approximation to the exact wave function for the molecule, one may explain the molecular properties in terms of its constituents. If the interaction between the molecules A and B is not small, then one would expect that a good trial function should contain, in addition to the ground states present in Eq. (6.9), a sum over the excited states of the constituents.

Since theories of chemical behavior have pictured the bond between atoms and molecules as composed of electron pairs it seems reasonable to attempt to describe a wide set of chemical systems using a particular case of Eq. (6.9), the antisymmetrized product of geminals wave function

$$\Psi(1, 2, \ldots, N) = \mathcal{N}\hat{\mathbf{A}}\Phi_1(1, 2)\Phi_2(3, 4) \cdots \Phi_{N/2}(N-1, N). \quad (6.10)$$

Here $\mathcal{N}$ is a normalization constant and $\hat{\mathbf{A}}$ is the antisymmetrization operator permuting all the variables $2n, 2n+1$. The basis $N/2$ geminals are referred to, for obvious reasons, as the "generating geminals" and contain correlated functions, which may be a function of the variable r_{ij} if necessary. On forming the 2-matrix, integrals of the class

$$I(1, 2, 1', 2') = \int \Phi_1(1, 3)\Phi_2(2, 4)\Phi_1^*(1', 3)\Phi_2^*(2', 4) \, d3 \, d4 \quad (6.11)$$

must be evaluated. Therefore, in addition to the $N/2$ generating geminals the density matrix will contain at least $[N(N-1)/2] - N/2$ additional geminals formed as a result of such integrals.

2. Strongly Orthogonal Geminals

As a result of the appearance of the functions $I(1, 2, 1', 2')$ in the 2-matrix the evaluation of expectation values and matrix elements becomes difficult. It is therefore useful to impose a restriction on the "generating geminals" which greatly simplifies the resulting 2-matrix. This is the condition of strong orthogonality

$$\int \Phi_N(1, 2)\Phi_M(1, 2')\,d1 = \delta_{NM}. \tag{6.12}$$

The satisfaction of this condition is equivalent to a partitioning of the orbital basis in which the N-particle wave function is expanded such that each generating geminal is expanded in only one nonintersecting subset of this single-particle basis. In the case of a geminal that is not a function of the interelectronic separation r_{ij} the condition is not difficult to apply, but it would appear to be nontrivial for more general geminals. The 2-matrix corresponding to an antisymmetrized product of strongly orthogonal geminals wave function has the following form:

(a) Its natural spin orbitals are partitioned into $N/2$ groups S_M such that the sum of the eigenvalues in each group is $2/N$. The natural spin orbitals are natural spin orbitals to the generating geminals and the "noncorrelated" geminals. [These noncorrelated geminals are the additional geminals formed from integrals of the type (6.11).]

(b) The $N/2$ generating geminals are also natural geminals and have occupation numbers of $2/[N(N-1)]$.

(c) The remaining natural geminals are uncorrelated and have associated weights less than $2/[N(N-1)]$

$$\lambda_n^{(2)} = \lambda_{(ij)}^{(2)} = \lambda_{iN}^{(1)}\lambda_{jM}^{(1)} \quad \begin{Bmatrix} i \in S_N \\ j \in S_M \end{Bmatrix}, \quad N \neq M \tag{6.13}$$

and are of the form

$$\Phi_n(1, 2) = [i, j : 1, 2] \quad \begin{Bmatrix} i \in S_N \\ j \in S_M \end{Bmatrix}, \quad N \neq M \tag{6.14}$$

where $\lambda_{iN}^{(1)}$ is the eigenvalue of the 1-matrix for the generating geminal Φ_N, corresponding to the ith natural orbital.

(d) The 2-matrix is block diagonal, the first block corresponding to the generating geminals and containing as nonvanishing terms their eigenvalues along the diagonal. The other block contains elements corresponding to the "noncorrelated" geminals. Relaxation of the condition of strong orthogonality will destroy the block-diagonal form of the 2-matrix.

It becomes possible to perform variation calculations using a parameterized 2-matrix of the form discussed above and it is to be expected that such calculations would be superior to those of the Hartree–Fock or other simple orbital approximation. The strong orthogonality condition, however, involves a separation of the basis geminals, and it is not clear to what extent this condition represents an undue restriction upon the density matrix or wave function. In the case of a system such as the beryllium atom in its ground state or the interaction of two helium or beryllium atoms, pairs of electrons are localized in different volumes of space and the approximation is expected to be succesful. Since the occupation numbers of the geminals in the strongly orthogonal geminal approximation may be calculated using value for the natural orbitals, according to Eq. (6.13), it is possible to compare the eigenvalues to the 2-matrix predicted by the strongly orthogonal geminal (SOG) approximation with the eigenvalues of an accurate 2-matrix. Such predicted values, in the case of the ground state of the beryllium atom (Kutzelnigg, 1965), are shown in the final column of Table III. It is found that the strongly orthogonal geminal condition does not present a severe restriction in the case of such a system. Indeed in this approximation a calculation of the ground-state energy variationally obtained 90% of the correlation energy (Miller and Rudenberg, 1968) although any further improvement in this calculation would necessitate some relaxation of the SOG condition.

Such an argument will not of course apply in the case of the carbon atom, for example, whose ground state $1s^2 2s^2 2p^2$ does not contain three distinct pairs of electrons; therefore in the case of more interesting chemical systems there is evidence that the strong orthogonality condition proves excessively unphysical.

3. *Antisymmetrized Geminal Power*

A particular form of the antisymmetrized product of geminals is to employ not $N/2$ generating geminals but one,

$$\Psi(1, 2, \ldots, N) = Ag(1, 2)g(3, 4) \cdots g(N-1, N). \qquad (6.15)$$

That is, the correlated behavior of N electrons is described by a single pair function.

It should be noted that a single determinantal wave function may always be written in this form. Those AGP functions which have been generated from a geminal whose natural orbitals are all associated with the same weight produce extreme 2-matrices.

As the rank r of this geminal becomes larger, the 2-matrix corresponding to (6.15) posesses an eigenvalue λ_1 which increases, according to

$$\lambda_1 = (N-1)^{-1}[1-(N-2)/r], \qquad (6.16)$$

to the limit $\lambda_1 \to (N-1)^{-1}$. The corresponding 2-matrix has been characterized by Coleman (1965).

It will be noted that the function (6.15) has the form of a BCS (Bardeen–Cooper–Schriffer) wave function used in superconductivity theory, which has been projected upon a configurational space having constant particle number N. It is not surprising, therefore, that this function and the corresponding 2-matrix are connected with condensation phenomena. A discussion of such phenomena is given in Section VIII.

4. General Geminal Wave Function

The functions discussed above have several disadvantages. For example, in addition to the generating geminals, others appear in the density matrix as a result of the antisymmetrization procedure. Therefore the structure of the density matrix is not simply related to the wave function, which is not a desirable situation from the point of view of understanding the density matrix. In addition if one selects a particular set of geminals as generating, it is desirable to express all matrix elements of an operator under investigation in terms of this set alone. Finally, the functions discussed above all involve some approximation and selection of a set (of geminals) whose size is small compared with the total number of pairs of electrons. A wave function for which the generating geminals are identical with all geminals occurring in the 2-matrix is

$$\Psi(1, 2, \ldots, N) = \sum_{}^{\infty} a_{n_1,n_2,\ldots,n_{N/2}} \Phi_{n_1}(1,2)\Phi_{n_2}(3,4) \cdots \Phi_{n_{N/2}}(N-1, N). \qquad (6.17)$$

If $\{\Phi_n\}$ is a complete set of functions, then any N-particle wave function may be expanded in the form (6.17). The 2-matrix corresponding to this function is simply

$$D^{(2)}(1, 2 : 1', 2') = \sum B_{nm} \Phi_n^*(1, 2)\Phi_m(1', 2') \qquad (6.18)$$

where the coefficients B_{nm} are given by

$$B_{nm} = \sum_{n_2,\ldots,n_{N/2}} a^*_{n,n_2,\ldots,n_{N/2}} a_{m,n_2,\ldots,n_{N/2}}. \qquad (6.19)$$

Given a complete set of geminals, therefore, it is possible to perform variation calculations using the wave function (6.17) or the density matrix (6.18) by varying the coefficients $\{a_{n_1,n_2,...,n_{N/2}}\}$ or $\{B_{nm}\}$. It is clear that, in addition to requiring normalization, the coefficients must satisfy additional restrictions to ensure that the total wave function is antisymmetric or the 2-matrix N-representable. A discussion of these conditions is given in Section VII.

5. Wave Function of Mixed Form

A form related to Eq. (6.17) which may be of use in approximation methods is

$$\Psi(1, 2, \ldots, N) = \sum_n \Phi_n(1, 2) f_n(3, \ldots, N). \qquad (6.20)$$

Again the 2-matrix is expanded in the same geminal basis but with coefficients

$$B_{nm} = \langle f_n | f_m \rangle. \qquad (6.21)$$

The $\{f_n\}$ are chosen to normalize the wave function and to ensure total antisymmetry; that is, they must be antisymmetric functions that satisfy

$$-f_n(3, \ldots, N) = \sum_m \int \Phi_n^*(1, 2) \Phi_m(1, 3) f_m(2, \ldots, 4)\, d1\, d2. \qquad (6.22)$$

If the $\{f_n\}$ are expanded in Slater determinants, then the coefficients may be chosen to minimize the energy, or the expectation value of some other operator, and satisfy Eq. (6.22). In the event that the expansion is not sufficiently large to satisfy all conditions, then the calculated result lies above a known lower bound to the energy.

D. The Grimley–Peat Approximation

The approximation discussed in this subsection expresses a 2-matrix built up from correlated pair functions much in the way that early theories of atomic structure attempted to build up atomic configurations by feeding electrons into orbitals. It is quite successful in treating atomic energy levels and provides a useful starting point for some extended approximation scheme.

The approximation owes its origins to an idea by Bopp (1959). Recalling that in a basis of eigenfunctions to the reduced Hamiltonian the energy expectation value becomes

$$E = (N/2)\, \text{tr}\, \hat{h}\hat{D} = (N/2) \sum B_{nn} \varepsilon_n, \qquad (6.23)$$

he suggested that a lower bound to the energy could be obtained. At the time he supposed, erroneously, that the upper bound to the eigenvalues of the 2-matrix was $2/[N(N-1)]$, and consequently replaced the terms in the sum (6.21) with the first $[N(N-1)]/2$ lowest energies. The "bounds" obtained lie slightly below the exact energies for a series of three-electron ions, the success of which is encouraging as an approximation method. However, as the number of electrons in the system increases the "bounds" fall further below the exact energies and cease to be of practical use. These are given in Table IV.

TABLE IV

GROUND-STATE ENERGIES OF ATOMIC IONS CALCULATED IN THE BOPP APPROXIMATION

Ion	Number of electrons	Experimental energy (cm^{-1})	"Bopp's approximation" (cm^{-1})
Be$^+$	3	$-3{,}144{,}100$	$-3{,}171{,}037$
C^{3+}	3	$-7{,}635{,}577$	$-7{,}654{,}244$
O^{5+}	3	$-14{,}104{,}969$	$-14{,}107{,}382$
C^{2+}	4	$-8{,}020{,}796$	$-8{,}969{,}757$
O^{3+}	5	$-15{,}648{,}668$	$-19{,}998{,}224$
Ne^{4+}	6	$-26{,}524{,}816$	$-36{,}283{,}775$
Mg^{5+}	7	$-41{,}115{,}222$	$-59{,}599{,}668$

While all calculated values, in the case of atoms, lie below the true ground-state energy, it must not be assumed that the method just mentioned *always* yields a lower bound for any many-body problem (except in the case $N=3$ in which $2/[N(N-1)]$ coincides with the exact upper bound to the eigenvalue, $1/N$.)

If the correct upper bounds to the eigenvalues of the 2-matrix are inserted into Eq. (6.21), the following exact bounds are obtained:

$$E \geq \tfrac{1}{2} \sum_{n=1}^{N} \varepsilon_n, \qquad N \text{ odd} \qquad (6.24)$$

$$E \geq [N/2(N-1)] \sum_{n=1}^{N-1} \varepsilon_n, \qquad N \text{ even}. \qquad (6.25)$$

The values corresponding to these bounds fall below those of the "Bopp approximation" given in Table IV. While these results are true lower bounds upon the energy, unlike the results first obtained by Bopp, many superior lower bound methods exist in the literature.

Inspection of Table IV shows the Bopp method to be quite successful in the case of three-electron ions and it would be attractive to extend it to more general systems. Obvious improvements such as requiring the B_{nn} to satisfy the spin sum rules [Eqs. (3.34) and (3.35)], do not result in any significant increase in the calculated energy. It is not difficult to see why such an approximation does not work. Reference to Table III shows that the $2s^2$ level is occupied in a four-electron atom. That is, the electrons in their orbitals obey the Pauli restriction and this is reflected in the occupation number of the geminals. It is therefore necessary for doubly excited geminals to appear in the 2-matrix of atoms, molecules, and solids but since such levels are associated with large energies they will never be included in a scheme which demands the inclusion of the first few lowest energy levels. In fact the doubly excited levels lie in the continuum of the reduced Hamiltonian; that is, they correspond to autoionizing levels (see Peat, Volume XIB, Chapter 12) and an infinite number of bound states lie below them; any approximation which is to prove successful must include them.

Grimley and Peat (1965) introduced the doubly excited levels into the density matrix in a simple fashion. The nature of their approximation is to expand the 2-matrix in a set of correlated geminals but using "zero-order coefficients" $B_{nm}^{(0)}$, that is, the coefficients that one would obtain using a wave function of the form

$$\Psi^0(x, y) = \sum_n \Phi_n^{(0)}(x) f_n^{(0)}(y), \qquad (6.26)$$

the eigenfunction of some zero-order Hamiltonian. Such functions may be expressed in terms of determinantal functions, in the case of a closed shell by a single Slater determinant and in other cases by the appropriate linear combination to give rise to the correct spin. Normalized Slater geminals g_k and functions of $(N-2)$ particles, U_{K-k}, may be projected from a determinant D_K

$$D_K = \{2/[N(N-1)]\} \sum_{k \in K} g_k U_{K-k} \qquad (6.27)$$

and these g_k expanded in terms of the zero-order eigenfunctions of the reduced Hamiltonian

$$g_k = \sum \phi_n^{(0)} a_{nk}. \qquad (6.28)$$

In terms of these expansion coefficients the $f_n^{(0)}$ are

$$f_n^{(0)} = \{2/[N(N-1)]\}^{1/2} \sum_{k \in K} a_{nk} U_{K-k}. \tag{6.29}$$

The weights of the 2-matrix corresponding to the wave function (6.26) are therefore

$$B_{nm}^{(0)} = \{2/[N(N-1)]\} \sum_{k \in K} |a_{nk}|^2. \tag{6.30}$$

The Grimley–Peat density matrix is expanded in a basis of the exact eigenfunctions of the reduced Hamiltonian but using these zero-order coefficients:

$$\mathbf{D}(x, x') = \sum B_{nm}^{(0)} \Phi_n{}^*(1, 2) \Phi_m(1', 2'). \tag{6.31}$$

That is, the required doubly excited levels are included in the density matrix and the success of the Bopp approximation is continued to larger systems.

Construction of a 2-matrix in this fashion ensures that the Pauli conditions are fulfilled. [It has been suggested (Kijewski and Percus, 1969, 1970) that the Grimley–Peat density matrix does not fulfill these conditions. However, the matrix referred to by Kijewski and Percus is a form of Bopp's density matrix modified by the spin sum rules, Eqs. (3.34) and (3.35), and not the Grimley–Peat 2-matrix. Examination of the numerical data referred to by these authors indicates that the Grimley–Peat density matrix does indeed satisfy the Pauli restrincions.]

Table V lists some calculations of the ground-state energies of atoms and ions calculated using the 2-matrix (6.31) that is, using

$$E = (N/2) \sum B_{nn}^{(0)} \varepsilon_n. \tag{6.32}$$

It will be noted that the calculated energies all lie below the exact non-relativistic energies. This is not surprising since a density matrix of the form (6.31) may not be N-representable. In the case of atomic calculations the Grimley–Peat approximation yields results which tend to overstimate the correlation energy, but it should be noted that the method does not necessarily yield a lower bound to the energy of all systems. For example, in a condensed system the Fermi surface has become unstable, resulting in a very large weight being associated with a single pair energy; energies corresponding to such cases may be expected to fall below the Grimley–Peat energy. However, as far as normal systems are concerned, the approximation may be expected to be consistently successful.

TABLE V

GROUND-STATE ENERGIES OF ATOMIC IONS CALCULATED IN THE GRIMLEY–PEAT APPROXIMATION

Ion	Number of electrons	Experimental energy (cm^{-1})	Grimley–Peat approximation
Be$^+$	3	−3,144,100	−3,177,462
C^{2+}	4	−18,020,796	−8,131,386
Ne^{4+}	6	−26,524,816	−26,922,230
Sc^{6+}	8	−59,675,557	−60,569,922
A^{8+}	10	−111,144,348	−113,042,358

The advantage of the method does not lie in any exceptional numerical accuracy but in its simplicity. It may also be applied to calculate a series of atomic excitation levels. Provided the appropriate pair energies ε_n are known, the excitations of a system may be quickly and accurately determined (Peat and Brown, 1966).

It would be of interest to apply this technique to the calculation of molecular energies. For example, if a set of two electron energies is calculated, and parameterized by nuclear charges and internuclear separations, the configurations of the homonuclear diatomic molecules could be built up. The bonding properties of these molecules would be expressed in terms of the relative weights of various molecular geminals. Preliminary calculations (Peat, 1967a; Bender *et al.*, 1968) indicate that such a program is worth pursuing.

The numerical success of the Grimley–Peat approximation lies in the use of correlated functions with approximately correct weights. Investigation indicates that the method is exact in first order and includes contributions from all higher orders of a perturbation sum in powers of the nuclear charge. It will also be noted that in the case $N = 3$ the method coincides with the modified Bopp approximation and yields an exact lower bound to the energy.

The Grimley–Peat approximation would prove a useful starting point for a perturbation method or some other approximation scheme. An example of such a scheme is to attempt a more accurate estimate of the correlation energy. The bonding energy of the lithium molecule is only

0.2% of the total energy of the molecule and it is therefore a severe test of an approximation scheme, even to determine the molecule as bound. In the Grimley–Peat approximation the molecule is found to be stable with a total energy of -15.2516 a.u. which overestimates the experimental value of -14.97 a.u. Wave functions of the form (Bender et al., 1968)

$$\Psi = a_0 \mid u_1\bar{u}_1 \cdots u_{N/2}\bar{u}_{N/2} \mid + \sum_{k=1}^{N/2} \sum_{l>N} a_{lk} \mid u_1\bar{u}_1 \cdots u_{N/2}\bar{u}_{N/2}; (u_l\bar{u}_l/u_k\bar{u}_k) \mid \quad (6.33)$$

are found to give accurate results. Here the second term in brackets is a Slater determinant in which the orbitals u_k ($=u_k\alpha$) and $\bar{u}_k$ ($=u_k\beta$) have been replaced by u_l and $\bar{u}_l$. Such a function suggests that the approximation

$$E = (N/2) \sum B_{nn}^{(0)} \varepsilon(\text{SCF}) + [N/2(N-1)] \sum B_{nn}^{(0)} \varepsilon_n \text{ (correlation)} \quad (6.34)$$

should be an improvement upon Eq. (6.32). Here the two-particle eigenfunctions of the reduced Hamiltonian ε_n have been divided into two parts

$$\varepsilon_n \text{ (correlation)} = \varepsilon_n - \varepsilon_n(\text{SCF}), \quad (6.35)$$

where $\varepsilon_n(\text{SCF})$ is the Self-Consistent Field or Hartree–Fock energy. Under this approximation the total energy of the system lies within 0.3% of the experimental energy and the bonding energy is 0.087 a.u., the correct order of magnitude when compared with the exact result of 0.03 a.u. Therefore an improvement in the Grimley–Peat method can be made without considerable complication of the calculation.

E. Hamiltonian-Dependent Conditions

Characterization of the full density matrix is made [Eqs. (2.3) and (2.4)] in terms of the expectation values of a series of operators. In the case of reduced density matrices much attention has been directed to the N-representability problem. However, there will also be a series of restrictions on trial p-matrices related to the operators which characterize the system. One such set of conditions is illustrated by Eq. (3.30) where the eigenvalue to an operator is known.

Condition which involve the Hamiltonian for the system have been examined and two examples are given below:

(a) Conditions may be imposed on the 2-matrix with the aid of a model Hamiltonian $\hat{H}_{\text{model}}$ (Kijewski and Percus, 1967), referring to a system with Hamiltonian $\hat{\mathbf{H}}$. If $\{\hat{\mathbf{D}}^{(2)}\}$ is varied over a wider class than the N-representable 2-matrices, one finds

$$\left(\min_{\hat{\mathbf{D}}^{(2)}}\right)(N/2)\,\text{tr}\,\hat{\mathbf{h}}^{(2)}\hat{\mathbf{D}}^{(2)} \leq E_0 \tag{6.36}$$

where E_0 is the ground-state energy of the N-particle system. In general this lower bound is too low to be of interest. Suppose, however, that the variation is performed over those 2-matrices $\{\hat{\boldsymbol{\Gamma}}^{(2)}\}$ which satisfy

$$(N/2)\,\text{tr}\,\hat{\mathbf{h}}_{\text{model}}\hat{\boldsymbol{\Gamma}}^{(2)} \geq E_0^{\text{model}}. \tag{6.37}$$

In this case

$$\left(\min_{\boldsymbol{\Gamma}^{(2)}}\right)(N/2)\,\text{tr}\,\hat{\mathbf{h}}^{(2)}\boldsymbol{\Gamma}^{(2)} \leq E_0 \tag{6.38}$$

may prove to be a more useful bound. If one neglects the possibility of pathological behavior and minimizes Eq. (6.38) subject to the inequality (6.37), then it may be supposed that the lower bound approaches the exact ground-state energy of the system as $\hat{\mathbf{H}}_{\text{model}}$ approaches $\hat{\mathbf{H}}$. In the case that equality in Eq. (6.37) is exactly fulfilled, then one has, in essence, performed a first-order perturbation calculation on the operator $(\hat{\mathbf{H}} - \hat{\mathbf{H}}_{\text{model}})$.

(b) The Virial theorem, that the sum of the potential and twice the kinetic energy vanishes for a system bound by Coulombic forces alone, may be applied as a condition the 2-matrix (Kijewski and Percus, 1969). Defining the operator

$$\hat{\mathbf{h}}'(1, 2) = 2\hat{\mathbf{h}}(1, 2) - \hat{\mathbf{H}}(1) - \hat{\mathbf{H}}(2) \tag{6.39}$$

the Virial theorem implies

$$\text{tr}\,\hat{\mathbf{h}}'\mathbf{D}^{(2)} = 0. \tag{6.40}$$

An alternative form of the condition may be derived as

$$\text{tr}_2\,[\hat{\mathbf{h}}'(1, 2), \hat{\mathbf{D}}(1, 2 : 1', 2')] = 0. \tag{6.41}$$

Equations (6.39) and (6.40) are derived using the properties of a totally antisymmetric wave function after the manner of Eq. (3.7). It is unfortunate that condition (6.41) may be accidentally satisfied by non-N-representable 2-matrices which do not satisfy the Virial theorem.

In terms of the eigenvalues ε_n of $\hat{\mathbf{h}}$ and the coefficients B_{nm} of the 2-matrix, Eq. (6.41) may be cast in the form

$$\sum B_{nm}(\varepsilon_n - \varepsilon_m) D_{nm}^{(1)}(1, 1') = 0. \tag{6.42}$$

$D_{nm}^{(1)}$ is the transition matrix of geminals

$$D_{nm}^{(1)}(1 : 1') = \int \Phi_n^*(1, 2) \Phi_m(1', 2) \, d_2. \tag{6.43}$$

Since these transition matrices are linearly related and the B_{nm} are not all positive, then the satisfaction of Eq. (6.42) by a trial 2-matrix is nontrivial.

VII. General Geminal Wave Functions

The statistics of pairs of electrons between geminal functions in a many-body system is both complicated and interesting. The difficulties of the situation are exhibited in the problem of N-representability of the second-order density matrix and the interrelationship of the geminals and their occupation numbers. Also if one wishes to apply any of the modern techniques of the many-body problem to a correlated pair function basis, then it will be necessary to work with pair creation and annihilation operators.

In the geminal basis $\{\Phi_n\}$ pair creation and annihilation operators may be defined; for example, the creation of an electron pair in the geminal Φ_n is performed with

$$A_n^+ = (1/\sqrt{2}) \int_{1,2} \Phi_n(1, 2) \psi^+(1) \psi^+(2). \tag{7.1}$$

The commutation relations

$$[A_n, A_m] = [A_n^+, A_m^+] = 0 \tag{7.2}$$

$$[A_n, A_m^+] = \delta_{nm} - 2 \int_{123} \Phi_n^*(1, 2) \Phi_m(1, 3) \psi^+(3) \psi(2) \tag{7.3}$$

indicate that the operators are not independent as is the case with single-particle creation and annihilation operators. Such dependent creation and annihilation operators are increasingly used in solid-state physics and chemistry where composite states appear as the elementary excitations of the system. Provided the concentrations of these excitations in the

7. Density Matrices

background fermion sea are very low, the second term on the right-hand side of an equation of the form (7.3) may be neglected. In such a case the elementary excitations are treated as "quasi bosons." However, it is important to derive a systematic treatment of composite excitations and electron pairs at all densities. In the present case the electron pairs are not considered as excitations out of a single-particle "sea" or "vacuum" but form the coordinate basis, as it were, of the representation.

Investigation of these problems is best begun with a consideration of the general geminal wave function introduced in Section VI (Girardeau, 1963; Peat and Brown, 1967). It will be recalled that the permutational symmetry (whether symmetric, antisymmetric, or some other representation) of the state function is governed by the coefficients $\{a_{n_1 n_2 \cdots n_{N/2}}\}$, where

$$\Psi(1, 2, \ldots, N) = \sum a_{n_1 \cdots n_{N/2}} \Phi_{n_1}(1, 2) \Phi_{n_2}(3, 4) \cdots \Phi_{n_{N/2}}(N-1, N). \tag{7.4}$$

In the case of fermions or bosons the coefficients $\{a_{n_1, n_2, \ldots, n_{N/2}}\}$ are symmetric with respect to interchange of index. To obtain the full permutational symmetry the following conditions must also be fulfilled

$$a^N_{n_1 n_2 \cdots n_{N/2}} = \varepsilon \sum_{m_1, \ldots, m_{N/2}} a^N_{m_1, \ldots, m_{N/2}} Q^N_{n_1, \ldots, n_2, m_1, \ldots, m_2},$$

$$\varepsilon = +1 \text{ for bosons}, \quad \varepsilon = -1 \text{ for fermions} \tag{7.5}$$

where

$$Q^N_{n_1 \cdots n_2 m_1 \cdots m_2} = \delta_{n_3 m_3} \cdots \delta_{n_{N/2} m_{N/2}}$$
$$\times \int_{1,\ldots,4} \Phi^*_{n_1}(1, 2) \Phi^*_{n_2}(3, 4) \Phi_{m_1}(1, 3) \Phi_{m_2}(2, 4). \tag{7.6}$$

These conditions clearly illustrate the interrelationship between the coefficients and the choice of basis. Since the weights B_{nm} of the 2-matrix are derived by contraction from the $\{a_{n_1, n_2, \ldots, n_{N/2}}\}$ it is seen that these weights are also interrelated and dependent on the choice of basis. Some of the difficulty of the N-representability problem is a result of this interconnection.

The indices of the coefficients may be ordered so that the $\{a_{n_1 \cdots n_{N/2}}\}$ may be represented by a vector $\mathbf{A}^N$; wave functions of the correct symmetry may therefore be characterized by eigenvectors of a matrix $\mathbf{Q}^N$ having associated eigenvalue ε. The eigenvectors of $\mathbf{Q}^N$ are highly degenerate since any wave function may be expanded as a suitable linear combination of them, provided that the geminal basis is complete.

Eigenvalues of $\mathbf{Q}^N$ lie between -1 and $+1$, the two extreme values corresponding to antisymmetric and symmetric particle functions, respectively. The other eigenvectors have a physical significance in that they characterize different representations of the permutation group; that is, they characterize the spatial part to a wave function for some particular spin symmetry and value of N.

Once given a geminal set, the $\mathbf{Q}^N$ matrix may be calculated, diagonalized, and its eigenvectors determined and in this sense a trivial solution to the N-representability problem is given. That is, the geminals of some 2-matrix may be used to calculate the matrix $\mathbf{Q}^N$. Once the eigenvectors corresponding to the appropriate eigenvalues are obtained they may be partitioned into sets according to the value N, into those which are internally symmetric when divided into N segments. It then remains to determine if the B_{nm} of that 2-matrix are contactable from one of the sets.

While this is a solution to the N-representability problem, it is equivalent to the construction of a wave function and therefore removes all advantages in working with the density matrix formalism. The form of the N-representability conditions desired in practice would be of constraints to be applied to the reduced density matrix.

In selecting the appropriate eigenvectors of $\mathbf{Q}^N$ from the set having $\varepsilon = \pm 1$, one may also be guided by certain "physical constraints." For example, the wave function characterized should have the correct symmetry properties with respect to rotations and translations; restrictions on total spin and other quantum numbers may be imposed once it is recognized that one presupposes a certain amount of information about the particular wave function desired. That is, eigenvectors of $\mathbf{Q}^N$ are selected which are simultaneous eigenvectors of a number of other operators and associated with desired eigenvalues.

Knowledge of the $\{a_{n_1,\ldots,n_{N/2}}^N\}$ coefficients also enables one to construct pair electron creation and annihilation operators which have desirable commutation properties.

A set of operators α_n, α_n^+ is introduced that satisfy Bose commutation laws

$$[\alpha_n, \alpha_m] = [\alpha_n^+, \alpha_m^+] = 0 \tag{7.7}$$

$$[\alpha_n, \alpha_m^+] = \delta_{nm}. \tag{7.8}$$

They operate on an ideal state space (Dyson, 1956) whose vectors are one-to-one correspondence with those defined by the $\{\mathbf{A}^N\}$. An N-particle wave function would therefore be

$$\Psi(1, 2, \ldots, N) = \sum_{n_1,\ldots,n_{N/2}} a_{n_1\cdots n_{N/2}} \alpha^+_{n_1} \alpha^+_{n_2} \cdots \alpha^+_{n_{N/2}} \mid 0\rangle \tag{7.9}$$

where $\mid 0\rangle$ is the vacuum of the ideal state space.

In the basis of eigenfunctions ε_n to the reduced Hamiltonian the N-particle Hamiltonian becomes

$$\hat{\mathbf{H}} = \sum \varepsilon_n \alpha_n^+ \alpha_n. \tag{7.10}$$

Similarly, a $\mathbf{Q}$ matrix operator may be defined:

$$\hat{\mathbf{Q}} = \sum_{n_1, n_2, m_1, m_2} Q^4_{n_1, n_2, m_1, m_2} \alpha^+_{n_1} \alpha^+_{n_2} \alpha_{m_1} \alpha_{m_2}. \tag{7.11}$$

The many-body problem, in the case of a geminal basis, is the determination of simultaneous eigenvectors of $\hat{\mathbf{H}}$, $\hat{\mathbf{Q}}$, and the number operator $\hat{\mathbf{N}} = \sum_n \alpha_n^+ \alpha_n$.

It should be emphasized that in working with "boson" creation and annihilation operators for particle pairs, the operators must be defined on an ideal state space. That is, the subsidiary conditions (7.5) must be fulfilled. It has been pointed out (Girardeau, 1963) that in certain investigations (Blatt and Matsubara, 1958) these conditions are not satisfied.

Since the construction of an N-representable 1-matrix requires constraints only on its eigenvalues, while the conditions on the 2-matrix involve interrelationships between the eigenvalues and natural geminals, it may be inquired if conditions can be set on the $\mathbf{Q}$ matrix elements alone such that a symmetric, or antisymmetric, wave function is always characterized. Further is it possible to derive useful necessary conditions that are independent of the geminal basis set used?

General properties of the matrix $\mathbf{Q}$ follow from the properties of general antisymmetric or symmetric pair functions. These take the form of bounds, inequalities, and symmetries relating the matrix elements (Peat, 1967b). With the aid of this information some characterization of the eigenvectors may be made and many of the properties of the p-matrices derived—for example, theorems concerning rank and bounds upon the eigenvalues. However, it will be noted that in discussing the general properties of the matrix elements independent of the basis, use is made only of the operations $P_{2n-1,2n}$ of the permutation group and therefore sufficient information to characterize a completely antisymmetric wave function of an N-representable 2-matrix is not forthcoming.

It is apparent therefore that conditions on the **Q**-matrix which ensure that some of the eigenvalues of the **Q**-matrix characterize symmetric (antisymmetric) wave functions must involve the basis set of geminals. The notion of N completeness has been introduced for the basis set (Ruskai, 1969, 1970); this is a condition on the basis, and therefore on **Q**, that antisymmetric (symmetric) wave functions may be characterized.

Since the calculation of all elements of the **Q**-matrix is tedious it would be desirable to limit their calculation by making certain approximations for the form of **Q**. For example, in the case of the geminal wave functions discussed in Section VI characteristic forms result. The antisymmetrized product of strongly orthogonal geminals (APSOG) wave function is a class for which the matrix **Q** reduces to block form with a consequent reduction in the number of nonvanishing matrix elements. A number of such forms have been classified together with suggested approximation methods (Peat, 1971).

In the simple case of a three-electron atom the **Q**-matrix has been obtained, diagonalized, and the 2-matrices for the ground and first two excited states calculated (Peat, 1968). Since a complete set of geminals was not employed in the calculation, the resulting wave function was not antisymmetric; that is, it was characterized by an eigenvector of **Q** which corresponded to an eigenvalue of $-(1-\lambda)$, where λ is small. Such a situation does not present a special problem since the value of λ determines a bound upon the energy calculated using this method; that is, wave functions which are only approximately antisymmetric, or symmetric, may be used in variational calculations (see also Simmons 1972).

Given a set of creation and annihilation operators for particle pairs one may discuss the elementary excitations of a system, in addition to its state functions. The difficulties inherent in the introduction of an ideal state space in the definition of these operators are illustrated when one wishes to describe simple excitations. In a single-particle framework the excitations are independent and may be represented by the annihilation of a particle in one state and the creation of a particle in another, higher, level. In the case of composite particles the excitations are no longer independent but interrelated. For example, an excitation of the beryllium atom $1s^22s^2$ ^1S–$1s^22s3s$ ^1S may be represented, in the Hartree–Fock approximation, by the notation 2s–3s. That is, an electron is excited from the 2s to the 3s orbital. In the case of electron pairs the excitation of the atom, in the Grimley–Peat approximation, is accompanied by the simultaneous promotions 1s2s ^{1}S–1s3s ^{1}S, 1s2s ^{3}S–1s3s ^{3}S, and 2s^2 ^{1}S–2s3s ^{1}S.

In the single-particle picture the excitation energy is approximately equal to the energy difference between single-particle levels; that is, the elementary excitation of the system may be identified with a single particle, or dressed particle. In the case of a composite-particle representation the elementary excitations of the composite particles themselves are not the elementary excitations of the system; rather a linear combination of such excitations gives the total excitation energy. Some simple excitations in the case of atoms are given in Table VI.

TABLE VI

EXCITATION ENERGIES OF ATOMIC IONS CALCULATED IN THE GRIMLEY–PEAT APPROXIMATION

Ion	Excitation	Energy (cm^{-1})	
		Experimental	Calculated
Be$^+$	1s^{2}2s–1s^{2}3s^{2}s	88,730	88,231
	1s^{2}2s–1s^{2}4d ^{2}D	119,966	114,422
	1s^{2}2s–1s^{2}8d ^{2}D	140,534	140,020
C^{3+}	1s^{2}2s–1s^{2}3s^{2}S	303,563	302,848
	1s^{2}2s–1s^{2}4p^{2}P	409,027	408,313
	1s^{2}s–1s^{2}6s^{2}S	469,400	468,765
	1s^{2}2s–1s^{2}8f ^{2}F	493,388	492,743

(The single-particle and composite-particle pictures of excitations are of course related. Excitations in an N-particle system involve the cooperative excitation of the order of $N/2$ pairs of particles; in a normal system the pair functions are weighted at the order of $2/(N^2)$ and therefore the weight associated with an elementary energy change in the system is $(N/2) \times (2/N^2)$, that is, $1/N$. In the single-particle picture the single elementary excitation is of course weighted at the order of $1/N$.)

The statistical behavior of electron pairs in geminal functions and, indeed, of more general composite particles may therefore be obtained from the ideal state space method outlined in this section. Composite particles generally give rise to statistics which are more complicated than Fermi-, Dirac-, or even Parastatistics.

VIII. Condensation Phenomena

A. INTRODUCTION

In this section the properties and applications of density matrices to the case of condensed systems that exhibit long-range correlated behavior are discussed. Under the title of correlated systems or long-range order one normally groups superfluidity, superconductivity, ferromagnetism, excitonic condensation, and the like. Such phenomena may not seem of direct interest to the physical chemist and a consideration of condensation rather out of place in a book of this nature. However, there is every likelihood that these ideas will be of increasing importance in the study of macromolecules and biological molecules. It is thought that some of the excitations in these molecules may exhibit correlations over macroscopic distances and it will therefore become necessary for the physical chemist to familiarize himself with theories of condensation. (Besides such a pragmatic excuse for introducing this section must be placed the observation that the erection of barriers between the sciences acts to the impoverishment of its practitioners. It should be the policy of the theoretical chemist to acquire the interesting and powerful tools of the theoretical physicist and of the physicist to acquaint himself with the considerable theoretical difficulties presented by chemical systems.)

A characteristic property of condensed systems is the ordering that persists over macroscopic distances. If one believes the elementary entities of the condensed system which are responsible for this ordering to be single-particle states, then the following property of the 1-matrix is exhibited:

$$\mathbf{D}^{(1)}(\mathbf{x}:\mathbf{x}') \to \text{constant} \quad \text{as} \quad |\mathbf{x}-\mathbf{x}'| \to \infty. \tag{8.1}$$

To bring our notation into conformity with that of Yang (1962), whose arguments on long-range order are given below, the 1-matrix is normalized to N and the 2-matrix to $N(N-1)$. Under periodic boundary conditions the 1-matrix may be written

$$\mathbf{D}^{(1)}(\mathbf{x}:\mathbf{x}') = (1/\Omega) \sum n_\mathbf{k} \exp(i k(\mathbf{x}-\mathbf{x}')), \tag{8.2}$$

and one has the following bound on the n_k:

$$n_\mathbf{k} \leq N \times (1/N) \leq 1 \quad \text{for fermions} \tag{8.3}$$

$$n_k \leq N \times 1 = N \quad \text{for bosons.} \tag{8.4}$$

7. Density Matrices

That is, for fermions

$$\mathbf{D}^{(1)}(x:x') \to 0 \quad \text{as} \quad |x-x'| \to \infty, \tag{8.5}$$

and no longer range ordering is possible in a fermionic system in which the elementary entities are single particles. For a boson system below the Bose–Einstein transition temperature one finds that $n_0 = \alpha N$, where α is referred to as the condensed fraction, and the corresponding density matrix possesses long-range ordering, for

$$\mathbf{D}^{(1)}(\mathbf{x}:\mathbf{x}') \to N\alpha/\Omega \quad \text{as} \quad |x-x'| \to \infty. \tag{8.6}$$

That is, there is a relationship between Bose–Einstein condensation and off-diagonal long-range order (ODLRO) in the 1-matrix.

Therefore condensation and long-range ordering in a boson system are associated with a large eigenvalue to the 1-matrix, ODLRO in the 1-matrix, and Bose–Einstein condensation. This ordering is also exhibited in all higher order density matrices; for example, the 2-matrix for a Bose condensed system is found to possess the following properties:

$$\mathbf{D}^{(2)}(\mathbf{x}_1, \mathbf{x}_2 : \mathbf{x}_1', \mathbf{x}_2') \to \text{constant} \quad \text{as} \quad x_1 \to \infty,\ x_2 \to \infty,\ x_1' \to \infty,\ x_2' \to \infty, \tag{8.7}$$

associated with an eigenvalue of the order of N^2 to $\mathbf{D}^{(2)}(\mathbf{x}_1, \mathbf{x}_2 : \mathbf{x}_1', \mathbf{x}_2')$.

Supposing that in the case of fermions, the elementary entities in the system are not single particles but particle pairs, it is then possible to have ODLRO in the 2-matrix without the corresponding ODLRO in the 1-matrix. In fact one finds that if an eigenvalue to the 2-matrix approaches its upper bound of N for fermions, then the 2-matrix is nonvanishing in the neighborhood of

$$\mathbf{x}_1 = \mathbf{x}_1' \quad \text{and} \quad \mathbf{x}_2 = \mathbf{x}_2' \tag{8.8}$$

$$\mathbf{x}_1 = \mathbf{x}_2' \quad \text{and} \quad \mathbf{x}_2 = \mathbf{x}_1' \tag{8.9}$$

$$\mathbf{x}_1 = \mathbf{x}_2 \quad \text{and} \quad \mathbf{x}_1' = \mathbf{x}_2', \tag{8.10}$$

and the individual coordinates may take on macroscopic separation.

Therefore macroscopic ordering of a fermionic system is possible if the elementary entities within the system are pairs of fermions and the corresponding 2-matrix exhibits a large eigenvalue with ODLRO. Clearly the term "Bose–Einstein condensation" may not be applied to this situation; the term "fermionic condensation" has been suggested.

Examples of ODLRO in the 2-matrix but not in the 1-matrix are superconductivity and excitonic condensation; an example of ODLRO in the 1-matrix is superfluidity. (It should be noted that several confusing discussions have appeared in the literature. These consider superconductivity or excitonic condensation as Bose–Einstein condensations and, by analogy, as superfluidity of the electron gas. Such statements are incorrect, the actual situation being more subtle.)

B. Superconductivity

The theory of superconductivity is well established; the lattice-mediated interactions between pairs of electrons become attractive for certain relative momenta. That is, "Cooper pairs" of electrons form which are stable, with respect to the Fermi sea, to disruption. The associated framework of Bardeen, Cooper, and Schrieffer utilizes a state function which, when projected onto a configurational space of constant particle number, is of the form

$$\Psi = \hat{\mathbf{A}} g(1, 2) g(3, 4) g(5, 6) \cdots g(N - 1, N). \tag{8.11}$$

This function is the AGP function that has been studied by Coleman (1965). The 2-matrix corresponding to this function has been characterized and, in the case of those geminals which have a 1-matrix possessing equal eigenvalues, corresponds to an extreme point of the convex set of N-representable density matrices. The eigenvalue of this extreme 2-matrix is

$$\lambda = N[1 - (N - 2)/r], \tag{8.12}$$

and for high 1-rank r this approaches N. Therefore a consequence of the BCS wave function may be a large eigenvalue to the 2-matrix with attendant macroscopic ordering of the system. It will be noted, however, that not all pairing of electrons, regardless of whether it can be expressed in the form of the function (8.11), leads to condensation and superfluidity—only those cases in which $g^{(12)}$ posesses a large natural orbital expansion with equal expansion coefficients. The Fourier transform of such a geminal represents an electron pair localized in momentum space, a Cooper pair.

In passing, it may be mentioned that the statistics of the electron pairs corresponding to the function (8.11), or to more general forms, is quite complicated. Some indication of the difficulties involved was given in Section VII. In the case of excitons whose concentration in a solid has reached such a proportion that interexcitonic interaction is no longer negligible, similar complications are to be expected. The treatment of excitonic condensation may be expected to be analogous to that for superconductivity, if the exciton is not associated with a particular lattice site, and more complicated otherwise.

C. Superfluidity

In the case of a free boson gas the occupation number of one of the eigenfunctions of the 1-matrix may approach N; that is, the condensed fraction is unity and perfect Bose–Einstein condensation occurs. In realistic systems this may never occur, since free boson states are no longer natural orbitals of the system. Therefore as a result of interactions, a number of excited states must be included in the 1-matrix, with non-vanishing weight, in the basis set of free boson states. In the case of the ground state of helium(II) it is believed that the condensed phase is not negligibly occupied, which suggests that the 1-matrix be factorized in the following manner:

$$\mathbf{D}^{(1)}(\mathbf{x}:\mathbf{x}') = \mathbf{\Phi}_1^*(\mathbf{x})\mathbf{\Phi}_1(\mathbf{x}') + \mathbf{\Lambda}^{(1)}(\mathbf{x}:\mathbf{x}') \qquad (8.13)$$

where $\mathbf{\Lambda}^{(1)}(\mathbf{x}:\mathbf{x}') \to 0$ when $|\mathbf{x} - \mathbf{x}'|$ becomes greater than a characteristic length for the system. $\mathbf{\Phi}_1$ therefore represents the macroscopically condensed part of the system and $|\mathbf{\Phi}_1|^2/\mathrm{tr}\,\mathbf{D}^{(1)}$ is therefore the condensed fraction α. A hierarchical set of density matrices can be derived for the p-matrices (Frochlich, 1969)

$$\hat{\mathbf{L}}_p \hat{\mathbf{D}}^{(p)} = \hat{\mathbf{W}}^{(p)} \qquad (8.14)$$

where

$$\hat{\mathbf{L}}_p = ih\,\partial_t + (h^2/2m) \sum_{i=1}^{p} \left\{(\partial_{\mathbf{x}_i}^2 - \partial_{\mathbf{x}_i'}^2) \right. $$
$$\left. - \sum_{i<j} (\hat{\mathbf{V}}(|\mathbf{x}_i - \mathbf{x}_j|) - \hat{\mathbf{V}}|\mathbf{x}_i' - \mathbf{x}_j'|)\right\}. \qquad (8.15)$$

$\hat{\mathbf{W}}^{(p)}$, which occurs on the right-hand side of Eq. (8.14), is a function of the $(p+1)$-matrix

$$\hat{\mathbf{W}}^{(p)}(\mathbf{x}_1, \ldots, \mathbf{x}_p : \mathbf{x}_1', \ldots, \mathbf{x}_p')$$
$$= \int \sum_{i=1}^{p} \{V(|\mathbf{x}_i - \mathbf{x}_{p+1}|) - V(|\mathbf{x}_p' - \mathbf{x}_{p+1}'|)\}$$
$$\times \hat{\mathbf{D}}^{(p+1)}(\mathbf{x}_1, \ldots, \mathbf{x}_{p+1} : \mathbf{x}_1', \ldots, \mathbf{x}_{p+1}')\,d\mathbf{x}_{p+1}. \qquad (8.16)$$

That is, the equation of motion of the pth-order reduced density matrix involves the $(p + 1)$th-order reduced density matrix. Faced with a hierarchy of this nature one is normally forced to decouple the equations of motion before any solution is attempted. However, if it is assumed that a necessary condition for the occurrence of superfluidity is the factorization of the 1-matrix, as in Eq. (8.13), and therefore the factorization of the 2-matrix as in Eq. (8.17) below, then it becomes possible to obtain exact equations of motion for the macroscopic parts of the 1-matrix uncoupled to higher order terms.

The 2-matrix is factorized as

$$\mathbf{D}^{(2)}(\mathbf{x}_1, \mathbf{x}_2 : \mathbf{x}_1', \mathbf{x}_2') = \mathbf{\Phi}_2^*(\mathbf{x}_1, \mathbf{x}_2)\mathbf{\Phi}_2(\mathbf{x}_1', \mathbf{x}_2') + \mathbf{\Lambda}_2(\mathbf{x}_1, \mathbf{x}_2 : \mathbf{x}_1', \mathbf{x}_2')$$
(8.17)

where again $\mathbf{\Phi}_2$ represents a macroscopic part to the density matrix. It is shown that this macroscopic part may be expressed in terms of the $\mathbf{\Phi}_1$ occurring in the 1-matrix. It therefore becomes possible to have an equation of motion for the $\mathbf{\Phi}_1$ function which does not involve macroscopic terms in higher order. Such a decoupling has not come about as a result of a usual approximation scheme but as a consequence of the structure of the 1-matrix for a condensed Bose system. Further investigation of the decoupled equation of motion (Lal and Terreaux, 1970) will of course involve certain approximations.

The elementary constituents of the superfluid system are considered to be helium atoms in their ground states, their correlated behavior being responsible for the particular superfluid properties of the helium(II) ground state as well as for the phonon and roton excitation spectrum. However, since a phenomenological potential is postulated between atoms in order that a fluid, let alone a superfluid, state may exist, the helium atoms in the liquid cannot be treated as structureless. The usual treatment is to derive some pair potential by expansion in virtually escited states of the atoms, but it has been suggested (Girardeau, 1970) that one should in fact consider the elementary constituents as electrons and alpha particles. Such a system of Bose and Fermi particles in interaction presents an interesting challenge since the statistics of the elementary states of the system will be far from trivial. Some controversy as to the correct behavior exists since the original treatment of Girardeau concluded that the condensed fraction of helium atom ground states vanished at absolute zero and that the superfluid behavior was a result of other causes. However, it has been suggested (Peat and Goswami, 1975) that this investigation did not involve the correct treatment of elementary states and that the condensed fraction is in fact nonvanishing, with the conventional form of statistics for the helium atoms being recovered in a limit.

References

Barrett, G. P., Lindenberg, G. P., and Shull, H. (1965). *J. Chem. Phys.* **43**, 580.
Bender, C. F., Davidson, E. R., and Peat, F. D. (1968). *Phys. Rev.* **174**, 75.
Benesch, R., and Smith, V. H. Jr. (1971). *Int. J. Quantum Chem.* **S5**, 35.
Blatt, J. M., and Matsubara, T. (1958). *Progr. Theor. Phys.* **20**, 553.
Boardman, A. D., and March, N. H. (1964). *J. Phys. Chem. Solids* **25**, 1435.
Bopp, F. (1959). *Z. Phys.* **156**, 348.
Breuckner, K. A., Buchler, J. R., Jorna, S., and Lombard, R. J. (1968). *Phys. Rev.* **171**, 1188.
Coleman, A. J. (1963). *Rev. Mod. Phys.* **35**, 668.
Coleman, A. J. (1965). *J. Math. Phys.* **6**, 1425.
Coleman, A. J. (1968). "Reduced Density Matrices with Applications to Physical and Chemical Systems" (A. J. Coleman and R. M. Erdahl, eds.), p. 2. Queen's Univ., Kingston, Ontario.
Coleman, A. J. (1972). *J. Math. Phys.* **13**, 214.
Davidson, E. R. (1963). *J. Chem. Phys.* **39**, 875.
Davidson, E. R. (1969). *J. Math. Phys.* **10**, 725.
Dyson, F. (1956). *Phys. Rev.* **102**, 1217.
Frohlich, H. (1969). *Phys. Kondes. Mater.* **9**, 350.
Garrod, C., and Percus, J. K. (1964). *J. Math. Phys.* **5**, 1756.
Girardeau, M. (1963). *J. Math. Phys.* **4**, 1096.
Girardeau, M. (1970). *J. Math. Phys.* **11**, 684.
Glauber, R. J. (1963). *Phys. Rev.* **131**, 2766.
Grimley, T. B., and Peat, F. D. (1965). *Proc. Phys. Soc.* **86**, 249.
Hilton, D., March, N. H., and Curtis, A. R. (1967). *Proc. Roy. Soc.* **A261**, 119.
Honenberg, P., and Kohn, W. (1964). *Phys. Rev.* **136**, B864.
Kijewski, L. J., and Percus, J. K. (1967). *Phys. Rev.* **164**, 228.
Kijewski, L. J., and Percus, J. K. (1969). *Phys. Rev.* **179**, 45.
Kummer, H. (1967). *J. Math. Phys.* **8**, 2063.
Kutzelnigg, W. (1963). *Theor. Chim. Acta*, **1**, 327.
Kutzelnigg, W. (1965). *Theor. Chim. Acta*, **3**, 241.
Kohn, W., and Sham, L. J. (1965). *Phys. Rev.* **140**, A1133.
Lal, P., and Terreaux, C. (1970). *Phys. Kondens. Mater.* **12**, 131.
Lang, N. D., and Kohn, W. (1970). *Phys. Rev. B* **1**, 4555.
March, N. H., and Murray, A. M. (1961). *Proc. Roy. Soc.* **A261**, 119.
Miller, K. J., and Rudenberg, K. (1968). *J. Chem. Phys.* **48**, 3414.
Peat, F. D. (1967a). *Can. J. Chem.* **45**, 847.
Peat, F. D. (1967b). *In* "Reduced Density Matrices with Applications to Physical and Chemical Systems" (A. J. Coleman and R. M. Erdahl, eds.), p. 315. Queen's Univ., Kingston, Ontario.
Peat, F. D. (1968). *Phys. Rev.* **173**, 69.

PEAT, F. D. (1971). *Phys. Rev.* **A4**, 2206.
PEAT, F. D. (1972). *Theor. Chem. Acta* **24**, 11.
PEAT, F. D., and BROWN, R. J. C. (1966). *Can. J. Phys.* **44**, 1349.
PEAT, F. D., and BROWN, R. J. C. (1967). *Int. J. Quantum Chem.* **S1**, 465.
PEAT, F. D., and GOSWAMI, D. (1975). To be published.
RUSKAI, M. B. (1969). *Phys. Rev.* **183**, 129.
RUSKAI, M. B. (1970). *Phys. Rev.* **A5**, 1336.
SIMMONS, J. (1972). *Int. J. Quantum Chem.* **6**, 439.
SMITH, D. W. (1966). *Phys. Rev.* **147**, 896.
SMITH, D. W. (1968). "Reduced Density Matrices with Applications to Physical and Chemical Systems" (A. J. Coleman and R. M. Erdahl, eds.), p. 169. Queen's Univ., Kingston, Ontario.
SMITH, D. W., and FOGEL, S. J. (1965). *J. Chem. Phys.* **43**, 591.
STOLER, D., and NEWMAN, S. (1972). *Phys. Lett.* **38A**, 433.
WEINHOLD, F., and WILSON, E. B. (1967). *J. Chem. Phys.* **47**, 2298.
YANG, C. N. (1962). *Rev. Mod. Phys.* **34**, 694.
YOUNG, W. H., and MARCH, N. H. (1960). *Proc. Roy. Soc.* **A256**, 62.

Chapter 8

The Green's Function Method

C. MAVROYANNIS

I.	Introduction	488
II.	Double-Time Temperature-Dependent Green's Functions	490
	A. Retarded, Advanced, and Causal Green's Functions	490
	B. Time Correlation Functions	494
III.	Spectral Representations	494
	A. Spectral Representation of the Time Correlation Functions	495
	B. Spectral Representation of the Green's Functions	496
IV.	Properties of the Green's Functions	500
	A. Dispersion Relations	500
	B. Energy Spectrum	501
	C. Symmetry Properties	502
	D. Invariance under Time Inversion	504
V.	The Reaction of a System to an External Perturbation	506
	A. Electrical Conductivity Tensor	510
	B. Absorption Coefficient	513
	C. Interband Optical Transitions	514
VI.	Calculation of the Green's Functions	517
	A. Systems with Direct Interactions	518
	B. One-Particle Green's Function	519
	C. Two-Particle Green's Function, Exciton Spectra	521
	D. Three-Particle Correlations	524
VII.	Charge-Transfer Spectra of Molecular Crystals	527
	A. Green's Functions	527
	B. Excitation Spectra	533
VIII.	Perturbation Theory for the Green's Functions	539
	A. High-Frequency Behavior	544
	B. Functional Derivatives	545
	C. Concluding Remarks	547
	References	548

I. Introduction

The many-body problem is of fundamental importance in quantum mechanics as well as in classical statistical physics. Its aim is to find effective methods of calculating equilibrium and nonequilibrium properties of systems consisting of a large number of particles. The most important aspect of problems dealt with in many-body theory is the necessity of taking into account the interactions between the particles, which alter the behavior of the isolated, noninteracting particles.

The main properties of a system are

(a) elementary excitation spectra. There are two kinds of low-lying excited states in a system of interacting particles, quasi particles and collective modes. Quasi particles correspond approximately to quasi-stationary states of the many-body system, having complex energies of excitation, whose small imaginary part (weak damping) determines the lifetime of the quasi particles. In general, the quasi particle spectrum is temperature dependent. Collective modes, on the other hand, arise from correlation of the particle motions brought about by particle interactions. The correlations responsible for the collective modes give rise to important coherence effects. Such correlations frequently take the form of screening, i.e., the alteration in the effective interaction of a pair of particles brought about by the remaining particles in the system;

(b) ground-state energy;

(c) mean values of the dynamical variables and their distribution functions, transition probabilities, electrical conductivity, thermodynamic and kinetic characteristics, phase transitions, and so on.

Considerable progress has been made in recent years in the investigation of the many-body problem. The progress is, to a considerable extent, due to the application of quantum field methods to statistical mechanics. One of the most effective developments is the double-time temperature-dependent Green's function. This method combines the effective techniques of modern quantum field theory and those of the density matrix approach, which was until recently used in statistical mechanics. The development of the Green's function formalism is a natural one since the basic problems in field theory and in statistical mechanics are similar. For example, in quantum field theory, the average of quantum mechanical operators is taken over the ground state of the system (zero temperature limit), while in statistical mechanics one is dealing with ensemble averages (finite temperatures). Also in both cases, one is interested in the asymptotic behavior of the system, which possesses a large number of degrees of freedom, in the limit $(N/V) =$ constant for $N \to \infty$ and $V \to \infty$, where N is the number of particles and V is the volume of the system.

The connection between the field theoretical concepts and those of statistical mechanics was first realized by Matsubara (1955), who put forward the basic ideas of the formulation. Since then a great deal of work has been done and the basic ideas of the formalism are described in the fundamental papers of Bogolyubov and Tyablikov (1959), Martin and Schwinger (1959), and Zubarev (1960). There are two general methods available for the calculation of the Green's functions: the use of perturbation theory (diagram technique) and the equation of motion method. The first method requires the summation of an infinite series of terms or Feynman diagrams in a perturbation expansion of the Green's function (Klein and Prange, 1958; Kohn and Luttinger, 1960; Luttinger, 1960; Luttinger and Ward, 1960; Luttinger, 1961; Abrikosov et al., 1963; Nozieres, 1964; Schultz, 1964; Mattuck, 1967). In the second method, on the other hand, the Green's functions are determined by the equations of motion and the boundary conditions (spectral representations) which they obey. It consists of solving an infinite hierarchy of differential equations connecting Green's functions of successively higher order (Bogolyubov and Tyablikov, 1959; Martin and Schwinger, 1959; Zubarev, 1960; Baym, 1962; Bonch–Bruevich and Tyablikov, 1962; Kadanoff and Baym, 1962).

In Section II we define the retarded, advanced, and causal double-time temperature-dependent Green's functions as well as the time correlation functions. We use the generalized version of the double-time Green's functions discussed by Zubarev (1960), Bonch–Bruevich and Tyablikov (1962), and Tyablikov (1967). The spectral representations for the time correlation functions and the Green's functions are discussed in Section III, while some general properties of the Green's functions are described in Section IV. The reaction of a system to an external perturbation is considered in Section V and, as an illustration, the expression for the absorption coefficient is derived corresponding to the physical process of direct optical transitions between one-electron bands.

In Section VI we deal with the calculation of the Green's functions. Considering systems with direct interactions, the one-particle Green's function is calculated in the first approximation by means of a simple decoupling procedure. The excitation spectrum is then compared with that derived by the Hartree–Fock self-consistent field approximation. In the same approximation, the two-particle (collective) Green's function is considered and the electron–hole pair (exciton) spectrum is discussed. The effect of the three-particle correlation on the one-particle spectrum is taken into account (Section VI.D), and is used in Section VII to discuss the charge-transfer spectra of molecular crystals. For the model under consideration, it is shown that the poles of the one-particle Green's function describe not only single-particle excitations but also collective modes, which arise from physical processes such as the simultaneous annihilation of a Frenkel exciton and a hole or the creation of a Frenkel exciton and the annihilation of an electron, respectively. The excitation spectrum of charge-transfer complexes is discussed and expressions for the absorption coefficient are derived corresponding to optical transitions between the states in question. In Section VI the decoupling approximation is used to truncate the infinite chain of the coupled equations, and the perturbation theory for the retarded and advanced Green's functions is described in Section VIII. The high-frequency behavior of the one-particle Green's function is examined and the justification for the asymptotic decoupling theorem is considered. A brief account of the functional derivative method is given.

No attempt has been made here to cover all the steadily growing literature on the subject or to describe all the various methods on Green's functions and their applications. Therefore, this chapter may be considered as an introduction to the subject in question. We hope that it will be useful to the graduate students and research workers in the field of theoretical chemistry as well as in the field of physics and chemistry of the organic solid state.

II. Double-Time Temperature-Dependent Green's Functions

A. Retarded, Advanced, and Causal Green's Functions

Let $A(t)$ and $B(t')$ be some operators in the Heisenberg representation

$$A(t) = \exp(i\mathcal{H} t)A(0)\exp(-i\mathcal{H} t)$$
$$B(t') = \exp(i\mathcal{H} t')B(0)\exp(-i\mathcal{H} t'), \qquad (2.1)$$

where $\mathcal{H}$ is the total Hamiltonian of the system $\mathcal{H} = H - \mu N$, where H is the time-independent Hamiltonian operator, N is the operator of the total number of particles in the system, and μ is the chemical potential. The system of units, where $\hbar = 1$, is used throughout. The operators

$A(t)$ and $B(t')$ are, in general, products of quantized wave functions or particle creation and annihilation operators. The double-time retarded (r), advanced (a), and causal (c) Green's functions are defined by the relations

$$G_r(t, t') = \langle\langle A(t); B(t')\rangle\rangle_r = -i\theta(t - t')\langle[A(t), B(t')]_{-\eta}\rangle \quad (2.2a)$$

$$G_a(t, t') = \langle\langle A(t); B(t')\rangle\rangle_a = i\theta(t' - t)\langle[A(t), B(t')]_{-\eta}\rangle \quad (2.2b)$$

$$G_c(t, t') = \langle\langle A(t); B(t')\rangle\rangle_c = -i\langle \hat{T} A(t)B(t')\rangle, \quad (2.2c)$$

where $\theta(t)$ is the step function

$$\theta(t) = 1, \quad t > 0; \quad \theta(t) = 0, \quad t < 0,$$

and $[A, B]_{-\eta}$ denotes the commutator or anticommutator

$$[A, B]_{-\eta} = AB - \eta BA. \quad (2.3a)$$

η is a disposal parameter which can be chosen to be either $+1$ or -1 according to what is more convenient for the problem. Usually the sign of η is taken to be positive or negative according to whether A and B are Bose or Fermi operators. $\hat{T}$ indicates Wick's chronological or $\hat{T}$ product of the two operators, i.e.,

$$\hat{T} A(t)B(t') = \theta(t - t')A(t)B(t') + \eta\theta(t' - t)B(t')A(t). \quad (2.3b)$$

The angular brackets $\langle \cdots \rangle$ indicate the average over a grand canonical ensemble

$$\langle U \rangle = (1/Z)\,\mathrm{tr}(\exp(-\beta\mathscr{H})U), \quad Z = \mathrm{tr}\exp(-\beta\mathscr{H}), \quad \beta = (K_B T)^{-1}, \quad (2.3c)$$

where K_B is Boltzmann's constant, T the absolute temperature, and Z the partition function for the grand canonical ensemble. The Green's functions (2.2) are generalized versions of those of field theory (Zubarev, 1960; Bonch–Bruevich and Tyablikov, 1962). The generalization consists of two parts. First, there is no restriction on the form of the operators A and B but the choice is determined by the conditions of the problem. They can be either Bose or Fermi operators or their products, Pauli operators and their products, density or current operators, and the like. In field theory, A and B are normally the second-quantized wave functions, which in turn are operators through their expansion in creation and annihilation operators. The average in Eqs. (2.2) is over a statistical ensemble, which could be chosen to be the canonical or grand canonical ensemble, i.e., the Green's functions in question are temperature dependent, while in field theory the average is over the ground (vacuum) state of the system. Another difference is that the Green's functions (2.2) depend on two-time arguments, while in field theory multiple-time Green's functions occur. It can easily be shown that the functions (2.2) depend on the difference of the time arguments. For instance, using the commutability of the operators under the trace, we have

$$
\begin{aligned}
G_r(t, t') &= -i\theta(t-t')\,\text{tr}\{\exp(-\beta\mathcal{H})\exp[i\mathcal{H}(t-t')] \\
&\quad \times A(0)\exp[-i\mathcal{H}(t-t')]B(0) \\
&\quad - \eta\exp[-\beta\mathcal{H}]\exp[i\mathcal{H}(t'-t)] \\
&\quad \times B(0)\exp[-i\mathcal{H}(t'-t)]A(0)\} \\
&= -i\theta(t-t')\,\text{tr}\{\exp(-\beta\mathcal{H})A(0)\exp[-i\mathcal{H}(t-t')] \\
&\quad \times B(0)\exp[i\mathcal{H}(t-t')] - \eta\exp(-\beta\mathcal{H}) \\
&\quad \times B(0)\exp[-i\mathcal{H}(t'-t)]A(0)\exp[i\mathcal{H}(t'-t)]\} \\
&= G_r(t-t'). \quad (2.4)
\end{aligned}
$$

From the definition (2.2) it follows that $G_r(t-t')$ and $G_a(t-t')$ vanish for $t < t'$ and $t > t'$, respectively, and the relation

$$
\begin{aligned}
&\langle\!\langle \alpha_1 A_1(t) + \alpha_2 A_2(t); B(t') \rangle\!\rangle_{(j)} \\
&= \alpha_1 \langle\!\langle A_1(t); B(t') \rangle\!\rangle_{(j)} + \alpha_2 \langle\!\langle A_2(t); B(t') \rangle\!\rangle_{(j)} \quad (2.5)
\end{aligned}
$$

holds, where α_1 and α_2 are arbitrary constants and $j = $ r, a, c. Notice that when $t = t'$ the Green's functions (2.2) are not defined because of the discontinuous factor $\theta(t-t')$. In the limit of zero temperature ($\beta \to \infty$), the Green's functions (2.2) correspond to those of field theory.

The operators $A(t)$ and $B(t')$ satisfy the equation of motion

$$i\,dA(t)/dt = [A(t), \mathcal{H}] = A(t)\mathcal{H} - \mathcal{H}A(t). \quad (2.6)$$

To derive the equations of motion for the Green's functions, we differentiate Eqs. (2.2) with respect to t, and using the relation

$$d\theta(t)/dt = -d\theta(-t)/dt = \delta(t), \quad (2.7)$$

we find

$$(id/dt)G_j(t - t') = \delta(t - t')\langle [A(t), B(t)]_{-\eta}\rangle + \langle\!\langle [A(t), \mathcal{H}]; B(t')\rangle\!\rangle_{(j)},$$
$$j = \text{r, a, c}, \quad (2.8)$$

where $\delta(t)$ is the Dirac delta function. The last term in Eq. (2.8) contains Green's functions which are of higher order than the initial one. When the equation of motion for the Green's function $\langle\!\langle [A(t), \mathcal{H}]; B(t')\rangle\!\rangle$ is considered, then an infinite set of coupled equations is obtained connecting Green's functions of successively higher order. The chain of equations (2.8) is the same for the retarded, advanced, and causal Green's functions. The infinite set of coupled equations (2.8) is characteristic for a system of interacting particles and stems from the fact that one particle or a group of interacting particles cannot be considered independently from the rest of the system.

Considering the Fourier transform of the Green's function $G_j(t - t')$

$$G_j(t - t') = \int_{-\infty}^{+\infty} G_j(\omega) \exp[-i\omega(t - t')] \, d\omega \quad (2.9a)$$

$$G_j(\omega) = (1/2\pi) \int_{-\infty}^{+\infty} G_j(t) \exp(i\omega t) \, dt, \quad (2.9b)$$

the chain of equations (2.8) takes the form

$$\omega G_j(\omega) = (1/2\pi)\langle [A(t), B(t)]_{-\eta}\rangle + \langle\!\langle [A(t), \mathcal{H}]; B(t')\rangle\!\rangle_{(j)(\omega)} \quad (2.10)$$

where the subscript ω indicates the Fourier transform of the Green's function in question. From now on, the Fourier transforms of the Green's functions will be referred to as Green's functions, whenever confusion does not arise, and the subscript ω will be suppressed for convenience. Analogous equations to that of (2.10) can be derived for the Green's functions $\langle\!\langle [A(t), \mathcal{H}]; B(t')\rangle\!\rangle$, $\langle\!\langle [[A(t), \mathcal{H}], \mathcal{H}]; B(t')\rangle\!\rangle$, and so on. To solve the equations of motion (2.8) or (2.10), one must use the necessary boundary conditions, the form of which may be established through the spectral representation for the Green's functions in question. The spectral theorems corresponding to the Green's functions are discussed in Section III. An exact solution of the chain of coupled equations (2.10) is impossible and, therefore, one must develop some approximate method of truncating the infinite hierarchy of the equations. Approximate methods for solving Eqs. (2.10) are considered in Section VI and VIII.

B. Time Correlation Functions

The Green's functions (2.2) are linear combinations of the time correlation functions $F_{AB}(t, t') = F_{AB}(t - t')$ and $F_{BA}(t, t') = F_{BA}(t' - t)$, defined as

$$F_{AB}(t - t') = \langle A(t)B(t')\rangle, \qquad F_{BA}(t' - t) = \langle B(t')A(t)\rangle, \quad (2.11)$$

where the average is over the statistical ensemble (2.3c). Many physical quantities of interest can be derived from the time correlation functions (2.11), which do not include the step function $\theta(t)$, and therefore they are also defined for $t = t'$. For $t = t'$, the functions $F_{AB}(0)$ and $F_{BA}(0)$ correspond to the usual distribution functions in statistical mechanics, which can be used to evaluate the average values of dynamical quantities.

Differentiating the correlation function $F_{AB}(t - t')$ with respect to the argument t, we derive the equation of motion

$$(id/dt)F_{AB}(t - t') = (id/dt)\langle A(t)B(t')\rangle = \langle [A(t), \mathcal{H}]B(t')\rangle, \quad (2.12a)$$

and similarly,

$$(id/dt)\langle [A(t), \mathcal{H}]B(t')\rangle = \langle [[A(t), \mathcal{H}], \mathcal{H}]B(t')\rangle, \quad (2.12b)$$

which leads to an infinite chain of coupled equations consisting of higher order time correlation functions, depending on the increasing number of variables. The chain of coupled equations (2.12) can be used to calculate the correlation functions of interest by truncating the series by some approximate method to obtain a finite set of equations. Then the correlation function in question can be calculated by direct integration of Eqs. (2.12), provided that the appropriate boundary conditions on t are taken into consideration. However, the mathematical formalism for the Green's function provides a more convenient evaluation of the time correlation functions than the direct one, because, as we will see later, the boundary conditions for the Green's functions are easily satisfied through the spectral theorems.

III. Spectral Representations

We consider in this section the spectral representations for the time correlation functions and the corresponding Green's functions. We follow the work of Bogolyubov and Tyablikov (1959), Zubarev (1960), Bogolyubov (1961), and Bonch–Bruevich and Tyablikov (1962).

A. Spectral Representation of the Time Correlation Functions

The time correlation functions (2.11) can be written as

$$F_{BA}(t - t') = \langle B(t')A(t) \rangle = (1/Z) \sum_{\nu} \exp(-\beta\omega_{\nu}) \langle \nu | B(t')A(t) | \nu \rangle$$
$$= (1/Z) \sum_{\nu,\mu} \exp(-\beta\omega_{\nu}) \langle \nu | B(t') | \mu \rangle \langle \mu | A(t) | \nu \rangle. \quad (3.1)$$

If we consider now the matrix elements (3.1) in the representation in which $\mathscr{H}$ is diagonal, i.e.,

$$\langle \nu | \mathscr{H} | \mu \rangle = \omega_{\nu} \delta_{\mu\nu}, \quad (3.2)$$

and keep in mind that $\exp(i\mathscr{H}t) | \nu \rangle = \exp(i\omega_{\nu}t) | \nu \rangle$, then from Eq. (3.1) we have

$$F_{BA}(t - t') = (1/Z) \sum_{\nu,\mu} \exp(-\beta\omega_{\nu}) \langle \nu | B(0) | \mu \rangle \langle \mu | A(0) | \nu \rangle$$
$$\times \exp[-i(t - t')(\omega_{\nu} - \omega_{\mu})]. \quad (3.3)$$

Similarly, for the time correlation function $F_{AB}(t - t')$, we find

$$F_{AB}(t - t') = \langle A(t)B(t') \rangle = (1/Z) \sum_{\nu,\mu} \exp[-\beta(\omega_{\mu} - \omega_{\nu}) - \beta\omega_{\nu}]$$
$$\times \langle \nu | B(0) | \mu \rangle \langle \mu | A(0) | \nu \rangle \exp[-i(t - t')(\omega_{\nu} - \omega_{\mu})]. \quad (3.4)$$

The Fourier transforms of the time correlation functions (3.3) and (3.4) can take the form

$$\langle B(t')A(t) \rangle = \int_{-\infty}^{+\infty} J(\omega) \exp[-i\omega(t - t')] \, d\omega, \quad (3.5)$$

$$\langle A(t)B(t') \rangle = \int_{-\infty}^{+\infty} J(\omega) \exp(\beta\omega) \exp[-i\omega(t - t')] \, d\omega, \quad (3.6)$$

where the function $J(\omega)$ is given by

$$J(\omega) = (1/Z) \sum_{\nu,\mu} \langle \nu | B(0) | \mu \rangle \langle \mu | A(0) | \nu \rangle \exp(-\beta\omega_{\nu}) \delta(\omega_{\nu} - \omega_{\mu} - \omega). \quad (3.7)$$

The expressions (3.5) and (3.6) are the spectral representations of the corresponding time correlation functions, while the expression for $J(\omega)$ [Eq. (3.7)] is the so-called spectral density or spectral function. It follows from Eqs. (3.5) and (3.6) that the spectral representations for the average value of the product of two operators taken in different order (BA and AB) differ from one another by a factor $e^{\beta\omega}$. When $t = t'$, the expressions (3.5) and (3.6) are reduced to

$$F_{BA}(0) = \langle B(t)A(t)\rangle = \int_{-\infty}^{+\infty} J(\omega)\,d\omega, \tag{3.8}$$

$$F_{AB}(0) = \langle A(t)B(t)\rangle = \int_{-\infty}^{+\infty} J(\omega)\exp(\beta\omega)\,d\omega, \tag{3.9}$$

which describe the corresponding distribution functions.

If we multiply Eqs. (3.5) and (3.8) by $\eta\,(= \pm 1)$ and then subtract from Eqs. (3.6) and (3.9), respectively, we have

$$\langle [A(t), B(t')]_{-\eta}\rangle = \int_{-\infty}^{+\infty} J(\omega)[\exp(\beta\omega) - \eta]\exp[-i\omega(t - t')]\,d\omega \tag{3.10a}$$

$$\langle [A(t), B(t)]_{-\eta}\rangle = \int_{-\infty}^{+\infty} J(\omega)[\exp(\beta\omega) - \eta]\,d\omega. \tag{3.10b}$$

The time correlation functions given by Eqs. (3.5) and (3.6) have been expressed in terms of the spectral function $J(\omega)$. Thus the problem of evaluating the equilibrium or nonequilibrium properties of a system is reduced to the determination of the corresponding spectral functions. This is what makes the Green's function formalism effective since, as it will be shown, the spectral functions can be immediately expressed in terms of the corresponding Green's functions.

B. Spectral Representation of the Green's Functions

We now consider the spectral representations for the retarded, advanced, and causal Green functions. Using the definition for the retarded Green's function, Eqs. (2.2a), (2.3a), (2.9b), and (3.10), we have

$$G_{\mathrm{r}}(\omega) = \langle\!\langle A(t); B(t')\rangle\!\rangle_{(\mathrm{r},\omega)}$$

$$= (1/2\pi i)\int_{-\infty}^{+\infty} d(t - t')\exp[i(\omega + i\varepsilon)(t - t')]$$

$$\times \int_{-\infty}^{+\infty} J(\omega')[\exp(\beta\omega') - \eta]\exp[-i\omega'(t - t')]\,d\omega'$$

$$= (1/2\pi)\int_{-\infty}^{+\infty}[\exp(\beta\omega') - \eta]J(\omega')[d\omega'/(\omega - \omega' + i\varepsilon)], \tag{3.11}$$

where use has been made of the relations

$$\theta(t) = \begin{cases} \exp(-\varepsilon t), & \varepsilon \to 0, \quad \varepsilon > 0 \quad \text{for } t > 0 \\ 0, & \text{for } t < 0, \end{cases} \quad (3.12a)$$

and the expression for the integral

$$\int_{-\infty}^{+\infty} dt \, \exp[it(\omega - \omega' + i\varepsilon)] = i/(\omega - \omega' + i\varepsilon). \quad (3.12b)$$

In the same fashion, we derive the following expression for the advanced Green's function

$$G_a(\omega) = \langle\!\langle A(t); B(t') \rangle\!\rangle_{(a,\omega)}$$
$$= (1/2\pi) \int_{-\infty}^{+\infty} [\exp(\beta\omega') - \eta] J(\omega')[d\omega'/(\omega - \omega' - i\varepsilon)]. \quad (3.13)$$

The expressions (3.11) and (3.13) describe the spectral representations for the retarded and advanced Green's functions, respectively.

Considering ω to be a complex quantity, we may combine Eqs. (3.11) and (3.13) in the form

$$G(\omega) = (1/2\pi) \int_{-\infty}^{+\infty} [\exp(\beta\omega') - \eta] J(\omega')[d\omega'/(\omega - \omega')]$$
$$= \begin{cases} G_r(\omega), & \text{Im } \omega > 0 \\ G_a(\omega), & \text{Im } \omega < 0. \end{cases} \quad (3.14)$$

$G_r(\omega)$ and $G_a(\omega)$ are analytic functions in the upper and lower half-planes, respectively, having singularities on the real axis (poles or cut lines). Assuming that ω is a complex quantity, then the Green's functions $G_r(\omega)$ and $G_a(\omega)$ will be referred to simply as $G(\omega)$. The left-hand side of Eq. (3.14) is a Cauchy-type integral. Bogolyubov and Parasyuk (1956) have shown that the function $G(\omega)$ defined by Eq. (3.14) is an analytic function in the complex ω plane and is equal to $G_r(\omega)$ and $G_a(\omega)$ everywhere in the upper and lower half-planes, respectively, provided that $G_{r,a}(t)$ is a generalized function in the Sobolev–Schwartz sense. If a cut is made along the real axis, the function $G(\omega)$ is an analytic function consisting of two branches, one of which is defined in the upper and the other in the lower half-plane for complex values of ω.

Using Eq. (3.14), we consider the expression

$$G(\omega + i\varepsilon) - G(\omega - i\varepsilon) = (2\pi)^{-1} \int_{-\infty}^{+\infty} [\exp(\beta\omega') - \eta] J(\omega')$$
$$\times [(\omega - \omega' + i\varepsilon)^{-1} - (\omega - \omega' - i\varepsilon)^{-1}] \, d\omega' \quad (3.15)$$

for real ω. To perform the complex integration in Eq. (3.15) we employ the well-known relation

$$\lim_{\substack{\varepsilon \to 0 \\ \varepsilon > 0}} (x \pm i\varepsilon)^{-1} = P(1/x) \mp i\pi \, \delta(x), \quad (3.16)$$

where x is real and P indicates that the principal value over the integral must be taken. Then the expression (3.15) becomes

$$i[G(\omega + i\varepsilon) - G(\omega - i\varepsilon)] = \int_{-\infty}^{+\infty} [\exp(\beta\omega') - \eta] J(\omega') \, \delta(\omega - \omega') \, d\omega'$$
$$= [\exp(\beta\omega) - \eta] J(\omega). \quad (3.17)$$

In deriving Eq. (3.17), no assumption has been made that the function $J(\omega)$ is real. Using Eq. (3.17), the time correlation functions (3.5) and (3.6) may be written as

$$\langle B(t')A(t) \rangle = i \int_{-\infty}^{+\infty} [G(\omega + i\varepsilon) - G(\omega - i\varepsilon)][\exp(\beta\omega) - \eta]^{-1}$$
$$\times \exp[-i\omega(t - t')] \, d\omega \quad (3.18a)$$

$$\langle A(t)B(t') \rangle = i \int_{-\infty}^{+\infty} [G(\omega + i\varepsilon) - G(\omega - i\varepsilon)][\exp(\beta\omega) - \eta]^{-1}$$
$$\times \exp(\beta\omega) \exp[-i\omega(t - t')] \, d\omega. \quad (3.18b)$$

Expressions (3.18) are of fundamental importance because they directly relate the time correlation functions $(t \neq t')$ or the distribution functions $(t = t')$ of two arbitrary dynamical variables with the Fourier transforms of the corresponding Green's functions. The problem of evaluating these functions is thus reduced to finding the Fourier transforms of the corresponding Green's functions. Thus the effectiveness of the Green's function is to a considerable extent due to the use of the spectral representations.

In the same way, we can derive the spectral representation for the causal Green's function. Considering the definition of the causal Green's function, Eqs. (2.2c) and (2.3b), its Fourier transform, Eq. (2.9b), and making use of Eqs. (3.5) and (3.6) as well as the relations (3.12) for the integration over the time, we find

$$G_c(\omega) = \left(\frac{1}{2\pi}\right) \int_{-\infty}^{+\infty} J(\omega') \left[\frac{\exp(\beta\omega')}{(\omega - \omega' + i\varepsilon)} - \frac{\eta}{(\omega - \omega' - i\varepsilon)}\right] d\omega'. \tag{3.19}$$

Employing the identity (3.16), Eq. (3.19) becomes

$$G_c(\omega) = \left(\frac{1}{2\pi}\right) \int_{-\infty}^{+\infty} J(\omega')[\exp(\beta\omega') - \eta]$$
$$\times \left[P\frac{1}{(\omega - \omega')} - i\pi \frac{[\exp(\beta\omega') + \eta]}{[\exp(\beta\omega') - \eta]} \delta(\omega - \omega')\right] d\omega'. \tag{3.20}$$

Separating the real and imaginary parts of Eq. (3.20), we obtain

$$\text{Re } G_c(\omega) = (1/2\pi)P \int_{-\infty}^{+\infty} [\exp(\beta\omega') - \eta]J(\omega')[d\omega'/(\omega - \omega')] \tag{3.21}$$

$$\text{Im } G_c(\omega) = -(\tfrac{1}{2})[\exp(\beta\omega) + \eta]J(\omega). \tag{3.22}$$

It follows from Eq. (3.19) that the function $G_c(\omega)$ is, in general, not an analytic function and cannot be extended into the complex plane. The lack of analyticity of the function $G_c(\omega)$ makes its use inconvenient. The causal Green's function can be analytically continued in the complex plane only in the limit of zero temperature ($\beta \to \infty$). Equations (3.21) and (3.22) were first derived by Landau (1958) for the single-particle Green's function.

We have only considered the spectral representation for the double-time Green's functions. Spectral representations for the many-time temperature-dependent Green's functions have been derived by Bonch–Bruevich (1960), to whom we refer the reader for details. Many-time Green's functions are important in many problems such as in the theory of electrical conductivity (Bonch–Bruevich, 1959).

Historically, spectral representations were first used in the theory of fluctuations and in statistical mechanics of irreversible processes by Callen and Welton (1951) and in quantum field theory by Cällen (1952) and Lehmann (1954). The spectral representations for the Green's functions in statistical mechanics were introduced by Bonch–Bruevich (1956),

Gor'kov (1958), Landau (1958), Martin and Schwinger (1958, 1959), Bogolyubov and Tyablikov (1959), Kogan (1959), and Bonch–Bruevich and Kogan (1960).

IV. Properties of the Green's Functions

In this section we discuss some general properties of the Green's functions which are independent of the particular form of the Hamiltonian of a system. These properties are extremely useful for the application of the Green's functions (Bonch–Bruevich and Tyablikov, 1962; Tyablikov, 1967).

A. Dispersion Relations

The expressions which relate the real and imaginary parts of the Green's functions for real values of ω are known as dispersion relations. By virtue of the relation (3.16), the expressions (3.11) and (3.13) may be transformed into

$$G_r(\omega) = (P/2\pi) \int_{-\infty}^{+\infty} [\exp(\beta\omega') - \eta][J(\omega')\,d\omega'/(\omega - \omega')]$$
$$- (i/2)[\exp(\beta\omega) - \eta]J(\omega) \qquad (4.1)$$

$$G_a(\omega) = (P/2\pi) \int_{-\infty}^{+\infty} [\exp(\beta\omega') - \eta][J(\omega')\,d\omega'/(\omega - \omega')]$$
$$+ (i/2)[\exp(\beta\omega) - \eta]J(\omega). \qquad (4.2)$$

From Eqs. (4.1), (4.2), (3.21), and (3.22), we obtain

$$\operatorname{Re} G_r(\omega) = (P/\pi) \int_{-\infty}^{+\infty} \operatorname{Im} G_r(\omega')\,d\omega'/(\omega' - \omega) \qquad (4.3)$$

$$\operatorname{Re} G_a(\omega) = -(P/\pi) \int_{-\infty}^{+\infty} \operatorname{Im} G_a(\omega')\,d\omega'/(\omega' - \omega) \qquad (4.4)$$

$$\operatorname{Re} G_c(\omega) = (P/\pi) \int_{-\infty}^{+\infty} [(\exp(\beta\omega') - \eta)/[\exp(\beta\omega') + \eta]]$$
$$\times [\operatorname{Im} G_c(\omega')\,d\omega'/(\omega' - \omega)]. \qquad (4.5)$$

The expressions (4.3)–(4.5) describe the dispersion relations for the retarded, advanced, and causal Green's functions, respectively. The importance of the dispersion relations (4.3)–(4.5) stems from the fact that if we know the imaginary part of one of the functions $G_r(\omega)$, $G_a(\omega)$, and $G_c(\omega)$, then they are all completely determined through the spectral function. It must be emphasized that Eqs. (4.3) and (4.4) follow from Eqs. (4.1) and (4.2), respectively, whether the spectral function $J(\omega)$ is real or complex. The proof is more complicated in the latter case.

B. Energy Spectrum

We now show that the singularities of the Green's functions determine the energy spectrum of the elementary excitations of a system.

Suppose that the poles of the Green's function in question are the points $\omega = \omega_i$ on the real axis. According to Eqs. (3.16) and (3.17), the spectral function $J(\omega)$ will have delta function singularities at the points $\omega = \omega_i$ and the correlation function (3.5) will oscillate at the frequencies ω_i. It is clear from Eq. (3.7) that when $T = 0$ the quantities ω_i are the exact energy eigenvalues of the Hamiltonian of the system. When $T \neq 0$, the quantities ω_i depend on the temperature and on the chemical potential of the system, and therefore they cannot be interpreted on purely quantum mechanical grounds. However, it is evident from Eqs. (3.5) and (3.6) that they represent undamped oscillations of the system. Thus the concept of a temperature-dependent energy spectrum has been introduced in a natural way through the formalism of the Green's functions. It is evident from Eq. (3.7) that the values of ω_i are of the form $\omega_\mu - \omega_\nu$, which indicates that when $T = 0$, the ω_i are the energies of excitation of the system calculated with respect to the ground-state energy. Thus the spectral functions determine the energy spectrum of the system.

For real systems with interactions having a continuous (or almost continuous) energy spectrum, there is always a finite damping. The latter is brought about by the exchange of energy, momentum, and so on, which takes place between the different degrees of freedom. From a mathematical point of view, the damping results in the displacement of the poles of the spectral function from the real axis into the complex plane. In this case, we may have, at least, approximate poles, whose imaginary parts represent the damping of the corresponding oscillations of the system. If the damping is sufficiently small, we may introduce the concept of quasistationary states. The correlation functions will then have the form of $f(t) \exp(-i\omega_i t)$, where $f(t)$ describes the damping of the states in question, $f(t) \to 0$ as $|t| \to \infty$. In such a case, the elementary excitations of the system are well defined, provided that the constant representing the damping is much smaller than the energies of excitation ω_i measured from the ground state.

Thus the study of the analytical properties of the spectral functions provides, in principle, the determination of the excitation spectrum, i.e., the appropriate energies of excitation, the damping constants, and the condition for the existence of quasi-stationary states. For further considerations concerning the relationship between elementary excitations and poles of the Green's functions, we refer the reader to the literature (Bolsterli, 1960; ter Haar and Parry, 1962; Pike, 1964).

C. Symmetry Properties

Using the definition of the retarded and advanced Green's functions, Eqs. (2.2a) and (2.2b), we may write the relation

$$\langle\!\langle A(t); B(t')\rangle\!\rangle_{(a)} = \eta \langle\!\langle B(t'); A(t)\rangle\!\rangle_{(r)}, \qquad (4.6)$$

and if we take the Fourier transforms with respect to time on both sides of Eq. (4.6), we obtain

$$\int_{-\infty}^{+\infty} \langle\!\langle A; B\rangle\!\rangle_{(a,\omega)} \exp[-i\omega(t-t')]\, d\omega$$
$$= \eta \int_{-\infty}^{+\infty} \langle\!\langle B; A\rangle\!\rangle_{(r,\omega)} \exp[-i\omega(t'-t)]\, d\omega. \qquad (4.7)$$

Therefore, for real values of ω

$$\langle\!\langle A; B\rangle\!\rangle_{(a,\omega)} = \eta \langle\!\langle B; A\rangle\!\rangle_{(r,-\omega)}. \qquad (4.8)$$

Considering Eqs. (3.13) and (3.11), we have

$$\langle\!\langle A; B\rangle\!\rangle_{(a,\omega)} = (1/2\pi) \int_{-\infty}^{+\infty} [\exp(\beta\omega') - \eta][J(\omega')\, d\omega'/(\omega - \omega' - i\varepsilon)] \qquad (4.9a)$$

$$\langle\!\langle B; A\rangle\!\rangle_{(r,-\omega)} = (1/2\pi) \int_{-\infty}^{+\infty} [\exp(\beta\omega') - \eta][J(\omega')\, d\omega'/(-\omega - \omega' + i\varepsilon)]. \qquad (4.9b)$$

Both expressions (4.9a) and (4.9b) are analytic in the lower half-plane. Therefore, for complex values of ω we have, instead of Eq. (4.8),

$$\langle\!\langle A; B\rangle\!\rangle_{(\omega)} = \eta \langle\!\langle B; A\rangle\!\rangle_{(-\omega)}. \qquad (4.10)$$

Considering now the complex conjugate of the expression (2.2a) for the retarded Green's function, we may write

$$\langle\!\langle A(t); B(t')\rangle\!\rangle^*_{(r)} = -\eta\langle\!\langle A^\dagger(t); B^\dagger(t')\rangle\!\rangle_{(r)} = -\langle\!\langle B^\dagger(t'); A^\dagger(t)\rangle\!\rangle_{(a)}. \quad (4.11)$$

Writing the Fourier transforms for the Green's functions in Eq. (4.11) and comparing their Fourier coefficients for real values of ω, we find

$$\langle\!\langle A; B\rangle\!\rangle^*_{(r,\omega)} = -\eta\langle\!\langle A^\dagger; B^\dagger\rangle\!\rangle_{(r,-\omega)} = -\langle\!\langle B^\dagger; A^\dagger\rangle\!\rangle_{(a,\omega)}. \quad (4.12)$$

From the Fourier transforms of the correlation functions

$$\langle A(t)B(t')\rangle = \int_{-\infty}^{+\infty} \exp[-i\omega(t-t')]J_{AB}(\omega)\exp(\beta\omega)\,d\omega, \quad (4.13)$$

$$\langle B^\dagger(t')A^\dagger(t)\rangle = \int_{-\infty}^{+\infty} \exp[-i\omega(t'-t)]J_{B^\dagger A^\dagger}(\omega)\exp(\beta\omega)\,d\omega$$

$$= \int_{-\infty}^{+\infty} \exp[-i\omega(t-t')]J_{A^\dagger B^\dagger}(\omega)\,d\omega, \quad (4.14)$$

the function (4.14) may be considered to be the conjugate of (4.13), and hence

$$\langle B^\dagger(t')A^\dagger(t)\rangle = \int_{-\infty}^{+\infty} \exp[-i\omega(t-t')]J^*_{AB}(\omega)\exp(\beta\omega)\,d\omega. \quad (4.15)$$

Comparing the Fourier coefficients of Eqs. (4.13)–(4.15), we derive the following relations for the spectral functions:

$$J^*_{AB}(\omega) = J_{B^\dagger A^\dagger}(\omega) \quad (4.16)$$

$$J^*_{AB}(\omega)e^{\beta\omega} = J_{A^\dagger B^\dagger}(-\omega) \quad (4.17)$$

$$J_{B^\dagger A^\dagger}(-\omega)e^{\beta\omega} = J_{A^\dagger B^\dagger}(\omega). \quad (4.18)$$

Taking $B^\dagger = A$, it then follows from Eq. (4.16) that $J_{AA^\dagger}(\omega)$ is real, i.e.,

$$J^*_{AA^\dagger}(\omega) = J_{AA^\dagger}(\omega). \quad (4.19)$$

The Green's function

$$\langle\!\langle A; A^\dagger\rangle\!\rangle_{(\omega)} = (1/2\pi)\int_{-\infty}^{+\infty}[\exp(\beta\omega') - \eta]J_{AA^\dagger}(\omega')[d\omega'/(\omega-\omega')] \quad (4.20)$$

is a function of the complex variable ω. Then using Eq. (4.19) we obtain

$$\langle\!\langle A; A^{\dagger}\rangle\!\rangle_{\omega}^{*} = -\langle\!\langle A; A^{\dagger}\rangle\!\rangle_{\omega}^{*}. \tag{4.21}$$

Considering the fact that $J_{AA^{\dagger}}(\omega)$ is real [Eq. (4.19)], we have

$$\langle AA^{\dagger}\rangle = \langle AA^{\dagger}\rangle^{*} = \int_{-\infty}^{+\infty} J_{AA^{\dagger}}(\omega)\exp(\beta\omega)\,d\omega \tag{4.22}$$

$$\langle A^{\dagger}A\rangle = \langle A^{\dagger}A\rangle^{*} = \int_{-\infty}^{+\infty} J_{AA^{\dagger}}(\omega)\,d\omega. \tag{4.23}$$

Using Eq. (3.10b),

$$\int_{-\infty}^{+\infty}[\exp(\beta\omega) - \eta]J_{AB}(\omega)\,d\omega = \langle AB - \eta BA\rangle. \tag{4.24}$$

Then for $B = A$ and $\eta = 1$, we obtain

$$\int_{-\infty}^{+\infty}[\exp(\beta\omega) - 1]J_{AA}(\omega)\,d\omega = 0. \tag{4.25a}$$

If we take $A = \alpha_f$, $B = \alpha_f^{\dagger}$ where $\alpha_f^{\dagger}$ and α_f are the creation and annihilation operators satisfying Bose or Fermi statistics, i.e.,

$$\alpha_f \alpha_{f'}^{\dagger} - \eta \alpha_{f'}^{\dagger}\alpha_f = \delta_{ff'}, \tag{4.25b}$$

then from Eq. (4.24), we have

$$\int_{-\infty}^{+\infty}[\exp(\beta\omega) - \eta]J_{\alpha_f \alpha_{f'}^{\dagger}}(\omega)\,d\omega = \delta_{ff'}. \tag{4.25c}$$

For Pauli operators, where

$$b_f b_{f'}^{\dagger} - b_{f'}^{\dagger}b_f = (1 - 2\eta_f)\,\delta_{ff'}, \qquad \eta_f = b_f^{\dagger}b_f, \tag{4.25d}$$

Eq. (4.24) leads to ($\eta = 1$)

$$\int_{-\infty}^{+\infty}[\exp(\beta\omega) - 1]J_{b_f b_{f'}^{\dagger}}(\omega)\,d\omega = \langle 1 - 2\eta_f\rangle\,\delta_{ff'}. \tag{4.25e}$$

D. Invariance under Time Inversion

Let the equation of motion for the operators A and B be invariant under time reversal, i.e.,

$$t \to -t, \qquad t' \to -t', \qquad \text{and} \qquad i = -i. \tag{4.26}$$

Then the left-hand side of Eq. (3.5) remains unchanged because of the interchange of i by $-i$, while on the right-hand side $J(\omega)$ is replaced by $J^*(\omega)$

$$\langle B(t')A(t)\rangle = \int_{-\infty}^{+\infty} J^*_{AB}(\omega) \exp[-i\omega(t-t')] \, d\omega. \tag{4.27}$$

Comparison between Eqs. (3.5) and (4.27) leads to

$$J^*_{AB}(\omega) = J_{AB}(\omega), \tag{4.28}$$

which indicates that the spectral function is real. Using Eq. (4.8), we derive from Eq. (3.14) the relation

$$[G(\omega)]^* = -G(\omega^*) \tag{4.29}$$

for complex values of ω. From Eqs. (4.13)–(4.15) and (4.28), we obtain

$$\langle A(t)B(t')\rangle = \langle B^\dagger(t)A^\dagger(t')\rangle \tag{4.30}$$

$$\langle B(t')A(t)\rangle = \langle A^\dagger(t')B^\dagger(t)\rangle. \tag{4.31}$$

Multiplying Eq. (4.31) by η and subtracting from Eq. (4.30), and then multiplying the result by $i\theta(t-t')$ or $-i\theta(t'-t)$, we obtain

$$\langle\!\langle A(t); B(t')\rangle\!\rangle_{(j)} = \langle\!\langle B^\dagger(t); A^\dagger(t')\rangle\!\rangle_{(j)}, \qquad j = \text{r, a} \tag{4.32}$$

and for their Fourier transforms with respect to time

$$\langle\!\langle A; B\rangle\!\rangle_{(j,\omega)} = \langle\!\langle B^\dagger; A^\dagger\rangle\!\rangle_{(j,\omega)} \tag{4.33}$$

$$\langle\!\langle A; B\rangle\!\rangle_{(\omega)} = \langle\!\langle B^\dagger; A^\dagger\rangle\!\rangle_{(\omega)}. \tag{4.34}$$

If A and B are Hermitian operators, then for $t = t'$, Eqs. (4.30) and (4.31) lead to $\langle AB - BA\rangle = 0$ and Eq. (4.24) becomes

$$\int_{-\infty}^{+\infty} [\exp(\beta\omega) - \eta] J_{AB}(\omega) \, d\omega = (1-\eta)\langle AB\rangle. \tag{4.35}$$

Formulas (4.28)–(4.35) are applicable only if the operators A and B are invariant under the transformation (4.26). The symmetry relations for the Green's functions as well as those for the spectral functions are useful in the application of the Green's function method to physical problems.

We now show that the diagonal elements of the retarded and advanced Green's functions satisfy an important inequality proved first by Bogolyubov (1961). Using Eqs. (2.9a), (2.9b), and (3.7), we may write for the Fourier transforms of the Green's functions

$$G_{AA^\dagger}(0) = \langle\!\langle A; A^\dagger \rangle\!\rangle_{(j,\omega=0)} = \left(\frac{1}{2\pi}\right) \frac{1}{Z} \sum_{\mu,\nu} \frac{|\langle \nu | A(0) | \mu \rangle|^2}{(\omega_\mu - \omega_\nu \pm i\varepsilon)} \quad (4.36)$$

$$G_{BB^\dagger}(0) = \langle\!\langle B; B^\dagger \rangle\!\rangle_{(j,\omega=0)} = \left(\frac{1}{2\pi}\right) \frac{1}{Z} \sum_{\mu,\nu} \frac{|\langle \nu | B(0) | \mu \rangle|^2}{(\omega_\mu - \omega_\nu \pm i\varepsilon)} \quad (4.37)$$

$$G_{AB^\dagger}(0) = \langle\!\langle A; B^\dagger \rangle\!\rangle_{(j,\omega=0)} = \left(\frac{1}{2\pi}\right) \frac{1}{Z} \sum_{\mu,\nu} \frac{\langle \nu | A(0) | \mu \rangle \langle \mu | B(0) | \nu \rangle}{(\omega_\mu - \omega_\nu \pm i\varepsilon)} \quad (4.38)$$

where $j = $ r, a. Multiplying the absolute magnitude of the expressions (4.36) and (4.37), comparing with (4.38), and making use of the Schwartz inequality, we find

$$|G_{AA^\dagger}(0)| |G^*_{BB^\dagger}(0)| \geq |G_{AB}(0)|^2. \quad (4.39)$$

Similarly, considering that $J_{AA^\dagger}(\omega) \geq 0$ and $J^*_{AB}(\omega) = J_{B^\dagger A^\dagger}(\omega)$ one can easily show that the spectral functions satisfy the inequality

$$J_{AA^\dagger}(\omega) J^*_{BB^\dagger}(\omega) \geq |J_{AB}(\omega)|^2. \quad (4.40)$$

V. The Reaction of a System to an External Perturbation

The Green's function method can be used to study not only the equilibrium properties of a system but also systems which are disturbed through an external action. In particular, it is possible to calculate the transport coefficients and the complex susceptibility of a system under the action of an external field corresponding to physical processes such as the electrical conduction, ferromagnetic resonance, optical absorption (see, for instance, Callen and Welton, 1951; Kubo and Tomita, 1954; Kubo, 1957; Zubarev, 1960; Bonch–Bruevich and Rozman, 1964), and so on. For general treatments of the thermodynamics of irreversible processes, we refer to the work of Mori (1956), Kubo et al. (1957), and Zubarev (1962).

We assume that when a system is in a state of statistical equilibrium it is represented by the Hamiltonian H_0, which does not depend on time.

When the time-dependent perturbation $V(t)$ is switched on, the total Hamiltonian of the system is described by

$$\mathcal{H} = H_0 + V(t). \tag{5.1}$$

We assume that when the perturbation is adiabatically switched on it has the form

$$V(t) = \sum_{\Omega} V_{\Omega} \exp(\varepsilon t - i\Omega t), \qquad \varepsilon > 0, \quad \varepsilon \to 0 \tag{5.2}$$

while for an instantaneous switching on at time $t = t_0$

$$V(t) = \begin{cases} 0, & t < t_0 \\ \sum_{\Omega} V_{\Omega} \exp(-i\Omega t), & t > t_0 \end{cases} \tag{5.3}$$

where V_{Ω} is an operator, that does not depend explicitly on time.

The average value of any dynamical variable $A(t)$ is given by

$$\langle A(t) \rangle = \text{tr}\{A(t)\varrho(t)\}, \qquad \text{tr } \varrho(t) = 1 \tag{5.4}$$

where $\varrho(t)$ is the statistical operator or the density matrix of the system described by the Hamiltonian (5.1) and satisfies the equation of motion

$$(id/dt)\varrho(t) = [\mathcal{H}, \varrho(t)] = [H_0, \varrho(t)] + [V(t), \varrho(t)], \tag{5.5}$$

with the initial condition that at $t = t_0$

$$\varrho(t)_{t=t_0} = \varrho_0 = (1/Z_0) \exp(-\beta H_0), \qquad Z_0 = \text{tr} \exp(-\beta H_0), \tag{5.6}$$

and ϱ_0 is the equilibrium density matrix of the system corresponding to the Hamiltonian H_0. We look for a solution of Eq. (5.5) in the form

$$\varrho(t) = \varrho_0 + \Delta\varrho, \tag{5.7}$$

where $\Delta\varrho$ is the increment of $\varrho(t)$ arising from the action of the perturbation $V(t)$. Inserting Eq. (5.7) into Eq. (5.5), we obtain

$$(id/dt)\Delta\varrho = [H_0, \Delta\varrho] + [V(t), \varrho_0] + [V(t), \Delta\varrho], \tag{5.8}$$

with the initial condition that $\Delta\varrho = 0$ at $t = t_0$. We introduce the operator

$$\tilde{\Delta\varrho} = \exp(iH_0 t)\, \Delta\varrho \, \exp(-iH_0 t). \tag{5.9}$$

Then the equation of motion for the operator $\tilde{\Delta}\varrho$ is given by

$$(id/dt)\,\tilde{\Delta}\varrho = [\tilde{V}(t), \varrho_0] + [\tilde{V}(t), \tilde{\Delta}\varrho] \qquad (5.10)$$

where

$$\tilde{V}(t) = \exp(iH_0t)V(t)\exp(-iH_0t), \qquad (5.11)$$

and $\tilde{\Delta}\varrho = 0$ for $t = t_0$. Assuming that $\tilde{\Delta}\varrho$ and $\tilde{V}(t)$ are small, we may solve Eq. (5.10) by iteration. The final result is given by

$$\Delta\varrho = \sum_{n=1}^{\infty}(-i)^n \int_{t_0}^{t}\int_{t_0}^{t_1}\cdots\int_{t_0}^{t_{n-1}} dt_1\,dt_2\cdots dt_n \exp(-iH_0t)$$
$$\times [\tilde{V}(t_1), [\tilde{V}(t_2), \ldots, [\tilde{V}(t_n), \varrho], \ldots]] \exp(iH_0t). \qquad (5.12)$$

Substituting Eq. (5.7) into Eq. (5.4), we obtain

$$\langle A(t)\rangle = \langle A(t)\rangle_0 + \Delta\langle A(t)\rangle \qquad (5.13a)$$

where

$$\langle A(t)\rangle_0 = \mathrm{tr}\{A(t)\varrho_0\}, \qquad (5.13b)$$

and the increment $\Delta\langle A(t)\rangle$ due to the action of the external perturbation $V(t)$ is given by

$$\Delta\langle A(t)\rangle = \mathrm{tr}\{A(t)\,\Delta\varrho\}. \qquad (5.13c)$$

Inserting Eq. (5.12) into Eq. (5.13c) and assuming that $\mathrm{tr}\,\Delta\varrho = 0$ by definition, we derive the expression

$$\Delta\langle A(t)\rangle = \sum_{n=1}^{\infty}\Delta^{(n)}\langle A(t)\rangle \qquad (5.14)$$

where

$$\Delta^{(n)}\langle A(t)\rangle = (-i)^n \int_{t_0}^{t}\int_{t_0}^{t_1}\cdots\int_{t_0}^{t_{n-1}} dt_1\,dt_2\cdots dt_n$$
$$\times \langle[[\tilde{A}(t), \tilde{V}(t_1)], \ldots, \tilde{V}(t_n)]\rangle, \qquad (5.15)$$

or using Eq. (5.2), Eq. (5.15) becomes

$$\Delta^{(n)}\langle A(t)\rangle = \sum_{\Omega_1,\ldots,\Omega_n}(-i)^n \int_{t_0}^{t}\int_{t_0}^{t_1}\cdots\int_{t_0}^{t_{n-1}} dt_1\,dt_2\ldots dt_n$$
$$\times \exp\left[\sum_{j=1}^{n}(\varepsilon t_j - i\Omega_j t_j)\right]$$
$$\times \langle[[\tilde{A}(t), \tilde{V}_{\Omega_1}(t_1)], \ldots, \tilde{V}_{\Omega_n}(t_n)]\rangle, \qquad (5.16)$$

where the operators $\tilde{A}(t)$ and $\tilde{V}_\Omega(t)$ are in the Heisenberg representation for the unperturbed system with the Hamiltonian H_0,

$$\tilde{A}(t) = \exp(iH_0 t) A \exp(-iH_0 t), \quad \tilde{V}_\Omega(t) = \exp(iH_0 t) V_\Omega \exp(-iH_0 t),$$

and the angular brackets denote the average over the density matrix ϱ_0. The expression (5.16) is known in the literature as the Kubo formula (Kubo, 1957).

We restrict ourselves to consider only terms linear with respect to the perturbation $V(t)$, i.e., we retain only the first nonvanishing term in the expansion given by Eq. (5.16)

$$\Delta^{(1)}\langle A(t)\rangle = \sum_\Omega (1/i) \int_{t_0}^t dt_1 \exp(\varepsilon t_1 - i\Omega t_1) \langle [\tilde{A}(t), \tilde{V}_\Omega(t_1)]\rangle$$

$$= \sum_\Omega (1/i) \int_{t_0}^t dt_1 \exp(\varepsilon t_1 - i\Omega t_1) \theta(t - t_1) \langle [\tilde{A}(t), \tilde{V}_\Omega(t_1)]\rangle. \tag{5.17}$$

Considering the definition of the retarded Green's function (2.2a), Eq. (5.17) or (5.13a) may be written as

$$\langle A(t)\rangle = \langle A(t)\rangle_0 + \sum_\Omega \int_{t_0}^t dt_1 \exp(\varepsilon t_1 - i\Omega t_1) \langle\!\langle \tilde{A}(t); \tilde{V}_\Omega(t_1)\rangle\!\rangle_{(r)}. \tag{5.18}$$

In the case where the perturbation is adiabatically switched on at time $t_0 = -\infty$, then using the Fourier transform for the retarded Green's function in Eq. (5.18) and integrating over t_1, we obtain

$$\langle A(t)\rangle = \langle A(t)\rangle_0 + (2\pi) \sum_\Omega \exp(\varepsilon t - i\Omega t) \langle\!\langle \tilde{A}; \tilde{V}_\Omega\rangle\!\rangle_{(r,\Omega)}. \tag{5.19}$$

The last term in Eq. (5.19) indicates that the change in the average value of an operator when a perturbation is switched on adiabatically can be expressed in terms of the Fourier transform of the retarded Green's function, which relates the perturbation operator and the operator for the observed quantity.

Similarly, when the perturbation is switched on instantaneously at time $t_0 > -\infty$ [Eq. (5.3)], then from Eq. (5.18), we find the expression

$$\langle A(t)\rangle = \langle A(t)\rangle_0 + i \sum_\Omega \exp(\varepsilon t_0 - i\Omega t_0)$$

$$\times \int_{-\infty}^{+\infty} \langle\!\langle \tilde{A}; \tilde{V}_\Omega\rangle\!\rangle_{(r,\omega)} \frac{\exp[-i\omega(t-t_0)]}{(\omega - \Omega - i\varepsilon)} d\omega, \tag{5.20}$$

which indicates that the reaction of the system can be also, in this case, expressed in terms of the corresponding retarded Green's function. The last term in Eq. (5.20) indicates that when the perturbation is switched on instantaneously at a time $t_0 > -\infty$, natural oscillations are created in the system which lead to additional changes in the increment of the average value of A.

Consider the case when the perturbation V_Ω has the form

$$V_\Omega = -F_0 B_\Omega \tag{5.21}$$

where F_0 is a c-number and represents the amplitude of the external field. It can be, for instance, the intensity of an electric field or that of an alternating magnetic field for physical processes corresponding to electrical conduction or ferromagnetic resonance, respectively; B_Ω is the operator part of the perturbation. Substituting Eq. (5.21) into Eq. (5.19), we obtain

$$\langle A(t) \rangle = \langle A(t) \rangle_0 + F_0 e^{-i\Omega t} \chi_{AB}(\Omega) \tag{5.22}$$

where $\chi_{AB}(\Omega)$ is the complex susceptibility of the system given by

$$\chi_{AB}(\Omega) = -(2\pi) \langle\!\langle \tilde{A}; \tilde{B}_\Omega \rangle\!\rangle_{(r,\Omega)}. \tag{5.23}$$

Equation (5.23) indicates that the complex susceptibility of a system is given by the Fourier transform of the corresponding retarded Green's function and, therefore, $\chi_{AB}(\Omega)$ satisfies the dispersion relations and all the properties of the Green's functions discussed in the preceding section. Equations (5.22) and (5.23) describe also the dissipation of the energy of the external field in the system in question. This is known in the literature as the fluctuation–dissipation theorem (Callen and Welton, 1951; Kubo, 1957).

Equations (5.22) and (5.23) describe the reaction of the system to the external perturbation of arbitrary physical nature. The only requirement is that the corresponding interaction Hamiltonian be written in the form of Eq. (5.2) or (5.3). As an example, we consider the relation between the electrical conductivity tensor and the Green's functions.

A. Electrical Conductivity Tensor

Consider the specific case when a uniform electric field of strength

$$\mathbf{E}(t) = \mathbf{E} \exp(-i\omega t + \varepsilon t) \tag{5.24a}$$

is switched on adiabatically with frequency ω. Then the perturbation assumes the form

$$V = -\mathbf{E} \cdot \mathbf{p}(t) \tag{5.24b}$$

where $\mathbf{p}(t)$ is the total dipole moment of the system. In this case the average operator $A(t)$ is the current density operator $\mathbf{j}$ and the function $\chi(\omega)$ is the complex electrical conductivity tensor denoted by $\sigma_{\alpha\beta}(\omega)$. Using Eq. (5.23), we have

$$\sigma_{\alpha\beta}(\omega) = -(2\pi)\langle\!\langle j_\alpha(t); p_\beta(t')\rangle\!\rangle_{(\mathbf{r},\omega)}$$

$$= i \int_{-\infty}^{+\infty} d(t-t')\theta(t-t')\langle[j_\alpha(t), p_\beta(t')]\rangle \exp[i\omega(t-t')]. \tag{5.25}$$

Integrating by parts, Eq. (5.25) assumes the form

$$\sigma_{\alpha\beta}(\omega) = (-1/\omega) \int_{-\infty}^{+\infty} d(t-t') \exp[i\omega(t-t')]\{\delta(t-t')\langle[j_\alpha(t), p_\beta(t')]\rangle$$
$$- \theta(t-t')\langle[j_\alpha(t), (d/dt')p_\beta(t')]\rangle\}. \tag{5.26}$$

If the volume of the system is taken to be equal to unity, then using the relation

$$(d/dt')p_\beta(t') = j_\beta(t'),$$

we have

$$\sigma_{\alpha\beta}(\omega) = (-1/\omega)\langle[j_\alpha(t), p_\beta(t)]\rangle + (2\pi i/\omega)\langle\!\langle j_\alpha(t); j_\beta(t')\rangle\!\rangle_{(\mathbf{r},\omega)} \tag{5.27}$$

where $\langle\!\langle j_\alpha(t); j_\beta(t')\rangle\!\rangle_{(\mathbf{r},\omega)}$ is the Fourier transform of the current–current retarded Green's function.

In order to calculate the commutator in the first term of Eq. (5.27), we need to specify the kinetic energy operator for the charged particles. For a parabolic and isotropic dispersion law, where the kinetic energy is given by

$$W(\mathbf{q}) = \mathbf{q}^2/2m, \tag{5.28}$$

where $\mathbf{q}$ and m are the momentum and the mass of the electron, respectively, we find

$$\langle[j_\alpha(t), p_\beta(t)]\rangle = (-ie^2n/m)\,\delta_{\alpha\beta}, \tag{5.29}$$

where $n = \langle\varrho(\mathbf{x})\rangle = \langle\psi_\mathbf{x}^\dagger\psi_\mathbf{x}\rangle$ is the average value of the particle density, which for a uniform system is constant and equal to the number of electrons per unit volume. $\psi_\mathbf{x}^\dagger$ and $\psi_\mathbf{x}$ are the Fermi creation and annihilation operators and **x** is the position vector of the electron. Then using Eq. (5.29), we obtain

$$\sigma_{\alpha\beta}(\omega) = (ie^2n/m\omega)\,\delta_{\alpha\beta} + (2\pi i/\omega)\langle\!\langle j_\alpha(t); j_\beta(t')\rangle\!\rangle_{(\mathbf{r},\omega)}. \quad (5.30)$$

For a more complicated dispersion law than that given by Eq. (5.28), especially for problems in the solid state, m in Eq. (5.30) has to be replaced by the average effective mass $\tilde{m}$ defined by the relation (see, for instance, Bonch–Bruevich, 1966a)

$$\tfrac{1}{2}\langle\psi^\dagger(\partial u_\alpha/\partial q_\beta)\psi + [(\partial u_\alpha/\partial q_\beta)\psi^\dagger]\psi\rangle = (n/\tilde{m})\,\delta_{\alpha\beta} \quad (5.31)$$

where $u_\alpha = \partial W(\mathbf{q})/\partial q_\alpha$ is the velocity operator. In such a case instead of Eq. (5.30), we have

$$\sigma_{\alpha\beta}(\omega) = (ie^2n/\tilde{m}\omega)\,\delta_{\alpha\beta} + (2\pi i/\omega)\langle\!\langle j_\alpha(t); j_\beta(t')\rangle\!\rangle_{(\mathbf{r},\omega)}. \quad (5.32)$$

When the dispersion law for $W(\mathbf{q})$ is given by Eq. (5.28), $\tilde{m}$ is equal to m, whereas when $W(\mathbf{q}) = \sum_{i=1}^{3} q_i^2/2m$, then $1/\tilde{m} = \tfrac{1}{3}\sum_{i=1}^{3}(1/m_i)$. When the dispersion law is not parabolic and the velocity operator is defined by the relation $u_\alpha(\mathbf{q}) = (\partial/\partial q_\alpha)W(\mathbf{q}) = q_\alpha f(\mathbf{q}^2)$, where $f(\mathbf{q}^2)$ is a scalar function (the particular case when Eq. (5.28) is applicable, then $f(\mathbf{q}^2) = 1/m$). In such a case, Eq. (5.31) becomes (Bonch–Bruevich and Tyablikov, 1962)

$$\sum_{\mathbf{q}} n(\mathbf{q})[f(\mathbf{q}^2) + \tfrac{2}{3}\mathbf{q}^2 f'(\mathbf{q}^2)] = 1/\tilde{m}, \quad (5.33)$$

where the prime on f indicates differentiation with respect to its argument and $n(\mathbf{q})$ is the electron distribution function, normalized to unity, i.e., $\sum_\mathbf{q} n(\mathbf{q}) = 1$. According to Eq. (5.33), the quantity $\tilde{m}$ may also be temperature dependent.

The average effective mass $\tilde{m}$ is also called the optical mass. The name is justified from consideration of the high-frequency behavior of the complex conductivity tensor. In the region of frequencies, where $\omega \to \infty$, we assume that the current–current correlations decay fast enough both in time and increasing distance between the particles in question; the high-frequency limit also indicates that $\omega\tau \gg 1$, where τ is the typical relaxation time. Therefore, for $\omega \to \infty$ the second term in Eq. (5.32) is small in comparison with the first one and $\sigma_{\alpha\beta}(\omega)$ assumes the form

$$\mathrm{Im}\,\sigma_{\alpha\beta}(\omega) \to (ne^2/\tilde{m}\omega)\,\delta_{\alpha\beta}, \quad \omega \to \infty. \quad (5.34)$$

The dielectric function is related to the complex conductivity tensor by

$$\varepsilon_{\alpha\beta}(\omega) = \varepsilon^0_{\alpha\beta} + (4\pi i/\omega)\sigma_{\alpha\beta}(\omega) \tag{5.35}$$

where $\varepsilon^0_{\alpha\beta}$ is the dielectric function of the lattice, which does not include charge carries. Then for a cubic crystal, we have

$$\text{Re } \varepsilon(\omega) = \varepsilon^0 - (4\pi n e^2/\tilde{m}\omega^2). \tag{5.36}$$

Thus at high frequencies the real part of the dielectric function is given by the corresponding expression for a system of free charges provided that the electron mass m is replaced by the average effective mass $\tilde{m}$. The effective mass of charge carries, which is determined by reflection experiments, is just the quantity $\tilde{m}$. The plasma frequency with mass $\tilde{m}$ is given by the solution of the equation $\text{Re } \varepsilon(\omega) = 0$.

Considering the integral of the quantity $\text{Re } \sigma_{\alpha\beta}(\omega)$ over all frequencies and using the fact that $\sigma_{\alpha\beta}(\omega)$ is an even function of ω, $\sigma_{\alpha\beta}(\omega) = \sigma_{\alpha\beta}(-\omega)$, we derive the following sum rule (Bonch–Bruevich and Tyablikov, 1962; Bonch–Bruevich, 1966a):

$$\int_0^\infty \text{Re } \sigma_{\alpha\beta}(\omega) \, d\omega = (\pi n e^2/2 \tilde{m}) \delta_{\alpha\beta}. \tag{5.37}$$

The expression (5.32) for the complex conductivity $\sigma_{\alpha\beta}(\omega)$ may be conveniently used to calculate the absorption coefficient describing electronic transitions.

B. Absorption Coefficient

The expression for the absorption coefficient $\alpha(\omega)$ and the index of refraction $\eta(\omega)$ at frequency ω can be derived from the well-known Maxwell equations in the form

$$\alpha(\omega) = (2\omega/c)(-(\varepsilon/2) + \{(\varepsilon^2/4) + [(2\pi/\omega) \text{ Re } \sigma(\omega)]^2\}^{1/2})^{1/2} \tag{5.38a}$$

$$\eta(\omega) = ((\varepsilon/2) + \{(\varepsilon^2/4) + [(2\pi/\omega) \text{ Re } \sigma(\omega)]^2\}^{1/2})^{1/2} \tag{5.38b}$$

where $\varepsilon = \text{Re } \varepsilon(\omega)$ is the real part of the dielectric function. In the optical region of frequencies the conditions (Bonch–Bruevich, 1966a,b)

$$\text{Re } \varepsilon(\omega) > 0, \qquad \omega \gg [4\pi/\text{Re } \varepsilon(\omega)] \text{ Re } \sigma(\omega) \tag{5.39}$$

are satisfied. Then the expressions (5.38) become

$$\alpha(\omega) = [4\pi/c\eta(\omega)] \operatorname{Re} \sigma(\omega) \qquad (5.40\text{a})$$

$$\eta(\omega) = \operatorname{Re} \varepsilon^{1/2}(\omega). \qquad (5.40\text{b})$$

In the range of frequencies where electronic transitions occur, $\eta(\omega)$ is a slowly varying function of ω and may be taken as a constant. Substitution of the real part of Eq. (5.32) into Eq. (5.40a) leads to

$$\alpha_{\alpha\beta}(\omega) = [8\pi^2/c\omega\eta(\omega)] \operatorname{Im} \langle\!\langle j_\alpha(t); j_\beta(t') \rangle\!\rangle_{(\mathbf{r},\omega)}. \qquad (5.41)$$

The expression (5.41), which relates the absorption coefficient with the imaginary part of the current–current retarded Green's function, is the most useful one for the investigation of optical properties of solids in particular, interband as well as intraband optical transitions.

C. Interband Optical Transitions

As an example, we use formula (5.41) to calculate the absorption coefficient describing the physical process of direct optical transitions between simple one-electron bands (band-to-band transitions) having relevant extrema at the center of the Brillouin zone. The current density operators can be expressed in the second quantization representation, taking as a basis system the Bloch functions describing an electron in an appropriate ideal lattice

$$\psi_\lambda(\mathbf{x}, s) = S_\sigma(s) u_{\mathbf{q}\nu} e^{i\mathbf{q}\cdot\mathbf{x}}$$

where $S_\sigma(s)$ is the spin wave function, $u_{\mathbf{q}\nu}(\mathbf{x})$ the usual periodic function, and s and σ the coordinate and spin quantum numbers, respectively. The index λ designates the set of the quantities $(\mathbf{q}, \nu, \sigma)$, where $\mathbf{q}$ is a reduced wave vector in the first Brillouin zone and ν is the band index. Then the current density operator takes the form

$$j_\alpha = \sum_{\lambda,\lambda'} \langle \lambda | j_\alpha | \lambda' \rangle a_\lambda^\dagger a_{\lambda'}, \qquad (5.42)$$

where in the absence of a magnetic field and spin–orbit interaction, the matrix elements of the current density operator are given by

$$\langle \lambda | j_\alpha | \lambda' \rangle = (e/V)\, \delta_{\lambda\lambda'} \mathbf{U}_\alpha(\mathbf{q}) + (e/V)(1 - \delta_{\nu\nu'})\, \delta_{\mathbf{q}\mathbf{q}'}\, \delta_{\sigma\sigma'} \mathbf{U}_{\nu\nu'}^\alpha(\mathbf{q})$$

$$\mathbf{U}_{\nu\nu'} = (1/im) \int d\mathbf{x}\, u_{\mathbf{q}\nu}^* \nabla u_{\mathbf{q}'\nu'}, \qquad \mathbf{U}(\mathbf{q}) = (\partial/\partial \mathbf{q}) W_\nu(\mathbf{q}). \qquad (5.43)$$

In Eq. (5.43), V is the volume of the crystal, e the electronic charge, and $W_\nu(\mathbf{q})$ the energy of an electron in the νth band with wave vector $\mathbf{q}$. $\alpha_\lambda^\dagger$ and α_λ are the Fermi creation and annihilation operators describing the λth state of an electron. Substituting Eqs. (5.42) and (5.43) into Eq. (5.41) and considering only interband transitions, we find (Bonch-Bruevich and Rozman, 1964)

$$\alpha_{\alpha\beta}(\omega) = [8\pi^2 e^2/c\omega\eta(\omega)V] \sum_{\lambda,\lambda'} \mathbf{U}^\alpha_{\lambda\lambda'}(\mathbf{q})\mathbf{U}^\beta_{\lambda'\lambda}(\mathbf{q}') \operatorname{Im}\langle\!\langle \alpha_\lambda^\dagger \alpha_{\lambda'} ; \alpha_{\lambda'}^\dagger \alpha_\lambda \rangle\!\rangle_{(\mathrm{r},\omega)}. \tag{5.44}$$

To proceed further, we shall consider only physical processes corresponding to electronic transitions between the isolated bands ν and ν'. This indicates that we exclude the possibility of the formation of bound states (collective excitations) arising from the interaction between the bands ν and ν'. In such a case, we may decouple the two-particle retarded Green's function as (Bonch-Bruevich and Rozman, 1964; Bonch-Bruevich, 1966a,b)

$$\langle\!\langle \alpha_{\lambda_1}^\dagger(t)\alpha_{\lambda_2}(t) ; \alpha_{\lambda_3}^\dagger(t')\alpha_{\lambda_4}(t') \rangle\!\rangle_{(\mathrm{r})}$$
$$\approx \langle \alpha_{\lambda_1}^\dagger(t)\alpha_{\lambda_4}(t') \rangle G_\mathrm{r}(\lambda_2, \lambda_3; t-t') + \langle \alpha_{\lambda_3}^\dagger(t')\alpha_{\lambda_2}(t) \rangle G_\mathrm{a}(\lambda_4, \lambda_1; t-t') \tag{5.45}$$

where $G_{\mathrm{r,a}}(\lambda, \lambda'; t-t') = \langle\!\langle \alpha_\lambda(t); \alpha_{\lambda'}^\dagger(t') \rangle\!\rangle_{(\mathrm{r,a})}$ denotes the retarded or advanced single-particle Green's function, respectively. The decoupling approximation (5.45) is valid asymptotically for large values of ω. It also indicates that we neglect quantities of the order $(\omega\tau)^{-2}$, where τ is of the order of the transport relaxation time. Therefore, the decoupling approximation (5.45) is sufficient only for the study of interband transitions, that is, those which take place between isolated bands. Using Eq. (3.18), we derive the Fourier transform of the imaginary part of the Green's function (5.45) in the form

$$\operatorname{Im}\langle\!\langle \alpha_{\lambda_1}^\dagger \alpha_{\lambda_2} ; \alpha_{\lambda_3}^\dagger \alpha_{\lambda_4} \rangle\!\rangle_{(\mathrm{r},\omega)}$$
$$= 2\int_{-\infty}^{+\infty} d\omega' \operatorname{Im} G_\mathrm{r}(\lambda_2, \lambda_3; \omega')[\operatorname{Im} G_\mathrm{r}(\lambda_1, \lambda_4; \omega'-\omega)$$
$$\times n_\mathrm{F}(\omega'-\omega) - \operatorname{Im} G_\mathrm{r}(\lambda_4, \lambda_1; \omega'-\omega)n_\mathrm{F}(\omega')], \tag{5.46}$$

where $n_\mathrm{F}(\omega)$ is the Fermi distribution function, $n_\mathrm{F}(\omega) = (e^{\beta\omega}+1)^{-1}$. Substitution of Eq. (5.46) into Eq. (5.44) leads to (for $\nu \neq \nu'$)

$$\alpha_{\alpha\beta}(\omega) = [32\pi^2 e^2/c\omega\eta(\omega)V] \sum_{\mathbf{q},\nu,\nu'} \mathbf{U}^\alpha_{\nu\nu'}(\mathbf{q})\mathbf{U}^\beta_{\nu'\nu}(\mathbf{q}) \int_{-\infty}^{+\infty} d\omega' \operatorname{Im} G_\mathrm{r}(\mathbf{q}, \nu; \omega')$$
$$\times \operatorname{Im} G_\mathrm{r}(\mathbf{q}, \nu'; \omega'-\omega)[n_\mathrm{F}(\omega'-\omega) - n_\mathrm{F}(\omega')]. \tag{5.47}$$

Consider the relations

$$|U^\alpha_{\nu\nu'}(\mathbf{q})|^2 = f^\alpha_{\nu\nu'}(\mathbf{q})\Delta_{\nu'\nu}(\mathbf{q})/2m, \qquad \Delta_{\nu'\nu}(\mathbf{q}) \equiv W_{\nu'}(\mathbf{q}) - W_\nu(\mathbf{q}), \quad (5.48)$$

and for a cubic crystal, we have

$$\tfrac{1}{3}\sum_{\alpha'}|U^{\alpha'}_{\nu\nu'}(\mathbf{q})|^2\delta_{\alpha\beta} = \frac{1}{2m}\Delta_{\nu'\nu}(\mathbf{q})\delta_{\alpha\beta}\sum_{\alpha'}f^{\alpha'}_{\nu\nu'}(\mathbf{q}) = \frac{\Delta_{\nu'\nu}(\mathbf{q})}{2m}\delta_{\alpha\beta}f_{\nu\nu'}(\mathbf{q}) \quad (5.49)$$

$$f_{\nu\nu'}(\mathbf{q}) = \tfrac{1}{3}\sum_{\alpha'=1}^{3}f^{\alpha'}_{\nu\nu'}(\mathbf{q})$$

where $f_{\nu\nu'}(\mathbf{q})$ is the oscillator strength for the transition $\nu \to \nu'$. Inserting Eqs. (5.48) and (5.49) into Eq. (5.47) and for a cubic crystal, we obtain (Bonch–Bruevich and Rozman, 1964)

$$\alpha(\omega) = [16\pi^2 e^2/c\omega\eta(\omega)mV]\sum_{\mathbf{q},\nu,\nu'}f_{\nu\nu'}(\mathbf{q})\Delta_{\nu'\nu}(\mathbf{q})\int_{-\infty}^{+\infty}d\omega'\,\mathrm{Im}\,G_\mathrm{r}(\mathbf{q},\nu;\omega')$$
$$\times\,\mathrm{Im}\,G_\mathrm{r}(\mathbf{q},\nu';\omega'-\omega)[n_\mathrm{F}(\omega'-\omega) - n_\mathrm{F}(\omega')]. \quad (5.50)$$

In order to proceed, one needs to know the form for the imaginary parts for the Green's functions appearing in Eq. (5.50) which describe the excitation spectra of the bands ν and ν', respectively. In the ideal case, which occurs in the absence of interactions, the imaginary part of the Green's function has a delta function distribution of the form

$$\mathrm{Im}\,G_\mathrm{r}(\mathbf{q},\nu;\omega) = \tfrac{1}{2}\delta(\omega - W_\nu(\mathbf{q})). \quad (5.51)$$

Then inserting Eq. (5.51) into Eq. (5.50) we derive the standard expression for the absorption coefficient describing the direct absorption process ($\omega > 0$) corresponding to the transition $\nu \to \nu'$

$$\alpha(\omega) = [4\pi^2 e^2/c\omega\eta(\omega)mV]\sum_\mathbf{q}f_{\nu\nu'}(\mathbf{q})\Delta_{\nu'\nu}(\mathbf{q})\delta(\omega - \Delta_{\nu'\nu}(\mathbf{q}))$$
$$\times\,\{n_\mathrm{F}(W_\nu(\mathbf{q})) - n_\mathrm{F}(W_{\nu'}(\mathbf{q}))\}. \quad (5.52)$$

Here ν and ν' may represent the valence and conduction bands, respectively. When interactions are taken into account the imaginary parts of the Green's functions in question do not have delta function distributions like that of Eq. (5.51) but instead have shape functions describing relaxation phenomena that occur within each band.

Notice that under the conditions of thermodynamic equilibrium the quantity in braces in Eq. (5.52) is of the order of unity. If the oscillator strength is a slowly varying function of **q** and can be taken as a constant, then

$$\alpha(\omega) \sim [4\pi^2 e^2/c\omega\eta(\omega)mV] f_{\nu\nu'} \Delta_{\nu'\nu} \varrho_1(\omega), \tag{5.53a}$$

where

$$\varrho_1(\omega) = \sum_{\mathbf{q}} \delta(\omega - \Delta_{\nu'\nu}(\mathbf{q})), \tag{5.53b}$$

and for a parabolic dispersion law $\alpha \sim (\omega - \Delta_{\nu'\nu})^{1/2}$, which is the well-known formula for the absorption coefficient. According to Eq. (5.50), $\alpha(\omega)$ cannot be expressed, in general, directly in terms of the expression for the density of states in question, which for a particular band is defined as

$$\varrho_\nu(\omega) = 2 \sum_{\mathbf{q}} \operatorname{Im} G(\mathbf{q}, \nu; \omega). \tag{5.54}$$

The expression for the density of states, Eq. (5.54), contains all the information needed to study the thermodynamic properties of the system in question.

The general expressions (5.27) and (5.41), which describe the electrical conductivity tensor (and consequently the dielectric tensor) and the absorption coefficient in terms of the Fourier transform of the current–current retarded Green's function, respectively, can be used to study the exciton spectra of molecular crystals with and without the participation of the phonon field (see, for instance, Mavroyannis, 1967b, 1970a).

VI. Calculation of the Green's Functions

As mentioned in Section II, the Fourier transform of the retarded Green's function satisfies the equation of motion

$$\omega \langle\!\langle A(t); B(t') \rangle\!\rangle = (1/2\pi) \langle [A(t), B(t)]_{-\eta} \rangle + \langle\!\langle [A(t), \mathcal{H}]; B(t') \rangle\!\rangle. \tag{6.1}$$

To proceed further, one has to consider analogous equations for the Green's functions $\langle\!\langle [A, \mathcal{H}]; B \rangle\!\rangle$, $\langle\!\langle [[A, \mathcal{H}], \mathcal{H}]; B \rangle\!\rangle$, and so on. Continuing this process, we obtain an infinite chain of coupled equations. Sometimes at one of the steps, it is convenient to differentiate with respect to the second time argument t'. In this case, the chain of equations consists of symmetrized Green's functions of the form $\langle\!\langle [A, \mathcal{H}]; [B, \mathcal{H}] \rangle\!\rangle$. An exact solution of such an infinite set of coupled equations is, as a rule, impossible, and therefore one must find some approximate method of decoupling the chain of equations. The usual technique is to decouple the hierarchy of equations by expressing some higher order Green's functions in terms of lower order ones. This is based on the assumption that the *i*th-order Green's function may be expressed in a well-defined manner in terms of the lower order Green's functions. It is at this point that various appproximations come into the act. Instead of considering these approximations in the general form, we discuss some specific examples. In particular, the decoupling approximation is used to calculate the excitation spectrum of one-particle and two-particle Green's functions for systems with direct interactions. In the same approximation, the effect of three-particle correlations on the excitation spectrum of one-particle Green's functions is considered. The derived results are used in Section VII to discuss the excitation spectra of charge-transfer complexes for a tight-binding model corresponding to a molecular crystal. In Section VIII, the retarded (advanced) double-time Green's function is calculated by means of perturbation methods, while the causal Green's function is considered through the functional derivative approach.

A. Systems with Direct Interactions

For a system of Fermi particles with direct interactions, the Hamiltonian is taken in the form

$$\mathcal{H} = \sum_{f,f'} L(f,f') \alpha_f^\dagger \alpha_{f'} + \tfrac{1}{2} \sum_{f,f',f_1,f_1'} \langle f, f_1 | V | f', f_1' \rangle \alpha_f^\dagger \alpha_{f_1}^\dagger \alpha_{f_1'} \alpha_{f'}. \tag{6.2}$$

For a crystal, the Hamiltonian (6.2) describes an ensemble of valence electrons in an undisplaced lattice of volume V. The compound index $f \equiv (\mathbf{n}, i, s)$, where $\mathbf{n}$ is the lattice site, and i and s designate the electron state and the spin component of an electron ($\pm \tfrac{1}{2}$), respectively. The creation and annihilation operators, $\alpha_f^\dagger$ and α_f, describe the fth electron state and satisfy Fermi anticommutation relations. The matrix elements of $L(f,f')$ are

$$L(f,f') = \langle f | L | f' \rangle = \int \psi_f^*(\mathbf{r_n}) [(-1/2m) \nabla^2 + \sum_{f'} V(\mathbf{r_n} - \mathbf{r_{n'}})] \\ \times \psi_{f'}(\mathbf{r_{n'}}) \, d\tau \tag{6.3a}$$

where $V(\mathbf{r_n} - \mathbf{r_{n'}})$ is the potential of an electron at the lattice site $\mathbf{n}$. The last term in Eq. (6.2) represents the electron–electron interaction, and the matrix elements of the coupling function are given by

$$\langle f, f_1 | V | f', f_1' \rangle \int \psi_f^*(\mathbf{r_n}) \psi_{f_1}^*(\mathbf{r_m}) V(\mathbf{r_n} - \mathbf{r_m}) \psi_{f'}(\mathbf{r_n}) \psi_{f_1'}(\mathbf{r_m}) \, d\tau_1 \, d\tau_2 \tag{6.3b}$$

and $V(\mathbf{r_n} - \mathbf{r_m}) = e^2 \mid (\mathbf{r_n} - \mathbf{r_m})$, where e^2 denotes the electronic charge divided by the static dielectric constant of the substance. The ψ's are the Wannier functions describing the electronic states, and $\mathbf{r_n}$ and $\mathbf{r_m}$ are the position vectors of an electron at the lattice site $\mathbf{n}$ and $\mathbf{m}$, respectively. The matrix elements in Eq. (6.3b) are taken to satisfy the symmetry relations

$$\begin{aligned}\langle f, f_1 | V | f', f_1' \rangle &= \langle f_1, f | V | f_1', f' \rangle \\ \langle f, f_1 | V | f', f_1' \rangle^* &= \langle f_1', f' | V | f_1, f \rangle.\end{aligned} \tag{6.3c}$$

Notice that if we consider the electron gas case, then f denotes the momentum and the spin component of the electron, and the last term in Eq. (6.2) represents the Coulomb interaction between the electrons.

B. One-Particle Green's Function

It will be shown that the one-electron spectrum can be obtained from the Hamiltonian (6.2) in the self-consistent field approximation. If we introduce the one-electron retarded Green's function $G(f, g; t - t') = \langle\!\langle \alpha_f; \alpha_g^\dagger \rangle\!\rangle$, then using Eqs. (6.1) and (6.2), we find the equation of motion

$$\omega G(f, g; \omega) = \delta_{gf}/2\pi + \sum_{f'} L(f, f') G(f', g; \omega)$$
$$+ \sum_{f_1, f_2, f_3} \langle f, f_1 | V | f_3, f_2 \rangle \langle\!\langle \alpha_{f_1}^\dagger \alpha_{f_2} \alpha_{f_3}; \alpha_g^\dagger \rangle\!\rangle. \tag{6.4}$$

We consider the approximate solution of Eq. (6.4) in the first approximation (Tyablikov and Bonch–Bruevich, 1962), and compare it with that obtained by the Hartree–Fock self-consistent field method (Martin and Schwinger, 1959).

To solve Eq. (6.4) we make the approximation

$$\langle\!\langle \alpha_{f_1}^\dagger \alpha_{f_2} \alpha_{f_3}; \alpha_g^\dagger \rangle\!\rangle \to \langle \alpha_{f_1}^\dagger \alpha_{f_2} \rangle \langle\!\langle \alpha_{f_3}; \alpha_g^\dagger \rangle\!\rangle - \langle \alpha_{f_1}^\dagger \alpha_{f_3} \rangle \langle\!\langle \alpha_{f_2}; \alpha_g^\dagger \rangle\!\rangle. \tag{6.5}$$

It can be shown (Tyablikov and Bonch–Bruevich, 1962) that the decoupling approximation (6.5) is asymptotically valid for large values of ω, provided that there are no bound states arising from the three-particle correlations. Equation (6.5) also indicates that all dynamic effects (various scattering processes) arising from electron–electron interactions are completely discarded and that the electron in question sees only the average field of all others (static effect), which is described by the distribution function of the form $\langle a_{f_1}^\dagger a_{f_2}\rangle$. Thus Eq. (6.5) is valid in the range of frequencies $\omega\tau \gg 1$, where τ is of the order of some average relaxation time. Substitution of Eq. (6.5) into Eq. (6.4) leads to the approximate equation

$$\omega G(f,g;\omega) - \sum_{f'}[L(f,f') + \sum_{f_1,f_2} J(f,f_1|f',f_2)\langle a_{f_1}^\dagger a_{f_2}\rangle]G(f',g;\omega)$$
$$= \delta_{fg}/2\pi, \tag{6.6}$$

where

$$J(f,f_1|f',f_2) = \langle f,f_1|V|f',f_2\rangle - \langle f,f_1|V|f_2,f'\rangle. \tag{6.7a}$$

If we introduce the expression for the mass operator in the first approximation as

$$M_1(f,f') = \sum_{f_1,f_2} J(f,f_1|f',f_2)\langle a_{f_1}^\dagger a_{f_2}\rangle, \tag{6.7b}$$

then Eq. (6.6) assumes the form

$$\omega G(f,g;\omega) - \sum_{f'}[L(f,f') + M_1(f,f')]G(f',g;\omega) = \delta_{fg}/2\pi. \tag{6.8a}$$

Let $\chi_{f\nu}$ and ω_ν be the eigenfunctions and eigenvalues of the equation

$$\omega_\nu \chi_{f\nu} - \sum_{f'}[L(f,f') + \sum_{f_1,f_2} J(f,f_1|f',f_2)\langle a_{f_1}^\dagger a_{f_2}\rangle]\chi_{f'\nu} = 0, \tag{6.8b}$$

with the normalization condition $\sum_f \chi_{f\nu}^* \chi_{f\nu'} = \delta_{\nu\nu'}$. Then we find

$$G(f,g;\omega) = (1/2\pi)\sum_\nu [\chi_{f\nu}\chi_{g\nu}^*/(\omega - \omega_\nu)]. \tag{6.8c}$$

Using Eq. (3.17), the spectral function

$$J(f,g;\omega) = \sum_\nu \chi_{f\nu}\chi_{g\nu}^*(\exp(\beta\omega_\nu) + 1)^{-1}\delta(\omega - \omega_\nu) \tag{6.9}$$

has a delta function distribution peaked at frequencies $\omega = \omega_\nu$. From Eq. (3.18a), we derive the distribution function

$$\langle a_g^\dagger a_f\rangle = \sum_\nu \chi_{g\nu}^*\chi_{f\nu}[\exp(\beta\omega_\nu) + 1]^{-1} \tag{6.10a}$$

and

$$n_f = \langle \alpha_f^\dagger \alpha_f \rangle = \sum_\nu |\chi_{f\nu}|^2 [\exp(\beta\omega_\nu) + 1]^{-1}. \quad (6.10b)$$

The subsidiary equation (6.8b) may be written as

$$\omega\chi_{f\nu} - \sum_{f'}[L(f,f') + \sum_{f_1,f_2} J(f,f_1|f',f_2)F(f_1,f_2)]\chi_{f'\nu} = 0 \quad (6.11a)$$

where

$$F(f_1,f_2) = \langle \alpha_{f_1}^\dagger \alpha_{f_2} \rangle = \sum_{\nu'} \chi_{f_1\nu'}^* \chi_{f_2\nu'}[\exp(\beta\omega_{\nu'}) + 1]^{-1}. \quad (6.11b)$$

The expression (6.11a) or (6.8b) is just the Hartree–Fock equation at finite temperature (Tyablikov and Bonch-Bruevich, 1962). In the limit, when $\beta \to \infty$, Eq. (6.11a) is reduced to the Hartree–Fock self-consistent field equation at zero temperature. Comparison of Eqs. (6.8a) and (6.11a) indicates that in this approximation the self-consistent field is represented by the mass operator. It can easily be shown that for spatially uniform systems, Eq. (6.8) is diagonal in the momentum representation.

C. Two-Particle Green's Function, Exciton Spectra

If we introduce the Fourier transform of the two-particle or collective Green's function $\langle\!\langle \alpha_{f_1}^\dagger \alpha_{f_2}; \alpha_{g_2}^\dagger \alpha_{g_1} \rangle\!\rangle$, then using Eq. (6.1) and the Hamiltonian (6.2), we derive the equation of motion

$$\omega\langle\!\langle \alpha_{f_1}^\dagger \alpha_{f_2}; \alpha_{g_1}^\dagger \alpha_{g_2} \rangle\!\rangle$$

$$= (1/2\pi)[\langle \alpha_{f_1}^\dagger \alpha_{g_2} \rangle \delta_{f_2 g_1} - \langle \alpha_{g_1}^\dagger \alpha_{f_2} \rangle \delta_{f_1 g_2}]$$

$$+ \sum_f L(f_2,f)\langle\!\langle \alpha_{f_1}^\dagger \alpha_f; \alpha_{g_1}^\dagger \alpha_{g_2} \rangle\!\rangle - \sum_f L(f,f_1)\langle\!\langle \alpha_f^\dagger \alpha_{f_2}; \alpha_{g_1}^\dagger \alpha_{g_2} \rangle\!\rangle$$

$$+ \tfrac{1}{2} \sum_{f,f',f''} J(f_2,f|f',f'')\langle\!\langle \alpha_{f_1}^\dagger \alpha_f^\dagger \alpha_{f''} \alpha_{f'}; \alpha_{g_1}^\dagger \alpha_{g_2} \rangle\!\rangle$$

$$+ \tfrac{1}{2} \sum_{f,f',f''} J(f,f'|f'',f_1)\langle\!\langle \alpha_f^\dagger \alpha_{f'}^\dagger \alpha_{f''} \alpha_{f_2}; \alpha_{g_1}^\dagger \alpha_{g_2} \rangle\!\rangle. \quad (6.12)$$

The last two terms in Eq. (6.12) contain three-particle Green's functions which have four particles on the left-hand side with the same time argument. To proceed further, one has to consider the equations of motion for the Green's functions that appear in the last two terms of Eq. (6.12). In this example, we limit ourselves to obtaining solutions of Eq. (6.12) in the first approximation only. Thus we consider the following decoupling approximation for the three-particle Green's function (Dzyub, 1959, 1961):

$$\langle\!\langle \alpha^\dagger_{f_1}\alpha^\dagger_{f_2}\alpha_{f_3}\alpha_{f_4}; \alpha^\dagger_{g_1}\alpha_{g_2} \rangle\!\rangle$$
$$\to \langle \alpha^\dagger_{f_1}\alpha_{f_3}\rangle\langle\!\langle \alpha^\dagger_{f_2}\alpha_{f_4}; \alpha^\dagger_{g_1}\alpha_{g_2}\rangle\!\rangle - \langle \alpha^\dagger_{f_2}\alpha_{f_4}\rangle\langle\!\langle \alpha^\dagger_{f_1}\alpha_{f_3}; \alpha^\dagger_{g_1}\alpha_{g_2}\rangle\!\rangle$$
$$+ \langle \alpha^\dagger_{f_1}\alpha_{f_4}\rangle\langle\!\langle \alpha^\dagger_{f_2}\alpha_{f_3}; \alpha^\dagger_{g_1}\alpha_{g_2}\rangle\!\rangle - \langle \alpha^\dagger_{f_2}\alpha_{f_3}\rangle\langle\!\langle \alpha^\dagger_{f_1}\alpha_{f_4}; \alpha^\dagger_{g_1}\alpha_{g_2}\rangle\!\rangle. \quad (6.13)$$

The approximation (6.13) is asymptotically valid for large values of ω, $\omega\tau \gg 1$, provided that no bound states are formed from the correlation of three or four particles with the same time argument which appear in the expressions for the Green's functions in question. Thus Eq. (6.13) implies that correlations of higher order than that of the two-particle ones are discarded. Substitution of Eq. (6.13) into Eq. (6.12) leads to the approximate expression

$$\omega\langle\!\langle \alpha^\dagger_{f_1}\alpha_{f_2}; \alpha^\dagger_{g_1}\alpha_{g_2}\rangle\!\rangle$$
$$= (1/2\pi)[\langle \alpha^\dagger_{f_1}\alpha_{g_2}\rangle \delta_{f_2 g_1} - \langle \alpha^\dagger_{g_1}\alpha_{f_2}\rangle \delta_{f_1 g_2}]$$
$$+ \sum_f \bar{L}(f_2,f)\langle\!\langle \alpha^\dagger_{f_1}\alpha_f; \alpha^\dagger_{g_1}\alpha_{g_2}\rangle\!\rangle - \sum_f \bar{L}(f,f_1)\langle\!\langle \alpha^\dagger_f\alpha_{f_2}; \alpha^\dagger_{g_1}\alpha_{g_2}\rangle\!\rangle$$
$$+ \sum_{f,f',f''} [J(f_2,f'|f,f'')\langle \alpha^\dagger_{f_1}\alpha_f\rangle - J(f,f'|f_1,f'')\langle \alpha^\dagger_f\alpha_{f_2}\rangle]$$
$$\times \langle\!\langle \alpha^\dagger_{f'}\alpha_{f''}; \alpha^\dagger_{g_1}\alpha_{g_2}\rangle\!\rangle, \quad (6.14)$$

where

$$\bar{L}(f_1,f_2) = L(f_1,f_2) + \sum_{f,f'} J(f_1,f|f_2,f')\langle \alpha^\dagger_f\alpha_{f'}\rangle. \quad (6.15)$$

The physical interpretation of the approximate expression (6.14) is that an excited electron interacts with a hole to form an electron–hole pair in the background of the occupied single-electron states. For each electron–hole pair, the remaining electrons create a self-consistent field which is determined by the electron density $\langle \alpha^\dagger_f\alpha_{f'}\rangle$. If we define the mass operator in the first approximation as

$$M_1(f_1,f_2;f',f'') = \sum_f [J(f_2,f'|f,f'')\langle \alpha^\dagger_{f_1}\alpha_f\rangle - J(f,f'|f_1,f'')$$
$$\times \langle \alpha^\dagger_{f_2}\alpha_f\rangle], \quad (6.16)$$

8. The Green's Function Method

then Eq. (6.15) assumes the form

$$\omega \langle\!\langle \alpha^\dagger_{f_1} \alpha_{f_2} ; \alpha^\dagger_{g_1} \alpha_{g_2} \rangle\!\rangle$$
$$= (1/2\pi)[\langle \alpha^\dagger_{f_1} \alpha_{g_2} \rangle \delta_{f_2 g_1} - \langle \alpha^\dagger_{g_1} \alpha_{f_2} \rangle \delta_{f_1 g_2}]$$
$$+ \sum_f \bar{L}(f_2, f) \langle\!\langle \alpha^\dagger_{f_1} \alpha_f ; \alpha^\dagger_{g_1} \alpha_{g_2} \rangle\!\rangle - \sum_f \bar{L}(f, f_1) \langle\!\langle \alpha_f{}^\dagger \alpha_{f_2} ; \alpha^\dagger_{g_1} \alpha_{g_2} \rangle\!\rangle$$
$$+ \sum_{f', f''} M_1(f_1, f_2 ; f', f'') \langle\!\langle \alpha^\dagger_{f'} \alpha_{f''} ; \alpha^\dagger_{g_1} \alpha_{g_2} \rangle\!\rangle. \tag{6.17}$$

We write the expressions for $\bar{L}(f_2, f)$ and $\bar{L}(f, f_1)$ as

$$\bar{L}(f_2, f) = L(f_2, f) + M_1(f_2, f) \tag{6.18a}$$
$$\bar{L}(f, f_1) = L(f, f_1) + M_1(f, f_1), \tag{6.18b}$$

where

$$M_1(f_2, f) = \sum_{f', f''} J(f_2, f' \,|\, f, f'') \langle \alpha^\dagger_{f'} \alpha_{f''} \rangle. \tag{6.18c}$$

Thus the expressions for $M_1(f_2, f)$, $M_1(f, f_1)$, and $M_1(f_1, f_2; f', f'')$ represent the mass operator in the first approximation for the electron, hole, and electron–hole pair, respectively.

In order to proceed with the analysis of Eq. (6.17), we consider the subsidiary equation

$$\omega_\nu \chi_\nu(f_1, f_2) - \sum_f \bar{L}(f_2, f) \chi_\nu(f_1, f) + \sum_f \bar{L}(f, f_1) \chi_\nu(f, f_2)$$
$$- \sum_{f', f''} M_1(f_1, f_2 ; f', f'') \chi_\nu(f', f'') = 0 \tag{6.19}$$

and the normalization condition $\sum_{f_1, f_2} \chi_\nu(f_1, f_2) \chi^*_{\nu'}(f_1, f_2) = \delta_{\nu \nu'}$. Then the solution of Eq. (6.17) assumes the form

$$\langle\!\langle \alpha^\dagger_{f_1} \alpha_{f_2} ; \alpha^\dagger_{g_1} \alpha_{g_2} \rangle\!\rangle = (1/2\pi) \sum_{\nu, \nu'} [c_{\nu \nu'}/(\omega - \omega_\nu)] \chi_\nu(f_1, f_2) \chi^*_{\nu'}(g_1, g_2) \tag{6.20}$$

where

$$c_{\nu \nu'} = \sum_{\substack{f, f' \\ g, g'}} [\langle \alpha_g{}^\dagger \alpha_{f'} \rangle \delta_{g'f} - \langle \alpha_f{}^\dagger \alpha_{g'} \rangle \delta_{gf'}] \chi_\nu{}^*(f, f') \chi_{\nu'}(g, g'). \tag{6.21}$$

Using Eq. (3.17) for the spectral function, we have

$$J(f_1, f_2 ; g_1, g_2 ; \omega) = \sum_{\nu, \nu'} c_{\nu \nu'} \chi_\nu(f_1, f_2) \chi^*_{\nu'}(g_1, g_2)(\exp(\beta \omega_\nu) + 1)^{-1}$$
$$\times \delta(\omega - \omega_\nu). \tag{6.22}$$

Thus the electron–hole (exciton) spectrum has a delta function distribution peaked at frequencies $\omega = \omega_\nu$ which are determined by the roots of Eq. (6.19) and are temperature dependent through the electron densities $\langle \alpha_f{}^\dagger \alpha_{f'} \rangle$. Equations (6.17) and (6.20) describe the bare exciton spectrum in solids (Dzyub, 1959, 1961). The solutions of Eq. (6.19) correspond to the Frenkel- or Mott-type exciton spectrum, depending on whether or not the electron and the hole are located at the same or different lattice sites, respectively. The expression (6.20) indicates that in this approximation, the electron–hole pairs behave approximately like Bose particles.

Notice that when the expression for the exciton Green's function is known, it can be used to derive the average or the ground-state energy of the crystal (Mavroyannis, 1965). When the coupling between the electron–hole pairs and the electromagnetic field is taken into account in the Hamiltonian (6.2) and the diagonalization is carried out in the self-consistent field approximation, then a new quasi particle is formed, the so-called polariton (Hopfield, 1958; Agranovich, 1960; Mavroyannis, 1967a). Polaritons consist of electron–hole pairs dressed by the photons of the electromagnetic field. For details concerning exciton and polariton spectra of molecular crystals as well as relevant literature, we refer the reader to the work of Mavroyannis (1970b, c).

It is pointed out that because of the use of the decoupling approximations given by Eqs. (6.5) and (6.13), both the one- and the two-particle excitation spectra have delta function distributions. Dynamic effects will appear only when the two- and three-particle Green's functions are taken into consideration, respectively.

D. Three-Particle Correlations

In Section VI.B, the one-electron spectrum is considered in the first approximation; i.e., the Green's function that appears in the last term of the equation

$$\omega G(f, g; \omega) = \delta_{gf}/2\pi + \sum_{f'} L(f, f') G(f', g; \omega)$$
$$+ \sum_{f_1, f_2, f_3} \langle f, f_1 | V | f_3, f_2 \rangle \langle\!\langle \alpha_{f_1}^\dagger \alpha_{f_2} \alpha_{f_3}; \alpha_g{}^\dagger \rangle\!\rangle \qquad (6.23)$$

has been decoupled and expressed in terms of the one-particle Green's function through Eq. (6.5). In the last term of Eq. (6.23) the two-particle Green's function occurs, which has on its left-hand side three particles with the same time argument. To improve the treatment for the one-particle spectrum derived in Section VI.B, let us proceed to calculate the Green's function $\langle\!\langle \alpha_{f_1}^\dagger \alpha_{f_2} \alpha_{f_3}; \alpha_g{}^\dagger \rangle\!\rangle$. Using Eqs. (6.1) and (6.2), we derive the equation of motion

$$\omega \langle\!\langle \alpha_{f_1}^\dagger \alpha_{f_2} \alpha_{f_3}; \alpha_g{}^\dagger \rangle\!\rangle$$

$$= (1/2\pi)[\langle \alpha_{f_1}^\dagger \alpha_{f_2} \rangle \delta_{gf_3} - \langle \alpha_{f_1}^\dagger \alpha_{f_3} \rangle \delta_{gf_2}]$$

$$+ \sum_{f'} L(f_3, f') \langle\!\langle \alpha_{f_1}^\dagger \alpha_{f_2} \alpha_{f'}; \alpha_g{}^\dagger \rangle\!\rangle + \sum_{f'} L(f_2, f') \langle\!\langle \alpha_{f_1}^\dagger \alpha_{f'} \alpha_{f_3}; \alpha_g{}^\dagger \rangle\!\rangle$$

$$- \sum_{f'} L(f', f_1) \langle\!\langle \alpha_{f'}^\dagger \alpha_{f_2} \alpha_{f_3}; \alpha_g{}^\dagger \rangle\!\rangle$$

$$+ \sum_{f', f'', f'''} \langle f_3, f' | V | f'', f''' \rangle \langle\!\langle \alpha_{f_1}^\dagger \alpha_{f_2} \alpha_{f'}^\dagger \alpha_{f'''} \alpha_{f''}; \alpha_g{}^\dagger \rangle\!\rangle$$

$$+ \sum_{f', f'', f'''} \langle f_2, f' | V | f'', f''' \rangle \langle\!\langle \alpha_{f_1}^\dagger \alpha_{f'}^\dagger \alpha_{f'''} \alpha_{f_3}; \alpha_g{}^\dagger \rangle\!\rangle$$

$$+ \sum_{f', f'', f'''} \langle f', f'' | V | f''', f_1 \rangle \langle\!\langle \alpha_{f'}^\dagger \alpha_{f''}^\dagger \alpha_{f'''} \alpha_{f_2} \alpha_{f_3}; \alpha_g{}^\dagger \rangle\!\rangle. \quad (6.24)$$

The three-particle Green's functions that occur in the last three terms of Eq. (6.24) have five particles on their left-hand side with the same time argument. To solve Eq. (6.24), we write the Green's functions of the form $\langle\!\langle \alpha_{g_1}^\dagger \alpha_{g_2}^\dagger \alpha_{g_3} \alpha_{g_4} \alpha_{g_5}; \alpha_g{}^\dagger \rangle\!\rangle$ approximately as

$$\langle\!\langle \alpha_{g_1}^\dagger \alpha_{g_2}^\dagger \alpha_{g_3} \alpha_{g_4} \alpha_{g_5}; \alpha_g{}^\dagger \rangle\!\rangle$$

$$\approx \langle\!\langle N_{g_2}[\delta_{g_2 g_3} \alpha_{g_1}^\dagger \alpha_{g_4} \alpha_{g_5} - \delta_{g_2 g_4} \alpha_{g_1}^\dagger \alpha_{g_3} \alpha_{g_5} + \delta_{g_2 g_5} \alpha_{g_1}^\dagger \alpha_{g_3} \alpha_{g_4}]$$

$$+ N_{g_1}[-\delta_{g_1 g_5} \alpha_{g_2}^\dagger \alpha_{g_3} \alpha_{g_4} + \delta_{g_1 g_4} \alpha_{g_2}^\dagger \alpha_{g_3} \alpha_{g_5} - \delta_{g_1 g_3} \alpha_{g_2}^\dagger \alpha_{g_4} \alpha_{g_3}]; \alpha_g{}^\dagger \rangle\!\rangle$$

(6.25)

and

$$\langle\!\langle N_{g'} \alpha_{g_1}^\dagger \alpha_{g_2} \alpha_{g_3}; \alpha_g{}^\dagger \rangle\!\rangle \approx \langle\!\langle N_{g'} N_{g_1}(\delta_{g_1 g_2} \alpha_{g_3} - \delta_{g_1 g_3} \alpha_{g_2}); \alpha_g{}^\dagger \rangle\!\rangle \quad (6.26)$$

where $N_g = \alpha_g{}^\dagger \alpha_g$. Then the decoupling procedure consists of the following approximation:

$$\langle\!\langle N_{g_2} \alpha_{g_1}^\dagger \alpha_{g_4} \alpha_{g_5}; \alpha_g{}^\dagger \rangle\!\rangle \approx n_{g_2} \langle\!\langle \alpha_{g_1}^\dagger \alpha_{g_4} \alpha_{g_5}; \alpha_g{}^\dagger \rangle\!\rangle \quad (6.27)$$

$$\langle\!\langle N_{g'} N_{g_1} \alpha_{g_3}; \alpha_g{}^\dagger \rangle\!\rangle \approx \langle N_{g'} N_{g_1} \rangle \langle\!\langle \alpha_{g_3}; \alpha_g{}^\dagger \rangle\!\rangle \quad (6.28)$$

where $n_{g_2} = \langle \alpha_{g_2}^\dagger \alpha_{g_2} \rangle$. The decoupling approximations (6.27) and (6.28) are equivalent to linearizing the equations of motion and they are correct in the Hartree–Fock self-consistent field approximation. Substituting Eqs. (6.25)–(6.28) into Eq. (6.24), we get the approximate expression

(Mavroyannis, 1972c)

$$\omega \langle\!\langle \alpha_{f_1}^\dagger \alpha_{f_2} \alpha_{f_3} ; \alpha_g^\dagger \rangle\!\rangle$$
$$= (1/2\pi) n_{f_1} (\delta_{f_1 f_2} \delta_{g f_3} - \delta_{f_1 f_3} \delta_{g f_2})$$
$$+ \sum_{f'} \bar{L}(f_3, f') \langle\!\langle \alpha_{f_1}^\dagger \alpha_{f_2} \alpha_{f'} ; \alpha_g^\dagger \rangle\!\rangle + \sum_{f'} \bar{L}(f_2, f') \langle\!\langle \alpha_{f_1}^\dagger \alpha_{f'} \alpha_{f_3} ; \alpha_g^\dagger \rangle\!\rangle$$
$$- \sum_{f'} \bar{L}(f', f_1) \langle\!\langle \alpha_{f'}^\dagger \alpha_{f_2} \alpha_{f_3} ; \alpha_g^\dagger \rangle\!\rangle$$
$$+ \sum_{f', f''} J(f_2, f' | f_1, f'')(n_{f_1} - n_{f_2}) \langle\!\langle \alpha_{f'}^\dagger \alpha_{f''} \alpha_{f_3} ; \alpha_g^\dagger \rangle\!\rangle$$
$$+ \sum_{f', f''} J(f_3, f' | f_1, f'')(n_{f_1} - n_{f_3}) \langle\!\langle \alpha_{f'}^\dagger \alpha_{f_2} \alpha_{f''} ; \alpha_g^\dagger \rangle\!\rangle$$
$$+ \sum_{f', f''} \langle f_3, f_2 | V | f', f'' \rangle (1 - n_{f_2} - n_{f_3}) \langle\!\langle \alpha_{f_1}^\dagger \alpha_{f'} \alpha_{f''} ; \alpha_g^\dagger \rangle\!\rangle$$
$$- \sum_{f', f'', f'''} [\langle f_2, f' | V | f'', f''' \rangle \delta_{f_1 f_3}$$
$$- \langle f_3, f' | V | f'', f''' \rangle \delta_{f_1 f_2}] n_{f_1}$$
$$\times \langle\!\langle \alpha_{f'}^\dagger \alpha_{f''} \alpha_{f'''} ; \alpha_g^\dagger \rangle\!\rangle + \sum_{g'} S(g') \langle\!\langle \alpha_{g'} ; \alpha_g^\dagger \rangle\!\rangle$$
$$+ L(f_2, f_1)(n_{f_1} - n_{f_2}) \langle\!\langle \alpha_{f_3} ; \alpha_g^\dagger \rangle\!\rangle \tag{6.29}$$

where

$$S(g') = \sum_{f', f''} J(f_3, f' | g', f'') \langle \alpha_{f_1} \alpha_{f_2}^\dagger \alpha_{f'}^\dagger \alpha_{f''} \rangle$$
$$+ \sum_{f', f''} J(f', f_2 | g', f'') \langle \alpha_{f_1}^\dagger \alpha_{f'}^\dagger \alpha_{f''} \alpha_{f_3} \rangle$$
$$+ \sum_{f', f''} \langle f', f'' | V | g', f_1 \rangle \langle \alpha_{f'}^\dagger \alpha_{f''}^\dagger \alpha_{f_2} \alpha_{f_3} \rangle - L(f_2, f_3) n_{f_1} \delta_{f_1 f_3}$$
$$= J(f_3, f_2 | g', f_1) \langle N_{f_1}(1 - N_{f_2}) - N_{f_1} N_{f_3}(1 - \delta_{f_1 f_3})$$
$$+ N_{f_2} N_{f_3}(1 - \delta_{f_2 f_3}) \rangle - L(f_2, f_3) n_{f_1} \delta_{f_1 f_3}. \tag{6.30}$$

The equation of motion (6.29) for the two-particle Green's function describes the three-particle correlations with the same time argument in the first approximation. Inspection of Eq. (6.29) reveals that the coupling functions that appear in the various terms can be expressed in terms of the mass operator in the first approximation corresponding to one, two, and three particles, respectively. For example, the coupling functions of the second, third, and fourth terms are of the form of Eqs. (6.18) and include the one-particle mass operators, those of the fifth, sixth, and seventh terms correspond to that of two-particle, while the coupling functions of the eighth term correspond to the three-particle mass operator. The last two terms in Eq. (6.29) indicate the coupling between the one-particle and two-particle Green's functions. Obviously, the existence of a bound state (if any), arising from the space correlation of three particles, depends entirely on the strength of the coupling functions of the eighth term in Eq. (6.29), provided that at least one of the arguments of the delta functions in question is satisfied and the electron state f_1 is occupied ($n_{f_1} \neq 0$). The problem is now reduced to finding solutions of the coupled equations (6.23) and (6.29), which are applicable for any interacting Fermi system.

VII. Charge-Transfer Spectra of Molecular Crystals

We make use of Eqs. (6.23) and (6.29) to discuss the excitation spectrum of charge-transfer (CT) complexes in molecular crystals (Mavroyannis, 1972c). We consider a tight-binding model for a molecular crystal having, for simplicity, one molecule (atom) per unit cell; there are N unit cells in the crystal volume V. We introduce the (Frenkel) exciton creation and annihilation operators $b^\dagger_{\mathbf{m}\nu} = (\alpha^\dagger_{\mathbf{m}\nu}\alpha_{\mathbf{m}0})$ and $b_{\mathbf{m}\nu} = (\alpha^\dagger_{\mathbf{m}0}\alpha_{\mathbf{m}\nu})$, respectively; $\mathbf{m}$ is the lattice site and 0 and ν denote the ground and excited electron state, respectively. In the model under consideration, electrons and tightly bound electron–hole pairs (excitons) describe excited states (excitation bands) in the crystal, while Fermi particles in the ground state represent holes in the valence band. Only spin-allowed transitions are considered and the spin indices are suppressed for convenience.

A. Green's Functions

Using the above-described model for a molecular solid, Eq. (6.23) may be written as

$$\omega G(\mathbf{n}\nu, \mathbf{n}_1\nu; \omega)$$
$$= \delta_{\mathbf{nn}_1} | 2\pi + \sum_{(\mathbf{n}'-\mathbf{n})}{}' \bar{L}(\mathbf{n}\nu, \mathbf{n}'\nu)G(\mathbf{n}'\nu, \mathbf{n}_1\nu; \omega)$$
$$+ \sum_{\substack{\mathbf{m},\mathbf{n}',\mu,\nu' \\ (\mathbf{m} \neq \mathbf{n})}} J(\mathbf{n}\nu, \mathbf{m}0 | \mathbf{n}'\nu', \mathbf{m}\mu)\langle\!\langle b_{\mathbf{m}\mu}\alpha_{\mathbf{n}'\nu'}; \alpha^\dagger_{\mathbf{n}_1\nu}\rangle\!\rangle$$
$$+ \sum_{\substack{\mathbf{m},\mathbf{n}',\mu,\nu' \\ (\mathbf{m} \neq \mathbf{n})}} J(\mathbf{n}\nu, \mathbf{m}\mu | \mathbf{n}'\nu', \mathbf{m}0)\langle\!\langle b^\dagger_{\mathbf{m}\mu}\alpha_{\mathbf{n}'\nu'}; \alpha^\dagger_{\mathbf{n}_1\nu}\rangle\!\rangle \quad (7.1)$$

where the prime on the summation indicates that only the term $\mathbf{n}' \neq \mathbf{n}$ has to be considered. The last two terms in Eq. (7.1) describe the interaction between an exciton at the lattice site $\mathbf{m}$ with either an electron ($\nu' \neq 0$) or a hole ($\nu' = 0$) at the lattice site $\mathbf{n}'$. In deriving Eq. (7.1), we have discarded terms describing electron–electron (or hole–hole) scattering processes. Writing Eq. (7.1) in $\mathbf{k}$ space, we have

$$(\omega - \omega_{\mathbf{k}\nu})G(\mathbf{k}\nu;\omega) = 1/2\pi + (1/\sqrt{N}) \sum_{\mathbf{R}_{\mathbf{mn}'},\mu,\nu'} J_{\nu 0,\nu'\mu}(\mathbf{k})$$
$$\times \sum_{\mathbf{q}} \exp(i\mathbf{q}\cdot\mathbf{R}_{\mathbf{mn}'})\langle\!\langle b_{\mathbf{q}\mu}\alpha_{\mathbf{k}-\mathbf{q}\nu'}; \alpha^{\dagger}_{\mathbf{k}\nu}\rangle\!\rangle + (1/\sqrt{N}) \sum_{\mathbf{R}_{\mathbf{mn}'},\mu,\nu'} J_{\nu\mu,\nu'0}(\mathbf{k})$$
$$\times \sum_{\mathbf{q}} \exp(i\mathbf{q}\cdot\mathbf{R}_{\mathbf{mn}'})\langle\!\langle b^{\dagger}_{-\mathbf{q}\mu}\alpha_{\mathbf{k}-\mathbf{q}\nu'}; \alpha^{\dagger}_{\mathbf{k}\nu}\rangle\!\rangle \quad (7.2)$$

where

$$J_{\nu 0,\nu'\mu}(\mathbf{k}) = \sum_{\mathbf{R}_{\mathbf{n'n}}} J(\mathbf{n}\nu, \mathbf{m}0/\mathbf{n}'\nu', \mathbf{m}\mu) \exp(i\mathbf{k}\cdot\mathbf{R}_{\mathbf{n'n}}) \quad (7.3\text{a})$$

$$\omega_{\mathbf{k}\nu} = {\sum_{\mathbf{R}_{\mathbf{n'n}}}}' \bar{L}(\mathbf{n}\nu, \mathbf{n}'\nu) \exp(i\mathbf{k}\cdot\mathbf{R}_{\mathbf{n'n}}), \quad \mathbf{R}_{\mathbf{mn}'} = \mathbf{r}_{\mathbf{m}} - \mathbf{r}_{\mathbf{n}'},$$
$$\mathbf{R}_{\mathbf{n'n}} = \mathbf{r}_{\mathbf{n}'} - \mathbf{r}_{\mathbf{n}} \quad (7.3\text{b})$$

$$G(\mathbf{k}\nu;\omega) = G(\mathbf{k}\nu,\mathbf{k}\nu;\omega) = \langle\!\langle \alpha_{\mathbf{k}\nu}; \alpha^{\dagger}_{\mathbf{k}\nu}\rangle\!\rangle. \quad (7.3\text{c})$$

The summations over $\mathbf{q}$ are all over the wave vectors in the first Brillouin zone. The coupling functions on the right-hand side of Eq. (7.2) consist of matrix elements involving three lattice sites. They are reduced to two-center integrals (matrix elements) when either $\mathbf{R}_{\mathbf{mn}'} = 0$ or $\mathbf{R}_{\mathbf{nn}'} = 0$. Notice that for $\mathbf{R}_{\mathbf{mn}'} \neq 0$, both terms have an exponential dependence on the wave vector $\mathbf{q}$ when the summation over $\mathbf{q}$ is taken. When $\mathbf{R}_{\mathbf{mn}'} = 0$, the three particles (two holes and one electron, or vice versa) are tightly bound together at the same lattice site $\mathbf{m}$. The Green's functions that appear in the last two terms of Eq. (7.2) will be calculated by means of Eq. (6.29).

In order to avoid intermediate steps consisting of complicated formulas, we shall spell out the approximations which are to be taken into consideration. The exciton energies are considered in the Heitler–London approximation; neglect of configuration mixing of different excitation bands and only diagonal contributions in the matrix elements are taken into account. All these assumptions are not necessary but they are useful in obtaining simple results. The interested reader is referred to the literature for details (Mavroyannis, 1972c). Using the assumptions just mentioned, Eq. (7.2) assumes the form

$$(\omega - \omega_{\mathbf{k}\nu})G(\mathbf{k}\nu;\omega) = 1/2\pi + (1/\sqrt{N}) \sum_{\mu,\mathbf{R}_{mn'}} J_{\nu 0, 0\mu}(\mathbf{k})$$
$$\times \sum_{\mathbf{q}} \exp(i\mathbf{q} \cdot \mathbf{R}_{mn'}) \langle\!\langle b_{\mathbf{q}\mu}\alpha_{\mathbf{k}-\mathbf{q}0}; \alpha^{\dagger}_{\mathbf{k}\nu}\rangle\!\rangle, \quad (7.4)$$

while the following expression for $\langle\!\langle b_{\mathbf{q}\mu}\alpha_{\mathbf{k}-\mathbf{q}0}; \alpha^{\dagger}_{\mathbf{k}\nu}\rangle\!\rangle$ is derived from Eq. (6.29) as

$$(\omega - \tilde{\omega}_{\mathbf{k}-\mathbf{q}0} - E_{\mathbf{q}\mu})\langle\!\langle b_{\mathbf{q}\mu}\alpha_{\mathbf{k}-\mathbf{q}0}; \alpha^{\dagger}_{\mathbf{k}\nu}\rangle\!\rangle + \sum_{\mu',\mathbf{R}_{m'm''}} J_{\mu 0,0\mu'}(\mathbf{k})(n_0/N)$$
$$\times \sum_{\mathbf{q}'} \exp(i\mathbf{q}' \cdot \mathbf{R}_{m'm''})\langle\!\langle b_{\mathbf{q}'\mu'}\alpha_{\mathbf{k}-\mathbf{q}'0}; \alpha^{\dagger}_{\mathbf{k}\nu}\rangle\!\rangle$$
$$= -(n_0/2\pi\sqrt{N})\,\delta_{\mu\nu}\exp(i\mathbf{k}\cdot\mathbf{R}_{mn'})$$
$$+ (1/\sqrt{N})J_{0\mu,\nu 0}(\mathbf{k})n_0\exp(-i\mathbf{q}\cdot\mathbf{R}_{mn'})G(\mathbf{k}\nu;\omega), \quad (7.5)$$

where $E_{\mathbf{q}\mu}$ is the exciton energy of the μth excitation band correct in the Heitler–London approximation, i.e.,

$$E_{\mathbf{q}\mu} = [\bar{L}(\mathbf{m}\mu,\mathbf{m}\mu) - \bar{L}(\mathbf{m}0,\mathbf{m}0)] + J_{\mu 0,0\mu}(\mathbf{q})(n_0 - n_\mu) \quad (7.6a)$$
$$J_{\mu 0,0\mu}(\mathbf{q}) = \sum_{\mathbf{R}_{m'm}}{}' J(\mathbf{m}\mu,\mathbf{m}'0/\mathbf{m}0,\mathbf{m}'\mu)\exp(i\mathbf{q}\cdot\mathbf{R}_{m'm}). \quad (7.6b)$$

$\tilde{\omega}_{\mathbf{k}-\mathbf{q}\nu'}$ is the energy of an electron in the νth excitation band:

$$\tilde{\omega}_{\mathbf{k}-\mathbf{q}\nu'} = \sum_{\mathbf{R}_{m'n'}} \tilde{L}(\mathbf{n}'\nu',\mathbf{m}'\nu')\exp[i(\mathbf{k}-\mathbf{q})\cdot\mathbf{R}_{m'n'}] \quad (7.6c)$$

$$\tilde{L}(\mathbf{n}'\nu',\mathbf{m}'\nu') = \bar{L}(\mathbf{n}'\nu',\mathbf{m}'\nu') - \sum_{\mathbf{m}'} J(\mathbf{n}'\nu',\mathbf{m}\mu/\mathbf{m}\mu,\mathbf{m}'\nu')(1 - n_\mu - n_{\nu'})$$
$$+ \sum_{\mathbf{m}'} J(\mathbf{n}'\nu',\mathbf{m}0/\mathbf{m}0,\mathbf{m}'\nu')(n_0 - n_{\nu'}). \quad (7.6d)$$

Physically, Eq. (7.4) implies that an electron in the νth excitation band ($\nu \neq 0$) interacts with a system where an exciton and a hole are annihilated simultaneously; the latter process, arising from the interaction between an electron and the density of the holes, is described by the expression (7.5). Thus for $\mu = \nu$, the coupled equations (7.4) and (7.5) describe interactions that take place in the νth excited state (excitation band).

Similarly, the following set of coupled equations

$$(\omega - \omega_{\mathbf{k}0})G(\mathbf{k}0;\omega) = 1/2\pi + (1/\sqrt{N})\sum_{\nu,\mathbf{R}_{mn'}} J_{0\nu,\nu 0}(\mathbf{k})$$
$$\times \sum_{\mathbf{q}} \exp(i\mathbf{q}\cdot \mathbf{R}_{mn'})\langle\!\langle b^{\dagger}_{-\mathbf{q}\nu}\alpha_{\mathbf{k}-\mathbf{q}\nu}; \alpha^{\dagger}_{\mathbf{k}0}\rangle\!\rangle, \quad (7.7)$$

$$(\omega - \tilde{\omega}_{\mathbf{k}-\mathbf{q}\nu} \mid E_{\mathbf{q}\nu})\langle\!\langle b^{\dagger}_{-\mathbf{q}\nu}\alpha_{\mathbf{k}-\mathbf{q}\nu}; \alpha^{\dagger}_{\mathbf{k}0}\rangle\!\rangle + \sum_{\mu,\mathbf{R}_{m'm''}} J_{0\mu,\mu 0}(\mathbf{k})$$
$$\times (n_{\nu}/N)\sum_{\mathbf{q'}} \exp(i\mathbf{q'}\cdot \mathbf{R}_{m'm''})\langle\!\langle b^{\dagger}_{-\mathbf{q'}\mu}\alpha_{\mathbf{k}-\mathbf{q'}\mu}; \alpha^{\dagger}_{\mathbf{k}0}\rangle\!\rangle$$
$$= -(n_{\nu}/2\pi\sqrt{N})\exp(i\mathbf{k}\cdot\mathbf{R}_{mn'})$$
$$+ (n_{\nu}/\sqrt{N})J_{\nu 0,0\nu}(\mathbf{k})\exp(-i\mathbf{q}\cdot\mathbf{R}_{mn})G(\mathbf{k}0;\omega) \quad (7.8)$$

describes interactions taking place in the valence band. Equation (7.7) indicates that a hole is coupled to a system where an exciton is created and an electron is annihilated simultaneously. Such a physical process, which is caused by the interaction between a hole and the electron density, is represented by Eq. (7.8). Obviously, the coupled equations (7.4)–(7.5) and (7.7)–(7.8) describe the distortion of the excited and ground state, respectively, which is caused by the corresponding interactions.

The solutions of Eqs. (7.5) and (7.8) can be written as

$$\left(\frac{1}{\sqrt{N}}\right)\sum_{\mu,\mathbf{R}_{m'm''}} J_{\nu 0,0\mu}(\mathbf{k})\sum_{\mathbf{q'}}\exp(i\mathbf{q'}\cdot\mathbf{R}_{m'm''})\langle\!\langle b_{\mathbf{q'}\mu}\alpha_{\mathbf{k}-\mathbf{q'}0}; \alpha^{\dagger}_{\mathbf{k}\nu}\rangle\!\rangle$$
$$= \frac{[\varepsilon_{\mathbf{k}\nu}(\omega) - 1]}{\varepsilon_{\mathbf{k}\nu}(\omega)}\left[\left(\frac{-1}{2\pi}\right)\exp(i\mathbf{k}\cdot\mathbf{R}_{mn'})\right.$$
$$\left. + J_{0\nu,\nu 0}(\mathbf{k})\exp(-\mathbf{q}\cdot\mathbf{R}_{mn'})G(\mathbf{k}\nu;\omega)\right] \quad (7.9)$$

$$\left(\frac{1}{\sqrt{N}}\right)\sum_{\mu,\mathbf{R}_{m'm''}} J_{0\mu,\mu 0}(\mathbf{k})\sum_{\mathbf{q'}}\exp(i\mathbf{q'}\cdot\mathbf{R}_{m'm''})\langle\!\langle b^{\dagger}_{-\mathbf{q'}\mu}\alpha_{\mathbf{k}-\mathbf{q'}\mu}; \alpha^{\dagger}_{\mathbf{k}0}\rangle\!\rangle$$
$$= \frac{[\varepsilon_{\mathbf{k}0}(\omega) - 1]}{\varepsilon_{\mathbf{k}0}(\omega)}\left[\left(\frac{-1}{2\pi}\right)\exp(i\mathbf{k}\cdot\mathbf{R}_{mn'})\right.$$
$$\left. + J_{\nu 0,0\nu}(\mathbf{k})\exp(-i\mathbf{q}\cdot\mathbf{R}_{mn'})G(\mathbf{k}0;\omega)\right] \quad (7.10)$$

where

$$\varepsilon_{\mathbf{k}\nu}(\omega) = 1 + \sum_{\mathbf{R}_{mn'}} J_{\nu 0,0\nu}(\mathbf{k})(n_0/N)\sum_{\mathbf{q}}[\exp(i\mathbf{q}\cdot\mathbf{R}_{mn'})/(\omega - \tilde{\omega}_{\mathbf{k}-\mathbf{q}0} - E_{\mathbf{q}\nu})] \quad (7.11)$$

$$\varepsilon_{\mathbf{k}0}(\omega) = 1 + \sum_{\nu,\mathbf{R}_{mn'}} J_{0\nu,\nu 0}(\mathbf{k})(n_{\nu}/N)\sum_{\mathbf{q}}[\exp(i\mathbf{q}\cdot\mathbf{R}_{mn'})/(\omega - \tilde{\omega}_{\mathbf{k}-\mathbf{q}\nu} + E_{\mathbf{q}\nu})]. \quad (7.12)$$

The expressions (7.11) and (7.12) have the form of dielectric functions. For instance, the last term in Eq. (7.11) represents the electronic polarizability arising from the interaction between the Frenkel exciton ($\mathbf{q}\nu$) and the hole ($\mathbf{k} - \mathbf{q}0$). Similar interpretation holds for $\varepsilon_{\mathbf{k}0}(\omega)$. Considering the approximations that have been made, the expressions for $\varepsilon_{\mathbf{k}\nu}(\omega)$ and $\varepsilon_{\mathbf{k}0}(\omega)$ are correct in the generalized random phase approximation. Because of the dependence on n_0 and n_ν, the existence of the last two terms in Eqs. (7.11) and (7.12) depends on whether or not the states 0 and ν, respectively, are occupied.

The coupling functions in Eqs. (7.11) and (7.12) have the form of Eq. (7.3a) and consist of three-center integrals. Considering only two-center integrals, the expressions (7.11) and (7.12) assume the form

$$\varepsilon_{\mathbf{k}\nu}(\omega) = 1 + \left(\frac{n_0}{N}\right) J_{\nu 0, 0\nu}(\mathbf{k}) \sum_{\mathbf{q}} \frac{1}{(\omega - \tilde{\omega}_{\mathbf{k}-\mathbf{q}0} - E_{\mathbf{q}\nu})}$$

$$+ \left(\frac{n_0}{N}\right) \sum_{\mathbf{q}} \frac{J_{\nu 0, 0\nu}(\mathbf{q})}{(\omega - \tilde{\omega}_{\mathbf{k}-\mathbf{q}0} - E_{\mathbf{q}\nu})} \quad (7.13)$$

$$\varepsilon_{\mathbf{k}0}(\omega) = 1 + \left(\frac{1}{N}\right) \sum_{\nu} J_{0\nu,\nu 0}(\mathbf{k}) n_\nu \sum_{\mathbf{q}} \frac{1}{(\omega - \tilde{\omega}_{\mathbf{k}-\mathbf{q}\nu} + E_{\mathbf{q}\nu})}$$

$$+ \left(\frac{1}{N}\right) \sum_{\mathbf{q},\nu} \frac{J_{0\nu,\nu 0}(\mathbf{q}) n_\nu}{(\omega - \tilde{\omega}_{\mathbf{k}-\mathbf{q}\nu} + E_{\mathbf{q}\nu})} \quad (7.14)$$

where

$$J_{\nu 0, 0\nu}(\mathbf{k}) = \sum_{\mathbf{R}_{mn}}{}' J(\mathbf{n}\nu, \mathbf{m}0 \mid \mathbf{m}0, \mathbf{m}\nu) \exp(i\mathbf{k} \cdot \mathbf{R}_{mn}) \quad (7.15a)$$

$$J(\mathbf{n}\nu, \mathbf{m}0 \mid \mathbf{m}0, \mathbf{m}\nu) = \langle \mathbf{n}\nu, \mathbf{m}0 \mid V \mid \mathbf{m}0, \mathbf{m}\nu \rangle - \langle \mathbf{n}\nu, \mathbf{m}0 \mid V \mid \mathbf{m}\nu, \mathbf{m}0 \rangle \quad (7.15b)$$

$$J_{\nu 0, 0\nu}(\mathbf{q}) = \sum_{\mathbf{R}_{mn}}{}' J(\mathbf{n}\nu, \mathbf{m}0 \mid \mathbf{n}0, \mathbf{m}\nu) \exp(i\mathbf{q} \cdot \mathbf{R}_{mn}) \quad (7.16a)$$

$$J(\mathbf{n}\nu, \mathbf{m}0 \mid \mathbf{n}0, \mathbf{m}\nu) = \langle \mathbf{n}\nu, \mathbf{m}0 \mid V \mid \mathbf{n}0, \mathbf{m}\nu \rangle - \langle \mathbf{n}\nu, \mathbf{m}0 \mid V \mid \mathbf{m}\nu, \mathbf{n}0 \rangle. \quad (7.16b)$$

The form of the coupling function (7.15b) indicates that two holes and one electron occupy the same lattice site, and hence are within the same molecule (atom). Thus the matrix elements of the first term in Eq. (7.15b) are of the CT type and describe the physical process where an electron undergoes the transition $(\mathbf{n}\nu) \to (\mathbf{m}0)$, i.e., upon deexcitation an electron is transferred from one molecule (atom) to another neighboring molecule, while the second electron undergoes the transition within the same molecule (atom). The second term in Eq. (7.15b) describes the exchange effect for the process under consideration. Similar interpretation holds for the coupling functions in Eq. (7.14). The matrix elements in Eq. (7.16b) describe physical processes where the electrons experience transitions within the same molecules. Therefore, from the point of view of coupling functions, the second terms in Eqs. (7.13) and (7.14) may be attributed as due to CT processes, while the last terms represent localized (excitonic) type transitions within the molecules.

The expressions for $\varepsilon_{k\nu}(\omega)$ and $\varepsilon_{k0}(\omega)$ may have two kinds of solutions: Scattering solutions exist when the conditions $\omega - \tilde{\omega}_{k-q0} - E_{q\nu} = 0$ and $\omega - \tilde{\omega}_{k-q\nu} + E_{q\nu} = 0$, respectively, are satisfied. On the other hand, when the conditions $\omega - \tilde{\omega}_{k-q0} - E_{q\nu} \neq 0$ and $\omega - \tilde{\omega}_{k-q\nu} + E_{q\nu} \neq 0$ are satisfied separately, the expressions for $\varepsilon_{k\nu}(\omega)$ and $\varepsilon_{k0}(\omega)$ may have bound states (collective excitations) with energies of excitation determined by the solutions of the equations $\varepsilon_{k\nu}(\omega) = 0$ and $\varepsilon_{k0}(\omega) = 0$, respectively. The exponential dependence on the distance $\mathbf{R}_{mn}$ of the last two terms in Eqs. (7.13) and (7.14) is very crucial for the existence of bound states. For example, when the sum over $\mathbf{q}$ is taken within the first Brillouin zone, then the last two terms in Eqs. (7.13) and (7.14) decay exponentially with distance and become negligibly small for distances $\mathbf{R}_{mn} \gg \lambda$, where λ is the wavelength of the energy of excitation determined by the roots of the equations $\varepsilon_{k\nu}(\omega) = 0$ and $\varepsilon_{k0}(\omega) = 0$, respectively. For distances $\mathbf{R}_{mn} \ll \lambda$, the exponential term may be replaced by unity, which is equivalent to expanding the exponential term in power series of $\lambda/\mathbf{R}_{mn}$ and retaining only the first nonvanishing term. Although the coupling function (7.16b) is, in general, larger than that of (7.15b), because of the term $\exp(i\mathbf{q}\cdot\mathbf{R}_{mn})$, the last two terms in Eqs. (7.13) and (7.14) contribute only at very small distances $\mathbf{R}_{mn}$, while the second terms predominate. Therefore, we may neglect contributions arising from the last two terms in Eqs. (7.13) and (7.14) and take $\varepsilon_{k\nu}(\omega)$ and $\varepsilon_{k0}(\omega)$ as

$$\varepsilon_{k\nu}(\omega) = 1 + (n_0/N)J_{\nu 0,0\nu}(\mathbf{k})\sum_q (\omega - \tilde{\omega}_{k-q0} - E_{q\nu})^{-1} \qquad (7.17)$$

$$\varepsilon_{k0}(\omega) = 1 + \sum_\nu J_{0\nu,\nu 0}(\mathbf{k})(n_\nu/N)\sum_q (\omega - \tilde{\omega}_{k-q\nu} + E_{q\nu})^{-1}. \qquad (7.18)$$

The last term in the expression for $\varepsilon_{k\nu}(\omega)$ arises from the physical process where an exciton and a hole are annihilated simultaneously, while the corresponding one of $\varepsilon_{k0}(\omega)$ is due to the creation of an exciton with the simultaneous annihilation of an electron. Notice that in both processes, the three particles are localized at the same lattice site; i.e., in each case the excitation takes place within the same molecule. In this approximation, using Eqs. (7.4), (7.9), (7.17) and (7.7), (7.10), (7.18), we derive the expressions for the electron and hole Green's functions in the form

$$G(k\nu; \omega) = [(2\pi)\varepsilon_{k\nu}(\omega)]^{-1}[\omega - \bar{\omega}_{k\nu} + J_{\nu 0, 0\nu}(\mathbf{k})/\varepsilon_{k\nu}(\omega)]^{-1} \quad (7.19)$$

$$G(k0; \omega) = [(2\pi)\varepsilon_{k0}(\omega)]^{-1}[\omega - \bar{\omega}_{k0} + \sum_\nu J_{0\nu, \nu 0}(\mathbf{k})/\varepsilon_{k0}(\omega)]^{-1} \quad (7.20)$$

where the renormalized energies of excitation for the electron and the hole are given by

$$\bar{\omega}_{k\nu} = \omega_{k\nu} + J_{\nu 0, 0\nu}(\mathbf{k}) \quad (7.21a)$$

$$\bar{\omega}_{k0} = \omega_{k0} + \sum_\nu J_{0\nu, \nu 0}(\mathbf{k}). \quad (7.21b)$$

Considering that all the expressions for the energies in Eqs. (7.19) and (7.20) are renormalized in the self-consistent field approximation, the Green's functions $G(k\nu; \omega)$ and $G(k0; \omega)$ are correct in the generalized random phase approximation.

B. Excitation Spectra

The Green's function $G(k\nu; \omega)$, given by Eq. (7.19), describes the spectrum of an electron in the νth excitation band. It consists of the product of two terms: the dielectric function $\varepsilon_{k\nu}(\omega)$, which describes collective or scattering modes, and the modified electron propagator with the CT interaction term screened by the dielectric function. Similar interpretation holds for the Green's functions of the hole $G(k0; \omega)$. The form of the excitation spectra described by Eqs. (7.19) and (7.20) is drastically dependent on whether the equations

$$\varepsilon_{k\nu}(\omega) = 0, \quad \text{for} \quad \omega - \bar{\omega}_{k-q0} - E_{q\nu} \neq 0 \quad (7.22)$$

$$\varepsilon_{k0}(\omega) = 0, \quad \text{for} \quad \omega - \bar{\omega}_{k-q\nu} + E_{q\nu} \neq 0 \quad (7.23)$$

have roots outside the frequency region $\bar{\omega}_{k-q0} + E_{q\nu}$ and $\bar{\omega}_{k-q\nu} - E_{q\nu}$, respectively.

Using a simple model for the exciton, electron, and hole spectra, it has been shown (Mavroyannis, 1972c) that Eqs. (7.22) and (7.23) may have solutions corresponding to bound states, provided that the conditions

$$|U_{k\nu}| \geq (1/3\sigma)\gamma\xi^2, \quad |U_{k0}| \geq (1/3\sigma)\gamma'\xi^2, \quad \sigma = (V\xi^3/6\pi^2 N) \quad (7.24)$$

are satisfied, respectively, where ξ is the radius of the sphere having the same volume as the Brillouin zone, and

$$U_{\mathbf{k}\nu} = n_0 J_{\nu 0, 0\nu}(\mathbf{k}), \qquad U_{\mathbf{k}0} = \sum_\nu J_{0\nu, \nu 0}(\mathbf{k}) n_\nu. \tag{7.25}$$

For large values of $U_{\mathbf{k}\nu}$ and $U_{\mathbf{k}0}$, Eqs. (7.22) and (7.23) have solutions corresponding to deep discrete levels (bound states) with energies $\Omega_{\mathbf{k}\nu}$ and $\Omega_{\mathbf{k}0}$ given by

$$\Omega_{\mathbf{k}\nu} = E_\nu + \omega_0 + (\alpha\beta/\gamma)\mathbf{k}^2 + \sigma U_{\mathbf{k}\nu}, \qquad \gamma = \alpha + \beta \tag{7.26}$$

$$\Omega_{\mathbf{k}0} = \omega_\nu - E_\nu - (\alpha\beta'/\gamma')\mathbf{k}^2 + \sigma U_{\mathbf{k}0}, \qquad \gamma' = \beta' - \alpha, \tag{7.27}$$

respectively. Here, E_ν, ω_0, and ω_ν are the energies of the exciton, hole, and electron, respectively, which are independent of wave vectors ($E_\nu \sim \omega_\nu - \omega_0$). The terms α, β, and β' are the effective masses for the exciton, hole, and electron, respectively.

For values of $|U_{\mathbf{k}\nu}|$ and $|U_{\mathbf{k}0}|$ slightly greater than $(1/3\sigma)\gamma\xi^2$ and $(1/3\sigma)\gamma'\xi^2$, respectively, Eqs. (7.22) and (7.23) have solutions corresponding to shallow levels (loosely bound states) with energies given approximately by

$$\Omega_{\mathbf{k}\nu} \simeq E_\nu + \omega_0 + (\alpha\beta/\gamma)\mathbf{k}^2 - (2/\pi)^2(\gamma\xi^2)[1 + \gamma\xi^2/(3\sigma U_{\mathbf{k}\nu})]^2 \tag{7.28}$$

$$\Omega_{\mathbf{k}0} \simeq \omega_\nu - E_\nu - (\alpha\beta'/\gamma')\mathbf{k}^2 - (2/\pi)^2(\gamma'\xi^2)[1 + \gamma'\xi^2/(3\sigma U_{\mathbf{k}0})]^2, \tag{7.29}$$

respectively. For $U_{\mathbf{k}\nu}$ and $U_{\mathbf{k}0}$ negative, the energies of excitation $\Omega_{\mathbf{k}\nu}$ and $\Omega_{\mathbf{k}0}$ given by Eqs. (7.26)–(7.29) are positive and correspond to bound states of positively and negatively charged molecular ions, respectively. Thus the existence of bound states as solutions of Eqs. (7.22) and (7.23) with energies $\Omega_{\mathbf{k}\nu}$ and $\Omega_{\mathbf{k}0}$ depends on the strength of the potentials $U_{\mathbf{k}\nu}$ and $U_{\mathbf{k}0}$, which should satisfy the conditions (7.24), respectively.

If $\Omega_{\mathbf{k}\nu}$ and $\Omega_{\mathbf{k}0}$ ar the energies of excitation satisfying Eqs. (7.22) and (7.23), respectively, we may expand $\varepsilon_{\mathbf{k}\nu}(\omega)$ and $\varepsilon_{\mathbf{k}0}(\omega)$ in power series and retain only the first nonvanishing term, i.e.,

$$\varepsilon_{\mathbf{k}\nu}(\omega) = 0 = (\omega - \Omega_{\mathbf{k}\nu})\lambda_{\mathbf{k}\nu}^{-1} \tag{7.30}$$

$$\varepsilon_{\mathbf{k}0}(\omega) = 0 = (\omega - \Omega_{\mathbf{k}0})\lambda_{\mathbf{k}0}^{-1} \tag{7.31}$$

where

$$\lambda_{\mathbf{k}\nu} = [(\partial/\partial\omega)\varepsilon_{\mathbf{k}\nu}(\omega)]^{-1}_{\omega=\Omega_{\mathbf{k}\nu}} = -J_{\nu 0, 0\nu}(\mathbf{k})n_0 \tag{7.32}$$

$$\lambda_{\mathbf{k}0} = [(\partial/\partial\omega)\varepsilon_{\mathbf{k}0}(\omega)]^{-1}_{\omega=\Omega_{\mathbf{k}0}} = -\sum_\nu J_{0\nu, \nu 0}(\mathbf{k})n_\nu. \tag{7.33}$$

Substituting Eqs. (7.30)–(7.33) into (7.19) and (7.20), we have (Mavroyannis, 1972c)

$$G(\mathbf{k}\nu;\omega) = \left(-\frac{1}{2\pi}\right) \frac{J_{\nu 0, 0\nu}(\mathbf{k})n_0}{[\omega - \Omega_+(\mathbf{k}\nu)][\omega - \Omega_-(\mathbf{k}\nu)]} \quad (7.34)$$

$$G(\mathbf{k}0;\omega) = \left(-\frac{1}{2\pi}\right) \frac{\sum_\nu J_{0\nu,\nu 0}(\mathbf{k})n_\nu}{[\omega - \Omega_+(\mathbf{k}0)][\omega - \Omega_-(\mathbf{k}0)]} \quad (7.35)$$

where the energies of excitation $\Omega_\pm(\mathbf{k}\nu)$ and $\Omega_\pm(\mathbf{k}0)$ are given by

$$\Omega_\pm(\mathbf{k}\nu) = \tfrac{1}{2}(\Omega_{\mathbf{k}\nu} + \bar{\omega}_{\mathbf{k}\nu}) \pm \tfrac{1}{2}[(\Omega_{\mathbf{k}\nu} - \bar{\omega}_{\mathbf{k}\nu})^2 + 4|J_{\nu 0, 0\nu}(\mathbf{k})|^2 n_0]^{1/2} \quad (7.36)$$

$$\Omega_\pm(\mathbf{k}0) = \tfrac{1}{2}(\Omega_{\mathbf{k}0} + \bar{\omega}_{\mathbf{k}0}) \pm \tfrac{1}{2}[(\Omega_{\mathbf{k}0} - \bar{\omega}_{\mathbf{k}0})^2 + 4\sum_\nu |J_{0\nu,\nu 0}(\mathbf{k})|^2 n_\nu]^{1/2}. \quad (7.37)$$

The splitting of the energies of excitation can be derived from Eqs. (7.36) and (7.37) as

$$\Omega_+(\mathbf{k}\nu) - \Omega_-(\mathbf{k}\nu) = [(\Omega_{\mathbf{k}\nu} - \bar{\omega}_{\mathbf{k}\nu})^2 + 4|J_{\nu 0, 0\nu}(\mathbf{k})|^2 n_0]^{1/2} \quad (7.38)$$

$$\Omega_+(\mathbf{k}0) - \Omega_-(\mathbf{k}0) = [(\Omega_{\mathbf{k}0} - \bar{\omega}_{\mathbf{k}0})^2 + 4\sum_\nu |J_{0\nu,\nu 0}(\mathbf{k})|^2 n_\nu]^{1/2}. \quad (7.39)$$

The splitting of the excitation spectrum given by Eq. (7.38) or (7.39) becomes substantial when the frequencies $\Omega_{\mathbf{k}\nu}$ and $\bar{\omega}_{\mathbf{k}\nu}$ (or $\Omega_{\mathbf{k}0}$ and $\bar{\omega}_{\mathbf{k}0}$) approach each other and for reasonably strong coupling functions. The coupling functions in Eqs. (7.38) and (7.39) are of the CT type, and are multiplied by the hole and electron occupation number, respectively. Therefore, the existence of the energy splitting depends on whether or not the hole state 0 or the electron state ν is occupied, provided that the coupling functions do not vanish by symmetry considerations.

The splitting of the excitation spectrum with energies $\Omega_\pm(\mathbf{k}\nu)$ [or $\Omega_\pm(\mathbf{k}0)$] is caused by the resonance interaction between the excited electron (or the positive hole) with energy $\bar{\omega}_{\mathbf{k}\nu}$ (or $\bar{\omega}_{\mathbf{k}0}$) and the positively (or negatively) charged bound state ion with energy $\Omega_{\mathbf{k}\nu}$ (or $\Omega_{\mathbf{k}0}$) being at a distance $\mathbf{R}_{mn}$ apart. We may identify Eq. (7.38) for $\Omega_\pm(\mathbf{k}\nu)$ with those of W_V and W_N derived by Mulliken (Mulliken and Person, 1969, Eqs. (2.10)) if we take $W_1 = \Omega_{\mathbf{k}\nu}$, $W_0 = \bar{\omega}_{\mathbf{k}\nu}$, $\beta_0\beta_1 = |J_{\nu 0, 0\nu}(2)|^2 n_0$ and for zero overlap between the wave functions. Following Mulliken, the energy splittings given by Eqs. (7.38) and (7.39) may be interpreted as the energies of the charge-transfer bands, i.e.,

$$\Omega_{CT}(\mathbf{k}\nu) = \Omega_+(\mathbf{k}\nu) - \Omega_-(\mathbf{k}\nu) \qquad (7.40)$$

$$\Omega_{CT}(\mathbf{k}0) = \Omega_+(\mathbf{k}0) - \Omega_-(\mathbf{k}0), \qquad (7.41)$$

corresponding to the donor and acceptor CT spectrum, respectively. Both CT complexes in the excited and in the ground state consist of four particles (two electrons and two holes), respectively, which are shared between two molecules. The CT complex in the excited state described by $G(\mathbf{k}\nu; \omega)$ is known in the literature as the excited dimer, or excimer.

The occupation numbers n_0 and n_ν are determined by Eq. (3.18), i.e.,

$$n_0 = 2 \int_{-\infty}^{+\infty} d\omega \, \mathrm{Im}\, G(\mathbf{k}0; \omega) n_F(\omega) \qquad (7.42)$$

$$n_\nu = 2 \int_{-\infty}^{+\infty} d\omega \, \mathrm{Im}\, G(\mathbf{k}\nu; \omega) n_F(\omega), \qquad n_F(\omega) = (1 + \exp(\beta\omega))^{-1} \qquad (7.43)$$

where $n_F(\omega)$ is the Fermi distribution function. For the unperturbed spectrum, $n_0^{(0)} = n_F(\omega_{\mathbf{k}0})$ and $n_\nu^{(0)} = n_F(\omega_{\mathbf{k}\nu})$. In all the expressions, derived so far n_0 and n_ν should be replaced by $n_0^{(0)}$ and $n_\nu^{(0)}$, respectively. Taking the imaginary part of Eqs. (7.36) and (7.37), we find

$$\mathrm{Im}\, G(\mathbf{k}\nu; \omega) = \tfrac{1}{2}\phi_{\mathbf{k}\nu}[\delta(\omega - \Omega_+(\mathbf{k}\nu)) - \delta(\omega - \Omega_-(\mathbf{k}\nu))] \qquad (7.44)$$

$$\mathrm{Im}\, G(\mathbf{k}0; \omega) = \tfrac{1}{2}\phi_{\mathbf{k}0}[\delta(\omega - \Omega_+(\mathbf{k}0)) - \delta(\omega - \Omega_-(\mathbf{k}0))] \qquad (7.45)$$

where

$$\phi_{\mathbf{k}\nu} = J_{\nu0,0\nu}(\mathbf{k})n_0^{(0)}/[\Omega_+(\mathbf{k}\nu) - \Omega_-(\mathbf{k}\nu)] = J_{\nu0,0\nu}(\mathbf{k})n_0^{(0)}/\Omega_{CT}(\mathbf{k}\nu) \qquad (7.46)$$

$$\phi_{\mathbf{k}0} = [\sum_\nu J_{0\nu,\nu0}(\mathbf{k})n_\nu^{(0)}]/[\Omega_+(\mathbf{k}0) - \Omega_-(\mathbf{k}0)] = \sum_\nu J_{0\nu,\nu0}(\mathbf{k})n_\nu^{(0)}/\Omega_{CT}(\mathbf{k}0). \qquad (7.47)$$

The spectral functions given by Eqs. (7.44) and (7.45) have delta function distributions and peak at frequencies $\Omega_\pm(\mathbf{k}\nu)$ and $\Omega_\pm(\mathbf{k}0)$, respectively. In this approximation, using Eqs. (7.42)–(7.45), we have

$$n_0^{(1)} = \phi_{\mathbf{k}0}[n_F(\Omega_+(\mathbf{k}0)) - n_F(\Omega_-(\mathbf{k}0))] \qquad (7.48)$$

$$n_\nu^{(1)} = \phi_{\mathbf{k}\nu}[n_F(\Omega_+(\mathbf{k}\nu)) - n_F(\Omega_-(\mathbf{k}\nu))]. \qquad (7.49)$$

The functions $\phi_{k\nu}$ and ϕ_{k0}, defined by Eqs. (7.46) and (7.47), indicate the strength of the resonance CT interaction between the two energy modes $\bar{\omega}_{k\nu}$ and $\Omega_{k\nu}$ or $\bar{\omega}_{k0}$ and Ω_{k0}, respectively. For example, when $\bar{\omega}_{k\nu} \sim \Omega_{k\nu}$ and $\bar{\omega}_{k0} \sim \Omega_{k0}$, which corresponds to the case of strong resonance coupling between the two structures, then

$$\phi_{k\nu} \sim n_F^{1/2}(\omega_{k0}), \qquad \phi_{k0} \sim n_F^{1/2}(\omega_{k\nu}). \tag{7.50}$$

For a weak resonance interaction, which occurs when the conditions

$$(\Omega_{k\nu} - \bar{\omega}_{k\nu})^2 \gg |J_{\nu 0, 0\nu}(\mathbf{k})|^2 n_0^{(0)} \tag{7.51}$$

$$(\Omega_{k0} - \bar{\omega}_{k0})^2 \gg \sum_\nu |J_{0\nu, \nu 0}(\mathbf{k})|^2 n_\nu^{(0)} \tag{7.52}$$

are satisfied, then we may expand Eqs. (7.46) and (7.47) in power series and retain only the first nonvanishing term

$$\phi_{k\nu} \sim J_{\nu 0, 0\nu}(\mathbf{k}) n_0^{(0)} / (\Omega_{k\nu} - \bar{\omega}_{k\nu}) - \cdots \tag{7.53}$$

$$\phi_{k0} \sim [\sum_\nu J_{0\nu, \nu 0}(\mathbf{k}) n_\nu^{(0)}] / (\Omega_{k0} - \bar{\omega}_{k0}) - \cdots . \tag{7.54}$$

In Mulliken's terminology, the energy difference $\Omega_{k\nu} - \bar{\omega}_{k\nu}$ or $\Omega_{k0} - \bar{\omega}_{k0}$ is denoted by the quantity Δ, which is defined as (Mulliken and Person, 1969)

$$\Delta = I_D^u - E_A^u + G_1 - G_0, \tag{7.55}$$

where I_D^u and E_A^u are the ionization potential of the donor and the electron affinity of the acceptor, respectively. G_1 is the Coulomb interaction responsible for the binding of the complex structure, while G_0 is the energy of the interaction, which modifies the energy of the D–A structure. It we take, in our case, Δ as $\Delta = \Omega_{k\nu} - \bar{\omega}_{k\nu}$ with $\Omega_{k\nu}$ and $\bar{\omega}_{k\nu}$ given by Eqs. (7.26) and (7.21a), then we may identify the quantities in Eq. (7.55) as

$$I_D^u = E_\nu + \omega_0 + (\alpha\beta/\gamma)\mathbf{k}^2, \quad E_A^u = \omega_{k\nu}, \quad G_1 = \sigma U_{k\nu}, \quad G_0 = J_{\nu 0, 0\nu}(\mathbf{k}) \tag{7.56}$$

$(E_\nu + \omega_0 \sim \omega_\nu)$. For a loosely bound state ion, G_1 is given by the last term of Eq. (7.28). Similar considerations hold for the energy difference $\Omega_{k0} - \bar{\omega}_{k0}$.

The spectral functions (7.44) and (7.45) can be used to calculate the density of states $\varrho_\nu(\omega)$ and $\varrho_0(\omega)$ given by

$$\varrho_\nu(\omega) = 2\sum_{\mathbf{k}} \text{Im } G(\mathbf{k}\nu;\omega), \qquad \varrho_0(\omega) = 2\sum_{\mathbf{k}} \text{Im } G(\mathbf{k}0;\omega),$$

as well as the expression for the absorption coefficient described by Eq. (5.50). We consider the physical process when CT complexes in both the excited and ground state are stable with energies $\Omega_\pm(\mathbf{k}\nu)$ and $\Omega_\pm(\mathbf{k}0)$, respectively. This is possible to occur when both states 0 and ν are occupied. Substituting Eqs. (7.44) and (7.45) into Eq. (5.50), retaining only the term that refers to the absorption process, and after integrating over ω', we obtain the expression for the absorption coefficient in the form

$$\alpha(\omega) = (\pi\omega_p^2/c\eta\omega N) \sum_{\mathbf{k}} f_{0\nu}(\mathbf{k})\omega_{\nu 0}(\mathbf{k})\phi_{\mathbf{k}0}\phi_{\mathbf{k}\nu}\{[\delta(\omega + \Omega_+(\mathbf{k}0) - \Omega_+(\mathbf{k}\nu))$$
$$- \delta(\omega + \Omega_+(\mathbf{k}0) - \Omega_-(\mathbf{k}\nu))][n_F(\Omega_+(\mathbf{k}0)) - n_F(\omega + \Omega_+(\mathbf{k}0))]$$
$$- [\delta(\omega + \Omega_-(\mathbf{k}0) - \Omega_+(\mathbf{k}\nu)) - \delta(\omega + \Omega_-(\mathbf{k}0) - \Omega_-(\mathbf{k}\nu))]$$
$$\times [n_F(\Omega_-(\mathbf{k}0)) - n_F(\omega + \Omega_-(\mathbf{k}0))]\} \quad (7.57)$$

where $f_{0\nu}(\mathbf{k})$ is the oscillator strength for the optical transition $0 \rightleftharpoons \nu$, $\omega_{\nu 0}(\mathbf{k}) = \omega_{\mathbf{k}\nu} - \omega_{\mathbf{k}0}$, and ω_p is the plasma frequency. The arguments of the delta functions in the expression (7.57) indicate that

$$\omega = \Omega_\pm(\mathbf{k}\nu) - \Omega_\pm(\mathbf{k}0); \quad (7.58)$$

i.e., in the absorption process, direct transitions take place between the bands $\Omega_\pm(\mathbf{k}0) \to \Omega_\pm(\mathbf{k}\nu)$. The reverse processes occur in emission. In the case when the CT complex exists only in the excited state, Eq. (7.57) assumes the form

$$\alpha(\omega) = (\pi\omega_p^2/c\eta\omega N) \sum_{\mathbf{k}} f_{0\nu}(\mathbf{k})\omega_{\nu 0}(\mathbf{k})\phi_{\mathbf{k}\nu}[\delta(\omega + \omega_{\mathbf{k}0} - \Omega_+(\mathbf{k}\nu))$$
$$- \delta(\omega + \omega_{\mathbf{k}0} - \Omega_-(\mathbf{k}\nu))] \times [n_F(\omega_{\mathbf{k}0}) - n_F(\omega + \omega_{\mathbf{k}0})], \quad (7.59)$$

which indicates that direct optical absorption $\omega_{\mathbf{k}0} \to \Omega_\pm(\mathbf{k}\nu)$ may occur, provided that the equation

$$\omega = \Omega_\pm(\mathbf{k}\nu) - \omega_{\mathbf{k}0}$$

is satisfied. On the other hand, when the CT complex is stable only in the ground state, then the expression for $\alpha(\omega)$ becomes

$$\alpha(\omega) = (\pi\omega_p^2/c\eta\omega N) \sum_{\mathbf{k}} f_{0\nu}(\mathbf{k})\omega_{\nu 0}(\mathbf{k})\phi_{\mathbf{k}0}[\delta(\omega + \Omega_+(\mathbf{k}0) - \omega_{\mathbf{k}\nu})$$
$$- \delta(\omega + \Omega_-(\mathbf{k}0) - \omega_{\mathbf{k}\nu})] \times [n_F(\omega_{\mathbf{k}\nu} - \omega) - n_F(\omega_{\mathbf{k}\nu})]. \quad (7.60)$$

In this case, direct absorption takes place between $\Omega_\pm(\mathbf{k}0) \to \omega_{\mathbf{k}\nu}$, provided that

$$\omega = \omega_{\mathbf{k}\nu} - \Omega_\pm(\mathbf{k}0),$$

and the reverse process holds for emission. It is obvious from this analysis that direct observation of the states $\Omega_\pm(\mathbf{k}0)$, if they exist, will be difficult. However, their existence may be deduced indirectly from the magnitude of the observed transition frequencies.

Under the conditions of statistical equilibrium, the functions $\phi_{\mathbf{k}\nu}$ and $\phi_{\mathbf{k}0}$ depend linearly on the hole and electron concentrations, respectively. Therefore, the expression (7.57) depends on the product of the electron and hole concentrations, while (7.59) and (7.60) vary linearly with respect to the hole and electron concentrations, respectively. However, considering the relations (7.50), which apply to the case of strong resonance coupling, and the fact that the energies of excitation $\Omega_\pm(\mathbf{k}\nu)$ and $\Omega_\pm(\mathbf{k}0)$ depend on the hole and electron concentrations, respectively, the discussion above on the dependence of $\alpha(\omega)$ on the electron and hole concentrations is applicable only in general.

For a discussion of the particular case when the dielectric functions $\varepsilon_{\mathbf{k}\nu}(\omega)$ and $\varepsilon_{\mathbf{k}0}(\omega)$ possess scattering solutions, we refer to the litarature (Mavroyannis, 1972c). Similar treatment to that of the CT spectra described here has been used to study the biexciton spectra of molecular crystals (Mavroyannis, 1972b).

VIII. Perturbation Theory for the Green's Functions

Systematic treatments of different forms of perturbation theory can be found in the works of Montroll and Ward (1959), Kostantinov and Perel (1961), Dzyaloshinskii (1962), and Baym and Sessler (1963); we follow Tyablikov and Bonch–Bruevich (1962).

Consider the Fourier transform of the Green's function $G(f, g; \omega) = \langle\!\langle A_f(t); A_g^\dagger(t')\rangle\!\rangle$, which satisfies the equation of motion with respect to the argument t

$$\omega G(f, g; \omega) = I_1 + \langle\!\langle [A_f(t), \mathscr{H}]; A_g^\dagger(t')\rangle\!\rangle, \tag{8.1}$$

where I_1 is defined as

$$I_1 = (1/2\pi)\langle [A_f(t), A_g^\dagger(t)]_{-\eta}\rangle. \tag{8.2}$$

Let us take the Hamiltonian of the system in the form

$$\mathcal{H} = H_0 + H' \tag{8.3}$$

where H_0, is the unperturbed Hamiltonian, for which

$$[A_f, H_0] = LA_f \tag{8.4}$$

where L is the energy operator for the free particles. H' is the Hamiltonian describing the interaction between the particles in the system. Upon substituting Eq. (8.3) into Eq. (8.1) and making use of Eq. (8.4), we obtain

$$G_0^{-1}(\omega)G(f,g;\omega) = I_1 + \langle\!\langle [A_f, H']; A_g^{\dagger}(t') \rangle\!\rangle \tag{8.5}$$

where $G_0(\omega) = (\omega - L)^{-1}$ is the unperturbed Green's function. Considering the equation of motion for the Green's function that appears on the right-hand side of Eq. (8.5) with respect to the argument t', we obtain

$$\langle\!\langle [A_f(t), H']; A_g^{\dagger}(t') \rangle\!\rangle = I_1 \tilde{M}(f,g;\omega)G_0(\omega) \tag{8.6}$$

where

$$\tilde{M}(f,g;\omega) = I_1^{-1}\{(2\pi)^{-1}\langle [[A_f(t), H'], A_g^{\dagger}(t)]_{-\eta}\rangle + \langle\!\langle [A_f(t), H']; [A_g(t'), H']^{\dagger} \rangle\!\rangle\}. \tag{8.7}$$

Substitution of Eq. (8.6) into Eq. (8.5) leads to

$$G_0^{-1}(\omega)G(f,g;\omega) = I_1\{1 + \tilde{M}(f,g;\omega)G_0(\omega)\}. \tag{8.8}$$

If we write now the expression (8.8) in the form of the Dyson equation

$$[G_0^{-1}(\omega) - M(f,g;\omega)]G(f,g;\omega) = I_1, \tag{8.9}$$

then the mass operator of the system is defined as

$$\begin{aligned} M(f,g;\omega) &= \tilde{M}(f,g;\omega)G_0(\omega)G^{-1}(f,g;\omega) \\ &= \tilde{M}(f,g;\omega)/[1 + \tilde{M}(f,g;\omega)G_0(\omega)]; \end{aligned} \tag{8.10}$$

i.e., the mass operator of the system is expressed in terms of the inverse Green's function. In fact, the poles of the Green's function determine the energies of the elementary excitations of the system, which depend on the interactions between the particles. The main idea is to develop a perturba-

8. The Green's Function Method

tion expansion (in terms of some small parameter indicating the strength of the interaction) of the inverse Green's function. The Green's function can then be obtained by inversion. Following this procedure one, at least, hopes to describe correctly the behavior of the Green's function in the region of the poles, which is important for many applications.

Considering that the function $\tilde{M}(f, g; \omega)$ is a complex quantity, then in the region of frequencies ω that satisfy the equation

$$1 + G_0(\omega) \operatorname{Re} \tilde{M}(f, g; \omega) = 0, \tag{8.11}$$

and from Eq. (8.8), we have that $\operatorname{Re} G(f, g; \omega) = 0$, while

$$\operatorname{Im} G(f, g; \omega) = G_0^2(\omega) I_1 \operatorname{Im} \tilde{M}(f, g; \omega) \tag{8.12}$$

where Re and Im stand for the real and imaginary parts. Notice that the frequencies ω satisfying Eq. (8.11) are the poles of the mass operator $M(f, g; \omega)$, which correspond to the zeros of the $\operatorname{Re} G(f, g; \omega)$. This indicates that when ω is near the poles of the mass operator, one has to use the expression (8.10). It physically implies that the formation of some collective type of excitations in the system is possible whose spectrum is determined by the solutions of Eq. (8.11).

For frequencies ω far from the zeros of the denominator in Eq. (8.10), we define $X \equiv \tilde{M}(f, g; \omega) G_0(\omega)$. Then symbolically we have

$$M = X G_0^{-1} / (1 + X). \tag{8.13}$$

Expanding the denominator of Eq. (8.13) in power series of X, we obtain

$$M \approx (X - X^2 + X^3 - \cdots) G_0^{-1}(\omega). \tag{8.14}$$

In the expansion (8.14), we have made the assumption that X is replaced by εX, where ε is a small parameter ($\varepsilon \ll 1$). This physically implies that the interactions involved in the interaction Hamiltonian H' are weak. In the expansion (8.14) the poles of the mass operator, which correspond to the zeros of $(1 + X)$, disappear. Equations (8.9) and (8.14) give a formal solution to the problem. For further details, we refer the reader to the work of Tyablikov and Bonch–Bruevich (1962).

Let us consider, as an example, only the first term in the expansion (8.14). Then Eq. (8.9) assumes the form

$$[G_0^{-1}(\omega) - \tilde{M}(f, g; \omega)] G(f, g; \omega) = I_1 \tag{8.15}$$

with $\tilde{M}(f, g; \omega)$ given by Eq. (8.7). We may rewrite Eq. (8.15) as

$$[G_1^{-1}(f, g; \omega) - M''(f, g; \omega)]G(f, g; \omega) = I_1 \quad (8.16)$$

where

$$G_1^{-1}(f, g; \omega) = G_0^{-1}(\omega) - M'(f, g) \quad (8.17)$$

$$M'(f, g) = (1/2\pi I_1)\langle[[A_f(t), H'], A_g^{\dagger}(t)]_{-\eta}\rangle \quad (8.18)$$

$$M''(f, g; \omega) = (1/I_1)\langle\!\langle[A_f(t), H']; [A_g(t'), H']^{\dagger}\rangle\!\rangle. \quad (8.19)$$

In this approximation, $M'(f, g)$ is the static part (independent of ω) of the mass operator, while $M''(f, g; \omega)$ is its dynamic part. It is easily shown that when the operator A_f is replaced by α_f or $\alpha_{f_1}^{\dagger}\alpha_{f_2}$ and the system is represented by the Hamiltonian given by Eq. (6.2), then making use of Eqs. (8.17) and (8.18), one recovers the results obtained in Section VI.B or VI.C for the mass operator in the first approximation.

The function $M''(f, g; \omega)$ can be calculated in the first approximation through the unperturbed Hamiltonian. Sometimes, instead of the unperturbed Hamiltonian, one can make use of an effective Hamiltonian which is diagonal and corresponds to the excitation spectrum described by $G_1(f, g; \omega)$ given by Eq. (8.17). In this case, the energy modes, which will appear in the expression for $M''(f, g; \omega)$, Eq. (8.19), are renormalized and they are correct in the self-consistent field approximation. For a more accurate calculation of the function $M''(f, g; \omega)$, one has to make use of the total Hamiltonian. In such a case, for the sake of consistency, care must also be given to the higher order terms in the expansion (8.14). As mentioned previously, there are physical processes for which the expansion (8.14) does not apply. For example, for systems of the hard-sphere type, the interaction is not weak at short distances, but the forces are short ranged. In this case, one considers Eq. (8.14) as an expansion in power series of the density, which is a small quantity, and the mass operator can be expressed in terms of the scattering amplitude (Tyablikov and Bonch–Bruevich, 1962). For systems with Coulomb interactions, collective excitations (plasma singularities) are important, but they are not included in Eq. (8.14). Such effects are of the order of $\varepsilon^{3/2}$, where ε is the expansion parameter in Eq. (8.14). In this case, one has to disregard successive expansions in powers of ε and, in general, it is preferable to use Eq. (8.10) or (8.13) and to perform partial summation of the series in perturbation theory. For details concerning the calculation of the mass operator with screening effects taken into consideration, we refer to the literature (Tyablikov and Bonch–Bruevich, 1962). The approach that has been developed in Section VI.D for the three-particle correlations and the application in Section VII is an illustration for systems consisting of charged fermions, for which polarization effects are important (Mavroyannis, 1972c).

Having derived the explicit form for the mass operator and considering that in general $M(f,g;\omega)$ is a complex quantity, then Eq. (8.9) assumes the form

$$G(f,g;\omega) = I_1\{G_0^{-1}(\omega) - \text{Re } M(f,g;\omega) + i \text{ Im } M(f,g;\omega)\}^{-1}. \quad (8.20)$$

Using Eqs. (3.17) and (8.20), we derive the expression for the spectral function (see, for instance, Zubarev, 1960)

$$J(f,g;\omega) = \frac{I_1 \gamma_{fg}(\omega)(\exp(\beta\omega) - \eta)^{-1}}{[G_0^{-1}(\omega) - \text{Re } M(f,g;\omega)]^2 + \gamma_{fg}^2(\omega)} \quad (8.21)$$

where $\gamma_{fg}(\omega) = \text{Im } M(f,g;\omega)$. From Eqs. (3.18) and (8.21), the correlation function is obtained as

$$\langle A_g^\dagger(t') A_f(t) \rangle = \int_{-\infty}^{+\infty} d\omega\, I_1 \frac{\gamma_{fg}(\omega)[\exp(\beta\omega) - \eta]^{-1} \exp[-i\omega(t-t')]}{[G_0^{-1}(\omega) - \text{Re } M(f,g;\omega)]^2 + \gamma_{fg}^2(\omega)}, \quad (8.22)$$

and for $t = t'$ the distribution function is given by

$$\langle A_g^\dagger(t) A_f(t) \rangle = \int_{-\infty}^{+\infty} d\omega\, I_1 \frac{\gamma_{fg}(\omega)[\exp(\beta\omega) - \eta]^{-1}}{[G_0^{-1}(\omega) - \text{Re } M(f,g;\omega)]^2 + \gamma_{fg}^2(\omega)}. \quad (8.23)$$

As $\gamma_{fg}(\omega) \to 0$, the spectral function (8.21) tends to a delta-shape distribution.

In the limiting case when $\gamma(\omega)$ is very small, the excitation spectrum is determined by the roots ω_ν of the equation

$$G_0^{-1}(\omega) - \text{Re } M(f,g;\omega) = 0. \quad (8.24)$$

Assuming that $\text{Re } M(f,g;\omega)$ and $\gamma_{fg}(\omega)$ are slowly varying functions of ω in the neighborhood of $\omega \sim \omega_\nu$, we expand $\text{Re } M(f,g;\omega)$ in power series of ω at $\omega = \omega_\nu$ and take $\gamma(\omega) \sim \gamma(\omega_\nu)$. Then the expression (8.21) for the spectral function takes the approximate form

$$J(f,g;\omega) \approx [\tilde{I}_1 \tilde{\gamma}_{fg}(\omega_\nu)(e^{\beta\omega} - \eta)^{-1}] / [(\omega - \omega_\nu)^2 + \tilde{\gamma}_{fg}^2(\omega_\nu)] \quad (8.25)$$

where

$$\tilde{I}_1 = I_1\{1 - [(d/d\omega) \operatorname{Re} M(f, g; \omega)]_{\omega=\omega_\nu}\}^{-1},$$
$$[(d/d\omega) \operatorname{Re} M(f, g; \omega)]_{\omega=\omega_\nu} \ll 1 \quad (8.26a)$$
$$\tilde{\gamma}_{fg}(\omega_\nu) = \gamma_{fg}(\omega_\nu)\{1 - [(d/d\omega) \operatorname{Re} M(f, g; \omega)]_{\omega=\omega_\nu}\}^{-1}. \quad (8.26b)$$

In this approximation, the spectral function given by Eq. (8.25) describes a Lorentzian line peaked at $\omega \sim \omega_\nu$ with a spectral width of the order of $\tilde{\gamma}_{fg}(\omega_\nu)$ (in energy units), provided that $\tilde{\gamma}_{fg}(\omega_\nu) \ll \omega_\nu$. The inverse of the damping function $\tilde{\gamma}_{fg}(\omega_\nu)$ describes the life time of the elementary excitation with energy ω_ν.

The perturbation approach of calculating Green's functions has been widely used in the literature (see, for instance, Tserkovnikov, 1962; Dharmawardana and Mavroyannis, 1970; Mavroyannis and Pathak, 1969; Mavroyannis, 1967b, 1970b,c,d, 1971, 1972a; Deverin and Mavroyannis, 1972).

A. High-Frequency Behavior

We discuss the asymptotic decoupling theorem, which holds at the high-frequency limit when $\omega \to \infty$ regardless of the strength of the interactions that might be present in the system. For a spatially uniform system, the Fourier transform of the retarded one-particle Green's function $G(\omega) = \langle\!\langle A_\mathbf{k}; A_\mathbf{k}^\dagger \rangle\!\rangle$ may be written as

$$[\omega - \omega_\mathbf{k} - M_\mathbf{k}(\omega)]G(\omega) = I/2\pi \quad (8.27)$$

where $\omega_\mathbf{k}$ is the energy of the particle at the state $\mathbf{k}$, $M_\mathbf{k}(\omega)$ is the mass operator, while

$$I = \langle [A_\mathbf{k}(t), A_\mathbf{k}^\dagger(t)]_{-\eta}\rangle. \quad (8.28)$$

$A_\mathbf{k}^\dagger$ and $A_\mathbf{k}$ are the one-particle creation and annihilation operators (Fermi, Bose, or Pauli). From Eq. (8.27), we obtain

$$-M_\mathbf{k}(\omega) = (I/2\pi)[G^{-1}(\omega) - G_0^{-1}(\omega)] \quad (8.29)$$
$$G_0(\omega) = (I/2\pi)(\omega - \omega_\mathbf{k})^{-1}. \quad (8.30)$$

Clearly, as $\omega \to \infty$

$$G_0^{-1}(\omega) \to (2\pi\omega/I), \quad |\omega| \to \infty. \quad (8.31)$$

From Eq. (3.14), we have that the function $G(\omega)$ is analytic in the upper half-plane and possesses a spectral representation, i.e.,

$$G(\omega) = (1/2\pi) \int_{-\infty}^{+\infty} [\exp(\beta\omega') - \eta] J_{A_\mathbf{k} A_\mathbf{k}^\dagger}(\omega')[d\omega'/(\omega - \omega')]. \quad (8.32)$$

Taking the limit $\omega \to \infty$ in Eq. (8.32) and using Eq. (3.10b), we have

$$G(\omega) = (1/2\pi) \int_{-\infty}^{+\infty} [\exp(\beta\omega') - \eta] J_{A_\mathbf{k} A_\mathbf{k}^\dagger}(\omega')(d\omega'/\omega) + O(\omega^{-2})$$
$$= (1/2\pi\omega)\langle [A_\mathbf{k}(t), A_\mathbf{k}^\dagger(t)]_{-\eta} \rangle + O(\omega^{-2}) = (I/2\pi\omega) + O(\omega^{-2}). \quad (8.33)$$

Thus the asymptotic form of $G(\omega)$ as $\omega \to \infty$ is given by

$$G(\omega) \to (I/2\pi\omega), \qquad \omega \to \infty. \quad (8.34)$$

Then by definition for the inverse function, we have

$$G^{-1}(\omega) \to (2\pi\omega/I), \qquad \omega \to \infty. \quad (8.35)$$

Comparison of Eqs. (8.29), (8.31), and (8.35) leads to

$$M_\mathbf{k}(\omega)/\omega \to 0, \qquad (1/\omega)(\partial/\partial\omega)M_\mathbf{k}(\omega) \to 0, \qquad |\omega| \to \infty. \quad (8.36)$$

This indicates that asymptotically at $|\omega| \to \infty$, the particles behave like free particles. In fact, large values of ω correspond to small time differences t. At such small time differences the operators $A_\mathbf{k}$, which are in the Heisenberg representation, do not substantially differ from those in the interaction representation. In other words, the interaction has no time to proceed during the small time interval $t \sim \hbar/|\omega|$. These considerations [Eq. (8.36)] justify the validity of the asymptotic decoupling approximation for the one-particle Green's function which has been used in Eq. (6.5). Rigorous treatments on this subject have been given by Bonch–Bruevich (1962, 1963) and Maleev (1962). We have also made use of the asymptotic decoupling theorem for the two-particle (collective) Green's functions in Eqs. (5.45) and (6.13). The proof of the theorem in this case has been given by Bonch–Bruevich and Rozman (1964); we refer to their paper for details.

B. Functional Derivatives

For the sake of completeness, we consider briefly the method of functional derivatives for determining the causal double-time Green's function. In accordance with Eqs. (2.2c) and (2.3b), we define the causal Green's functions

$$G_c(f, t; g, t') = -i\langle \hat{T}\alpha_f(t)\alpha_g{}^\dagger(t')\rangle \tag{8.37a}$$

$$G_c(f_1, f_2, f_3, t; g, t') = -i\langle \hat{T}\{\alpha_{f_1}^\dagger(t)\alpha_{f_2}(t)\alpha_{f_3}(t)\alpha_g{}^\dagger(t')\}\rangle. \tag{8.37b}$$

Then using the Hamiltonian (6.2), we derive the equation of motion

$$(id/dt)G_c(f, t; g, t') - \sum_{f_1} L(f_1, f)G_c(f_1, t; g, t')$$
$$= \delta_{fg}\,\delta(t - t') + \sum_{f_1, f_2, f_3} \langle f, f_1 | V | f_3, f_2\rangle G_c(f_1, f_2, f_3, t; g, t'). \tag{8.38}$$

The functional derivative method consists in writing the last term on the right-hand side of Eq. (8.38) in the form of a functional derivative of the one-particle Green's function.

We introduce the generalization of the Green's function in the functional form

$$G_c(f, t; g, t'; Q) = (-i/\langle S\rangle)\langle \hat{T}\{\alpha_f(t)\alpha_g{}^\dagger(t')S\}\rangle \tag{8.39}$$

where S is given by

$$S = \hat{T}\{\exp[i\sum_{f,g}\int dt\, Q(f, g; t)\alpha_f{}^\dagger(t)\alpha_g(t)]\}, \tag{8.40}$$

and Q is the function that we will use to take functional derivatives. If we define the functional derivative

$$\psi(f_1, f_2; t) = -i[\delta \ln\langle S\rangle/\delta Q(f_1, f_2; t)], \tag{8.41}$$

then performing the variation for the function

$$G_c(f_1, f_2, f_3, t; g, t'; Q) = (-i/\langle S\rangle)\langle \hat{T}\{\alpha_{f_1}^\dagger(t)\alpha_{f_2}(t)\alpha_{f_3}(t)\alpha_g{}^\dagger(t')S\}\rangle, \tag{8.42}$$

we find

$$G_c(f_1, f_2, f_3, t; g, t'; Q)$$
$$= \frac{-i}{\langle S\rangle}\,\frac{\delta\{\langle S\rangle G_c(f_3, t; g, t'; Q)\}}{\delta Q(f_1, f_2; t)}$$
$$= (-i)\frac{\delta G_c(f_3, t; g, t'; Q)}{\delta Q(f_1, f_2; t)} + \psi(f_1, f_2; t)G_c(f_3, t; g, t'; Q). \tag{8.43}$$

The equation of motion [Eq. (8.39)] for the Green's function $G_c(f, t; g, t'; Q)$ takes the form

$$\frac{id}{dt} G_c(f, t; g, t'; Q) - \sum_{f_1} L(f_1, f) G_c(f_1, t; g, t'; Q)$$
$$+ \sum_{f_1} Q(f, f_1; t) G_c(f_1, t; g, t'; Q) = \delta_{fg}\, \delta(t - t')$$
$$+ \sum_{f_1, f_2, f_3} \langle f, f_1 | V | f_3, f_2 \rangle \Big\{ -i\, \frac{\delta G_c(f_3, t; g, t'; Q)}{\delta Q(f_1, f_2; t)}$$
$$+ \psi(f_1, f_2; t) G_c(f_3, t; g, t'; Q) \Big\}. \tag{8.44}$$

Using Eqs. (8.40) and (8.41), the functional derivative can be written as

$$\psi(f_1, f_2; t) = (1/\langle S \rangle) \langle \hat{T}\{\alpha_{f_1}^\dagger(t) \alpha_{f_2}(t) S\} \rangle, \tag{8.45}$$

which in the limit when $Q \to 0$, becomes

$$\psi(f_1, f_2; t) \xrightarrow[Q \to 0]{} \langle \alpha_{f_1}^\dagger(t) \alpha_{f_2}(t) \rangle. \tag{8.46}$$

Similarly,

$$G_c(f, t; g, t') = \lim_{Q \to 0} G_c(f, t; g, t'; Q). \tag{8.47}$$

Thus the function $\psi(f_1, f_2; t)$ has a clear physical meaning; i.e., as $Q \to 0$, $\psi(f_1, f_2; t)$ is equal to the electron distribution function. From Eq. (8.44) one can generate solutions to arbitrary order by means of iteration procedures. In this fashion, one can derive general forms for the Green's functions involving mass operators. The function Q has only formal meaning and it must be taken equal to zero after the functional derivatives have been calculated. For further details we refer to the literature (Fredkin, 1959; Martin and Schwinger, 1959; Baym and Kadanoff, 1961; Bonch-Bruevich and Tyablikov, 1962; Engelsberg, 1962). The functional derivative technique has also been applied to study problems in lattice dynamics (Wehner, 1966, 1967; Kwok, 1967).

C. Concluding Remarks

We have outlined the properties and the methods by which the retarded and advanced double-time temperature-dependent Green's functions can be calculated. Their usefulness in the study of many-body problems stems from the fact that they can be analytically continued in the complex plane. However, the method has one shortcoming, namely, the absence of a convenient diagrammatic representation. For recent attempts in this direction, we refer to the papers by Morita (1970), Mattuck and Theumann (1971), and Tserkovnikov (1971).

REFERENCES

NRCC No. 12741.
ABRIKOSOV, A. A., GOR'KOV, L. P., and DZYALOSHINSKII, I. E. (1963). "Methods of Quantum Field Theory in Statistical Mechanics." Prentice-Hall, Englewood Cliffs, New Jersey.
AGRANOVICH, V. M. (1960). *Sov. Phys.-JETP* **10**, 307.
ALEKSEEV, A. I. (1961). *Sov. Phys.-Usp.* **4**, 23.
BAYM, G. (1962). *Phys. Rev.* **127**, 1391.
BAYM, G., and KADANOFF, L. P. (1961). *Phys. Rev.* **124**, 287.
BAYM, G., and SESSLER, A. M. (1963). *Phys. Rev.* **131**, 2345.
BOGOLYUBOV, N. N. (1961). Quasi-Averages in Statistical Mechanics Problems. Rotaprint Rep. OIYaI, D-781, Dubna.
BOGOLYUBOV, N. N., and PARASYUK, O. S. (1956). *Dokl. Akad. Nauk SSSR* **109**, 717.
BOGOLYUBOV, N. N., and TYABLIKOV, S. V. (1959). *Sov. Phys.-Dokl.* **4**, 589.
BOLSTERLI, M. (1960). *Phys. Rev. Lett.* **4**, 82.
BONCH-BRUEVICH, V. L. (1956). *Sov. Phys.-JETP* **4**, 456.
BONCH-BRUEVICH, V. L. (1959). *Sov. Phys.-JETP* **9**, 653.
BONCH-BRUEVICH, V. L. (1960). *Sov. Phys.-Dokl.* **4**, 1275.
BONCH-BRUEVICH, V. L. (1962). *Phys. Lett.* **2**, 146.
BONCH-BRUEVICH, V. L. (1963). *Sov. Phys.-Dokl* **7**, 1108.
BONCH-BRUEVICH, V. L. (1966a). The Optical Properties of Solids. *Proc. Int. School Phys. "Enrico Fermi," Varenna, Italy*, p. 331. Academic Press, New York.
BONCH-BRUEVICH, V. L. (1966b). "The Electronic Theory of Heavily Doped Semiconductors." Amer. Elsevier, New York.
BONCH-BRUEVICH, V. L., and KOGAN, SH. M. (1960). *Ann. Phys. N. Y.* **9**, 1275.
BONCH-BRUEVICH, V. L., and ROZMAN, R. (1964). *Sov. Phys.-Solid State* **5**, 2117.
BONCH-BRUEVICH, V. L., and TYABLIKOV, S. V. (1962). "The Green Function Method in Statistical Mechanics." North-Holland Publ., Amsterdam.
CÄLLEN, G. (1952). *Helv. Phys. Acta* **25**, 417.
CALLEN, H. B., and WELTON, T. A. (1951). *Phys. Rev.* **83**, 34.
DEVERIN, J.-A., and MAVROYANNIS, C. (1972). *Helv. Phys. Acta* **45**, 1005.
DHARMAWARDANA, M. W. C., and MAVROYANNIS, C. (1970). *Phys. Rev. B* **1**, 1166.
DZYALOSHINSKII, I. E. (1962). *Sov. Phys.-JETP* **15**, 778.
DZYUB, I. P. (1959). *Sov. Phys.-Dokl.* **5**, 125.
DZYUB, I. P. (1961). *Sov. Phys.-JETP* **12**, 429.
ENGELSBERG, S. (1962). *Phys. Rev.* **126**, 1251.
FREDKIN, E. S. (1959). *Nucl. Phys.* **12**, 465.
GOR'KOV, L. P. (1958). *Sov. Phys.-JETP* **1**, 505.

HOPFIELD, J. J. (1958). *Phys. Rev.* **112**, 1555.
KADANOFF, L. P., and BAYM, G. (1962). "Quantum Statistical Mechanics." Benjamin, New York.
KLEIN, A., and PRANGE, R. (1958). *Phys. Rev.* **112**, 994.
KOGAN, SH. M. (1959). *Sov. Phys.-Dokl.* **4**, 604.
KOHN, W., and LUTTINGER, J. M. (1960). *Phys. Rev.* **118**, 41.
KOSTANTINOV, O. V., and PEREL, V. I. (1961). *Sov. Phys.-JETP* **12**, 142.
KUBO, R. (1957). *J. Phys. Soc. Japan* **12**, 570.
KUBO, R., and TOMITA, K. (1954). *J. Phys. Soc. Japan* **9**, 888.
KUBO, R., YOKOTA, M., and NAKAJIMA, S. (1957). *J. Phys. Soc. Japan* **12**, 1203.
KWOK, P. C. K. (1967). *Solid State Phys.* **20**, 213.
LANDAU, L. D. (1958). *Sov. Phys.-JETP* **7**, 182.
LEHMANN, H. (1954). *Nuovo Cimento* **11**, 342.
LUTTINGER, J. M. (1960). *Phys. Rev.* **119**, 1153.
LUTTINGER, J. M. (1961). *Phys. Rev.* **121**, 942.
LUTTINGER, J. M., and WARD, J. C. (1960). *Phys. Rev.* **118**, 1417.
MALEEV, S. V. (1962). *Sov. Phys.-JETP* **14**, 1191.
MARTIN, P. C., and SCHWINGER, J. (1958). *Bull. Amer. Phys. Soc.* **3**, 202.
MARTIN, P. C., and SCHWINGER, J. (1959). *Phys. Rev.* **115**, 1342.
MATSUBARA, T. (1955). *Progr. Theor. Phys. Japan*, **14**, 351.
MATTUCK, R. D. (1967). "A Guide to Feyman Diagrams in the Many-Body Problem." McGraw-Hill, New York.
MATTUCK, R. D., and THEUMANN, A. (1971). *Advan. Phys.* **20**, 721.
MAVROYANNIS, C. (1965). *J. Chem. Phys.* **42**, 1772.
MAVROYANNIS, C. (1967a). *J. Math. Phys.* **8**, 1515.
MAVROYANNIS, C. (1967b). *J. Math. Phys.* **8**, 1522.
MAVROYANNIS, C. (1970a). *J. Math. Phys.* **11**, 491.
MAVROYANNIS, C. (1970b). *Phys. Rev. B* **1**, 2706.
MAVROYANNIS, C. (1970c). *Phys. Rev. B* **1**, 3439.
MAVROYANNIS, C. (1970d). Optical Properties of Solids. *Proc. Int. Conf., Chania, Crete (Greece)*, p. 349. Gordon and Breach, New York.
MAVROYANNIS, C. (1971). *Phys. Rev. B* **3**, 2750.
MAVROYANNIS, C. (1972a). *Chem. Phys. Lett.* **14**, 264.
MAVROYANNIS, C. (1972b). *Chem. Phys. Lett.* **14**, 497.
MAVROYANNIS, C. (1972c). *Phys. Rev. B.* **6**, 2463.
MAVROYANNIS, C., and PATHAK, K. N. (1969). *Phys. Rev.* **182**, 872.
MONTROLL, E. W., and WARD, J. C. (1959). *Physica* **25**, 423.
MORI, H. (1956). *J. Phys. Soc. Japan* **11**, 1029.
MORITA, T. (1970). *J. Phys. Soc. Japan* **28**, 1128.
MULLIKEN, R. S., and PERSON, W. B. (1969). "Molecular Complexes." Wiley, New York.
NOZIERES, P. (1964). "Theory of Interacting Fermi Systems." Benjamin, New York.
PIKE, E. R. (1964). *Proc. Phys. Soc.* **84**, 83.
SCHULTZ, T. D. (1964). "Quantum Field Theory and the Many-Body Problem." Gordon and Breach, New York.
TER HAAR, D., and PARRY, W. E. (1962). *Phys. Lett.* **1**, 145.
TSERKOVNIKOV, YU. A. (1962). *Sov. Phys.-Dokl.* **7**, 322.
TSERKOVNIKOV, YU. A. (1971). *Theor. Math. Phys.* **7**, 511.

TYABLIKOV, S. V. (1967). "Methods in Quantum Theory of Magnetism." Plenum Press, New York.
TYABLIKOV, S. V., and BONCH-BRUEVICH, V. L. (1962). *Advan. Phys.* **11**, 317.
WEHNER, R. K. (1966). *Phys. Status Solidi* **15**, 725.
WEHNER, R. K. (1967). *Phys. Status Solidi* **22**, 527.
ZUBAREV, D. N. (1960). *Sov. Phys.-Usp.* **3**, 320.
ZUBAREV, D. N. (1962). *Sov. Phys.-Dokl.* **6**, 776.

Author Index

Numbers in italics refer to the pages on which the complete references are listed.

A

Abramowtiz, M., 222, 237, 245, *259, 335, 357, 372*
Abrikosov, A. A., 489, *548*
Agranovich, V. M., 524, *548*
Ahlberg, J. H., 346, *372*
Ahlfors, L. V., 169, 194, *259*
Alekseev, A. I., *548*
Amemiya, A., *427*
Apostol, T. M., 129, *150*
Arfken, G., *335*
Arsac, J., 83, *149*

B

Barrett, G. P., 449, *485*
Bateman, H., *335*
Baxter, R. J., 83, *149*
Baym, G., 489, 539, 547, *548, 549*
Beltrami, E. J., 83, *149*
Bender, C. F., 471, 472, *485*
Benesch, R., 458, *485*
Blatt, J. M., 477, *485*
Boardman, A. D., 454, *485*
Bogolyubov, N. N., 489, 494, 497, 500, 506, *548*
Bolsterli, M., 502, *548*
Bonch-Bruevich, V. L., 489, 491, 494, 499, 500, 506, 512, 513, 515, 516, 519, 521, 539, 541, 542, 543, 545, 547, *548*
Bopp, F., 467, *485*
Bouix, M., 83, *149*
Bremermann, H., 83, 124, *149*
Breuckner, K. A., 457, *485*
Brown, R. J. C., 471, 475, *486*
Buchler, J. R., *485*
Buerger, M. J., *427*
Bunn, C. W., *427*

C

Cällen, G., 499, *548*
Callen, H. B., 499, 506, 510, *548*
Capri, A. Z., 83, *149*
Carslaw, H. S., *335*
Churchill, R. V., 167, *259,* 263, 328, 332, *335*
Coleman, A. J., 442, 444, 445, 446, 482, *485*
Cotten, F. A., *426*
Courant, R., *335*
Cross, P. C., *427*
Curtis, A. R., 454, *485*

D

Davidson, E. R., 446, 448, 471, 472, *485*
Davis, H. F., *335*
Decius, J. C., *427*
Dennery, P., *78*
Deverin, J.-A., 544, *548*
Dharmawardana, M. W. C., 544, *548*
Dirac, P. A. M., *79,* 82, *149*
Donoghue, W. F., Jr., 83, *149*
Duff, G. F. D., *335*
Dyson, F., 476, *485*
Dzyaloshinskii, I. E., 489, 539, *548*
Dzyub, I. P., 522, 524, *548*

E

Edmonds, A. E., *427*
Edwards, R. W., 83, *149*
Emde, F., *335*
Engelsberg, S., 547, *548*
Erdélyi, A., 83, 103, *149,* 222, 232, *259*

F

Fano, U., *427*
Feshbach, H., *78,* 268, 316, 332, *335*
Fogel, S. J., 461, *486*
Fredkin, E. S., 547, *548*
Friedman, A., 83, *149,* 316, *335*
Frohlich, H., 483, *485*

G

Gårding, L., 83, *149*
Garrod, C., 445, *485*
Gel'fand, I. M., 83, 118, 143, *150*
Girardeau, M., 475, 477, 484, *485*
Glauber, R. J., *485*
Goldstein, J., *78*
Gor'kov, L. P., 489, 500, *548*
Goswami, D., 484, *486*
Graev, M. I., 83, 118, 143, *150*
Griffith, J. S., *427*
Grimley, T. B., 469, *485*

H

Hadamard, J., 124, *150*
Hall, G. G., *426*
Hall, L. H., *426*
Halmos, P. R., *78*
Halperin, I., 83, *150*
Hammermesh, M., *426*
Heaviside, O., 82, *150*
Heine, V., *426*
Herzberg, G., *427*
Hilbert, D., *335*
Hilton, D., 454, *485*
Hobson, E. W., 166, *259, 335*
Holloway, T. T., 256, *260*
Honenberg, P., 455, *485*
Hopfield, J. J., 524, *549*
Hoskins, W. D., 350, *372*

I

Irving, J., *78*
Ishiguro, E., *427*

J

Jackson, J. D., 316, *335*
Jaeger, J. C., *335*
Jaffe, H. H., *427*
Jager, E. M. de, 83, *150*
Jahnke, E., *335*

Jantscher, L., 83, *150*
Jeffreys, B., *335*
Jeffreys, H., *335*
Jorna, S., *485*

K

Kadanoff, L. P., 489, 547, *549*
Kay, K. G., 241, 245, 247, *259, 260*
Kijewski, L. J., 470, 472, 473, *485*
Kimura, T., *427*
King, P. R., 350, *372*
Klein, A., 489, *549*
Kogan, Sh. M., 500, *548, 549*
Kohn, W., 455, 457, *485,* 489, *549*
Korevaar, J., 83, *150*
Kostantinov, O. V., 539, *549*
Kotani, M., *427*
Krzywicki, A., *78*
Kubo, R., 506, 509, 510, *549*
Kummer, H., 445, *485*
Kuo, S. S., *79*
Kutzelnigg, W., 449, 461, 465, *485*
Kwok, P. C. K., 547, *549*

L

Lal, P., 484, *485*
Landau, L. D., 499, 500, *549*
Lang, N. D., 457, *485*
Lass, H., *78*
Lebedev, N. N., *335*
Lehmann, H., 499, *549*
Lighthill, M. J., 83, 97, 141, *150,* 214, 251, *260*
Lindenberg, G. P., 449, *485*
Lions, J. L., 83, *149*
Liverman, T. P. G., 83, 97, 148, *150*
Lombard, R. J., *485*
Luttinger, J. M., 489, *549*

M

MacRobert, T. M., *335*
McWeeny, R., *426*
Maleev, S. V., 545, *549*
March, N. H., 453, 454, 460, *485, 486*
Marchand, J.-P., 83, 103, *150*
Margenau, H., *335*
Martin, P. C., 489, 500, 519, 547, *549*
Matson, F. A., *78*
Matsubara, T., 477, *485,* 489, *549*

Author Index

Mattuck, R. D., 489, 548, *549*
Mavroyannis, C., 517, 524, 526, 527, 529, 533, 535, 539, 543, 544, *548, 549*
Merzbacher, E., *79*
Messiah, A., 83, *150*
Mikusinski, J. G., 83, 103, *150*
Miller, K. J., 465, *485*
Milne, W. E., 364, *372*
Montroll, E. W., 539, *549*
Mori, H., 506, *549*
Morita, T., 548, *549*
Morse, P. M., *78,* 268, 316, 332, *335*
Mulliken, R. S., 535, 537, *549*
Mullineux, N., *78*
Murphy, G. M., *335*
Murray, A. M., 453, *485*

N

Nakajima, S., 506, *549*
Naylor, D., *335*
Nehari, Z., 166, 167, *260*
Newman, S., 433, *486*
Nilson, E. N., 346, *372*
Nozieres, P., 489, *549*

O

Orchin, M., *427*

P

Panofsky, W. K. H., *78,* 316, *335*
Papoulis, A., 83, *150*
Parasyuk, O. S., 497, *548*
Parry, W. E., 502, *549*
Pathak, K. N., 544, *549*
Peat, F. D., 441, 469, 471, 472, 475, 477, 478, 484, *485, 486*
Percus, J. K., 445, 470, 472, 473, *485*
Pere'l, V. I., 539, *549*
Person, W. B., 535, 537, *549*
Phillips, M., *78,* 316, *335*
Pike, E. R., 502, *549*
Powell, M. J. D., 346, *372*
Prange, R., 489, *549*

R

Racah, G., *427*
Reid, J. K., 371, *372*
Reinsch, C., 369, 370, *372*
Rose, M. E., *427*

Rozman, R., 506, 515, 516, 545, *548*
Rudenberg, K., 465, *485*
Ruskai, M. B., 478, *486*

S

Sack, R. A., 256, *260*
Sagan, H., *335*
Saltzer, C., 83, *150*
Schmiedler, W., *79*
Schoenberg, I. J., 346, *372*
Schonland, D., *426*
Schultz, T. D., 489, *549*
Schwartz, L., 82, *150*
Schweber, S. S., 83, *150*
Schwinger, J., 489, 500, 519, 547, *549*
Sessler, A. M., 539, *548*
Sham, L. J., 457, *485*
Shilov, G. E., 83, 118, 143, *150*
Shull, H., 449, *485*
Silverstone, H. J., 138, *150,* 241, 244, 245, 247, 256, *259, 260*
Simmons, J., 478, *486*
Slater, L. J., 232, *260*
Smith, D. W., 445, 461, *486*
Smith, V. H., Jr., 458, *485*
Smythe, W. R., *335*
Sneddon, I. N., 129, *150, 335*
Sommerfeld, A., *335*
Stanton, R. G., 364, *372*
Stegun, I. A., 222, 237, 245, *259, 335,* 357, *372*
Stoler, D., 433, *486*
Streater, R. F., 83, *150*

T

Temple, G., 83, 97, *150*
ter Haar, D., 502, *549*
Terreaux, C., 484, *485*
Theumann, A., 548, *549*
Titchmarsh, E. C., 129, *150*
Todd, H. D., 138, *150,* 241, 245, 247, *260*
Tomita, K., 506, *549*
Tranter, C. J., *336*
Treves, F., 83, *150*
Tserkovnikov, Yu. A., 544, 548, *549*
Tyablikov, S. V., 489, 491, 494, 500, 512, 513, 519, 521, 539, 541, 542, 543, 547, *548, 550*

U

Urch, D. S., *427*

V

Vilenkin, N. Ya., 83, 118, 143, *150*

W

Walsh, J. L., 346, *372*
Ward, J. C., 489, 539, *549*
Watson, G. N., 222, 232, *260, 336*
Webster, A. G., *336*
Wehner, R. K., 547, *550*
Weinhold, F., 461, *486*
Welton, T. A., 499, 506, 510, *548*
Weyl, H., *427*
Wheatley, P. J., *427*

Whittaker, E. T., 222, 232, *260, 336*
Wightman, A. S., 83, *150*
Wigner, E. P., *426*
Wilkinson, J. H., 369, 370, *372*
Wilson, E. B., *427,* 461, *486*
Wohlers, M. R., 83, *149*
Wylie, C. R., *78, 79,* 263, *336*

Y

Yang, C. N., 480, *486*
Yokota, M., 506, *549*
Young, W. H., 460, *486*

Z

Zemanian, A. H., 83, 148, *150*
Zubarev, D. N., 489, 491, 494, 506, 543, *550*

Subject Index

A

Abelian groups, 390
 defined, 376
 symmetry operations in, 388-391
Absorption coefficient, Green's function and, 513-514
Abstract vector space, 284-285
Addition theorem, for spherical harmonics, 310-312
Airy function, 235
Algebra, fundamental theorem of, 181-183
Algebraic eigenvalue methods, 75-78
Algebraic functions, 194-198
Analyticity
 in hypergeometric functions, 222
 in power series, 187-189, 192
 in transcendental functions, 215-216
Antilinear operator, 33
Averaging operator, 339

B

Band matrix, 369
Basis, transformation of, 45-47, 60
Basis functions, in group theory, 395-398
Basis matrices, of small dimension, 63-64
Basis vectors, 6
 relationships for, 15
Benzene molecule, twofold axes in, 380
Beryllium atom, ground state of, 448-449
Bessel functions, 235
 in cylindrical coordinates, 292-294
 integral-order, 302
 modified, 295-296
 orthogonality of, 298-300
 relations for, 296-298
 spherical, 235-237, 300
Beta function, 217-222
Bilinear transformation, 195
Binary operation, in group theory, 375

Bloch density matrix, 450-455
Bloch wave functions, 425-426
Block diagonalization, in non-Abelian groups, 386, 391-392, 398, 405
Boosting operator, 77
Born-Oppenheimer approximation, 377
Bose-Einstein condensation, 481, 483
Boundary conditions, 262
 matching, 266
 periodic, 266
Boundary-value problems, 261-334
 conformal mapping in, 331-334
 in cylindrical coordinates, 291-304
 defined, 262
 eigenvalues, eigenfunctions, and expansion problems in, 280-291
 Green's functions in, 316-327
 Laplace transform methods in, 327-331
 potential problems in, 264-265
 separation of variables in, 269-280
 in spherical coordinates, 304-316
 typical, 263-268
Branch points and cuts, 196-197
Bravais lattice, 422
Bra vector, 10
Brillouin-Wigner series, 257

C

Calculus, fundamental theorem of, 172-175
Carson-Keller theorem, 439
Cauchy-Riemann equations, 164-165, 173, 196, 332
Cauchy's integral formula, 179-181
Cauchy's theorem, 167, 175-179, 191, 256
 extension of, 209-210
 in multiply connected domain, 178-179
 in simply connected domain, 175-178
Cauchy test for convergence, 185
Cayley-Hamilton method, 57
Cayley-Hamilton theorem, 61-63

Cayley-Klein parameters, 67-68
Central difference operator, 339
Character tables, in group theory, 393-395
Classes
  computation of, 384
  in symmetry operations, 383-384
Clebsch-Gordon coefficients, 421
Closure relation, 9
Column matrix, 37
Commutation, vector, 34
Commutative property, multiplication as, 376
Commutator, defined, 34
Commuting operators, 44
Complementary function, 266
Completeness relation, 9
Complex analysis, open sets and domains in, 172
Complex conjugate, 157
Complex integral
  defined, 168-171
  existence of, 170-171
Complex integration, 167-184
Complex numbers, 153-160
  additive and multiplicative identity in, 154-155
  analytic functions of, 160-167
  associative, distributive, and commutative laws in, 154
  defined, 153
  field of, 154-155
  geometric representation of, 158
  notation and conventions in, 155-158
  powers and roots of, 159-160
  functions of, 160-162
Complex variable theory, 151-259
  analytic functions in, 165-166
  complex integration in, 167-184
  conjugate coordinates in, 166
  defined, 152
  derivative in, 162-163
  elementary functions in, 193-205
  Fourier transforms in, 238-241
  higher transcendental functions and, 214-238
  inequalities in, 171-172
  integral evaluation in, 242
  integration and differentiation of power series in, 187-189
  limits and continuity in, 161-162
  real definite integrals in, 205-214
  real line integrals and, 183-184
  paths in, 168-171
  power series in, 184-193
  quantum chemistry integrals in, 241-255
  as scientific tool, 153
  substitution formula in, 183
  uniqueness theorem in, 192-193
Condensation phenomena, 480-485
Confluent hypergeometric functions, 231-234
  spherical Bessel functions and, 235-237
Conformal mapping, in boundary-value problems, 331-334
Conjugate coordinates, in complex variable theory, 166
Conservative force field, 25
Conservative vector field, 25
  irrotational, 26-27
Continuous groups, in group theory, 417-421
Continuous linear functional, 91
Contraction process, tensors and, 69-70
Convergence, strong and weak, 72
Convergence tests, 185
Convergent series, 189
Convolution product, 102
Convolution quotients, theory of, 84
Convolution theorem, 244
  special form of, 240-241
Cooper pairs, 482
Coordinate axes, rotation of, 64-68
Cosine operator, 35
Creation operator, 77
Cross product, 16
Crystal-field splitting, 423-425
Crystallographic point groups, 422
Crystal symmetry, in group theory, 421-423
Cubic splines, in numerical analysis, 346-350
Curl, 22
  defined, 29
Cyclic group, 376
Cylindrical coordinates
  boundary-value problems in, 291-304
  Laplace's equation in, 292

D

D'Alembert solution, of wave equation, 268-269, 279-280

Subject Index

Definite integral, 116-118
  real, 205-214
Del operator, 21-23, 264
Delta function, 82, 102, 214, 316
De Moivre's theorem, 159
Density matrix(-ices), 50, 429-485
  Dirac and Bloch types, 450-455
  effective potential method in, 454
  energy functional method in, 455-457
  full, 431-434
  geminals in, 462-465
  Grimley-Peat approximation in, 467-472
  Hamiltonian-dependent conditions in, 472-474
  N-representability problem in, 442-447
  philosophy of, 430
  reduced, 435-441
  second-order reduced, 458-472
  single-particle reduced density matrix, 447-458
  thermodynamic quantities in, 454-455
  transition matrix in, 441
  2-matrix in, 459-467
  X-ray form factor in, 457-458
Derivative, in complex variable theory, 162-164
Destruction operator, 77
Determinants, 55-56
Diagonal matrix, 54
Differencing, interval of, 338
Differential equations, in numerical analysis, 362-364
Differential operator, 21-23, 264
Differentiation
  of generalized functions, 108-114
  rules of, 109
Differentiation operator, 339
Dirac delta function, 82, 102, 214, 316
Dirac density matrix, 450-455, 458
Directional derivative, 26
Direct product, of two representations, 399-400
Dirichlet boundary conditions, 266
Dispersion relations, Green's functions and, 500-501
Displacement operator, 339
Divergence, defined, 22
Divergence theorem, 29-32
Division, of generalized functions, 104-108
Domain, defined, 33
Dot product, 13
Double factorial, 235
D space, in test functions, 85-87

E

Eigencolumns, 60
Eigenfunctions, in boundary-value problems, 280-291
Eigenket, 38
Eigenvalue problem, 280
  for matrices, 59-63
Eigenvalues, 38
  in boundary-value problems, 280-291
  of natural orbitals, 447-449
Eigenvectors, 38-39
Elastic string, in boundary-value problems, 263-264
Electrical conductivity tensor, 510-513
Electron pairs, in geminal functions, 476-479
Elementary functions, in complex variable theory, 193-205
Energy functional method, in density matrix analysis, 455-457
Energy spectrum, in Green's function, 501-502
Entire function, 193
Euclidean vector, 2
  three-dimensional, 10-32
Euclidean vector spaces, tensors and, 69-70
Euler angle, 66-67
Euler approximation formula, 363
Euler's integral, 214-216
Excitation spectra, Green's function and, 533-539
Exciton spectra, 521-524
Exponential functions, 198-205
  definition and properties of, 198-200
Exponential operator, 35
Exponential-type integral, series for, 237-238
External perturbation, reaction of system to, 506-517

F

Fermi-Dirac distribution function, 455
Fermi distribution function, 536
Fermi-Thomas approximation, 452-453

Subject Index

Forward difference operator, 338
Fourier coefficients, 140
Fourier integral, 278
Fourier inversion theorem, 128, 131
   generalized function counterpart of, 134
Fourier series, 140, 271
   orthogonality property of, 281
Fourier transform(s), 128-143, 238-241
   defined, 238
   dimensions of, 142-143
   discrete, 141-142
   of generalized function in S', 134-135
   for Green's functions, 539
   particular results in, 136-138
   properties of, 135
   of retarded Green's functions, 517
   of Slater-type atomic orbital, 243
   table of, 130
   in three dimensions, 240
   of two-center product, 246-250
Fourier transform convolution theorem, 239
Fourier transform inversion formula, 238
Fourier transform pair, 129
Free ion, crystal-field splitting and, 424
Frenkel exciton, 490, 527
Function(s)
   see also Generalized function(s); Green's function(s)
   algebraic, 194-198
   beta, 217-222
   complementary, 266
   entire, 193
   exponential, 198-205
   Fourier transform of, 128-130
   gamma, 214-218, 221
   generalized, *see* Generalized function(s)
   Laplace transform of, 143-144
   locally integrable, 88
   multiple-valued, 196-198
   orthogonal sets of, 281-284
   products of, 101-103
   slow growth, 132
   trigonometric and hyperbolic, 200-202
Functions of a complex variable, 160-161
Function space, 76
Fundamental sequence, in generalized functions, 96-98
Fundamental theorem of the calculus, 172-175

G

Gamma function, 214-217
   beta function and, 218
Gamma function reflection formula, 221
Gaussian Elimination, 367-368, 370
Gauss-Jordan reduction method, 55-57
Gauss's hypergeometric series, 224
Gauss's law, 30-31
Geminal basis, pair creation and annihilation in, 476-477
Geminal power, antisymmetrized, 465-466
Geminals
   antisymmetrized product in, 462-463
   strongly orthogonal, 463-465
Geminal wave functions, 474-479
   general, 466-467
Generalized function(s), 81-149
   see also Function(s)
   algebra of, 98-108
   calculus of, 108-118
   convolution of, 102-103
   differentiation of, 108-114
   direct product of, 103
   division of, 104-108
   equality of, 92
   Fourier transforms of, 134-135
   fundamental sequences in, 96-98
   integration of, 106, 114-116
   Laplace transform and, 145-148
   parametric, 95-96
   periodic, 139-142
   products of, 101
   regular, 88-93
   in S', 134-135
   sequences of, 93-96
   series of, 94-95
   of several variables, 112-113
   singular, 90-92, 118-127
   of slow growth, 132
   spatial transformations in, 99-101
   translation in, 100
GF theory, *see* Generalized function(s)
Gradient, defined, 22
Green's first identity, 32
Green's function(s), 316-327, 487-548
   absorption coefficient and, 513-514
   calculation of, 517-527
   and charge transfer spectra of molecular

Subject Index

crystals, 527-539
density matrix and, 434
direct interactions and, 518
double-time temperature-dependent, 490-494
electrical conductivity tensor and, 510-513
energy spectrum and, 501-502
excitation spectra and, 533-539
Fourier transform of, 539
functional derivatives and, 545-547
high-frequency behavior and, 544-545
interband optical transitions and, 514-517
invariance under time inversion, 504-506
one-particle, 519-521
perturbation theory for, 539-548
properties of, 500-506
reaction of system to external perturbation and, 506-517
retarded, advanced, and causal, 490-494
solution of potential problems by, 324-326
spectral representations and, 494-500
for sphere, 326-327
for Sturm-Liouville operator, 319-324
symmetry properties and, 502-504
three-particle correlations in, 524-527
two-particle, 521-524
Green's second identity, 32
Grimley-Peat approximation, 467-472
Group(s)
  Abelian, 376, 390
  defined, 375
  non-Abelian, 383, 391-393
  order of, 376
  types of, 376
Group multiplication tables, 376
Group representation theory, 385-401
Group theory, 373-426
  application of in molecular quantum mechanics, 401-411
  basis functions and projection operators in, 395-398
  Bloch wave functions and, 425-426
  character tables in, 393-395
  continuous groups in, 417-421
  crystal symmetry in, 421-423
  direct product of two representations in, 399-400
  matrix elements in, 401-404
  Pauli principle in, 415-417
  permutation group in, 411-414
  permutation symmetry in, 381-383
  point groups in, 380-381
  solid state and, 421-426
  spin functions in, 414-415
  symmetry operators in, 376-388
  Young diagrams in, 412-413
  wave functions in, 385-388

H

Hadamard finite part of integral, 124
Hadamard's formula, 186, 188
Hamiltonian operator, 576-577
  in density matrix systems, 432-433, 472-474
  Green's function and, 506-507
  permutation symmetry and, 381-382
  and reduced density matrix, 435
  two-body reduced, 435
Hankel functions, 298-299
Hankel's formula, 216-217
Harmonic function, 166-167
Hartree-Fock approximation, 438, 478
Heat conduction problems, 275-280
  in boundary-value problems, 290-291
  in finite rod, 329-331
  in semiinfinite rod, 328-329
Heaviside step function, 89
Hermite polynomial, 235
Hermitian adjoint, 40
Hermitian conjugate, 53
Hermitian matrices, diagonalization of, 60-61
Hermitian operator, 39-45, 432, 443
  commuting, 43-45
  eigenvalues and eigenvectors of, 40
  noncommuting, 45
Hilbert space, 2, 75, 285, 443
  completeness in, 72-75
Homogeneous function, 100
Hyperbolic function, 200-202
  inverse in, 204
Hypergeometric function(s), 221-225
  confluent, 231-234
  generalized, 233
  Legendre polynomials and functions related to, 226-231

I

Idempotent projection operators, 50
Identity operator, 10, 34
Imaginary number, 156
Indefinite integral, 114-116
Inertia tensor, 70-71
Inner product, 285
Integral(s)
  arbitrary multicenter, 255
  in complex variable theory, 173-174
  Hadamard finite part of, 124
  real definite, 205-224
  three-center, 250-255
  two-center overlap type, 243-246
Integral calculus, fundamental theorem of, 117
Integral powers, 118-119
Integration, approximate, in numerical analysis, 359-364
Interband optical transitions, 514-517
Interval of differencing, 338
Invariance, 47
Inverse Fourier transform, 247-250
Inverse operator, 35
Inverse powers, definitions of, 118-123
Inverse trigonometric and hyperbolic functions, 204-205
Inversion theorem, 136, 142
Irrotational vector field, 26-27
Isolated singularity, 193
Isomorphic group, 376

J

Jacobi identity, 18, 34
Jordan's lemma, 209

K

Ket vector, 10
Kronecker delta, 7, 272
Kubo formula, 509
Kummer's differential equation, 232
Kummer's transformations, 234

L

Lagrange formula, 256-257
Lagrange interpolation polynomial, 344-346
Laguerre polynomial, 235
Laplace equation, 264
  harmonic function and, 166-167
  two-dimensional, 166-167
Laplace integrals, 227
Laplace inversion theorem, 144
Laplace transform(s), 143-148
  in boundary-value problems, 327-331
  properties of, 147
Laplacian operator, 22
Laurent series, 190-191, 256
Laurent series expansion, 221
Least squares, approximation by, 352-355
Left inverse, of operator, 34
Legendre equation, associated, 305
Legendre function(s), 228-231
  in boundary-value problems, 307-308
  orthogonality of, 309-310
Legendre polynomials, 226-231, 306
  hypergeometric functions and, 236
Linear equations, systems of, 366-369
Linear fractional transformation, 195
Linear manifold, of set, 6
Linear operator, concept of, 32-36
  see also Operator(s)
Linear transformations, 32, 195
Linear vector spaces, 1-3
  see also Vector(s)
Line integral, 23-24
Liouville equation, 289
Liouville's theorem, 182
Logarithm, as multiple-valued function, 202
Logarithmic branch point, 203-204
Lowering operator, 77
Lorentz force, 18

M

Many-body problem, 477, 488
Mapping
  conformal, 331-334
  defined, 33
Matrix(-ices), 36-37, 51-64
  addition of, 52
  column type, 36-37
  complex conjugate and, 53
  density, see Density matrix(-ices)
  determinant of, 55-56
  eigenvalues of, 59-63
  equality of, 51

inverse of, 56-59
multiplication of, 52-53
null, 51
scalar multiplication of, 53
special, 54
square, 36
transmutations of, 53
transpose of, 53
Matrix elements, 36-37
  in group theory, 401-404
Matrix operation, basic, 51-53
Matrix representations, 37
Maximum modulus theorem, 182-183
Minor, in determinant calculations, 55
Möbius transformation, 195
Molecular crystals, charge-transfer spectra of, 527-539
Molecular quantum mechanics
  normal modes of vibration in, 406-411
  transition probabilities in, 404-405
Molecular wave functions, computation of, 405-406
Moments of inertia, 70
Morera's theorem, 182, 188
Multiple-valued function, 196-198
Multiplication, associative property in, 376

N

Nabla differential operator, 21-23, 264
Natural orbitals, eigenvalues and, 447-449
Neumann boundary conditions, 266
Newton's advancing difference formula, 342, 355
Newton's method or formula, 363-366
Non-Abelian group, 376, 391-393
Nondegenerate perturbation theory, 257-259
Nonintegral powers, 119-122
Nonsingular operator, 35
N-representability problem, 474-475
  density matrices and, 442-447
  for 1-matrix and 2-matrix, 444-447
nth normal mode of vibration, 272
Null matrix, 51
Null operator, 34
Numerical analysis, 337-372
  approximate integration or quadrature in, 359-364
  approximation by least squares in, 352-355

approximation by spline interpolation in, 346-352
differential equations in, 362-364
equations in a single unknown in, 364-366
Gaussian elimination in, 367-368, 370
numerical differentiation and, 355-358
Simpson's rule in, 361
sparse sets of equations in, 369-371
systems of linear equations in, 366-369
Numerical differentiation, 355-358

O

ODLRO (off-diagonal long-range order), 481-482
Operator(s)
  addition of, 37
  commuting, 44
  defined, 33
  functions of, 35
  Hamiltonian, *see* Hamiltonian operator
  Hermitian, *see* Hermitian operator
  m-fold degeneracy of, 39
  product of, 37
  projection, 49-50
  unitary, 46
Operator algebra, 33-35
Ordinary functions, Laplace transforms and, 143-145
  see also Function(s); Generalized function(s)
Orthogonality condition, 272
  defined, 6
Orthogonalization process, 7

P

Parseval's theorem, 283
Partial fractions, 195
Paths, in complex variable theory, 168-171
Pauli matrices, 63-64, 67
Pauli principle, 415-417
Periodic parametric spline, 352
Permutation group, in group theory, 411-414
Permutation symmetry, 381-382
Plane wave expansion, 236
Plucked string, in separation of variables, 269-273
Pochhammer's contour, 218, 220, 222
Point groups, 380-381
  crystallographic, 422

Point group symmetry, crystal-field splitting and, 423-425
Polynomial interpolation, approximation by, 338-345
Polynomials, as entire functions, 195
Position vector, 19
Postfactor, in scalar product, 4
Potential problems, Green's functions and, 324-326
Power series
  analyticity in, 187-189, 192
  in complex variable theory, 184-193
  integration and differentiation of, 187-189
  representation of analytic functions by, 189-190
  uniqueness theorem and, 192-193
Power series convergence
  basic theorem on, 186-187
  tests for, 185
Prefactor, in scalar product, 4
Products, of generalized functions, 101-103
Projection operators, 49-50

Q

Quadrature, in numerical analysis, 359-364
Quantum chemistry integrals, 241-255

R

Radius vector, 19
Raising operator, 77
Rayleigh's formula, 236
Rayleigh-Schrödinger perturbation theory, 256
Rayleigh-Schrödinger series, 258
Real definite integral(s)
  Cauchy principal value in, 212-214
  evaluation of, 205-214
  residue theorem in, 205-211
Rectangular arrays, matrices and, 51
Recurrence and reflection formulas, 215, 233
Reduced density matrix
  properties of, 436-441
  second-order, 458-474
  single-particle, 447-458
Regular generalized function, 92
  of slow growth, 132
Removable singularity, 194

Residue theorem, 205-211
Resonance, 324
Riemannian integral, 23-24, 96
Riemann-Lebesgue lemma, 128, 131
Riemann sheets, 198
Riemann surface, 198
Right inverse, 34
Rodriques's formula, 307, 309
Rotation matrix, 65
Rotation symmetry operator, 377-378
Runge-Kutta process, 363-364

S

Scalar multiplication, 3, 12-13
Scalar product, 4, 13-14, 16
  defined, 285
Scalars, invariants as, 47
Schmidt orthogonalization process, 7, 42
Schrödinger's equation, 265, 374
  for n-electron system, 415
Schwartz's inequality, 5, 14, 72-73
Second-order reduced density matrix, 458-472
Secular equation, 39
Semiinfinite rod, heat problem in, 328-329
Separation of variables, 269-280
Sequences
  differentiation of, 114
  of generalized functions, 93-96
Set D, in test functions, 85-87
Shift operator, 339
Simpson's Rule, 361-362
Sine operator, 35
Single-particle basis, 2-matrix in, 459-467
Singular generalized functions, 90-92, 118-127
  applications of, 123-127
  extension to complex values, 133
  fundamental sequences in, 126-127
  physical examples in, 125-126
  spaces S and S' in, 130-133
  in elementary functions, 193
Singular operator, 35
Slater determinant, 405, 468
Slater-type atomic integral, Fourier transform of, 243
Slater-type atomic orbitals, 241-242
Slow growth function, 132
Solenoidal field, 30

Solid state, group theory and, 421-426
Space groups, in group theory, 421-423
Sparse sets of equations, solution of, 369-371
Spatial symmetry, 376-381
  classes in, 384-385
Spectral representation, of Green's functions, 494-500
Sphere, Green's function for, 326-327
Spherical Bessel functions, 300
  modified, 246
Spherical coordinates
  boundary-value problems in, 304-316
  Laplace's equation in, 304-305
Spherical harmonics, 241-242
  addition theorem for, 310-312
  in group theory, 414-415
Spline(s)
  cubic, 346-350
  periodic parametric, 352
Spline interpolation, approximation by, 346-355
Square matrix, 36-37
  as sum of two matrices, 54
Statistical matrix, 431-434
  see also Density matrix
Stokes's theorem, 27-29, 32
Strong convergence, 72
Sturm-Liouville operator, Green's function for, 319-324
Sturm-Liouville problem, 285-290
Substitution formula, in complex variable theory, 183
Superconductivity, 482-483
Superfluidity, 483-485
Superposition principle, 266, 270
Surface integral, 23
  defined, 27
Symmetry operator(s), 374, 377
  classes of, 383-384
  in group theory, 376-385
  wave functions and, 385-388

T

Taylor series, 340
Taylor's theorem, 340
Tensor(s), 69-71
  inertia, 70
Test function, 84
  defined, 85

set D in, 85-87
Thomas-Fermi approximation, 456
Three-center integral, evaluation of, 250-255
Three-dimensional Euclidean vector (TDEV), 10-32
  example of, 64-71
  vector product for, 16-18
Three-particle correlations, Green's functions and, 524-527
Time-correlation functions, spectral representation of, 495-496
Time derivative, 21
Transcendental functions, 214-238
Transformation, conformal, 331-334
Transformation matrices, orthogonality of, 64
Transpose conjugate, 53
Trigonometric functions, 200-202
  inverse, 204
Two-center overlap type integral, 243-246
Two-center product, Fourier transform of, 246-250
Two-matrix, 459-467
  Grimley-Peat approximation in, 470

U

Uniqueness theorem, in power series theory, 192-193
Unitary operators, 46-49

V

Value at a point, in generalized function, 112
Vector(s)
  basis, 6
  concept of, 2
  Euclidean, 2
  linear independence of, 5-7
  in physical sciences, 2
  position or radius, 19
  realizations of, 2
  scalar coefficients of, 8
  scalar product of, 16
  three-dimensional Euclidean, 10-32
Vector addition, 3, 11-12
Vector analysis, 10
Vector components, 8-10
Vector derivatives, 20

Vector fields, 18-32
  conservative, 25
  defined, 19
Vector function, 19
Vector line integration, 23
Vector operations, 11-14
Vector product, 16-18
Vector rotations, 68
Vector space, 2
  of infinite dimensions, 71-78
  orthonormal basis for, 15-16
Vector surface integrals, 27
Vector transformations, 32-50
Virial theorem, 455, 473

W

Water molecule, configurations for, 377, 389

Wave equation, 263
  D'Alembert solution of, 268-269
Wave functions, symmetry operators and, 385-388
Weak convergence, 72
Weierstrass M test, 185-187
Wronskian operator, 320, 323

X

X-ray form factor, 457-458

Y

Young diagrams, in group theory, 412-413, 416

QD
453
.P55
v.11a

80284

Physical Chemistry

East Texas Baptist College
Library
Marshall, Texas 75670